From Chromosomes to Mobile Genetic Elements

This biography of Nobel Laureate Barbara McClintock (1902–1992) places her life and work in its social, scientific, and personal context. The author examines the development of Barbara McClintock's scientific work and her influence upon individuals and upon the fields of cytogenetics and evolutionary biology in the period from 1902 to the present. The history documents years of McClintock's notable and lauded scientific work long before she discovered and named transposable elements in the mid-1940s for which she ultimately received the Nobel Prize. The biography employs documented evidence to expose, demystify, and provide clarity for legends and misinterpretations of McClintock's life and work.

Key Features

- Exposes and demystifies myths and legends told about McClintock's time in Missouri
- Clarifies the changing language of genes and genetics
- Places in perspective the history of McClintock's research
- Documents McClintock's family and early life before college
- Provides documented details of McClintock's time in Nazi Germany

From Chromosomes to Mobile Genetic Elements

The Life and Work of Nobel Laureate Barbara McClintock

Lee B. Kass

CRC Press
Taylor & Francis Group
Boca Raton London New York

CRC Press is an imprint of the
Taylor & Francis Group, an **informa** business

Designed cover image of McClintock: Courtesy of E. H. Coe
Designed cover image of corn ear: Courtesy of Ed Gerson©2023

First edition published 2024
by CRC Press
2385 NW Executive Center Drive, Suite 320, Boca Raton FL 33431

and by CRC Press
4 Park Square, Milton Park, Abingdon, Oxon, OX14 4RN

CRC Press is an imprint of Taylor & Francis Group, LLC

ISBN: 9781032365329 (hbk)
ISBN: 9781032365350 (pbk)
ISBN: 9781003332527 (ebk)

DOI: 10.1201/9781003332527

Typeset in Times
by codeMantra

To the memories of
William B. Provine,
For inspiring this project

&

Royse P. Murphy,
Mentor and Friend

&

To
My husband, Robert E. Hunt,
For his constant encouragement and support of my efforts

Contents

Preface: Records and Recollections x
About the Author xii
Acknowledgments xiii

Introduction Holding Court: How I Came to Know Barbara McClintock 1
Literature Cited 4

Chapter 1 An Unconventional Childhood 6
Growing Up in an Unconventional Family 7
Children of Many Names 8
Formative Years in Brooklyn, New York 9
Literature Cited 12

Chapter 2 Genetics in the Roaring '20s 14
The Broad Undergraduate Curriculum at the New York State College of Agriculture, Cornell University 14
Friends and Cornell Women in Science 16
Extracurricular Activities 18
Antisemitism in Academia 18
Educating Cornell Students in Agriculture 19
AAAS Toronto Meeting, December 1921 22
McClintock's Continuing Studies of Genetics and First Course in Cytology 24
Literature Cited 27

Chapter 3 From Botany Scholar to Maize Cytologist: Fact, Fiction, and Faulty Memories 30
Women in Plant Breeding at Cornell University 30
Botany Scholar 31
Master's Degree with Professor Sharp 32
Research and Teaching Assistant in Botany 32
Cornell Alumna and Instructor in Botany 41
Literature Cited 42

Chapter 4 The Golden Age of Corn Genetics: Cornfests, Cornfabs, and Cooperation 46
Postgraduate Instructor at Cornell University 46
Recognition and Discrimination, 1928 48
Significant Contributions, 1928–1931 49
Emerson and His School at Cornell 49
R. A. Emerson, Maize, and Genetics 50
Emerson's Maize Cooperation 52
The Genetics of Maize 53

Cytology Meets Corn Genetics at Cornell 54
Assigning Linkage Groups to Cytologically Identifiable Chromosomes 60
Cytogenetics Helps Genetic Mapping 65
Literature Cited 66

Chapter 5 Awards and Recognition: Chromosome Structure and Behavior 72

Separation, Cooperation, and Reunion 72
The "Push" to Publish the Cytological Proof for the Theory of Crossing-over 75
Searching for the Cause of Mutations 79
National Research Fellow, California Institute of Technology 81
The International Congress of Genetics (ICG), 1932 82
Barbara McClintock from Caltech to Germany 82
Curt and Evelyn Stern meet Barbara McClintock in America in 1932 84
McClintock's Guggenheim Research Proposal 86
McClintock, Stern, and the Rise of National Socialism in Germany, 1932–1933 87
Cornell University, 1934–1936 92
Literature Cited 95

Chapter 6 Missouri Compromise, Tenure or Freedom?: Barbara McClintock Leaves Academe 100

Assistant Professor, University of Missouri 100
The University of Missouri with Stadler's Genetics Research Group 103
Invitations to Cold Spring Harbor, 1941 108
McClintock Considers a Leave of Absence 110
Reactions to McClintock's Request for a Leave 112
Considerations for Changing Jobs 113
General Milieu 113
Problems with University Administration 115
Teaching and Advising Responsibilities 117
Research and Publication Responsibilities 118
Promotion, Tenure, and the AAUP at the University of Missouri 119
Summer 1941 – Cold Spring Harbor 121
Guest Investigator, Department of Genetics, Cold Spring Harbor, 1941–1942 123
Negotiating a Permanent Appointment at Department of Genetics, CIW/CSH, 1943 125
Missouri Compromise 127
Records and Recollections 129
Literature Cited 131

Chapter 7 The Road to Transposition 136

Studies with Broken Chromosomes 143
Presentations and Collaborations 145
Scientific Societies, Committee Memberships, and Recognition 149
Literature Cited 151

Chapter 8 Resignation, Renewal, and Reorganization ... 156

Resignation ... 156
Renewal ... 159
Reorganization ... 167
Literature Cited ... 175

Chapter 9 Coming Home Again: Andrew Dickson White, Professor-at-Large, Cornell University ... 182

Andrew Dickson White, Professor-at-Large ... 182
The Lecture Circuit and Other Obligations ... 188
Cornell A. D. White Professor Reappointment: Coming Home Again ... 192
Recognition by Geneticist Colleagues ... 192
Acceptance and Rejection of McClintock's Ideas ... 195
Literature Cited ... 201

Chapter 10 Golden Age of Corn Genetics: Reprise ... 208

The Name Game: Reconciling Prokaryotic and Eukaryotic Genetic Systems ... 208
Recognition and Awards ... 216
Literature Cited ... 223

Chapter 11 Unshared Nobel Prize and Inclusive Recognition ... 229

The Nominations for the 1983 Nobel Prize ... 229
Who Decides? ... 231
The Dynamic Genome and Cell Biology ... 232
Molecular Studies Identify Transposable Elements in Maize ... 233
McClintock's Endeavors During 1983 ... 234
The 1983 Nobel Prize in Physiology or Medicine ... 236
McClintock's Nobel Lecture ... 239
Rationale for Delay of Her Recognition ... 240
"Time for the Molecular Analysis" of Transposable Elements ... 243
A Nobel Laureate ... 244
Lessons from Stories and Legends ... 247
Literature Cited ... 249

Index ... 257

Preface
Records and Recollections

All history, so far as it is not supported by contemporary evidence, is romance

– Samuel Johnson

Distinguished scientists are often asked to recount their life and work as memoirs or oral histories. These chronicles are usually recorded from memory at an age far removed from the events and therefore subject to memory's shortcuts and subjective interpretations. So it has been with the memories of Nobel Laureate Barbara McClintock, a towering figure of twentieth-century genetics. McClintock is reputed to have worked alone for most of her life and achieved international acclaim late in her career, only then attracting intense attention and interviews. While existing biographies retell her stories from such interviews, an examination of papers and correspondence written at the time makes it possible to assemble a more complete picture of the environment at the various institutions where McClintock lived and worked. My effort has been devoted to analyzing these documents in order to understand the scientific, institutional, and societal context in which McClintock made momentous discoveries. These records have allowed me to assemble an inevitably more complex picture of her life and work than memories reveal.

This biography of Barbara McClintock (1902–1992) examines the development of her scientific work and its influence upon individuals and upon the fields of cytogenetics and evolutionary biology. McClintock contributed years of notable and lauded scientific work long before she discovered transposable elements in the mid-1940s, the achievement that resulted in her being awarded the 1983 Nobel Prize in Physiology or Medicine. Previous biographies, as described in more detail below, relied broadly on interviews conducted with McClintock in the last years of her life. In contrast, contemporaneously written records not only have substantially added to her story but have also helped to revise and clarify the anecdotes she and others told from memory years after the recalled events.

As a graduate student at Cornell University in the early 1970s, I observed first-hand how McClintock had long been honored and revered by faculty, students, and staff. In 1976, at the urging of Cornell geneticist Adrian Srb, William B. Provine, an historian of science at Cornell, interviewed Barbara McClintock at her Cold Spring Harbor Laboratory, Long Island, New York, about her life and work. Later, in 1980, Provine and then graduate student Paul Sisco conducted a day-long confidential interview with Dr. Barbara McClintock at her Cold Spring Harbor Laboratory. The interview was recorded, professionally transcribed, and deposited in the Cornell University Archives. At McClintock's request, the interview remained restricted until after her death.

The year following McClintock's death, Dr. Provine invited me to collaborate on conducting research for a biography of Barbara McClintock. Provine provided copies of his interview transcripts from the Cornell Archives. These restricted interviews with McClintock served as leads for further investigations and resulted in my discovery of the documents that provide the factual basis for this biography. With support from the National Science Foundation and other organizations, I searched for evidence to corroborate or clarify interviewees' recollections. I located contemporaneous documents revealing background on McClintock's family, her life in Brooklyn, New York, her high school years, and her experiences as a student at Cornell University, as a research fellow in Germany, and as a faculty member at the University of Missouri, thus illuminating for the first time a different view of her early life and work. I also used this same approach in the chapters devoted to the more widely reported portions of her later life.

During this time, in 1996, we shared the transcribed interview with then graduate student N. Comfort, who extensively cited it in his Ph.D. dissertation, focusing on the discovery and reception of McClintock's controlling elements. The Provine and Sisco interview with McClintock subsequently became available to all scholars in 1999.

The first biography of McClintock, entitled *A Feeling for the Organism*, was published in 1983, by Evelyn Fox Keller. Coming out in the same year McClintock won her Nobel Prize, Keller's story was mostly based on her interviews with McClintock and her colleagues. Keller's book was not intended as a documented biography of McClintock's life and work. Then in 2001, expanding on his doctoral dissertation, Comfort published a biography of Dr. McClintock entitled *The Tangled Field,* relying broadly on the transcripts of the 1980 interview of Provine and Sisco and Keller's interviews with McClintock, among other oral history sources, to reconstruct with limited confirmation, McClintock's early and later career.

Because these two biographies depended so heavily on oral interviews conducted late in the lives of Dr. McClintock and her colleagues, Dr. Provine and I saw a clear need for an expanded and critically documented biography. Archived documents and records led to my expanded and refined interpretations, which were previously founded on oral histories, memoirs, or autobiographical recollections. Frequently, secondary sources did not provide an accurate representation of McClintock's scientific contributions as reported in her primary papers. Hence, reading the seminal papers from the beginning of her scientific career is essential for an understanding of her insights and appreciation of her approach to research. Consequently, I published a collection of McClintock's early papers with accompanying perspectives to assist those who might be hesitant to use McClintock's primary papers for teaching. That *Companion Volume* provides readers of this biography with more technical details of McClintock's early scientific contributions, which led to her Nobel Prize–winning work on mobile genetic elements.

Research on memory shows that stories people tell about their past are shaped by beliefs they hold in the present and are often reexamined in terms of current experiences. The quest for precision can lead historians to criticize the factual inconsistencies, exaggerations, falsehoods, or self-deceit often found in autobiographies. The latter is not my goal.

This biography uses contemporaneous documents to place in historical perspective the autobiographical reminiscences, recollections, and stories told by and about McClintock. I further call attention to unsubstantiated stories that are often repeated but without supporting evidence; such legends are treated as anecdotes rather than factual accounts of events.

About the Author

Dr. Lee B. Kass received her Ph.D. in Botany and Genetics from Cornell University after earning a B.S. in Biology from The City University of New York. She did postdoctoral research at the University of Cambridge-UK, and Vanderbilt University. She has served on the faculties of the University of Cambridge, University of Tennessee-Nashville, Elmira College, The College of the Bahamas, Cornell University, and West Virginia University-Morgantown. Kass has authored, edited, or co-edited 12 books, and authored or co-authored numerous book chapters, proceedings papers, and articles in scientific journals. She is a member of the Botanical Society of America, the Bahamas National Trust, and the Maize Genetics Cooperation, and a former member of many botanical organizations. Kass was a chair of the Historical Section of the Botanical Society of America for many years. She established the Elmira College Herbarium and currently serves on the Science Advisory Committee of the Bahamas National Trust. Among her awards are the Josef Stein Award for excellence in teaching and scholarly achievement, various research grants, and a Fulbright Scholar Award, during which time she and her spouse, Dr. Robert E. Hunt, established the National Herbarium of the Bahamas. Currently, she is recognized in *American Men of Science* and *Marquis Who's Who.* She is Visiting/Adjunct Professor at Cornell University and at West Virginia University-Morgantown. Her research focuses on history of botany, and biodiversity and reproductive biology of Bahamian plants.

Acknowledgments

You are not obligated to complete the task, but neither are you free to refrain from it

– Pirke Avot 2:21

There are so many colleagues and friends who have supported my efforts in this project that I fear I may have overlooked some in my acknowledgments. If so, please know that I appreciated your guidance, documents, and direction, as I often tried to refrain from completing this task.

I am grateful to my colleague and friend Professor William B. Provine* for inviting me to assume responsibility and expand upon his project to examine the work of Barbara McClintock, for unlimited use of his vast library and reprint collection, and for his unfailing support and encouragement. I acknowledge partial research support from the National Science Foundation (grants SBR 9511866 and SBR 9710488); American Philosophical Society Library, Mellon Resident Research Fellowship; and Lilly Library, Helm Fellowship. I appreciate permissions granted from publishers, colleagues, and the Cold Spring Harbor Laboratory Archives (Clare Clark, Stephanie Satalino, and the Estate of Barbara McClintock) to reproduce images and reprints for this volume and the *Companion Volume* that accompanies this book (see acknowledgments therein). For logistical support, I thank my colleagues in L. H. Bailey Hortorium, Department of Plant Biology (now Section), Cornell University [CU] (William Crepet, Kevin Nixon, Robert Dirig, Peter Fraissinet, Sherry Vance*, Pat Garrett*); Department of Plant Breeding & Genetics (now Section), CU (W. Ronnie Coffman, Elizabeth Earle, Mark E. Sorrells, Jeffrey Doyle, Michael Gore, Leslie Larson, Judy Singer, Royse P. Murphy*, Ron A. Anderson*); and Division of Plant & Soil Sciences, West Virginia University (Barton Baker, Joginder Nath*, Vagner Benedito, Sven Verlinden, William Bryan). Special thanks to colleagues who supported and encouraged my efforts from the beginning (Kathleen Gale, Len Simons, William Lindsay*). I am most grateful to colleagues who spent extraordinary time helping me to understand McClintock's experiments (Edward H. Coe, James A. Birchler, Paul Chomet, Paul Sisco, Nina V. Fedoroff, James A. Shapiro, Harriet B. Creighton*, Charles H. Uhl*), and to those who granted me interviews, as cited in my chapters. Special thanks to those who donated major time and effort reviewing draft manuscript chapters (Robert E. Hunt, J. A. Birchler, E. H. Coe, N. V. Fedoroff, J. A Shapiro, Allan P. Shatkin, V. Benedito), although I take final responsibility for any remaining errors.

I am grateful to faculty, staff, and students whose special efforts assisted my research at national and international archives and libraries: AAUP Archives, Gelman Library, George Washington University, Washington, DC (Jennifer King); A. D. White Professor-at-Large Program, CU (Uri Bronfenbrenner*, Cindy K. Greco); American Philosophical Society Library, Philadelphia, PA (Charles Greifenstein, Joe DiLullo); Archives and Special Collections, University of California Davis Library, Davis, CA (Michael Dietrich); Brookhaven National Laboratory, Upton, NY (Benjamin & Francis F. Burr); Brooklyn Public Library, Center for Brooklyn History, Brooklyn, New York; Cold Spring Harbor Laboratory Archives, Cold Spring Harbor, NY (C. Clark, S. Satalino); Division of Rare and Manuscripts Collections, CU Library (Gould Coleman*, Elaine Engst, Eileen Keating, Laura Linke, Nancy L. Dean, David Corson, Evan F. Earl, Corey R. Earle); Dolph Briscoe Center for American History, the University of Texas at Austin (Marisa Jefferson); California Institute of Technology Archives, Pasadena, CA (Judith R. Goodstein); Carnegie Institution of Washington Archives, Washington, DC (Maxine Singer); Columbia University Archives, Manhattan, NY (Jocelyn Wilk); Emerson College Archives, Boston, MA (Rosalie Gartner); Guggenheim Foundation Archives, the Solomon R. Guggenheim Museum & Archives, Manhattan, NY (Mary Kiffer); Harvard University Archives, Cambridge, MA; Hunt Institute for Botanical Documentation, Pittsburgh, PA (Anita L. Karg, Angela Todd); John Innis Horticultural

Society Archives, UK (Sarah Wilmot); Lilly Library, Indiana University (Drew Schwartz*, Tom Kauffman); L. H. Bailey Hortorium, CU (R. Dirig, S. Vance*, P. Fraissinet, Jeff Kirby, Rebecca Waldron); Mann Library, CU (Janet McCue, Mary Ochs, Michael N. Cook); Manuscript and Map Collections, Bancroft Library, University of California, Berkeley, CA; Max Planck Society Archives, Berlin, Germany (Simon Nobis, Hans Jörg Rheinberger, Bretislav Friederich); Missouri Botanical Garden Archives, St. Louis, MO (Kim Kleinman, Doug Holland); National Academy of Sciences Archives, National Research Council, Washington, DC (Daniel Barbiero); National Archives and Records Administration, Washington, DC (Joseph Schwartz); National Museum of American History, Smithsonian Institution, Washington, DC (Patricia L. Gossel*, Ann Seeger); New York Academy of Sciences, New York, NY (Douglas Braaten); North Carolina State University Archives, Raleigh, NC (Major M. Goodman; Muquarrab A. Qureshi); Olin Library, CU, Ithaca, NY; Office of Alumni Affairs and Development, College of Agriculture and Life Sciences, CU [formerly Public Affairs Office] (Barbara Littlefair, Patti A. Weldon, Sharon L. Detzer); Plant Breeding Department Archives, CU, Ithaca, NY (R. P. Murphy*, Henry Munger*, J. Singer, Leslie Larson); Provine/Cain Library and Reprint Collections, CU (W. B. Provine*); Rockefeller Archive Center, Sleepy Hollow, NY (Thomas Rosenbaum*); Special Collections and Archives, University of Michigan Library (Beverly J. Rathcke*); Stanford Libraries, University Archives, Special Collections, Stanford, CA; the State Historical Society of Missouri Research Center-Columbia [formerly the State Historical Society of Missouri, Columbia, MO, & Western Historical Manuscript Collection, Columbia, MO] (Laura Jolley); University of Illinois Archives, University Library, University of Illinois, Urbana-Champaign, IL (Jameatris Y. Rimkus); University of Minnesota Archives, Minneapolis, MN (Katelyn Morke); University of Missouri Archives, Columbia, MO (E. H. Coe, J. A. Birchler); University of Washington, University Libraries, Special Collections, Seattle, Washington (K. Gale); University of Wisconsin Archives, Special Collections, Steenbock Memorial Library, Madison, WI (Bernard Schermetzler); West Virginia University Library and Archives, Morgantown, WV (Martha Yancy); Williams College Archives and Special Collections (Naomi Pasachoff, Sylvia Brown); Yale University Archives, New Haven, CT (Genevieve Coyle).

For leads and document sharing, in the spirit of maize cooperation, I thank Pnina G. Abir-Am, Sankar Adhya, Bruce Alberts, Garland E. Allen*, Gregory J. Anderson, Edward G. Barry, Harlan P. Banks*, V. Benedito, Marjorie McKinley Bhavnani, J. A. Birchler, Steven E. Bloom, James L. Brewbaker*, Patricia Brown*, Christophe Bonneuil, Edward S. Buckler, B. & F. Burr, Allan M. Campbell*, Neil Campbell*, P. Chomet, Edward Cobb, E. H. Coe, N. L. Comfort, Dale Corson*, Susan L. G. Cooper, Denise E. Costich, Harriet B. Creighton*, James F. Crow*, William deJong-Lambert, M. Dietrich, John Doebley, J. Doyle, E. Earle, B. Friederich, Joseph Gall, Mel Green*, Susan Gabay-Laughnan, M. M. Goodman, Margaret Gordon, Alexander Grobman, Michael Gutensohn, N. V. Fedoroff, Edith Fine, Jim Haber, Rada Demerec Dyson-Hudson*, Maureen R. Hanson, Zlata Demerec Hartman*, William Haut, Hugh Iltis*, Andre Jagendorf*, Gina B. Kolata, Elizabeth Jones*, Burke H. Judd*, Alfred Kahn*, Thomas Kaufmann, Evelyn Fox Keller*, Matthew Kessler, Barbara Kimmelman, John M. Kingsbury*, K. Kleinman, Robert E. Kohler, Jean Kreizinger, Joshua Lederberg*, Edward B. Lewis*, Kent Loeffler, Ross MacIntyre, Daniel R. Marshak, Robert Metzenberg*, Ulises Moreno, Rosalind Morris*, H. Munger*, Peter McKinley, June & Mike Nasrallah, M. Gerald Neuffer*, Erling Norrby, William Pardee*, Dom Paolillo, Naomi Pasachoff, Wojtek Pawlowski, Peter A. Peterson*, David Perkins*, Ronald L. Phillips*, Gale Provine, Hope Punnett, George Rèdei*, Virginia Rhoades*, Margaret Rossiter, D. Schwartz*, Maria Scurra, Nancy Slack*, J. A. Shapiro, M. Singer, P. Sisco, Margaret E. Smith, V. Betty Smocovitis, M. Sorrells, Josetta & Adrian Srb*, H. David Thurston*, David & Marian Timothy*, Natalie W. & C. H. Uhl*, Randy O. Wayne, David (Burt) Walden, Clifford F. Weil, Richard H. Whalen*, Ann & Allan* Witztum, Roberta, Miriam* & Bruce* Wallace, Takeo Angel Kato Yamakake.

Thanks to family, friends, and colleagues, who hosted me during archival sojourns, Rochelle & N.* Campbell, E. H. Coe, K. Gale, Ellen Dorfman & K. Kleinman, Rebecca & Bud Dolch, Nancy Elliott, Marie Knowlton, Melissa Luckow, Nancy Potter & George Gull, Raana & M. A. Qureshi, B. J. Rathcke*, M. & Dan Singer, Sandra* & Nick Tsantillis, Florina Tseng, Lorraine & Gene Traynor.

I owe special recognition to my editors Chuck Crumly and Kara Roberts for encouraging my efforts to make this project a reality. Lastly, I gratefully acknowledge my husband Dr. Robert E. Hunt for his continued and tireless backing of my efforts to complete this project.

*deceased

Introduction

Holding Court

How I Came to Know Barbara McClintock

Nobel Laureate Barbara McClintock discovered transposable elements (mobile genetic elements, also called transposons and popularly called jumping genes) in corn (maize) in the mid-1940s, published her initial findings in the *Carnegie Institution of Washington Year Book* (McClintock 1946), and reported her results in the scientific journal *PNAS* in 1950 (McClintock 1950; Weil 2013). At that point in her career, she had been studying the genetics and cytology of the corn plant for about 25 years. Her experience in teaching and studying her subject, since the mid-1920s, provided the insights for observing and recording unexpected results. More than 30 years later (1983), she was recognized for her significant discovery with an unshared Nobel Prize in Physiology or Medicine. Only five months previously, her biography was published (Keller 1983) and the award "instantly transformed her into a scientific heroine and a darling of the media" (Keller 1993, p. vii).

This biography offers a different historical picture of McClintock's life and contributions, as previously reflected in published literature and legends. For example, readers will learn here that: McClintock was born in Hartford, Connecticut (not Brooklyn, NY); she participated in many extracurricular social and sports activities in high school (she was not antisocial or a loner); she graduated high school at age 17, in June 1919 (not mid-year at age 16); Professor Lester W. Sharp (not R. A. Emerson) was Barbara McClintock's major professor in graduate school; not until she had obtained her Ph.D. (not as an undergraduate) did she morphologically describe the ten chromosomes of corn (at mitosis, not meiosis); the first maize chromosome map was published in 1922 by her genetics professor C. B. Hutchison (not by McClintock in 1931); she did not discover ring chromosomes, but reported finding the first *maize* ring chromosome in 1931; she was not denied tenure at the University of Missouri, but rejected a forthcoming offer of tenure and instead accepted a research position elsewhere; and Barbara McClintock is *not buried* at Huntington Rural Cemetery, on Long Island, NY, etc. These and many other legends will be demystified in the chapters that follow.

Indeed, as a woman who wished to pursue a career in science, Barbara McClintock did confront many obstacles before settling into a full-time job where she felt the "female element" would not be a barrier. Yet, years before McClintock became a public figure, she was recognized by her scientific peers in Plant Biology and in Plant Breeding (Genetics) and by her alma mater, Cornell University (Fedoroff & Botstein 1992). As a graduate student at Cornell University in the early 1970s, I observed first-hand how McClintock had long been honored and revered by faculty, students, and staff (Kass 2003; Murphy & Kass 2007, 2011). McClintock was a Cornell University A. D. White Professor-at-Large (1965–1974, Cornell 2022a,b), when I first met her in the spring of 1972. The details of this distinguished visiting professorship are described in Chapter 9, Coming Home Again. She was hosted by the Section of Genetics, Development and Physiology, where I was a graduate student. The Section had absorbed the former Botany Department from which she had received her undergraduate and graduate degrees (B.S. 1923, M.A. 1925, Ph.D. 1927; Kass 2003).[1]

[1] Most of the faculty in Cornell's former Department of Botany were subsumed into the Section of Genetics, Development and Physiology, where they hosted Professor-at-Large Barbara McClintock from 1965 to 1974 (Kass & Cobb 2007).

DOI: 10.1201/9781003332527-1

I learned about McClintock's early scientific contributions in the Cytology course that I completed in 1971, during my first year as a Cornell graduate student. In the Cytology laboratory, on the second floor of the Plant Science Building, we had examined the maize microscope slides that McClintock had prepared for her research and for the course while she was a graduate student and later an Instructor.[2] In addition to a current cytology textbook, one of our reference texts was Lester W. Sharp's *Introduction to Cytology* (1934, 3rd edition), which included summaries of many of McClintock's published papers and unpublished photographs (Coe & Kass 2005). Dr. Sharp had been McClintock's major professor for both her Master's and doctoral research, in the Department of Botany. We also studied microscope slides prepared by Harriet Creighton (Ph.D. 1934), a former student of Professor Dr. Sharp and Instructor Dr. McClintock. We were most excited to examine many corn chromosome slides prepared by graduate student George Beadle (Ph.D. 1930), a Nobel Laureate, who learned cytology from Professor Sharp and Instructor McClintock, while majoring in Genetics with R. A. Emerson, Head of Cornell's Department of Plant Breeding.[3] When Sharp retired in 1947, our Cytology Professor, Dr. Charles H. Uhl, also Sharp's former student, was appointed to his faculty position upon completing his Ph.D. that year.

While enrolled in Cytology during my first term at Cornell, I soon learned that the pedagogy for teaching graduate students, at least in Plant Biology and in Plant Breeding (Genetics), did not rely on textbooks. Students received a mimeographed handout for each lecture, listing primary sources that were on reserve in the library; so-called further reading if you will. From reading primary sources and observing laboratory demonstrations of original materials, we learned to be skeptical of review articles and other secondary sources that misrepresent original research results. The class and laboratory notebooks of Professor Henry Munger, who took Cytology (1939) and Advanced Cytology (1941) with Dr. Sharp at Cornell as a graduate student, demonstrate this same teaching pedagogy, and many illustrations in his lab book were made directly from the microscope slides prepared by McClintock for her early pioneering research publications (Munger 1934–1941; Kass 2013ff.). When I later found McClintock's teaching documents for her cytogenetics lectures given first at the University of Missouri, it became clear that she had used these same techniques to teach her students.[4]

On Sunday afternoon, April 2, 1972, Professor Harry T. Stinson Jr., Chair of the Section of Genetics, Development and Physiology, was pacing the second floor of Bradfield Hall, in anticipation of Dr. Barbara McClintock's annual visit.[5] As I left my Emerson Hall laboratory, on the way to Bradfield Hall to pick up some deionized water for my experiment, I passed him in the hall. I asked if something was wrong, which seemed apparent from the expression on his face. He told me that Dr. McClintock should have arrived already and he was very concerned that something might have happened to her.

[2] Unfortunately, the Cytology lab, Room 219, Plant Science Building, and the adjoining office that McClintock and Harriet Creighton had once occupied were renovated around 2010 (Kass 2013) as was the building during 2021–2022 (see Chapter 9); L. B. Kass (2018) deposited the Cornell Department of Botany cytology and cytogenetics microscope slide collection at the Hunt Institute for Botanical Documentation, and also the Department of Plant Breeding photograph collection; see *Bulletin of the Hunt Institute for Botanical Documentation*. Vol. 31 (No. 2) Fall 2019, News from the Archives, p. 5, p. 9 (recognition: a gift of Cornell's Department of Plant Breeding photographic slides & glass lantern slides), https://www.soc-botanical-artists.org/wp-content/uploads/hibd-bulletin-31-2.pdf

[3] L. W. Sharp *Roll Book* 1924–1931, "S-6 (cytology) summer 1927, Beadle, G. W. (Plant Breeding)"; "A. Marshak," the only undergraduate in this summer school class of eight students, began his master's degree with Sharp. McClintock became a good friend and supporter of his work in later years. Note that by the summer of 1927, Barbara McClintock had earned her Ph.D. and was an Instructor in the Cornell Botany Department. See Murphy and Kass (2007, 2011) for a list of Professor R. A. Emerson's students.

[4] McClintock's 1941 Cytogenetics lecture notes were given to Marcus Rhoades and are in the Rhoades' Collection at the Lilly Library, University of Indiana. The microscope slides used in Cornell's Cytology and Advanced Cytology courses (the latter named Cytogenetics in 1948) were deposited at Hunt Institute in Pittsburgh, PA (Hunt Institute 2018).

[5] H. Stinson to Max Black, May 10, 1972, A. D. White Professors' File, CALS Dean's office.

To Stinson's relief, McClintock arrived mid-afternoon. She had a flat tire on the drive to Ithaca from Cold Spring Harbor Laboratory, Long Island, New York. She apologized for being late. It took her a while, she said, to change the tire on the turquoise and white station wagon that belonged to the Laboratory. I don't recall that either Stinson or I were surprised that at age 69, McClintock had changed the tire herself on this large estate car.

McClintock's annual visits to Cornell were more intimidating to the faculty than to the students as I had observed. Years later, in anticipation of writing this documented biography of the life and work of Barbara McClintock, I interviewed Stinson and other faculty in our Division of Biological Science about their memories of McClintock's visits to Cornell, which reinforced my recollection and is detailed and documented in Chapter 9. McClintock preferred visiting with students in their labs and joining us for dinners, and walks in the woods, rather than meeting with faculty. She told us that students were more receptive to new ideas, while faculty held preconceived notions.

Graduate and undergraduate students were encouraged to make appointments to meet with McClintock, and since she was so honored and respected, we fondly remembered those meetings as her "Holding Court." Our reverence was justified by her recent award of the 1970 National Medal of Science, which had been presented on May 21, 1971, by the President of the United States (Nixon 1971). She had been recognized for "establishing the relations between inherited characters in plants and the detailed shapes of their chromosomes, and for showing that some genes are controlled by other genes within chromosomes." Having a distinguished female scientist in our midst was so welcome to the women graduate students in our Section, since only male faculty had been hired for teaching and research in Botany, Physiology, or Genetics at that time.

I can also add that all lab technicians were women, as were the departmental secretaries. There was even a statement in the Cornell University 1970 *Graduate School*, *Biological Sciences* course announcements as follows: "EMPLOYMENT FOR WIVES OF STUDENTS. Cornell University offers many nonacademic positions for working wives ... Types of work include secretarial and clerical work, work for technicians in the various laboratories, library work, limited nursing positions, and some administrative positions" (Cornell University 1970, p. 17). Even though the Sections of Microbiology and of Biochemistry, within Cornell's Division of Biological Sciences, had three female faculty members by 1971 (*Agriculture Handbook* 1971, p. 133), it would be over a decade later (1985) that the first female would be hired and awarded tenure in the Section of Plant Biology (Kass & Cobb 2007, p. 96: Table 6, p. 101).

Perhaps because I was a first-year graduate student and did not know of her reputation for intimidation, I accepted the invitation (extended to all graduate students) to meet and discuss my research with Dr. McClintock. I was quite impressed with her depth of knowledge in my subject area of plant chloroplast development. As we talked, we drew pictures together on the blackboard and talked about the recent literature in this field. She asked me about a paper recently published on the subject. When I said I had not read it, she stopped talking and looked at me. "Come back after you have read the paper," I recall her saying, "and we will continue the conversation." So that is exactly what I did. Years later, when I told this story to Harry Stinson, he responded that this was an example of her daunting personality. I, however, felt no threat, but was grateful for her lead to the literature, which I eventually cited in my doctoral dissertation (Kass & Paolillo 1974; Kass 1975). Possibly because we both grew up in Brooklyn, or because I was just too naive to know better, I truly enjoyed talking with Visiting Professor Barbara McClintock about my research, attending her seminars, and especially joining her for dinner and drinks, with my graduate student colleagues at local restaurants (Fig. 0.1) or in our homes.

She taught us to be open-minded and well informed, and to think independently – a great legacy for a great mind (Kass 2013ff., Vol. I, p. 0.17).

FIGURE 0.1 Barbara McClintock, postdoc Michael Vodkin and Cornell University students at a restaurant in Ithaca, New York, ca. 1973. (Photograph by Janet Sawicki, used with permission.)

LITERATURE CITED

Agriculture Handbook No. 305. 1971. *1971–1972: Professional Workers in State Agricultural Experiment Stations and other Cooperating State Institutions*, Revised December 1971. Cooperative State Research Service, United States Department of Agriculture, Washington, D.C.

Coe, E. and L. B. Kass. 2005. Proof of physical exchange of genes on the chromosomes. *Proceedings of the National Academy of Science* Vol. 102 (No. 19, May 10), pp. 6641–6656; online May 2, 2005: http://www.pnas.org/content/102/19/6641.full

Cornell. 2022a. Cornell University Program for Andrew D. White Professor-at-Large; Program description and history. https://adwhiteprofessors.cornell.edu/program-info/ (accessed May 4, 2022).

Cornell. 2022b. Cornell University Program for Andrew D. White Professor-at-Large; All Professors at Large 1965 to 2027. https://adwhiteprofessors.cornell.edu/all-professors-at-large-1965-to-2027/ (accessed May 4, 2022).

Cornell University. 1970. *Cornell University Announcements: Graduate School, Biological Sciences, 1971–72.* Volume 62 (No. 22, November 23, 1970). Cornell University, Sheldon Court, 420 College Avenue, Ithaca, NY 14850. https://ecommons.cornell.edu/handle/1813/37718

Fedoroff, N. and D. Botstein. 1992. Introduction, p. 1–4, *In*, N. Fedoroff and D. Botstein (Eds.), *The Dynamic Genome: Barbara McClintock's Ideas in the Century of Genetics,* Cold Spring Harbor Laboratory Press, Plainview, NY.

Hunt Institute for Botanical Documentation. August 15, 2018. Institute receives cytology microscope slide collection. News: https://www.huntbotanical.org/news/show.php?id=91

Kass, L. B. 1975. *A study of chloroplast replication in spores of polytrichum.* Thesis, Ph.D. Cornell Univ. August 1975 [Ithaca, NY].

Kass, L. B. 2003. Records and recollections: A new look at Barbara McClintock, Nobel Prize-Winning geneticist. *Genetics* Vol. 164 (No. 4, August 1), pp. 1251–1260. https://www.genetics.org/content/164/4/1251

Kass, L. B. 2013. "Cornell University: the McClintock Years, 1919–1939 and beyond," Department of Plant Biology Centennial Celebration, June 29, 2013, Invited Symposium Presentation; Plant Science Building, College of Agriculture and Life Science, Cornell University; video recording. https://www.ecommons.cornell.edu/handle/1813/33663

Kass, L. B. 2013ff. Preface, pp. 0.17–0.19, *In*, L. B. Kass (Ed.), *Perspectives on Nobel Laureate Barbara McClintock's Publications (1926–1984): A Companion Volume.* Vol. I, The Internet-First University Press. http://hdl.handle.net/1813/34897 [online December 30, 2013]

Kass, L. B. and D. J. Paolillo, Jr. 1974. The effect of darkness and inhibitors of protein synthesis on the replication of chloroplasts in the moss. *Polytrichum. Zeitshrift für Pflanzenphysiologie* Vol. 73 (No. 3, August 1974), pp. 198–207.

Kass, L. B. and E. Cobb. 2007. Landmarks and milestones in American plant biology, the Cornell connection. *Plant Science Bulletin* Vol. 53 (No. 3, fall), pp. 90–101. Available online, September 2007.

Keller. E. F. 1983. *A Feeling for the Organism, The Life and Work of Barbara McClintock*, 1st edition, W. H. Freeman & Co., San Francisco, CA.

Keller. E. F. 1993. *A Feeling for the Organism, The Life and Work of Barbara McClintock*; reprinted with an author's note on the Tenth Anniversary edition, W. H. Freeman & Co., San Francisco, CA.

McClintock, B. 1946 [July 1, 1945–June 30, 1946]. Maize genetics: Continuation of the study of the induction of new mutants in chromosome 9. Modification of mutant expression following chromosomal translocation. The unexpected appearance of a number of unstable mutants. *Carnegie Institution of Washington Year Book No. 45*, 1945–1946 [Issued 13 December 1946, submitted June 1946], pp. 176–186.

McClintock, B. 1950. The origin and behavior of mutable loci in maize. *Proceedings of the National Academy of Sciences* Vol. 36 (No. 6) [June 15, 1950, communicated April 8, 1950], pp. 344–355.

Munger, H. 1934–1941. Class note books and laboratory note books on loan to L. B. Kass, will be deposited in the Cornell University Archives.

Murphy, R. P. and L. B. Kass. 2007. *Evolution of Plant Breeding at Cornell University: A Centennial History, 1907–2006*. Department of Plant Breeding & Genetics, Cornell University, Ithaca, NY, pp. 1–98, Appendices A1–A98, Photo Section P1–P38. (July 2007).

Murphy, R. P. and L. B. Kass. June 27, 2011. *Evolution of Plant Breeding at Cornell University: A Centennial History, 1907–2006*. Rev. edition i–x, 178 p., Appendices, Photo Section. e-book, The Internet-First University Press, Ithaca, NY. http://hdl.handle.net/1813/23087

Nixon, R. M. 1971. 37th President of the United States: 1969–1974, Remarks on Presenting the National Medal of Science Awards for 1970, May 21, 1971, https://www.presidency.ucsb.edu/documents/remarks-presenting-the-national-medal-science-awards-for-1970

Sharp, L. W. 1934. *Introduction to Cytology*, 3rd edition, McGraw-Hill Book Co., New York.

Weil, C. 2013. Transposable controlling elements step out onto the broader scientific stage, pp. 3.11–3.21, *In*, L. B. Kass. (Ed.), 2013ff. *Perspectives on Nobel Laureate Barbara McClintock's publications (1926–1984): A Companion Volume*. Vol. II, The Internet-First University Press. http://hdl.handle.net/1813/34897

1 An Unconventional Childhood

The individuality of the student is of the first importance

—*Emerson College of Oratory, 1896*

While many legends and stories of Barbara McClintock's early family life have been repeated, not much verifiable information has been published about her or her family concerning her upbringing or the family deliberations that led her parents to support her decision to attend college. During an interview Professor William Provine and a graduate student Paul Sisco conducted with McClintock in 1980 at her Cold Spring Harbor Laboratory, she avoided answering such questions, preferring instead to discuss the history of agriculture (Provine & Sisco 1980). Evelyn Fox Keller's (1983, 1993) biographical memoir, *A Feeling for the Organism*, introduced McClintock's family and her life and work to a wider audience just five months prior to her Nobel Prize Award (1983). Interviews cited in Keller's biography provided many wonderful recollections and photographs of McClintock and her extended family. The year before her death, a book for young adults (Kittridge 1991) appeared and included additional recollections and photographs from her youth, the latter again provided by McClintock's niece, Marjorie (Mitz) Bhavnani, the youngest child of her sister Marjorie. A second biography for adults (Comfort 2001) repeated these same stories of McClintock's family history and her early and later years (Kass 2002; Fedoroff 2002). Additional books or book chapters for juveniles continued to preserve the family lore (Heiligman 1994; Gardner 1997; Fine 1998; Cullen 2003; Pasachoff 2006; Spangenburg & Moser 2008).

While seeking to corroborate the recollections of McClintock's early life, I came across documents that confirm, expand upon, and also seem to provide another perspective for some of the family legends. While documents can also deceive due to gaps in the written record, as Nobel Laureate Joshua Lederberg reminded me in a 2003 letter he wrote to me commending my *Genetics* Perspective about Barbara McClintock (Kass 2003),[1] I found that many contemporaneous records of the same event can substantiate or clarify recollections.

Barbara McClintock's early family life, as first summarized in biographies by Keller (1983) and Kittridge (1991), was mostly stories repeated from recollections by her and her family many years after the events had transpired.[2] The stories were then repeated in subsequent biographies without further documentation. When I first began searching for documents to confirm those stories, I asked Professor William Provine, who gave me this project, why I should continue to interview people about McClintock's life and work, since memories are so ephemeral. He said, "Interviews lead to documentation. Don't worry about stories not being true - get the documentation."[3] For recollections that can be documented, I provide sources in the literature cited and notes.

1 Letter Joshua Lederberg, University Professor Emeritus, The Rockefeller University to Prof. Lee B. Kass, Cornell University, September 9, 2003, in LBK Lederberg file, January 1998–September 2003.

2 Keller (1983/1993), interviews with McClintock (1978 & 1979), and with her longtime friends Marcus Rhoades, George Beadle, Harriet Creighton, and other researchers in the field, e.g., Lottie Auerbach (1981); Keller's interviews are deposited at the APS Library in Philadelphia.

3 Kass, L. B. *McClintock Research Notebook I,* p. 13, May 9, 1996.

DOI: 10.1201/9781003332527-2

GROWING UP IN AN UNCONVENTIONAL FAMILY

Nobel Laureate Barbara McClintock was born into an educated family. Why then would her mother have discouraged her children, specifically her youngest daughter Barbara, from seeking higher education? A newspaper article published in 1948 precipitated my search to answer this question. My findings also helped clarify and verify certain family history up to the time that Barbara McClintock entered University.

The local *Brooklyn Daily Eagle* newspaper's account of the 50th wedding anniversary of McClintock's parents provides a story that her parents married while "Dr. McClintock was in his senior year at Boston University and his bride was a student at Emerson College."[4] Thirty years later, the family recalled that McClintock's mother at first discouraged her daughter Barbara's college aspirations, possibly because of concerns about money and a fear that "too much education was likely to make a young woman unmarriageable" (Keller 1993, p. 26). Barbara McClintock recalled that her mother was "afraid" that she would become "a college professor" (Keller 1993, p. 26). Did this fear stem from her mother's own experiences at Emerson College of Oratory (now Emerson College), in Boston, which she attended for two years (1896–1898) but did not graduate.[5] While attending college, "the former Sarah Handy" met and married her future husband, Thomas Henry McClintock,[6] who was enrolled at Boston University School of Medicine (*Boston University Year Book* 1894; Sutherland 1905, p. 204). Keller (1993, p. 19) reported that they married shortly before he graduated, and Porter (1898, p. 603) records his graduation date as June 1, 1898. Did "Sara Ryder" Handy's time at College open her eyes to the future for female students or women faculty during this time? Or was she concerned that coeducational college life, away from the oversight of family, might preclude her daughters from completing college, as had happened to her? The McClintock's celebrated their golden wedding anniversary on January 3, 1948, which would have placed the birth of their first child ten months after their marriage. The legal documents record their marriage on March 30, 1898, "Sara R. Handy, student to Thomas H. McClintock student (med) in Essex" (Massachusetts Marriages, 1841–1915). Their first child was born on October 23, 1898, seven months after their legally recorded marriage.

Sara Handy McClintock seems to have aspired to the aims of her college alma mater, Emerson College of Oratory. This coeducational institution was established as a private school in 1880, and incorporated in 1886 as a College by the Commonwealth of Massachusetts. The college qualified students to "become professors and teachers of elocution and oratory in the colleges and schools of the land ..." (Emerson College of Oratory [ECO] 1896, p. 8). Sara seems to have believed in and followed the Colleges' goals that "The individuality of the student is of the first importance" (ECO 1896, p. 9). She taught independence to her young children, and aimed to make these students "not only readers but thinkers as well" (ECO 1896, p. 8): "The pupil is not to be an imitator of his teacher ..." (ECO 1896, p. 9). The stories McClintock told to Keller (1983, 1993), among others, regarding her mother's different approaches to parenting from those of her friends, and her mother's defense of her four children from "... pressure to conform to social expectations ..." (Keller 1993, p. 28), reflect the importance her mother must have felt for her children's individuality – at least in their early childhood years. It also may have given Barbara McClintock the confidence to pursue new ideas and to be comfortable with non-conformity regarding societal norms.

McClintock's father, Thomas Henry McClintock M.D., knew the value of an education for women. His 1898 class from Boston University School of Medicine included 46 graduates – more than one-third women (Porter 1898, p. 603). The institution evolved from the Boston Female Medical School (1848) to the New England Female Medical College (1852), and in 1873 merged

[4] The *Brooklyn Daily Eagle* (henceforth *BDE*), January 4, 1948, p. 16.

[5] Emerson College Archives, Archivist Rosalie Gartner email to L. Kass, September 14, 2016; Emerson College of Oratory (1897, p. 46; 1898, p. 42), Handy, Sara Ryder, Record Card, 1896–1897, 1897–1898, College Archivist Rosalie Gartner, to L Kass, September 16, 2016.

[6] *BDE*, January 4, 1948, p. 16.

with the newly established Boston University to become its homeopathic medical school, known as Boston University School of Medicine (BUSM). The detailed history of the Medical School is reported by Sutherland (1905, pp. 159–213): "... from its inception Boston University freely welcomed women to all the opportunities it offered to men" (Sutherland 1905, p. 165), admitting students of "both sexes on uniform terms and conditions" (Sutherland 1905, p. 183). The school taught homeopathic medicine, which was introduced into America around 1825. The new medical practice was soon embraced by the Massachusetts Homeopathic Medical Society, which had encouraged the establishment of a Homeopathic Medical College. At his quarter-century address (1898), President Warren reminded attendees that Boston University was the *first university ever organized, logically and from the start, on principle of no discrimination in privilege on the ground of sex* (Sutherland 1905, p. 187).[7] Clearly, Dr. McClintock M.D. would have had no difficulties understanding his daughter's wish for higher education, or illuminating that view to his wife.

CHILDREN OF MANY NAMES

Barbara McClintock was born on June 16, 1902, in Hartford, Connecticut, the third child of [Grace] Sara Handy and Thomas Henry McClintock. Originally given the name Elenore, her parents called her Barbara, which they felt was better suited to her independent personality. She recalled that her father had officially changed her name in 1943, for fear she might not be able to obtain a passport.[8] It was more likely a passport renewal, since she had provided her passport number to the Guggenheim Foundation in 1933, thus permitting the foundation to transfer funds to her overseas account, while on her fellowship in Germany.[9] The family recalled that a little over a year following the birth of their younger brother "Tom" on December 3, 1903, Barbara went to live with her paternal Aunt and Uncle, to relieve the strain on her mother, who was looking after "four small children with little help" (Keller 1993, p. 20). By December 1904, Barbara was just over two years old, and her older sister, [Sara] Marjorie (born on October 23, 1898, in Searsport, Maine),[10] age 6, may have been attending a Connecticut elementary school. But [Edna] Mignon (born on November 13, 1900, Hillsboro, New

[7] Cornell University admitted the first woman undergraduate student in 1870, two years after its founding.

[8] E. F. Keller oral history interview September 24, 1978, APS; Spectre (1993, p. 19), Name change notarized Affidavit, May 27, 1943, Series II, Box 8. Comfort (2001, p. 20, note 9, p. 283) reported the name change as of June 1943. Note: The "Second War Powers Act" passed by the U.S. Congress on March 27, 1942, temporarily repealed the confidentiality of census data, permitting the U.S. Secret Service [established 1865] to obtain this information. Title XIII of this act permitted the "INSPECTION AND AUDIT OF WAR CONTRACTORS" (Second War Powers Act, 56 Stat. 176, Senate Bill 2208, March 27, 1942, https://uscode.house.gov/view.xhtml?req=granuleid%3AUSC-1999-title50a-node230&edition=1999). Members of the Department of Genetics at Cold Spring Harbor had discussed working on war-related research. Knowing that Barbara McClintock officially became a member of this Department beginning January 1943 may have inspired McClintock's father to officially change her name, which he did on May 27, 1943, in New York City. McClintock explained in a letter for the Mayer Prize that her father officially changed her name, McClintock to R. Courrier, December 6, 1982, File: Charles Leopold Mayer Prize, APS, Spectre (1993, p. 13).

[9] Barbara McClintock Guggenheim Fellow, Guggenheim Foundation Files, 1932–1935.

[10] Marjorie McClintock b. October 1900 Census Record, age 1; "Maine Births and Christenings, 1739-1900," https://www.wikitree.com/wiki/McClintock-587, b. "25 October 1923" (*sic*), but a Passenger Search (accessed June 22, 2022) for married name Marjorie McKinley provided a 1932 Ship's manifest: b. October 23, 1898, traveling with children Howard [McKinley], b. April 13, 1925, Peter [McKinley], b. October 20, 1926, Lewis (1925, p. 10, #131), recorded Sara Marjorie, b. October 23, 1898, but misdates marriage of "Sara Grace Ryder" to "Thomas Henry McClintock," and she recorded the marriage of "Sara Marjorie" [McClintock] to Howard Alexander MacLean, and his death on September 6, 1924, but makes no mention of their child born on April 13, 1925, which might indicate that Lewis' (1925) records were made before the birth of Howard Alexander MacLean [Jr.] to the widow Marjorie [McClintock] MacLean; also, Lewis does not record the second marriage of Marjorie McClintock MacLean to L. P. McKinley, September 4, 1925, Cert. no. 13440, Kings County [Brooklyn], New York (German Genealogy Group, Brides and Grooms index, http://www.germangenealogygroup.com/records-search/grooms.php), or the name change of her son to Howard Alexander Maclean McKinley, following the marriage to L. P. McKinley.

Hampshire),[11] age 4, and not yet of school age, would have remained at home with Barbara, and her baby brother Tom (christened Malcolm Ryder), at least during the day, before her sister returned from school. So at about two years of age, Barbara was sent to her father's sister's family to live, "on and off until she reached school age" (Keller 1993, p. 20). McClintock recalled that she enjoyed living with her Aunt and Uncle and was not homesick (Keller 1993, p. 20). She recalled traveling by horse and carriage and later by motor car with her fish dealer Uncle to sell his wares to the families of Massachusetts. Observing the constant repairs required of his car caused her to wish for tools of her own. Her father recalled that she asked him for a set of tools when she was age 5 (Keller 1993, p. 22), which might offer a clue that she may have been with his brother's family "on and off" for about three years. By that time (1907), both of her older sisters, by then ages 10 and 8, would have been attending school, and only Barbara and Tom would have been at home, possibly relieving the stress on her mother.

FORMATIVE YEARS IN BROOKLYN, NEW YORK

The McClintock family moved to Brooklyn in 1908 (Sutherland & Watters 1908, p. 236), and the children attended Public School 139 in Flatbush.[12] Dr. T. H. McClintock, M.D., was at first in private practice at his home, which was a common venue for medical doctors at that time (*American Medical Directory* 1906ff.). His Army National Guard unit had been called to service on the Mexican border in June 1916 (Ryan 1921a: Chapter 2). He then worked with the U.S. Public Health Service until October 1920, when he was hired by the Marine Medical Department of the Standard Oil Company of New Jersey, from where he retired in 1940, as Chief Medical Officer (U.S. Census 1920, January 9, 1920; Anonymous 1941; *American Medical Directory* 1942, p. 1261).

After graduating from the eighth grade, the children went on to the local coeducational high school, Erasmus Hall, where they had an opportunity to take academic courses to prepare for college. In addition to high school academics, extracurricular activities engaged Barbara McClintock's attention (1915–1919). Contrary to reports that McClintock was a "loner," local newspapers, along with the third Erasmus Hall High School Year Book, *The Flying Dutchman III* (1920), provides many accounts of the extracurricular high school activities of Barbara McClintock and her three siblings. I found no record that Barbara engaged in the politics of women's suffrage or theater, both of which her older sisters Marjorie and Mignon were much involved,[13] although she did take part in many adolescent girls' sports and other high school social organizations. In the spring term of 1917, during Barbara's sophomore year at Erasmus Hall High School, she was elected "second year captain of the Girls Swimming Club."[14] In January 1917, she and 19 other girls in the swimming club qualified for the Public Schools Athletic League (P.S.A.L; *The Flying Dutchman III* 1920, p. 135). She was also among the first members of the Girls' [horseback] Riding Club organized in 1917 (*The Flying Dutchman III* 1920, p. 75).

Upon graduating high school in January 1916, the McClintock's oldest daughter [Sara] Marjorie had declined an invitation to attend Vassar College and instead pursued her life's dream as an

[11] Mignon McClintock, 1928, Ship Manifest, Carmania, arriving from France, line 9, lists b. "13 II 1900" [November 13, 1900], D [Divorced] (Passenger Search). Birth certificate for "Edna Mignon McClintock," November 13, 1900, Father T. H. McClintock Physician, Mother Sara R. Handy (New Hampshire Births and Christenings, 1714–1904). See also Lewis (1925, p. 10), Mignon b. November 13, 1900, "Malcom Rider" (*sic*?), born in Hartford, Ct. December 3, 1903. Many records use "Ryder" for S. Handy's mothers' name, e.g., Lewis (1925, p. 10); Massachusetts Marriages, 1841–1915. In contrast, Keller (1993, pp. 18–19) spelled the name "Malcom Rider," with no source citation. The 1940 Census reports the name as "Malcom Ryder McClintock," Sixteenth Census of the United States, 1940, NARA digital publication T627. Records of the Bureau of the Census, 1790–2007, RG 29. Washington, D.C.: National Archives and Records Administration, 2012, roll 2611.

[12] *BDE*, Friday, January 26, 1912, p. 6, Graduates from Brooklyn Public Schools, Public School 139, "Sara Marjorie McClintock"; *BDE*, Friday, March 08, 1912, p. 13, Public School 139, "Barbara McClintock" [name listed for Grade 3 is possibly an error-she would have been in fifth grade at this time].

[13] *BDE*, September 24, 1915, p. 19; September 29, 1915, p. 27; April 15, 1916, p. 7.

[14] *BDE*, February 11, 1917, p. 29.

actress.[15] At that time, it seems Marjorie had rejected a personal invitation from Vassar College President H. N. MacCracken to attend his institution, which was dedicated to women's education. She resisted her parents' encouragement to accept the offer from this notable proponent of women's suffrage who gave the mid-year graduation speech at Erasmus Hall High School. In fact, by February 1919, Miss Marjorie McClintock "who gave up a college career" had become a success on Broadway.[16] It is not known whether Marjorie's younger sister Barbara, then in her first year of high school, was aware of the family deliberations regarding her older sister's future to return for a college education, if her dream did not succeed after a two-year trial. Within a year of graduation (January 1916), however, Marjorie McClintock was in a 1917 production at The Winter Garden on Broadway, and by 1918, she became a member of the Washington Square Players, staring in plays written by Theodore Dreiser and Eugene O'Neill, thus precluding a return to academic life.[17]

By April 1917, while the three youngest McClintock children were still attending school, America was drawn into the Great War, which began in Europe three years previously. Within a week of the U.S. declaration of war on Germany, on April 6, 1917, Barbara McClintock's father was commissioned Major in the U.S. National Guard (23rd Infantry, New York) on April 11, 1917. On October 1, he was ordered to transfer to Federal Service in the 12th Infantry from the 23rd NY upon arrival at the 27th Division Headquarters, Camp Wadsworth, Spartanburg, South Carolina (Ryan 1921b, p. 604; Dearborn 1923, p. 324). He was still at Camp Wadsworth in January 1918,[18] and was transferred to Surgeon of the Second Pioneer Infantry, Camp Wadsworth, in February 1918 (Dearborn 1923, p. 324), which would have precluded his attendance at Mignon's 1917–1918 mid-year graduation – delayed until February 6, 1918, due to a coal shortage that winter.[19] By September 1918, as his youngest daughter Barbara was entering her senior year in high school, Major McClintock was overseas with the American Expeditionary Forces (A.E.F.), Combat Operations, Camp Hospital No. 8.

In her senior year (1918–1919), Barbara McClintock was present at the first meeting, held on Wednesday, September 18, 1918, of the Erasmus Hall High School girls' debating society, *The Witan*.[20] Along with other seniors in the class of 1919, she played and won (2-0) the Erasmus Hall junior-senior championship game in girls' field hockey held at Prospect Park, Brooklyn, on Saturday, November 23, 1918.[21] Soon after, a photograph and listing of names for the Erasmus Hall High School January 1919 mid-year graduating class were depicted in the local newspaper.[22] Barbara McClintock's photo and name are *not included* (my emphasis), signifying that she did not graduate early in 1919.

During this time, the McClintock family also weathered the 1918 flu pandemic, which arrived in New York Harbor in August 1918. Fortuitously, New York City fared much better with respect to the rest of the county, thanks to the efforts of the experienced staff of the New York City Health Department (Toole 2018). The city's schools were not closed by decree of the city's Board of Health,

---

15 *BDE*, January 28, 1916, p. 4: "Vassar President Talks to Erasmians. Dr. Henry Nobel MacCracken ...;" speech to the Erasmus Hall High School graduation class, January 27, 1916; "Sara Marjorie McClintock" listed among 145 of "the graduates" from Erasmus Hall H. S. Flatbush, Brooklyn.

16 *BDE* "Flatbush Girl Wins Success on Stage: Miss Marjorie McClintock, Former Erasmus student, Gave Up College Career for Footlights," Sunday, February 23, 1919, p. 15; see also *BDE*, January 28, 1916, p. 4.

17 The Washington Square Players was forced to disband in 1918 due to wartime economic pressure and personnel shortages, https://en.wikipedia.org/wiki/Washington_Square_Players#Dissolution; Marjorie McClintock acted in Broadway productions of *The Girl in the Coffin* Comedy Theatre (October 31, 1917–April 27, 1918), Produced by The Washington Square Players, Written by Theodore Dreiser; and *In The Zone* (October 31, 1917–April 27, 1918), Comedy Theatre [Comedy Theatre (1913–1937), now Mercury Theater], Produced by The Washington Square Players, written by Eugene O'Neill, https://www.ibdb.com/broadway-production/in-the-zone-8588; she also is listed for the 1917–1918 theater for two additional plays, *The Avenue* by Fenimore Merrill, *Neighbors* by Zona Gale; https://www.ibdb.com/broadway-production/the-girl-in-the-coffin-8597; she was featured in the musical in two acts, *Flo-Flo*, Cort Theatre [Broadway] (December 20, 1917–June 29, 1918), produced by John Cort, 220 performances [Now James Earl Jones Theatre, 138 W. 48th St., New York, NY, https://www.ibdb.com/broadway-production/flo-flo-8636].

18 *BDE*, January 24, 1918, p. 3.

19 *BDE*, January 11, 1918, p. 15; January 24, 1918, p. 4, February 6, 1918, p. 3.

20 *BDE*, September 19, 1918, p. 8.

21 *BDE*, November 26, 1918, p. 5.

22 *BDE*, January 30, 1919, p. 19.

led by Health Commissioner Royal S. Copeland.[23] Like McClintock's father, he too had training in homeopathic medicine, and had just stepped down as a dean of the New York Medical College, formerly New York Homeopathic Medical College, to become the president of the city's Board of Health. The City experienced fewer deaths by utilizing public education (explaining how disease is spread), quarantine, and decentralized medical care, employing visiting nurses to provide care at home. Meanwhile, McClintock's father had been fighting the flu in field hospitals on the battlefields of France. Between 1918 and 1920, the flu infected 300 million people worldwide, 1/3 of the world's population. It claimed the lives of approximately 50 million people globally and 675,000 in the United States. Furthermore, no vaccines or antibiotics were yet available to treat secondary bacterial infections associated with the virus. We do not know whether any of McClintock's family was affected, but fortunately, all members survived the pandemic.

Barbara McClintock graduated at age 17 from Erasmus Hall High School on June 26, 1919.[24] Her photo appeared in the local newspaper along with all of her classmates (Figure 1.1). Only students graduating with distinction or receiving medals were individually named in the news article; graduating senior Barbara McClintock's name is not among those so honored. In contrast, when her sisters, Sara Marjorie and Mignon, had mid-year graduations in January 1916 and January 1918, respectively, their entire graduating classes were named in the local newspapers.[25] The family seems to have conflated her sister Mignon's early graduation from high school (January 1918) with that of Barbara (Keller 1993, p. 30), who – from period reports – neither graduated a semester early in the academic years 1917–1918, nor 1918–1919. Furthermore, had she graduated in mid-year, at age 16, she would not have been eligible to attend Cornell University. The entry requirement for women was age 17, in contrast to age 16 for male students (Cornell University 1920, p. 45). She would have achieved age eligibility on June 16, 1919, only ten days before graduating from high school.

McClintock's father's extended stay overseas as Post Surgeon for the post-war Advanced Ordnance Depot No. 4 (A.E.F.), commencing in February 1919 through August 1919 (Dearborn 1923, p. 324), would have prevented his attendance at his daughter Barbara's high school graduation

FIGURE 1.1 Barbara McClintock (second from left, second row from top) Erasmus Hall High School graduation photo, June 27, 1919, *Brooklyn Daily Eagle*, p. 22. McClintock graduated high school on June 26, 1919. (Public domain; image courtesy of Brooklyn Public Library, Center for Brooklyn History.)

[23] R. S. Copeland was later elected to three terms as New York's Democratic U.S. Senator, in 1922, and his term was cut short by his death on June 17, 1938. See also Kolata (1999) for a perspective on the search for the virus that caused the 1918 flu.

[24] *BDE*, Friday, June 27, 1919, Erasmus Hall High School class photo, p. 22 [Section 2, p. 4]; McClintock is included in the photograph of the class of 1919 (upper right); see Figure 1.1.

[25] *BDE*, January 28, 1916, p. 4; January 11, 1918, p. 15.

in June 1919. He had remained in France following the Armistice of November 11, 1918, and was discharged from his appointment as Post Surgeon, with the A.E.F., France [Joncery-Villers-le-Sec (north of Dijon)],[26] on August 4, 1919 (Dearborn 1923, p. 324).

Only upon his return home, toward the end of summer 1919, would Major McClintock have learned of his daughter Barbara's wish to attend coeducational Cornell University. On June 9, 1919, a local newspaper had announced the names of "Erasmians going to College."[27] Among the dozen students reported to be heading to Cornell University was Barbara McClintock. When interviewed years later, McClintock did not recall when she had made the decision to attend Cornell (Keller 1993, p. 30), but she did remember that her mother was not in favor of the idea.

Major McClintock's presence at home seems to have influenced his wife's change of mind to sanction their third daughter, Barbara, to attend college (Keller 1993, p. 30). Sara McClintock finally agreed with her husband's decision to support their daughter's wishes. Barbara recalled that just before leaving for work one morning, her mother told her they decided she would go to college. Her mother went to the high school to see about credentials for Barbara's entry to Cornell. Barbara recalled heading to Ithaca on the train, but Harriet Creighton, McClintock's former graduate student, told me that Barbara said she got a ride with the parents of a friend, who was returning to classes for the fall of 1919. Either way, in her memory she arrived without papers to present to the registrar, but somehow they appeared when her name was called. Her older sister, Marjorie, believed that her resourceful mother would have "done anything necessary to see that Barbara was admitted, once the decision was made" (Keller 1993, p. 31).

By the time Barbara McClintock was considering attending college in early June 1919, three and a half years after Marjorie's decision not to enroll at Vassar College, it is not completely clear why Mrs. Sara Handy McClintock would have discouraged her youngest daughter from pursuing higher education at a college of her choice. Despite these barriers, in September 1919, Barbara McClintock entered the New York State College of Agriculture, at Cornell University, Ithaca, New York, as a freshman.[28] The Erasmus Hall High School Year Book (*The Flying Dutchman III* 1920, p. 105) promptly reported that "Barbara McClintock, class of '19 [1919], is president of the freshman class at Cornell."

LITERATURE CITED

American Medical Directory. 1906(1st edition)–1992. American Medical Association, Chicago, IL.

Anonymous. 1941. Dr. Thomas H. McClintock Retires. *The Medical Bulletin* (Standard Oil Co.) N.Y. 1941–1942, Vol. 5, pp. 85–87. [Issued by the Standard Oil Company (New Jersey) and affiliated companies, 1932–72; by Exxon Corporation and affiliated companies, 1973–]

Boston University Year Book. 1894. Vol. XXL, Boston University Offices, Somerset Street, School of Medicine, First Year Class, Thomas Henry McClintock, p. 166.

Cornell University. 1920. *The Register of Cornell University* 1919–1920, September 1, 1920. [Admission to the Undergraduate Colleges, Admission of Freshmen (p. 45); Students, 1919–1920: Barbara McClintock (p. 290)]. https://ecommons.cornell.edu/handle/1813/22309.

Comfort, N. 2001. *The Tangled Field*, Harvard University Press, Cambridge, MA.

Cullen, J. H. 2003. *Barbara McClintock*, Chelsea House, New York, NY, Juvenile Nonfiction, 122 pp.

Dearborn, F. M. (Ed.). 1923. *American Homeopathy in the World War*, American Institute of Homeopathy, Chicago, IL. https://archive.org/details/americanhomeopat00fred/page/n7/mode/2up

Emerson College of Oratory. 1896. *Annual Catalogue of Emerson College of Oratory, 1895 and 1896*, The Barta Press, Boston, MA. https://books.google.com/books?id=6ghEAQAAMAAJ&printsec=frontcover#v=onepage&q&f=false

Emerson College of Oratory. 1897. *Annual Catalogue of Emerson College of Oratory, 1896 and 1897*, The Barta Press, Boston, MA. [SRH, p. 46, freshmen year]

[26] National Archives, Records of the A.E.F. WWI, RG 120.76.

[27] *BDE*, June 9, 1919, p. 8.

[28] Barbara McClintock Record Card. Division of Rare and Manuscript Collections, Cornell University Library. Cornell University (1920): CALENDAR, First Term 1919–1920, registration of new students September 29 & 30 (p. 5), "Students, 1919–1920" (p. 251), "McClintock, Barbara, 1923 Ag [Agriculture], Brooklyn" (p. 290).

Emerson College of Oratory. 1898. *Annual Catalogue of Emerson College of Oratory, 1897 and 1898*, The Barta Press, Boston, MA. [SRH, p. 42, junior year]
Fedoroff, N. 2002. The well mangled McClintock Myth. *Trends in Genetics* Vol. 18 (No. 7, July 1, 2002), pp. 378–379.
Fine, E. F. 1998. *Barbara McClintock: Nobel Prize Geneticist*, Enslow Publishers, Springfield, NJ.
Gardner, A. L. 1997. Barbara McClintock, geneticist, 1902–1992, pp. 307–324, *In*, M. L. Matyas and A. E. Haley-Oliphant (Eds.), *Women Life Scientists: Past, Present, and Future – Connecting Role Models to the Classroom Curriculum*, American Physiological Society, Bethesda, MD.
Handy, S. R., Record Card, 1896–1897, 1897–1898, Emerson College Archives, RG 032.01. [Documentation from Emerson College Archivist Rosalie Gartner, September 16, 2016]
Heiligman, D. 1994. *Barbara McClintock, Alone in Her Field* (Science Superstars), illustrated by Janet Hamlin. W. H. Freeman & Co. (August 1, 1994) 64 pp. [Children biology book, 9–12 years]
Kass, L. B. 2002. Book review, *The Tangled Field*, by N. Comfort. *Isis* Vol. 93 (No. 4), pp. 729–730.
Kass, L. B. 2003. Records and recollections: A new look at Barbara McClintock, Nobel-Prize-Winning geneticist, *Genetics* Vol. 164 (August), pp. 1251–1260. https://www.ncbi.nlm.nih.gov/pmc/articles/PMC1462672/
Keller, E. F. 1983. *A Feeling for the Organism, The Life and Work of Barbara McClintock*, W. H. Freeman & Co., San Francisco, CA.
Keller, E. F. 1993. *A Feeling for the Organism, The Life and Work of Barbara McClintock*; reprinted with an author's note on the Tenth Anniversary edition, W. H. Freeman & Co., San Francisco, CA.
Kittridge, M. 1991. *Barbara McClintock: Biologist* (American Women of Achievement series). Chelsea House, New York.
Kolata, G. B. 1999. *Flu: The Story of the Great Influenza Pandemic of 1918 and the Search for the Virus That Caused It*, Farrar, Straus and Giroux, New York, xi, 330 pp., illustrated.
Lewis, E. M. (Mrs. Seth Ames Lewis). 1925. *Handy Family, Descendants of Richard of Sandwich, Mass.* Springfield, MA. https://archive.org/details/handyfamilydesce00lewi/page/10
Massachusetts Marriages, 1841–1915. Database with images, FamilySearch, (https://familysearch.org/ark:/61903/1:1:N4H2-493: 17 Februaty 2016), Thomas H. McClintock and Sara R. Handy, 30 March 1898; citing, Boston, Massachusetts, United States, State Archives, Boston; FHL microfilm 1,843,709.
Pasachoff, N. E. 2006. *Barbara McClintock: Genius of Genetics*, Enslow Publishers, Berkeley Heights, NJ.
Passenger Search. The Statue of Liberty-Ellis Island Foundation, Inc. https://heritage.statueofliberty.org/passenger
Porter, E. H. (Ed.). 1898. Boston University School of Medicine Commencement, June 1, 1898. p. 603, T. H. McClintock. *Transactions of the Fifty-Fourth Session of the American Institute of Homeopathy, held at Omaha, Neb.* June 23, 1898. Volume 54. New York: Trow Directory, Printing and Bookbinding Co.
Provine, W. B. and P. Sisco. 1980. Interview with Barbara McClintock at Cold Spring Harbor Laboratory, Long Island, New York, August 28, 1980. Division of Rare and Manuscript Collections, Cornell University Library, Ithaca, NY.
Ryan, J. F. 1921a. *The Story of the 27th Division*, Table of Contents, 562 pp. [Vols. I & II, Total 1157 pp.] https://museum.dmna.ny.gov/unit-history/conflict/world-war-1-1914-1918/27th-infantry-division/story-27th-division-john-f-oryan-table-contents. [For Exhibits listed as unavailable see Ryan 1921, Vol. II Appendix A, Exhibits, beginning p. 569]
Ryan, J. F. 1921b. *The Story of the 27th Division*, Vol. II, Appendix A, Exhibits, Wynkoop Hallenbeck Crawford Co., New York, pp. 569–1157, https://archive.org/details/cu31924086494394
Spangenburg, R. and D. K. Moser. 2008. *Barbara McClintock: Pioneering Geneticist*, Chelsea House/Infobase Publishing, New York.
Spectre, M. 1993. *Barbara McClintock Papers. Guide to the Collection.* American Philosophical Society Library, Philadelphia, PA. (Online: Barbara McClintock Papers Mss.Ms.Coll.79, https://search.amphilsoc.org/collections/view?docId=ead/Mss.Ms.Coll.79-ead.xml) [Also see selected papers and summary of APS Library papers of McClintock available for the National Library of Medicine, Profiles in Science: https://profiles.nlm.nih.gov/spotlight/ll; The collection also reprints papers/letters from other institution's archives]
Sutherland, J. P. 1905. Chapter V. Boston University School of Medicine, pp. 159–213, *In*, W. H. King (Ed.), *History of Homeopathy, and Its Institutions in America*, Vol. III, The Lewis Publishing Co., Chicago, IL.
Sutherland, J. P. and W. H. Watters. 1908. "Persons and General Items; Dr. Thomas H. McClintock Class of 1898B.U.S.M. has removed to 1802 Church Ave, Brooklyn, NY." *The New England Medical Gazette* Vol. 28, p. 236.
The Flying Dutchman III. 1920. Year Book of Erasmus Hall High School, Brooklyn, NY.
Toole, P. 2018. *The Flu Epidemic of 1918*. NYC Department of Records & Information Services. https://www.archives.nyc/blog/2018/3/1/the-flu-epidemic-of-1918 (accessed, March 17, 2020).

2 Genetics in the Roaring '20s

If I have seen further it is by standing on the shoulders of Giants

– Sir Isaac Newton, 1675

Barbara McClintock arrived at Cornell at a time when genetics was developing rapidly after the rediscovery (1900) of Mendel's experiments on plant hybrids, and its teachings, and when Darwin's evolutionary theory was being challenged in the courts of the land (1923). The end of World War I (1918) and the passage of the 19th Amendment (1919, ratified 1920) combined to make the 1920s a decade where women could indeed become scientists. Documents in the Cornell University Archives indicate that McClintock was part of the group of 222 women out of 272 in her graduating class who studied subjects other than domestic economy. The challenge was the social expectations faced by these educated women and their limited opportunities for careers in science (Rossiter 1982, pp. 163–168; Kass 2005; Plant Breeding Records 1906–).

The comprehensive undergraduate curriculum required by the Cornell University College of Agriculture ("Ag School") was taught by distinguished faculty who had a strong commitment to both teaching and research. McClintock's undergraduate curriculum and the encouragement she received from the faculty provided a strong foundation for her future development as a cytogeneticist (Kass & Provine 1997; Kass 1999, 2003). By February 1923, McClintock had completed all course requirements for the Bachelor of Science in Agriculture and was "permitted to register grad," while completing the residency and one PT [Physical Training] class required for her B.S. degree, awarded June 1923.

THE BROAD UNDERGRADUATE CURRICULUM AT THE NEW YORK STATE COLLEGE OF AGRICULTURE, CORNELL UNIVERSITY

Barbara McClintock received a very comprehensive undergraduate education while enrolled in the College of Agriculture at Cornell University. In 1923, she graduated with a B.S. in Agriculture, with a concentration in Plant Breeding and Botany.[1] A review of the required courses for students majoring in Agriculture shows the broad background required by students studying in the College at this time (Table 2.1).[2] These documents, among many others, refute the legends that McClintock was narrowly focused on one organism and thus knew little about subjects other than genetics of maize.[3]

New students registered at the New York State College of Agriculture at Cornell University on Monday and Tuesday, September 29 and 30, 1919. Barbara McClintock registered with the class of 1923 to pursue a Bachelor of Science in Agriculture, with a concentration in Plant Breeding and Botany.[4] She had graduated from high school soon after reaching her 17th birthday on June 16, 1919. Thus, she would not have been granted permission to enter Cornell University any earlier because female candidates for admission to the regular four-year undergraduate course had to be at least

[1] McClintock Registrar's Card, McClintock Graduate School File, Division of Rare and Manuscript Collections, Cornell University Library (henceforth CU).

[2] Announcements of the *New York State College of Agriculture* from 1919 through 1923; *The Register of Cornell University 1919–1920.*

[3] J. Lederberg recollections of McClintock's narrow educational background from an interview with N. Comfort, March 23, 1996, which Lederberg shared with LBK January 22, 1998.

[4] McClintock's Graduate School File includes an application for graduate school listing her undergraduate course of study as "Plant Breeding and Botany"; McClintock Graduate School File.

 DOI: 10.1201/9781003332527-3

TABLE 2.1
Required Courses [45 hours], Chosen from Those Listed, Leading to the Degree of Bachelor of Science at Cornell University as Detailed in the *New York State College of Agriculture Announcement 1919–1920* (Issued June 15, 1919)

English	6 hours
Botany, Biology, or Zoology	6 hours
Chemistry or Physics	6 hours
Physiology, one of the following:	
Physiology of Domestic Animals	3 hours
Human Physiology	3 hours
Plant Physiology	4 hours
Political Science	6 hours
Botany, Zoology, Bacteriology, Chemistry, Physics, Geology, Physical Training, Mathematics, Drawing	18 hours

"Not less than 24 hrs. of the required work is to be taken in the freshman year, including English; botany, biology, or zoology; and physics or chemistry" (p. 26).
"When an option of required courses is offered, consideration should be given to the prerequisites demanded by the elective courses to be taken subsequently" (p. 27).
"Botany 1 is prerequisite for further work in botany, ... for courses in plant breeding ..." (p. 27).
"Botany 20 [Plant Physiology] is prerequisite for ... courses in plant breeding ..." (p. 27).
Required courses not in the College of Agriculture are described on pp. 81–83.
All students were required to register for a minimum of 12 hours, and new students could register for no more than 18 credits (p. 26). Additionally, freshmen and sophomore women were required to take, "throughout the year, three periods a week" (p. 23) of "Physical Training for Women" consisting of "outdoor games and exercises from the beginning of the year to Thanksgiving, and from the Easter vacation to the end of the year. From Thanksgiving to Easter the work is in large part indoors, and consists of floor exercises, folk and aesthetic dancing, and indoor games, in all of which certain prescribed tests must be met at the end of each period. At the beginning of the college year, six lectures on health and hygiene are given attendance upon which is compulsory for all first year students" (p. 83).
Students receiving grade C are to receive normal credit toward graduation; grade B, 10% additional credit; grade A, 20% additional credit; grade D, credit reduced 10%; and grade P, 20% reduced credit. No student may be graduated in less than eight terms unless his work in the College of Agriculture averages 15% excess credit (p. 28).

17 years of age; male candidates were admitted at age 16.[5] One hundred and twenty hours of required and elective work toward graduation were outlined in the *New York State College of Agriculture Announcement* published in June 1919.[6] Students were expected to take 45 hours of required courses (Table 2.1), and the remaining 75 hours were chosen with the help of an advisor from a description of elective courses listed under various departments in the College of Agriculture ("Unless otherwise noted all classes were given in the buildings of the College of Agriculture," p. 29). During their required eight terms of residence, students were permitted to take up to 20 hours of elective credits

[5] *New York State College of Agriculture Announcement 1919–1920*, p. 2 (hereafter *NYSCOAA 1919–1920*).
[6] *NYSCOAA 1919–1920*, pp. 25, 26.

chosen from courses "in other colleges than Agriculture."[7] Students were permitted to graduate early (in fewer than eight terms) if their work in the College of Agriculture averaged 15% excess credit (Table 2.1). McClintock was permitted to enroll as a graduate student after seven terms, but she did not have the excess credit to obtain residece credit; thus, she was awarded her bachelor's degree after four full years in residence at Cornell (McClintock, Barbara, Registrar's Card). In the "*Announcement*," students could also find a list of the "Staff of Instruction and Extension Work,"[8] which included the names and titles of all the Professors, Assistant Professors, Instructors, and Assistants involved in teaching in the College of Agriculture.[9] Also described was a history of the founding of the College of Agriculture, its buildings and farms, Department's facilities and equipment, costs of attending and availability of scholarships, and a description of the honor system.[10]

Cornell University was comprised of eight colleges in 1919: Arts and Sciences; Law; Architecture; Medical; Mechanical Engineering and the Mechanic Arts; Civil Engineering; Agriculture; and Veterinary. The latter two were state-funded and free to residents of New York State. McClintock said that she registered at the "Ag School" because she did not have to pay tuition (Provine & Sisco 1980). The engineering colleges were combined into the College of Engineering in 1921–1922. The College of Home Economics, the third New York State–funded college, did not open until 1925. Of the 840 Cornell graduates (men and women) from the class of 1923, 25% were women, and of these, 222 (82%) were not in home economics.[11]

FRIENDS AND CORNELL WOMEN IN SCIENCE

As an undergraduate, McClintock befriended women who had declared concentrations in Chemistry, Engineering, English, Agriculture, Botany, and Entomology, among other subjects taken in the various colleges at Cornell. She was friends with foreign students and students of different faiths. One Jewish friend, Emma Weinstein, later became a director of sales promotion for *Forbes Magazine*, and was active in Cornell Alumni affairs.[12] Another Jewish friend in their group was Laura Keane Zametkin, class of 1921, notable by her married name Laura Hobson, an award-winning author of *Gentleman's Agreement* (1947), which depicted "polite" antisemitism. Her parents were refugees from Czarist Russia.[13] Other friends of the class of 1923, also Russian refugees, were Helen Alexander Ziegler (Chemistry) and her older sister Mary Alexander Ziegler (Mechanical Engineering), a winner of The Cornell Scholarship.[14] Mary was courted by and later married R. A. Emerson's graduate student Milislav Demerec (Figure 2.1). Helen, who was reputed to be brilliant according to her niece Rada Demerec, had switched her concentration to Plant Breeding and was enrolled in classes with McClintock.[15] Helen was elected vice president of the International Agricultural Society, graduated a year early (1922), and continued on for a Master's degree in the Department of Plant Breeding, under the direction of the Department Chairman R. A. Emerson (Figure 2.2; Trajkovich 1923, 1924).[16]

---

7 *NYSCOAA 1919–1920*, p. 27.

8 NYSCOAA 1919–1920, pp. 3–7.

9 No rank of Associate Professor was established at Cornell at this time.

10 *NYSCOAA 1919–1920*, pp. 3–22.

11 Report of the Registrar in the Annual President's Reports, 1919–1920 through 1922–1923, and the College of Home Economics Records in CU.

12 *Cornell Alumni News*, July 1961, p. 35.

13 Zametkin, Laura Keane. Class of 1921. *The Cornellian* 1921, p. 567. *The Register 1921*, degree conferred September 21, 1921. Winner of the Barnes Shakespeare Prize. Married 1930, name changed to Laura Z. Hobson. Famous for writing *Gentlemen's Agreement* (1947). Laura Z. Hobson, June 19, 1900–February 28, 1986, by Julia Wolf Mazow; *The Shalvi/Hyman Encyclopedia of Jewish Women*, https://jwa.org/encyclopedia/article/hobson-laura. Note: Keller (1993, p. 33) reports that Hobson was not Jewish (sic) and was a friend of Barbara McClintock.

14 The *Register of Cornell University 1919–1920*, p. 191.

15 Prof. L.W. Sharp *Roll Book*, Course 10 (Cytology) Spring 1922, "Ziegler, H. (Mrs. B. Trajkovich)"; Helen Z. Trajkovich Registrar's Card, and Graduate School File, CU; LBK interview with Rada Demerec Dyson-Hudson, October 1, 2003.

16 *The Register 1922–1923*, Commencement June 18, 1923, Bachelor of Science, Helen Alexander Ziegler, p. 218, conferred September 23, 1922; Masters of Science, conferred June 18, 1923, Helen Alexander Trajkovich, B.S. Plant Breeding, Plant Pathology, p. 22. See Murphy and Kass (2011, p. 151), or Kass et al. (2005, Figure 2) for identifications of all persons in Figure 2.2.

FIGURE 2.1 Milislav Demerec joined R. A. Emerson's maize genetics group upon emigrating to the U.S. in 1919. He married Mary Alexander Ziegler in 1921 and would play a major role in Barbara McClintock's future career. (Courtesy of Rada Demerec Dyson-Hudson.)

FIGURE 2.2 Cornell University Synapsis Club, 1922–1923. Mrs. Helen A. Z. Trajkovich, Master's student of R. A. Emerson. Trajkovich, front row second from right; Emerson second row center. (Permission of Plant Breeding & Genetics, Cornell University.)

Both sisters married during this time and graduated under their married names, Mrs. Helen A. Z. Trajkovich and Mrs. Mary A. Z. Demerec.[17] Along with Barbara McClintock, they were also active in the Women's Cosmopolitan Club.[18] The Women's Club was founded in 1921, on similar principles as the seminal organization established at Cornell in 1904: the Club's motto was "Above all nations is humanity," and the organization hoped "to bring intelligent thinking [people] of different nations in such contact that they might find the best in one another; that they might learn to love, to live on common bases side by side; ... to understand the spirit of nations so that in difficult international conflicts they might have a clear judgment and correct reasoning" (Engst & Friedlander 2014). Carlton Gajdusek, 1976 Nobel Laureate in Physiology or Medicine, recalled that his aunt "Irene Dobroczki" [Dobroscky] was a friend of Barbara McClintock (Hargittai 2002, p. 125). She was also a member of the College of Agriculture, class of 1923,[19] going on to obtain an M.S. in Entomology and Bacteriology in September 1924, and in September 1928 was awarded a Ph.D. in Entomology, Plant Pathology, and Botany.[20] While pursuing their doctorates, she and McClintock attended the 1926 International Congress of Plant Sciences held in Ithaca (see Chapter 3), and Dobroscky learned cytology from McClintock and Professor Sharp in the spring of 1927.[21] After graduation, she continued research in entomology and plant diseases; and following her marriage, she was still prominent in *American Men of Science* (1938, p. 366) as "Dr. Irene D. Dobroscky (Mrs. Carleton van deWater)."

Science laboratories were often taught by graduate student Assistants or Instructors, the latter going on to successful careers. Male Instructors were often recommended for faculty appointments within the University. If permitted to remain as Instructors, women might not be promoted for decades.[22] Women Instructors had fewer opportunities, as McClintock would later learn (see Chapters 4–6). These women often obtained jobs teaching in women's colleges or as technical assistants to university professors. Helen Trajkovich and Irene Dobroscky may have been exceptions. Mrs. Mary Ziegler Demerec initially held a good engineering job with AT&T but was passed over for promotion all too often. She eventually switched fields and became a teacher of science.[23]

EXTRACURRICULAR ACTIVITIES

Respected, as well as popular, Barbara McClintock was elected President of the [Women's] Freshmen Class and thus a representative on the Women's Self-Government Association Council.[24] She recalled playing with a banjo group, but unfortunately, there is no listing for individual members of the Banjo Club, in the *1920 Cornellian* (p. 626).

ANTISEMITISM IN ACADEMIA

In her first year at Cornell, McClintock pledged and became a member of the Delta Zeta (Beta Chapter) Sorority.[25] Upon learning of its discriminatory (antisemitic; excluded Jewish students) policy, she broke her pledge (Keller 1993, p. 33).[26] McClintock also was not initiated into the Graduate Women's Scientific Honor Society, *Sigma Delta Epsilon*, founded at Cornell in 1921. A

---

[17] *The 1923 Cornellian*, photos, pp. 570, 590.

[18] *The 1923 Cornellian*, p. 650; *The 1922 Cornellian*, p. 661.

[19] *The 1923 Cornellian*, "Irene Dobroscky," photo p. 570.

[20] *The Register for 1928–1929*, p 50.

[21] L.W. Sharp, *Roll Book* 1924–1931 "124(10) Cytology 1927."

[22] See Cornell University *Registers*, and *College Announcements*, which list faculty and their ranks.

[23] L. Kass interview with Mary's daughter Rada Demerec Dyson-Hudson, October 1, 2003. See Rossiter 1982: Chapter 7, and 1995 for much detail on this subject.

[24] 1920 *Cornellian*, p. 468, McClintock is included in the photograph of the council on p. 469.

[25] *Cornellian* 1920, p. 449; *Brooklyn Daily Eagle*, October 30, 1919, p. 25.

[26] *Cornellian* 1920, p. 449, Delta Zeta Freshmen members include Barbara McClintock; later year books, did not include her in their memberships.

letter in their records, regarding inviting a student to membership, seems to reflect their antisemitic policy of excluding Jewish students.[27]

E. G. Anderson, R. A. Emerson's graduate student, was an Instructor in Plant Breeding during the 1919–1920 academic year (*Directory Cornell ... 1919–1920, The Register 1919–1920*), and McClintock's affiliation with him would loom large in her later career. He completed his doctorate in June and worked both at Cold Spring Harbor and at Columbia University, from where he wrote disparaging remarks about Jewish students in his letters to Professor Emerson.[28] In August 1922, he told Emerson that he would be teaching at the "Yiddish Emporium" [CCNY]; they offered him the same amount that he would get at Cold Spring Harbor. And that T. H. Morgan encouraged him to take the job but not for more than two years. By February 1923, along with a discussion of submitting a paper for publication to *Botanical Gazette*, he wrote, "Teaching American brand of ethics to the Lord's Chosen people is quite an interesting diversion, at least it has proven so to me." By March 1923, Anderson wrote to Emerson that his finances were "OK and the Jews still interesting." Anderson would later be appointed to the faculty at Caltech when Morgan's group moved there in the fall of 1928 from Columbia University.

Morgan had Jewish students and postdoctoral associates at Columbia University (e.g., H. J. Muller, Jack Schultz, Alexander Weinstein, graduate students; Curt Stern, postdoc), and it's not clear what Anderson's attitude was toward these colleagues. Whether Morgan was antisemitic (prejudice against persons of Jewish heritage) is controversial – he had protected one of his graduate students by suggesting that he change his name to one that would not be recognized as Jewish, since it was difficult for Jewish scholars to obtain jobs in academia during this time.[29] In 1928, Morgan's insulting offer of low rank and salary for a faculty appointment to one of his former Jewish students, Alexander Weinstein, appears to have attained its goal when the offer was refused (Goodstein 1991, pp. 203–204). Weinstein continued to have difficulty finding permanent employment (Glass 1986). Years later, Barbara McClintock received a National Research Council Fellowship to work with Anderson's maize genetics research group at the California Institute of Technology from 1931 to 1933 (see Chapter 5). It appears that by then Anderson's attitude had changed, since I found no disparaging comments about Jewish faculty or students in Anderson's papers from that time. Nevertheless, by 1934, Morgan was desperate to find someone who was not Jewish to fill a faculty appointment in his Division of Biology.[30] Furthermore, questions about attitudes concerning Jewish students continued to surface in correspondence with Cornell faculty and students, as documented in this and in forthcoming chapters.

EDUCATING CORNELL STUDENTS IN AGRICULTURE

In her first two years as an undergraduate, McClintock enrolled in the requisite courses as outlined in *The College of Agriculture Announcement for 1919–1920*. Within the required subjects, students had a great deal of choice in the classes they could take; two-thirds of the required courses were offered in colleges outside the College of Agriculture. One hundred and twenty [120] hours were required for graduation, and of those, 54% could be taken in Colleges outside of Agriculture.

[27] Letter Helen Craig to Beatrice Bolton, November 26, 1930, "no one seemed to think her Jewish," SDE Box 2, File 2-5 1930-32, CU.

[28] *The Register 1919–1920*, p. 247, E. G. Anderson, degree conferred June 23, 1920; Communications of E. G. Anderson with R. A. Emerson in File Anderson, E. G., Box 1, Plant Breeding Records 1906-.

[29] At Morgan's suggestion, Albert Titlebaum, Caltech Ph.D. 1929, changed his name to Albert Tyler when he was appointed to the embryology faculty at Caltech (*Commencement 1929*, Caltech; *Bulletin*, Annual Catalogue, December 1929 Caltech). I thank Maxine Singer for alerting me to the name change; see also Goodstein 1991, p. 205, for documentation on name change for Albert Tyler.

[30] Goodstein 1991, pp. 205–206; for documentation, see her note on p. 296, citing H. M. Miller Diary, June 7, 1934, Rockefeller Foundation Archives Center. See Engst (2006) for the status of Jewish students and faculty at Cornell University, 1865–2005.

McClintock recalled that she enrolled in many classes but only attended the ones she liked, causing her to accumulate Zs [indicating not completing classes/receiving no grade] and preventing her from graduating early. Her undergraduate record card shows that it was only in her first year that McClintock did not complete classes in which she had enrolled and that she completed her requirements for graduation one term early, and no "Zs" are recorded on her Registrar's Card. As mentioned above, only students with excess credit toward graduation were permitted to graduate early. In her first year, McClintock enrolled for 37 credit hours and completed 31. Most of her classes were in the basic sciences, and all were offered in the College of Arts and Sciences. She registered for required classes in English, Inorganic Chemistry, Geology, and Human Physiology, in addition to electives. She did not complete her electives in Music or Agricultural Chemistry. She was given permission to drop music, and she received no grade in Agricultural Chemistry. It is possible that she stopped attending the former class because she did not have the recommended prerequisite of Organic Chemistry (which she would later take in her junior year and receive outstanding grades).

In her first year, McClintock also registered for General Zoology [Zoology 1, two terms] with Professor Hugh Daniel Reed, Head of the Department of Zoology.[31] Reed earned his Ph.D. at Cornell in 1903. In 1910, he went to study with the eminent zoologist August Weismann, who in 1863 had joined the medical faculty of the University of Freiburg. In 1886, Weisman published his theory of the continuity of the germplasm, in which he noted that germ cells can be distinguished from somatic cells early in embryonic development. From this, he visualized the protoplasm of the germ cell as being passed on unchanged through the generations, and therefore, it was responsible for inheritance. Professor Reed, upon returning to Cornell, organized and was appointed Head of the Department of Zoology. There he offered the first complete course in Zoology. Reed's scientific field of interest was Zoology and Comparative Anatomy, and he taught courses in those areas. McClintock recalled that she gained an understanding of development from her courses in Zoology. Reed later became a member of McClintock's Ph.D. committee, and he required his class in Morphology of Invertebrates (Zoology 2, perquisite Zoology 1) to fulfill her minor in Zoology.

By the middle of McClintock's second undergraduate year, she completed all the required coursework toward the B.S. in Agriculture, including Political Science, Physics, and an elective in German. During this year, she elected her first course in Botany [Botany 1, General Botany, two terms], which was the prerequisite for all further coursework in Botany and in Plant Breeding. She also was introduced to her first course in Meteorology, a subject that kept her interest for the rest of her life. During this second year, 1920–1921, she enrolled for and completed 39 hours, and accumulated a total of 68 credits toward graduation.

In September 1921, the beginning of McClintock's third undergraduate year, in addition to classes in Organic Chemistry, Rural Education, and Biology of the Human Species, she took her first course in Genetics. This course, titled Plant Breeding 1, was the introductory course in Plant Breeding and was taught by Claude Burton Hutchison, Professor of Plant Breeding, and an outstanding teacher. Hutchison had earned a Master's in Agriculture from Cornell in 1913 and was immediately appointed Professor of Farm Crops at the University of Missouri, his alma mater. Because of Hutchison's renowned reputation as a teacher and researcher, he was invited to apply for a faculty position in the Plant Breeding Department at Cornell. He returned to Cornell in 1916 as Professor of Plant Breeding. The Department Head, R. A. Emerson, requested that he enroll for a doctorate at Harvard University, where Emerson had been awarded his ScD. in 1913. Hutchison immediately took a leave of absence to complete a Ph.D. at Harvard's Bussey Institution, with Edward Murray East and William Ernest Castle, two of the most prominent plant and animal geneticists of the time. Hutchison returned to Cornell in 1917 after obtaining a Master's degree from Harvard. I found no record of why he did not complete his Ph.D. At Cornell, Hutchison soon broadened the introductory and advanced genetics courses in the Plant Breeding Department to include relevant zoological research. McClintock recalled that Hutchison's class was one of the best courses she had taken at Cornell, and it inspired her continued interest in genetics (McClintock 1987; Provine & Sisco 1980).

---

[31] Faculty Folder, Reed, Hugh D. CU.

Hutchison's introductory genetics course (Plant Breeding 1) was open to all students at Cornell. Men and women interested in the biological sciences enrolled in the class as an elective. The course catalogue described the course as "A general introductory course and survey of the field of genetics designed to acquaint the student with the laws of variation and heredity, with some suggestions of their application to the improvement of plants and animals, and their bearing upon eugenics. Laboratory studies of variations in certain species, of hybridization and selection, and the laws of segregation and recombination illustrated by breeding experiments with plant and animal materials." The course was offered in first term and prerequisites were Botany 1 [General Botany] and 20 [Plant Physiology] or equivalents in zoology and physiology. The lectures were held on Tuesdays, Thursdays, and Saturdays at 8 am in the Forestry Building [later named Fernow Hall in 1922], one conference period to be arranged. The labs were held on Tuesday or Thursday afternoons or Saturday mornings for 2.5 hours. Hutchison's assistants were Assistant Professor [Allan] Fraser and Mr. [Ernest] Dorsey. This course was the prerequisite course for Plant Breeding 13, "Genetics, Advanced Course," offered in the second term, and was intended primarily for graduate students.

Hutchison believed that it was advisable for students specializing in any phase of biology to have a course in Genetics, but he did not advocate it as a requirement. At the close of the term in December 1921, he presented his rationale for Universities and especially Agricultural Colleges to offer an elementary course in Genetics, along with the details of the course content, to the Conference of Geneticists interested in Agriculture at the Toronto, Canada, meeting of the American Association for the Advancement of Science (AAAS) (Hutchison 1922a). This meeting was notable not only for Hutchison's advocacy of teaching genetics but also because of the other prominent invited speakers, who held different views of the mechanism of evolution. I will discuss below the importance of this meeting to the status of genetics and evolution in the early twentieth century, after first providing the details and significance of his course, which apparently influenced the future of McClintock's career (McClintock 1987).

A review of Hutchison's (1922a) published paper provides us with an opportunity to learn of his pedagogy and is a window into the first course in Genetics that McClintock took in her junior year at Cornell. Hutchison considered Genetics a subdivision or phase of biology not "confined to one or the other kingdom." His course gave "less consideration to the presentation of information and more to the development of habit and desire for real thinking." … "Better teaching would result, if courses were designed for the purpose of real mental training rather than for … giving out a lot of half-digested facts … to be handed back often in the same undigested form at examinations" (Hutchison 1922a). His course was a survey of the entire field of heredity. He wanted the course to "acquaint the student not only with a knowledge of the principles and facts of heredity but of how the science of genetics has been developed and is being developed, and give an elementary knowledge of the modes of genetic research." The course should present the subject "from the research point of view and demonstrate how human knowledge is advanced." His publication included the course syllabus for "Plant Breeding 1-Genetics, consisting of three lectures and one laboratory period a week for a term of sixteen weeks" (Hutchison 1922a, pp. 420–421). The first lectures provided methods, problems, scope, and relationships of genetics; their relation to evolution, to plant and animal breeding, and to eugenics. He presented historical background on revisions of early theories leading to modern concepts in genetics. After introducing the essential features of Mendel's studies of inheritance and the methods of testing and calculating his hypothesis, he lectured on the mechanism of Mendelian heredity, providing evidence for the chromosome theory of heredity. This was followed by the principle of linkage, first proposed by William Bateson and extended by Thomas Hunt Morgan and his students with *Drosophila*. The course closed with details of the mutation concept. "The DeVriesian view" of mutation, which held that new species arose by discontinuous variation [saltation], had been replaced by "The modern view" of mutations as "point or factor [gene] mutations" and multiple allelomorphs, and included regional mutations, chromosome aberrations, bud variations, and attempts to induce mutation. This was followed by a discussion of the mode of evolution from the mutation point of view [continuous variation, gradual variations, and selection], in contrast to the former view as defined by DeVries (and at the time still promoted by Bateson). The full course syllabus is available in Hutchison's (1922a) publication.

What resources were available for students to read about the subject at this time? Historian of science Will Provine asked McClintock what textbook she used for this course. She had no recollection of any textbook, but Hutchison's paper again provides us with some insights. He wrote, "In my judgement genetics falls in the same category as cytology and evolution in which the best instruction cannot be given without the free and unrestricted choice of illustrative material and evidence from both plant and animal life." The illustrative materials were discussed in the lectures and provided in the laboratory by experiments using *Drosophila* and by observing and reading papers on Mendelian heredity (ratios) and interaction between allelomorphic factors [genes] in fowl and plant color in maize; the appearance of new or old characters with modified Mendelian ratios, in *Mirabilis* [four o'clocks], carnations, snapdragons, and yellow mice; and sex inheritance in plants (mosses, ferns, and liverworts) and in animals (gypsy moth and *Drosophila*, among others).

The College of Agriculture Library in the Agronomy Building, and the Department of Plant Breeding Library, in the Forestry Building, had extensive reference books, journals, and reprints to supplement the teaching library and laboratory materials. A few examples of the materials available to the students were the first edition of Castle's (1916) textbook, *Genetics and Eugenics*; multiple copies of Bateson (1913, third corrected edition), *Mendel's Principles of Heredity*; Morgan's (1919) monograph, *The Physical Basis of Heredity*; DeVries (1907, reprinted 1919), *Plant Breeding*; Morgan et al. (1915), *The Mechanism of Mendelian Heredity;* and East and Jones (1919), *Inbreeding and Outbreeding*, among others. These libraries also included journals relevant to the study of plant breeding, genetics, and evolution: *Genetics*, *Journal of Heredity*, *American Naturalist*, *Genetica*, *Journal of Experimental Zoology*, *Journal of Morphology*, *Proceedings of the National Academy of Sciences*, *American Journal of Botany*, *Proceedings of the American Philosophical Society*, *British Journal of Experimental Biology*, *Science*, *Biological Bulletin*, *Botanical Gazette*, *Proceedings of the Society of Experimental Biology and Medicine*, and *Annals of Botany*, in addition to science journals and books published in France and Germany. It also included *Bulletins* and *Memoirs* published by U.S. Agricultural Experiment Stations and College of Agriculture Experiment Stations, as well as contributions by independent foundations and societies such as the *Year Books* of the Carnegie Institution of Washington, and the *Journal of the Washington Academy of Sciences*. The *Anatomical Record* also included abstract of papers presented at conferences, thus alerting researchers to the most current work conducted in the field of heredity.

AAAS TORONTO MEETING, DECEMBER 1921

As I mentioned earlier, the most current views on heredity and evolution were featured in Symposia at the AAAS annual meeting in December 1921. The meeting opened on the evening of December 28, 1921, with an invited presentation by Professor William Bateson, Director of the John Innes Horticultural Institution, Merton Park, Surrey, England. He had been welcomed by joint invitation of the American Society of Zoologists and the AAAS. His address on "Evolutionary Faith and Modern Doubts" emphasized that students of evolution hold no doubts as to the fact of evolution, but the exact mode of evolution is still an unsolved problem (Bateson 1922).[32] Darwin's (1859) "Abstract" *On the Origin of Species by Means of Natural Selection or the Preservation of Favoured Races in the Struggle for Life* had inspired many theories for how new species arose and how organic evolution could proceed. Darwin had proposed that new species arose by small random variations among organisms, and those variations were selected by the natural environment for their adaptive advantage (natural selection of small continuous variations).[33] This idea was also independently

[32] Address of William Bateson, Director of the Johns Innes Horticultural Institution, Merton Park, Surrey, England, was presented on Wednesday evening, December 28, 1921, and published in *Science*, January 20, 1922 (Bateson 1922). The University of Toronto conferred an honorary doctorate of science on Professor Bateson, "News and Notes," 1922, *Science* Vol. LV (No. 1410, January 6), p. 15.

[33] Darwin also published revisions of his 1859 essay, the last of which was issued in 1872, the 6th edition; see Introduction by E. Mayr, *In*, Darwin 1859, Facsimile 1964, p. xxiv.

recognized by Alfred Russell Wallace, who's communication with Darwin precipitated the publication of Darwin's "Abstract" after decades of accumulating proof for Darwin's novel idea. He recognized Wallace for his discovery and attributed credit to those colleagues who provided a resolution for both of their contributions to be presented together (Darwin 1859, Introduction; Kass 2001).

On December 29 and 30, three AAAS Symposia would address the issues raised by Bateson regarding the mode [mechanism] of evolution. These issues were some of the "special topics" of consideration in Hutchison's advanced course in Genetics (Plant Breeding 13) that McClintock would take in the spring term of 1922 (see below). The symposium on "The Species Concept" was held under the auspices of AAAS Section G [Botany], and featured the viewpoints of botanical systematists, geneticists, bacteriologist/physiologists, and pathologists (Wylie 1921, 1922), and presentations were published in the *American Journal of Botany* (1923). The symposium on "The origin of variations" was held under the auspices of the American Society of Naturalists (Rand 1922); their presentations were published in the *American Naturalist* (1922). This symposium seems to have held sway with Bateson, and he wrote to his wife on December 29, 1921, that on that day, "I had to announce my conversion on the main point: that chromosomes are definitely associated with transferable characters [morphological characters associated with factors/genes of heredity] ... much enthusiasm of this, of course, but as a candid man I don't see how any other view can be maintained" (Rushton 2014, p. 157). A third symposium on "Orthogenesis" was held under the auspices of the American Society of Zoologists, and some presentations were published in *American Naturalist*. Bateson gave a "General Discussion" to sum up this symposium, but it was "unpublished" (*Summarized Proceedings* 1925). Bateson was also scheduled to give an invited address on "The Outlook in Genetics" at the dinner following the Program for the American Society of Naturalists, on Friday, December 30, 1921, as announced in *Science* (December 16, 1921). Bateson's dinner address was unpublished, but his notes have since become available among his papers (Rushton 2014). We learn from his lecture notes for "The Outline of Genetics" that apparently three days after his first address, he was leaning toward accepting the evidence that chromosomes played a part in heredity, which was a mode for understanding evolution:

> Bateson accepted the mechanism for heredity, but still had doubts about whether crossing over and recombination were actual events ... Bateson agreed that the chromosome theory [of heredity] 'had proven to be an extraordinary efficient symbolism ... and is a fine service as any theory could perform'.
>
> *(Rushton 2014, p. 158)*

Just prior to attending the Toronto meeting, Bateson had visited the laboratory of Thomas Hunt Morgan at Columbia University and Calvin Bridges showed him microscope slides of *Drosophila* chromosomes, in which the variations in the morphology of the flies [phenotype] could be associated with abnormal *Drosophila* chromosomes [genotype] ... And he noted that his "Course of Conversion was begun" (Rushton 2014, p. 157). On December 24, 1921, he wrote to his wife that "the chromosomes must in some way connect with our transferable characters, but the details of the linkage theory strike me still as improbable" (Rushton 2014, p. 157).

Researchers from Cornell University, who participated in the AAAS Toronto meeting, invited Bateson and other geneticist with an interest in their work on maize genetics to visit them at Cornell. And on January 1, 1922, they met in Ithaca to discuss and demonstrate their work (Murphy & Kass 2007, 2011, photos p. 150). R. A. Emerson, Head of the Department of Plant Breeding, and Professor Hutchison demonstrated their maize breeding studies to Bateson and to the other visiting zoologists, Charles Metz, Sewell Wright, and C. B. Bridges, who were also working on problems of genetics and heredity in *Drosophila* and other animal species (Murphy & Kass 2011, Figure 18, p. 150). On that same day, a group photo of Emerson's former and current students and the visiting colleagues was taken at the Forestry Building, where the Cornell researchers conducted their work (Figure 2.3).[34]

[34] The Forestry building would officially be renamed Fernow Hall on October 5, 1922 (Anon. 1923).

FIGURE 2.3 R. A. Emerson with Cornell faculty, former and current students, hosting William Bateson and visitors at the Forestry Building (later named Fernow Hall), Cornell University, January 1, 1922, following the AAAS meeting in Toronto, where Hutchison presented his course syllabus for teaching introductory genetics. Back row from left to right: Milislav Demerec, Sterling Emerson, Ernest G. Anderson, Charles Metz; front row from left to right: Maxwell J. Dorsey, Sewall Wright, Rollins A. Emerson, William Bateson, Claude Burton Hutchison, Calvin Bridges, Frank B. Bussell, Lewis A. Eyster. (Permission of Plant Breeding & Genetics, Cornell University.)

Earlier that year, Hutchison (1921) had established that genes affecting maize kernel plumpness (full vs. shrunken, *Sh*, *sh*) and starch type (non-waxy vs. waxy, *Wx*, *wx*) were linked to kernel endosperm color (color vs. no color, *C*, *c*). Hutchison's (1921) report included the first map for this linkage group. Emerson and Hutchison (1921) had also published a genetic study of the development of maize microspores (pollen) and megaspores (embryo sac). They found that crossing over did not greatly differ during development of these maize reproductive structures. Hutchison (1922b) then provided additional linkage data for color and endosperm characters in the maize kernel, which might have further alleviated Bateson's doubts about the theory of heredity. Hutchison's studies of the genetic factors (genes) for kernel characteristics in maize would play an important part in the future work on the cytogenetics of maize conducted by Barbara McClintock and her colleagues. McClintock was introduced to the most recent maize research in the advanced course in Genetics, where students were required to read the primary literature.

McCLINTOCK'S CONTINUING STUDIES OF GENETICS AND FIRST COURSE IN CYTOLOGY

By the close of the 1921 term, Hutchison recognized McClintock's abilities in the subject of genetics. And at the end of her first course in Genetics, for which she received an A grade, he encouraged her to take the advanced class in Genetics. It was not unusual for undergraduates to take this course, and its description has provisions for qualified undergraduates to register with permission. Hutchison's advanced course dealt with "special topics in genetics with particular reference to the physical basis of heredity," including consideration of the "more important problems in Mendelism, linkage, sex inheritance and mutation."[35]

[35] *Announcement of the New York State College of Agriculture*, 1921–1922, p. 44.

In February of her junior year, 1922, McClintock accepted Hutchison's invitation to take the Advanced Genetics class, Plant Breeding 13, obtained the required permission to register, and simultaneously enrolled in the prerequisite course Botany 10, Cytology, offered by Professor Sharp. His Cytology course was "planned to give instruction in the morphology and physiology of the cell, and in the cytological aspects of reproduction and inheritance." Her teachers in genetics and cytology were distinguished faculty who had a strong commitment to both teaching and research. She did not take the second prerequisite, Botany 20, Plant Physiology, until the second term of her senior year. She received an A grade in the advanced course in Genetics. She also enrolled in Principles of Plant Breeding (Plant Breeding 8), Rural Education, and a second class in Meteorology. Professor Allen Cameron Fraser taught the Advanced Genetics class and Plant Breeding 8, because Hutchison left Cornell to become Director of the Agriculture Farm at Davis, California. McClintock recalled that Fraser did an excellent job teaching Advanced Genetics. He was well qualified for a number of reasons. First, after receiving his B.S. from Cornell in 1913, he was an Instructor in Plant Breeding at Cornell between 1914 and 1918, while pursuing his Ph.D. awarded in 1918. Second, in 1919, the year following his graduation, he was appointed Assistant Professor in Plant Breeding, and he assisted Hutchison in the introductory and advanced classes in Genetics, as well as independently teaching the course in Plant Breeding Methods. When McClintock took classes with Fraser, he was involved with maize research and had published an important paper on the inheritance of aleurone color in corn (Fraser 1924). He would later become a member of her graduate school committee.

Lester Wyland Sharp had received his Ph.D. from the University of Chicago in 1912. In 1914, he was appointed an Instructor of Botany in the recently established Department of Botany of the New York State College of Agriculture at Cornell University, and was promoted to Assistant Professor of Botany in 1915. In September 1920, he had completed the manuscript for the first edition of his notable textbook, *An Introduction to Cytology*, published by the McGraw-Hill Book Co, in early 1921. That same year, he was promoted to Professor of Botany. Sharp's research interests were in studies of the embryo sac, spermatogenesis, and chromosome structure in plants. In addition to teaching and directing graduate students in cytology, he taught courses in Plant Comparative Morphology. Sharp was also known for his excellent sense of humor, and many may remember him for his satiric monograph on *Eoörnis pterovelox gobiensis* or the Whiffenpoof, published under a pseudonym by the Buighleigh Press, London (Fotheringham 1928; Kass 2005).[36] Professor Sharp was starred in the 1927 edition of *American Men of Science*, indicating that his colleagues recognized him to be an outstanding contributor to the field of Botany (*American Men of Science* 1921–1944; Kass & Cobb 2007; Kass 2005).

In the first term of her senior year (1922–1923), McClintock completed her final classes as an undergraduate. She enrolled in two Botany courses, Histology and Plant Physiology. She also conducted original investigations for an advanced class in Meteorology, a subject that fascinated her throughout her career, and completed a general education requirement with a class in Harmony. She had completed all the requirements for graduation except residency. On February 9, 1923, she was granted permission to register in the graduate school for the second term of her senior year and submitted an application for the doctoral program that same day.[37] Sharp had written to Dean Creighton of the graduate school recommending that she be allowed to register as a candidate for the degree of Doctor of Philosophy. She applied for admission as a Ph.D. candidate and chose to major in Cytology with Sharp and to minor in Plant Breeding with R. A. Emerson, and in Zoology with Reed.[38] For her Graduate School File, Sharp wrote that McClintock was "in residence

[36] Fotheringham was the pseudonym for Cuthbert V. Fraser who co-authored with Lester W. Sharp the monograph on *Eoörnis*, published in 1928. Fraser wrote the first draft monograph in 1926; he gave a presentation on the work at a Cornell Botany Department seminar in January 1927. C. V. Fraser was enrolled in Sharp's Cytology class in the spring term of 1919 and was awarded his Bachelor of Science from Cornell on June 23, 1919. He was not a graduate student at Cornell but continued his friendship with Sharp. On a trip to the Adirondacks, they were inspired to write this scientific spoof. The background history was published in February 1928, along with the distributed copy of their monograph. C. V. Fraser became ill the following September (*Cornell Alumni News* April 15, 1929, p. 354), and by May 5, 1929, he had died.

[37] McClintock Undergraduate Record Card and Graduate School File.

[38] *The Register* 1922–1923, section on the Graduate School.

FIGURE 2.4 Barbara McClintock graduated with a Bachelor of Science in Agriculture from the New York State College of Agriculture at Cornell University on June 18, 1923. (Photo reproduced from 1923 *Cornellian*, p. 580, public domain.)

during the second term of the academic year 1922–23, and completed satisfactorily the work of that term."[39] On McClintock's undergraduate record, there is only one course listed for the spring term, P.T. [physical training]. There are no remarks on the Graduate Work Report that Sharp filed on June 12, 1923. It is not clear what type of academic work she was engaged in during the second term of her senior year, but it seems clear that she had neither identified the ten chromosomes of maize nor learned how to make chromosome squashes of maize spore mother cells to observe pachytene chromosomes. That would come much later in her graduate student career.

On June 18, 1923, two days after her 21st birthday, Barbara McClintock was awarded the degree of Bachelor of Science [in agriculture] from the New York State College of Agriculture at Cornell University (*The Register* 1922–1923, p. 219; *Cornellian* 1923, p. 580; Figure 2.4). At that time,

[39] In McClintock's Graduate School File for Ph.D.

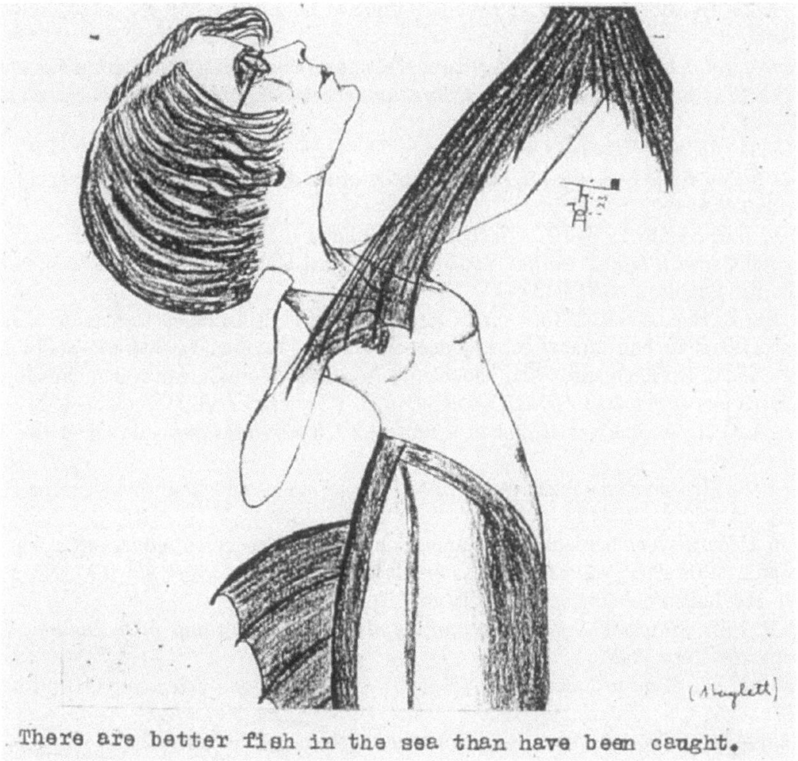

FIGURE 2.5 Sketch of Barbara McClintock drawn for the Synapsis Club Razzberry meeting held on January 28, 1924, where students and faculty poked fun at each other. (Courtesy of Plant Breeding & Genetics, Cornell University.)

she had completed her first term of graduate work at Cornell (*The Register* 1922–1923, p. 73) and had been inducted into the *Phi Kappa Phi* College of Agriculture Honor Society. Students so honored were admitted by invitation at the nomination and approval by the chapter. She was one of twelve from the Agriculture class of 1923 to be so honored (*Cornellian* 1923, p. 382). She soon became an active member of the Department of Plant Breeding Student/Faculty Synapsis Club, which met weekly for seminars and social events (see Chapter 3). In her first year in graduate school (1923–1924), she helped organize their 1924 annual Razzberry meeting, where both student and faculty poked fun at each other (Kass 2003; Murphy & Kass 2011, p. 31; Figure 2.5).[40]

McClintock's comprehensive undergraduate curriculum and the encouragement she received from the faculty provided a strong foundation for her future development as a cytogeneticist.

LITERATURE CITED

American Men of Science, A Biographical Directory. 1938, 6th edition, J. McKeen Cattell and J. Cattell (Eds.). The Science Press, New York.

Anon. 1923. Forestry building named for Dr. Fernow. *Journal of Forestry* Vol. 21 (No. 4, April 1923), pp. 316–318. https://doi.org/10.1093/jof/21.4.316

Bateson, W. 1913 (3rd corrected edition). *Mendel's Principles of Heredity*. With a translation of Mendel's original paper on hybridization. Preface, November 1912; Cambridge University Press, Cambridge, UK. [3rd impression 1913, with corrections in appendices.]

[40] LBK Scanned on February 17, 1997, from the Synapsis Club Record Book Razzberry Meeting, January 28, 1924, then in the Plant Breeding Department Archives, later deposited with the Plant Breeding Records in the Cornell Archives.

Bateson, W. 1922. Evolutionary faith and modern doubts. *Science* Vol. 55 (No. 1412, 20 January 1922), pp. 55–61.

Cornellian. 1921. Vol. 1 to present (1868)ff. Ithaca, NY. https://newcatalog.library.cornell.edu/catalog/314011

Cornellian. 1923. *The Nineteen Twenty Three Cornellian, The Annual of Cornell University and Class Book of the Senior Class*, Cornell University, Ithaca, NY.

Darwin, C. 1859. *"Abstract" On the Origin of Species by Means of Natural Selection, or the Preservation of Favored Races in the Struggle for Life*. Facsimile of the first edition, 1964, Harvard University Press. Cambridge, MA

Directory Cornell University 1919-1923. Second Term. (March 1, 1920) Cornell University, Ithaca, NY.

Engst, E. D. 2006. *Jewish Life at Cornell 1865–2005*, Cornell University Library, Ithaca, NY. https://ecommons.cornell.edu/handle/1813/13793

Engst, E. D. and B. Friedlander. 2014. Cornell Rewind: 'Above all nations is humanity.' *Cornell Chronicle*, November 20, 2014. https://news.cornell.edu/stories/2014/11/cornell-rewind-above-all-nations-humanity

Emerson, R. A. and C. B. Hutchison. 1921. The relative frequency of crossing over in microspores and in megaspore development in maize. *Genetics* Vol. 6 (No. 5, 1 September 1921), pp. 417–432.

Fotheringham, A. C. [L. W. Sharp and Cuthbert V. Fraser]. 1928. *Eoörnis Pterovelox Gobiensis*. The Buighleigh Press, London.

Fraser, A. C. 1924. Heritable characters in maize XVII. Intensified red and purple aleurone color. *Journal of Heredity* Vol. 15, pp. 110–123.

Glass, B. 1986. Geneticists embattled: Their stand against rampant eugenics and racism in America during the 1920's and 1930s. *Proceedings of the American Philosophical Society* Vol. 130 (No. 1, March 1986), pp. 130–154. https://pubmed.ncbi.nlm.nih.gov/11611930/

Goodstein, J. R. 1991. *Millikans' School: A History of the California Institute of Technology*. W. W. Norton & Company, Inc., New York.

Hargittai, I. 2002. *The Road to Stockholm: Nobel Prizes, Science, and Scientists*. Oxford University Press, Oxford.

Hobson, L. Z. 1947. *Gentleman's Agreement*. Simon and Schuster, New York.

Hutchison, C. B. 1921. Heritable characters of maize. VII. Shrunken endosperm. *Journal of Heredity* Vol. 12 (No. 2, February), pp. 76–83.

Hutchison, C. B. 1922a. The elementary course in genetics. *Science* Vol. 55 (No. 1425, 21 April 1922), pp. 416–421.

Hutchison, C. B. 1922b. The linkage of certain aleurone and endosperm factors in maize, and their relation to other linkage groups. *Cornell University Agricultural Experiment Station Memoir No. 60,* June 1922, pp. 1419–1473. https://digital.library.cornell.edu/catalog/chla7251474_8544_009

Kass, L. B. 1999. Barbara McClintock and the 1926 International Botanical Congress. *XVI International Botanical Congress, Abstracts*, p. 474. [Poster available in the Cornell Archives. https://newcatalog.library.cornell.edu/catalog/4031434]

Kass, L. B. 2001. Ethics in science: preparing students for their career. *Plant Science Bulletin* Vol. 47 (No. 2, Summer 2001), pp. 42–48.

Kass, L. B. 2003. Records and recollections: A new look at Barbara McClintock, Nobel Prize-Winning geneticist. *Genetics* Vol. 164 (No. 4, August 1, 2003), pp. 1251–1260. https://www.genetics.org/content/164/4/1251

Kass, L. B. 2005. Harriet B. Creighton: Proud botanist. *Plant Science Bulletin* Vol. 51 (No. 4, Winter 2005), pp. 118–125. Available online, December 2005: https://www.botany.org/psbarchive/issue/2005-v51-no-4.html

Kass, L. B., C. Bonneuil, and E. Coe. 2005. Cornfests, cornfabs and cooperation: The origins and beginnings of the Maize Genetics Cooperation News Letter. *Genetics* Vol. 169 (No. 4, April), pp. 1787–1797; online May 6, 2005: https://www.genetics.org/content/169/4/1787

Kass, L. B. and E. Cobb. 2007. Landmarks and milestones in American plant biology, the Cornell connection. *Plant Science Bulletin* Vol. 53 (No. 3, fall), pp. 90–101).

Kass, L. B. and W. B. Provine. 1997. Genetics in the roaring 20s: The influence of Cornell's professors and curriculum on Barbara McClintock's development as a cytogeneticist. *American Journal of Botany Abstracts* Vol. 84 (No. 6, Supplement), p. 123.

McClintock, B. 1987. *The Discovery and Characterization of Transposable Elements*. The Collected Papers of Barbara McClintock. Genes Cells and Organisms; Great Books in Experimental Biology, J. A. Moore, series editor, Garland Publishing Co., New York. [Published December 1, 1987; Currently in the series Genes Cells and Organisms, Routledge, Taylor & Frances Group.]

Murphy, R. P. and L. B. Kass. 2007. *Evolution of Plant Breeding at Cornell University: A Centennial History, 1907–2006*, pp. 1–98, Department of Plant Breeding & Genetics, Cornell University, Ithaca, NY, Appendices A1–A98, Photo Section P1–P38 (July 2007).

Murphy, R. P. and L. B. Kass. 2011 (27 June). *Evolution of Plant Breeding at Cornell University: A Centennial History, 1907–2006*. Rev. edition i–x, 178 pp., Appendices, Photo Section. e-book, The Internet-First University Press, Ithaca, NY. https://hdl.handle.net/1813/23087

New York State College of Agriculture Announcement 1919-1920. Cornell University, Ithaca, NY [Issued June 15, 1919]. https://ecommons.cornell.edu/items/b7ffd934-8b4b-4b91-9d95-f917460f73d8

Plant Breeding Records, Cornell University Archives: Cornell University Department of Plant Breeding Records, 1906–1970. Collection Number: 21-28-889 Division of Rare and Manuscript Collections. Cornell University Library. [Boxes 56-65 (student files) are restricted to Plant Breeding and Genetics Section Chair or their designate]. https://rmc.library.cornell.edu/EAD/htmldocs/RMA00889.html

Provine, W. B. and P. Sisco. 1980. Interview with Barbara McClintock at Cold Spring Harbor Laboratory, Long Island, New York, August 28, 1980. Division of Rare and Manuscript Collections, Cornell University Library. Ithaca, NY.

Rand, H. W. 1922. Section F – Zoological sciences and associated societies. *Science* Vol. 55 (No. 1430, May 26, 1922), pp. 572–573.

Rossiter, M. W. 1982. *Women Scientists in America. Struggles and Strategies to 1940*. Johns Hopkins University Press, Baltimore, MD.

Rushton, A. R. 2014. William Bateson and the chromosome theory of heredity: A reappraisal. *The British Journal for the History of Science* Vol. 47 (No.1, March), pp. 147–171. https://doi.org/10.1017/S0007087413000320

Sharp, L. W. 1921. *An Introduction to Cytology*, 1st edition, McGraw-Hill Book Company, Inc. NY.

Summarized Proceedings from the American Association for the Advancement of [AAAS] for the period from June 1921 to June 1925, Reporting the second Toronto Meeting, 74th meeting, December 27 to 31, 1921, Published by the American Association for the Advancement of Science, Smithsonian Institution Building, Washington, DC, 1925, pp. 96–100. Report includes Symposia, speakers, and where their presentations were published. [AAAS Proceedings were initially published in *Science*.]

The Register 1919–1920 (September 1, 1920) Cornell University, Ithaca, NY. https://ecommons.cornell.edu/items/83939039-b736-4787-ab1c-4bd3a5365b4e

The Register 1922–1923 (September 1, 1923). Cornell University, Ithaca, NY. https://ecommons.cornell.edu/handle/1813/22316

Trajkovich, H. 1924. Inheritance of *Xantha* seedlings in maize. *Cornell University Agricultural Experiment Station Memoir 82* [December], pp. 1–13. [Publication of her Master's thesis]. Digital document, Cornell University: https://digital.library.cornell.edu/catalog/chla7251474_8565_002

Trajkovich, H. A. 1923. *Yellow-virescent seedling in maize*, M.S. Cornell Univ. June 1923 [Master's], Thesis 1923 T768. https://newcatalog.library.cornell.edu/catalog/5018346

Wylie, R. B. 1921. The program of the section of Botany for the Toronto meeting. *Science* Vol. 54 (No. 1406, 9 December 1921), pp. 576–577.

Wylie, R. B. 1922. Section G – Botanical sciences and associated societies. *Science* Vol. 55 (No. 1430, 26 May 1922), pp. 573–574.

3 From Botany Scholar to Maize Cytologist
Fact, Fiction, and Faulty Memories

WOMEN IN PLANT BREEDING AT CORNELL UNIVERSITY

Before McClintock arrived at Cornell and throughout her undergraduate years, the College of Agriculture had employed 18 female Assistant and Full Professors (1913–1923) – half in departments other than Home Economics (Faculty Files, Cornell University; Kass 2003). The first woman Full Professor was appointed in 1920 – five years prior to the establishment of its College of Home Economics (Kass 2003). Yet, while McClintock was a graduate student, many women were enrolled in both undergraduate and graduate classes in Plant Breeding, and a few of them also were awarded Master's degrees from the Plant Breeding Department (Registrar's Cards, Cornell University; Kass & Murphy 2007, 2011). As mentioned in the previous chapter, Helen Trajkovich's graduate school file reveals that she majored in plant breeding with R. A. Emerson, who was head of the department. She was awarded her Master's degree on the very day that McClintock received her Bachelor's degree, in June 1923. Her thesis was revised and published the following December (Trajkovich 1924). She was employed by the New Jersey Agricultural Experiment Station, New Brunswick, New Jersey, as a research botanist (Rudolfs 1924, p. 4; Rudolfs & Trajkovich 1924a,b). She and her husband were then employed by an Agricultural Experiment Station in "Beograd Jugoslavia."[1] Another woman, Elizabeth M. Bodger, received a Master's degree with A. C. Fraser in the Plant Breeding Department the same year as McClintock was awarded her Ph.D. from the Botany Department, in 1927. Upon graduation from Cornell, E. Bodger joined her father's family business, a flower seed company, Bodger Seeds Co./Bodger Seeds, in Lompoc California, where she became a world famous plant hybridist.[2]

By the early 1930s, Emerson had accepted a number of women graduate students to work with him toward the doctorate. They did not, however, complete their Ph.D.'s, although they were awarded Master's degrees (Murphy & Kass 2007, 2011). Since Emerson and other Plant Breeding faculty did accept female graduate students, can we clarify why McClintock did not apply to major in Emerson's department? Can we explain why the story of discrimination against women in Cornell's Plant Breeding Department has been widely accepted?

Former head of the Cornell Plant Breeding Department, Professor R. P. Murphy, provided an insight regarding admitting women in their department. Murphy joined Cornell's Plant Breeding faculty in 1946 (Department Head 1953–1964; Murphy & Kass 2007, 2011). His understanding is that before he arrived at Cornell, it was accepted practice in the Plant Breeding Department, and

[1] Letter Helen Trajkovich to Prof. Emerson, February 11, 1925, File Helen Trajkovich, Plant Breeding Records, 21/28/889, Box 7, Division of Rare and Manuscript Collections, Cornell University Library (henceforth CU).

[2] *New York Times* Obituary, March 18, 1943, p. 20: "ELIZABETH BODGER, PLANT EXPERT, DIES; Hybridist of Bodger Seeds Co., Who joined Concern in 1927, Stricken in Los Angeles DICTATED FLORAL STYLES 'America's No. 1 Flower Girl' Set Fashions for Thousands of Gardeners in U.S." https://lompocrecord.com/news/local/bodger-seeds-closing-friday/article_fcfe2d92-42d0-11df-8c03-001cc4c002e0.html (April 8, 2010): The company was sold in 2010, Elizabeth Bodger was the "Daughter of company founder Walter Bodger." *Lompoc Review*, Volume XIII, Number 1, May 26, 1931, p. 1: Obituary of Walter Bodger, father of Elizabeth M. Bodger. *The Los Angeles Times*, Los Angeles, California, March 18, 1943, Thursday. p. 23: Mrs. E. Baertschinger, considered a world-famous plant hybridist, known by her maiden name, died at age 39 of a heart attack five days after giving birth to a son.

DOI: 10.1201/9781003332527-4

elsewhere, to admit students into their programs only if professors thought they could place students in jobs after graduation. This practice was common during the 1970s when I was a Cornell graduate student. My first advisor discouraged me from majoring in my chosen subject and encouraged me to switch to a field with job opportunities (Kass 2003). Harriet Creighton, McClintock's friend, student, and collaborator in Cornell's Botany Department during the late 1920s and early 1930s (see Chapter 4), suggested that Emerson may have discouraged McClintock from majoring in the Plant Breeding Department given this rationale (see Creighton 1992). This image says more about societal customs and the availability of jobs for women in the field of Plant Breeding during this era (Rossiter 1982) than it does about Emerson's presumed anti-female attitude. Emerson's letters in the Cornell Archives demonstrate that he dealt with similar circumstances in placing Jews and other minorities during this time (see Chapter 4). Stories of discrimination in Cornell's Plant Breeding Department have been interpreted and embellished in the context of contemporary views. Furthermore, Cornell's Department of Botany included male and female Instructors, which in the 1920s were considered the first faculty rank toward tenure, but the first woman was not given Professorial rank in Cornell's Department of Plant Biology until 1985 (Kass 2003; Kass & Cobb 2007). Likewise, during the 1920s Cornell's Department of Plant Breeding had only male Instructors, and only in 1958, did they hire their first female Assistant Professor (Murphy & Kass 2007, 2011).

BOTANY SCHOLAR

While an undergraduate in the College of Agriculture at Cornell University, Barbara McClintock began graduate school during the last term of her senior year, as briefly mentioned in the previous chapter. On February 9, 1923, she applied for admission as a Ph.D. candidate and chose to major in Cytology, and minor in "Genetics in Plant Breeding" and Zoology.[3] The major and minor professors for her proposed degree were listed under the Graduate School section of "*The Register 1922–1923*" (p. 73),

> McClintock, Barbara, B.S. (Cornell) 1923 Brooklyn
> Botany, Plant Breeding, Zoology. (Sharpe [*sic*], Emerson, Reed Ph.D.

She registered for the first full year of graduate school beginning with the fall term of 1923 (September 28, 1923). Her graduate school file shows that within a month of registering she had applied to be a Master's candidate. Entries on the cover of the file were erased from Ph.D. to A.M. [Master's degree], and dated October 3, 1923; also erased was the second minor in Zoology. On October 9, 1923, she accepted an appointment to a "Graduate Scholarship in Botany" [Botany Scholar], and on October 31, 1923, she officially applied to candidacy for a Master of Arts in Cytology and Genetics (McClintock Graduate School File; *The Register 1923–1924*, p. 96). There is no documentation on why McClintock decided to change her status as a graduate student. She told Professor Provine during an interview in 1980 that she was concerned that she would not have funding to complete a doctorate (Provine & Sisco 1980). Neither do we know why Professor A. C. Fraser replaced Dr. R. A. Emerson as a member of her graduate school committee. Since Emerson was scheduled to leave the country at the end of 1923 for a maize-collecting trip to South America, one can only speculate that this may have been why Dr. Fraser assumed his responsibilities as a mentor to Barbara McClintock. Fortunately, after completing her Master's degree, she would go on for a Ph.D. with the support of research/teaching assistantships in Cornell's Department of Botany.

McClintock may have considered majoring in Genetics, but documents in Cornell's archives and libraries make clear that she never applied for admission to the Plant Breeding Department. There is no record in her graduate school file of an application to major in Genetics in the Plant Breeding Department. Neither can it be shown that the Department refused to accept women, although the legend has often been repeated (Keller 1983; Berg & Singer 2003).

[3] McClintock Graduate School File, CU.

MASTER'S DEGREE WITH PROFESSOR SHARP

During McClintock's first term as a Botany Scholar (1923–1924), Sharp submitted a "Report Regarding Graduate Work" to her file (February 1, 1924; McClintock Graduate School File). He wrote that she spent the time on work that was "assigned her as a basis for her Master's thesis and had made substantial progress." She also "completed Botany 14a" [Morphology of Bryophytes and Vascular Plants Part I].[4] At the end of the second term, Sharp reported that she had completed Botany 14b [Morphology of Bryophytes and Vascular Plants Part II][5] and that she had spent most of the term on work "in connection with her Master's thesis."

During the 1924–1925 academic year, she held an appointment as "assistant in the Department of Botany" (McClintock Graduate School File, Sharp, 7 February 1925), working with L. F. Randolph (Kass & Bonneuil 2004, and see below). The bound copy of McClintock's Master's thesis had been submitted to the Graduate School on December 1, 1924. By early December 1924, she passed the exam for her Master's thesis and was recommended for the degree of A.M. [Master of Arts], which was conferred on February 4, 1925.[6]

McClintock's Master's thesis research, begun in early 1923, was a literature review of cereals with particular attention to wheat. She did no experimental work during that time, and the review excluded studies being made with the corn plant.

> Corn has been omitted from this paper as the earlier work has been summarized by Weatherwax (1923) in his book, *The Story of the Maize Plant*, to which the reader is referred. Also much cytological investigation is now being conducted on the corn plant which will cover the situation more completely than it is possible to do at the present time.
>
> *(McClintock 1925, p. 2)*[7]

Why did Sharp "assign" this topic "for her Master's thesis"? The purpose of her thesis was to "gather together from the scattered sources the results of cytological work in cereals with particular reference to wheat." Sharp was at the time mentoring Ernest Dorsey who was an Instructor and a graduate student in Plant Breeding and Botany. He had submitted a draft of his doctoral thesis on studies of the reduction division [meiosis] in cereal hybrids, especially wheat and oats, and it seems that further cytological work was necessary to complete his dissertation, which was revised in June 1924 (Dorsey 1923, 1924). The studies reviewed in McClintock's thesis seem to have contributed to this work, since she cited his thesis draft many times in her thesis.

RESEARCH AND TEACHING ASSISTANT IN BOTANY

At the time McClintock completed the writing of her Master's thesis (December 1, 1924), she held a teaching Assistantship in Cornell's Department of Botany and was a research assistant to L. F. Randolph.[8] Randolph had studied cytology with Lester W. Sharp in the Botany Department at

---

[4] Botany 14a, taught by Professor Eames, *Announcement of the New York State College of Agriculture, 1923–1924.*

[5] Botany 14b, taught by Professor Sharp, *Announcement of the New York State College of Agriculture, 1923–1924.*

[6] Cover of McClintock's Graduate School File, CU, date thesis received, December 1, 1924, deposited in Library February 28, 1925. Exam Report December 6, 1924, Miss Barbara McClintock, signed by Lester W. Sharp and Allan C. Fraser; *The Register 1924–1925*, p. 133, Masters of Arts, conferred June 15, 1925, Barbara McClintock, B.S.: Cytology, Genetics. Thesis: *A Resume of Cytological Investigations of the Cereals with Particular Reference to Wheat.*

[7] Citations for the publications of McClintock do not follow the chronology in Kass 2013ff.: Table I.

[8] L. F. Randolph to F. D. Richey May 21, 1924, National Archives and Records Administration, U.S. (henceforth NARA), Box no 121, Quisenberry, K. S. to Randolph, L. F., Rg 54, records of the Bureau of Plant Industry, Soils & Agricultural Engineering (henceforth BPI/SAE). Correspondences 1917–35. File Randolph, L. F. 1922–1926. The graduate school received McClintock's thesis on December 1, 1924; she was examined on her thesis on December 6, 1924; after she was awarded her Master's degree on February 4, 1925, the thesis was deposited in the library on February 28, 1925; all documentation in McClintock Graduate School File, Cornell Archives; when McClintock completed her Master's thesis (February 1925), she had been working as an assistant to L. F. Randolph since September 1924.

Cornell and minored in Plant Breeding with R. A. Emerson. The latter recommended a United States Department of Agriculture (U.S.D.A.) appointment for Randolph at Cornell to investigate corn chromosome cytology. Emerson knew that a cooperative effort with a cytologist should prove valuable because of the work that was coming from the laboratories of T. H. Morgan at Columbia University and A. F. Blakeslee, working with J. Belling at Cold Spring Harbor, New York. Emerson's former students, E. G. Anderson and M. Demerec, worked with both Morgan's and Blakeslee's groups and wrote to Emerson at Cornell about the exciting advances in those laboratories (Kass & Bonneuil 2004; Chapter 4). Anderson and Demerec often visited Cornell and gave invited seminars at the Plant Breeding Department's Synapsis Club, a student/faculty organization that met weekly for social and intellectual get-togethers.[9] Emerson, therefore, encouraged Randolph's pursuit of corn chromosome numbers to correlate with the linkage groups that students of maize genetics were finding. Cornell's Botany Department provided research space, laboratory equipment, and some funding for their project. The Plant Breeding Department provided field space for growing plants. When Randolph began his cytological investigations in 1923, the base number for chromosomes in corn was still controversial. Randolph would soon learn more about techniques for the study of chromosomes from notable researchers who visited Cornell during these years.

The Cornell Synapsis Club played an important role in introducing faculty and students to current research before it was published. The club was established in 1907, as the Plant Breeding Department's student/faculty organization, which met weekly for seminars and social events (Murphy & Kass 2007, 2011). The organization invited leading researchers in the field to present their findings at their weekly evening seminars. Following a dinner prepared by the "feed committee" (students and faculty), invited guests would present their research and hold discussions with those attending. The meetings were open to all those interested in the topics, and Dr. Sharp, his colleagues and students, attended quite regularly. Occasionally, Emerson and Sharp would receive a visit from a special guest and hold "a special meeting called on short notice." Of note was such a visit in early September 1923, from the Director of the "Cold Spring Harbor Station," Dr. C. B. Davenport, who presented the "work of the station."[10] Emerson, Fraser, and only twelve others were in attendance. But more importantly, the Plant Breeding faculty were most interested in the current research of Davenport's "Department of Genetics," where A. F. Blakeslee, J. Belling, M. Demerec, and C. W. Metz were studying the role of chromosomes, mutations, heredity, and evolution in plants and insects.[11] Within the month, another special meeting was called on short notice, and speaking to 24 in attendance, Blakeslee reported on "Mutants in *Datura*-their explanation from the chromosomal standpoint." Their "mutants" were plants with extra chromosomes, $(2n+1)$ trisomics, that were identified by the plant's morphology and its chromosome constitution [genome]. Randolph, Emerson, and Fraser were in attendance, and learned of the new findings in *Datura* (Jimson weed) plants and the continuation of Belling's techniques for studying their chromosomes, even before their report appeared in the 1922–1923 *Carnegie Year Book*.[12] By 1925, Randolph, with the assistance of McClintock, would find a triploid maize plant among Dr. Fraser's corn stocks, and this work would provide offspring (F_1 progeny) for McClintock's doctoral research on plants with extra chromosomes (trisomics) (Randolph & McClintock 1926; McClintock 1927).

McClintock attended her first Synapsis meeting in November 1923.[13] Within a week, the Synapsis Club's invited guests were "C[harles] R[upert] Stockard, Raymond Pearl & James W[atts] Mavor."[14] The Johns Hopkins University Professor Pearl gave a short talk on "Length of Life from geneticists'

[9] For instance, see E. G. Anderson's (then at Columbia University) letters to Emerson at Cornell, quoted in Chapter 4; *Synapsis Club Record Book* (henceforth *SCRB*) *1920–1927*, now in Plant Breeding Records, CU.

[10] *SCRB 1920–1927*, September 17, 1923, p. 70.

[11] *Carnegie Institution of Washington Year Book No.* 22, November 1, 1922–October 31, 1923, published February 1924.

[12] *Carnegie Institution of Washington Year Book No.* 22, November 1, 1922–October 31, 1923, published February 1924, pp. 89–92.

[13] *SCRB* 1920–1927, November 7, 1923, p. 76.

[14] *SCRB* 1920–1927, November 12, 1923, p. 78.

point of view," and Cornell Medical School Professor Stockard spoke on "The developmental factors [genes] in relation to inheritance." Professor Mavor of Union College in nearby Schenectady, New York, did not present a talk on his much publicized X-ray work. His papers had been published in *Science* during the previous two years and heralded in the newspapers (Hook 1986). Mavor was one of the first researchers to report that X-rays induced mutations in *Drosophila* (Mavor 1921, 1922, 1947, p. 738; *Science News-Letter* 1922; Sharp 1921; Hook 1986). The X-ray work was known to Dr. Sharp, who cited it in the first and subsequent editions of his *Cytology* texts (Sharp 1921, 1926, 1934). Sharp, Randolph, and McClintock attended this meeting with over 30 in attendance. The maize variations (mutants) that they studied were not induced by X-rays, but Randolph soon had an interest in the methods. At this time, the term mutation as reported in the scientific literature was often confused with respect to a DeVriesian mutation (caused by chromosome aberrations; salutations) and point mutations as observed by gradual variation, of which the physical basis at the level of cytology and genetics was of interest (Mayr 1980; Morgan et al. 1915; Morgan 1919).[15] By 1927, Mavor's work on X-rays and mutation causing chromosome aberrations (e.g., loss of chromosomes or parts of chromosomes, deletions) would be overshadowed by H. J. Muller's 1927 publication in *Science* reporting the effects of X-rays on the rapid induction of point mutations in *Drosophila* (Muller 1927a). Muller's X-ray work would also be immediately heralded and would also overshadow L. J. Stadler's contributions to plants, which he conducted and reported simultaneously (Kass 2018; Stadler unpublished;[16] see below and Chapter 4).

By spring of 1924, Randolph had applied John Belling's recently described (1921a,b, 1923) "iron-acetocarmine" smear technique to clarify plant chromosome numbers reported in the literature (Longley 1924), and in August 1924, Randolph and Professor Sharp consulted with Belling at Cold Spring Harbor, New York, about applying his new smear technique to corn sporocytes.[17] The diploid number of chromosomes in *Zea mays* ($2n=20$) was in question when L. F. Randolph commenced his intensive cytological investigations in 1924.[18] Randolph realized the value of Belling's technique for associating extra-chromosome plants with genetic characters, and would soon share it with his new assistant Barbara McClintock.[19] It was in Randolph's laboratory that McClintock learned Belling's iron-acetocarmine smear technique (Randolph & McClintock 1926; Kass & Bonneuil 2004). And years later, only after completing her Ph.D. (McClintock 1927), she would modify the technique to identify the ten chromosomes of maize (McClintock 1929b,c; Kass & Bonneuil 2004) and apply it to further the understanding of the cytogenetics of maize (Emerson et al. 1935; Rhoades & McClintock 1935; Kass 2013ff.; also see Chapter 4).

Assigning each linkage group to cytologically identifiable chromosomes was on Emerson's mind when he and Randolph began collaborative studies supported by the U.S.D.A. Randolph's usual method of preparing root tip sections, and observing their division figures with the microscope afforded limited identification of chromosome morphology. Randolph quickly adopted Belling's "aceto-carmine smear technique," but the methods or materials for morphologically identifying the chromosomes had not yet been devised. In his cultures (strains), late prophase (diplotene and diakinesis stage) chromosomes were visible and countable, but each chromosome was not readily

[15] For a discussion of the term mutation during this era, see Mayr (1980), Protologue.

[16] Stadler's unpublished [1927] presentation on X-rays was given at the AAAS meeting in Nashville, Tennessee, December 1927, and mentioned in Stadler (1928b); see Chapter 4 for details.

[17] L. F. Randolph to F. D. Richey, May 21, 1924. L. F. Randolph to F. D. Richey, August 14, 1924, August 20, 1924 – on August 16, 1924, Randolph drove to Cold Spring Harbor with Dr. Sharp, all letters in NARA, Box no 121, Quisenberry, K. S. to Randolph, L. F., Rg 54, BPI/SAE. Correspondence 1917–35. File Randolph, L. F. 1922–1926.

[18] A. E. Longley (1924) and L. F. Randolph (1924) (unpublished) had studied a limited number of corn varieties and showed that the basic chromosome number in *Zea* "appeared" to be 10. See L. F. Randolph to F. D. Richey, May 21, 1924, August 14, 1924, July 1, 1924, all letters in NARA, Box 121, Quisenberry to Randolph, NARA, Box 121, Quisenberry, K. S. to Randolph, L .F., RG 54, BPI/SAE. Correspondence 1917–35. File Randolph, L. F. 1922–1926.

[19] L. F. Randolph to F. D. Richey, February 5, 1925, NARA, Box no 121, Quisenberry, K. S. to Randolph, L. F., Rg 54, BPI/SAE. Correspondence 1917–35. File Randolph, L. F. 1922–1926. Note: McClintock became an Assistant in Botany in the fall of 1924.

distinguishable. While working as Randolph's research assistant, in February 1925, McClintock began research for her Ph.D. Her interest was in the "B" or accessory chromosomes of corn – a project in which Randolph was much involved. In early February 1925, Randolph reported on the chromosome studies he and McClintock were planning:

> A number of problems suggest themselves which seem to me to be of considerable importance. not only to learn about the behavior of these extra chromosomes, but also of characters [genes, factors] which may be associated with them. Also the origin of the extra chromosome is still open for definite solution. ... The examination of pollen-mother-cells will begin soon.[20]

However by late summer of 1925, McClintock discovered, in A. C. Fraser's corn cultures [stocks], the first triploid maize plant, a plant that had three complete sets of chromosomes.[21] They applied Belling's smear technique to study the chromosomes in the "microsporocytes" of this plant, reported its cytology, and hypothesized the origin of polyploidy in maize (Randolph & McClintock 1926).[22] They described the chromosomes in late meiotic prophase. Note that M. Navashin's (1925) cytogenetic studies in *Crepis*, published a few months previously, had similar implications for understanding the origin of polyploidy and were of "extreme interests for genetics," because progeny of polyploid individuals "should permit a quick solution of the question as to the locations of different groups of genes ... in the haploid set."[23]

Still, neither Randolph nor McClintock could identify each chromosome individually (Randolph & McClintock 1926; Rousseau-Gueutin & Wendel 2020; McClintock 1927; Randolph 1928). While examining offspring from this triploid plant for her doctoral research, McClintock found a trisomic plant ($2n + 1$), which she used to correlate with a particular linkage group (McClintock 1927, 1929a), yet she had not determined the identity of the extra chromosome (McClintock 1927, p. 20). Contrary to persistent legend, her material and techniques at this time did not yet permit her to morphologically distinguish each of corn's ten chromosomes or observe these chromosomes at the pachytene stage [early prophase] of meiotic cell division:

> In the triploid individual the presence of three homologous chromosomes is functionally indicated by the genetic evidence of trisomic inheritance for 4 factors ... Little can be said in estimating homology from the form [morphology] of the chromosomes in the diploid or triploid complement. No exact studies have been made as yet, since the chromosomes show no clearly distinguishable characters and since they fall into a closely graded series. In comparing diakinesis [late prophase] and metaphase I figures of diploid and triploid individuals it can safely be stated that the three chromosomes of a trivalent are morphologically equivalent as far as size and general form are concerned.
>
> *(McClintock 1927, p. 20)*

She believed, however, that it would not be long before she would find a technique to do so (McClintock 1927; Kass & Bonneuil 2004; see Chapter 4).

McClintock's collaboration with Randolph did not last. She was displeased that Randolph's name appeared first "on their 1926 publication." The reason is recorded in a letter written by Emerson in

[20] L. F. Randolph to F. D. Richey, February 5, 1925, NARA, Box 121, Quisenberry, K. S. to Randolph, L. F., RG 54, BPI/SAE. Correspondence 1917–35. File Randolph, L.F. 1922–1926.

[21] L. F. Randolph to F. D. Richey, August 4, 1925, NARA, Box 121, Quisenberry to Randolph, Rg 54, BPI/SAE. Correspondence 1917–35. File Randolph, L. F. 1922–1926.

[22] McClintock's culture card, no 56 [= F1270, F is Fraser's culture number], McClintock MS Col 79, Series V, Box 68, APS. See also Provine and Sisco (1980), Interview with Barbara McClintock 1980, CU [interview restricted until 1998 – see Kass and Provine (1998, 1999)].

[23] LBK read an annotated copy of Navashin (1925) in Sharp's reprint collection, then in the possession of W. B. Provine; the quoted statements are underlined in red pencil on Sharp's copy. Both Randolph and McClintock had free access to Sharp's reprints. Following Dr. Provine's death in 2015, his reprint collection was disbursed; some reprints are in the Cornell Archives, History of Science Collections, and others were given away; see also Smocovitis (2009), *The Plant Drosophila* for contributions by Navashin in the genus *Crepis*. M. Navashin also presented an exhibit on chromosomes of *Crepis* at the 1926 International Botanical Congress (IBC 1926a).

1930: McClintock "was quite displeased that her name appeared second on their 1926 publication when she had done most of the work ...."[24] Additionally, their approach to experimental work was very different. Randolph was extremely careful, meticulous, and cautious. McClintock was notably bright and liked to explore and to modify new experimental techniques, which apparently caused discomfort for her employer. Upon finding the triploid maize plant, McClintock changed the focus of her dissertation project and decided to investigate the cytology and genetics of this unusual plant, but her material did not yet permit her to identify corn's ten chromosomes morphologically (McClintock 1927). Upon ending her service with Randolph, Cornell continued to employ McClintock in the Botany Department as a teaching and research assistant to Professor Sharp. McClintock taught Cytology labs with her Major Professor, who was at the time revising his *Cytology* text and hence had access to the latest references in the field. Sharp referred both Botany and Plant Breeding graduate students and postdocs to McClintock for advice and consultation on their research. Approximately 50 years later, Provine interviewed her about working with Randolph, but she was uneasy and preferred not to discuss specifics (Provine & Sisco 1980). Her experience with Randolph seems to have played a part in her decision to encourage students to work independently. It was during this time, in 1926, after distancing herself from Randolph, that McClintock met both Lewis John Stadler and George Wells Beadle (Kass & Bonneuil 2004).

While an Assistant Professor of Farm Crops at the University of Missouri, Lewis J. Stadler had first come to Cornell in Summer term 1919 to work towards a Ph.D. on the genetics of oats in the Plant Breeding Department with Dr. Love, but was not a successful candidate. Stadler was advised to first take classes in the Department of Botany with his minor professor, the plant morphologist Arthur J. Eames, who was uncomfortable with his attitude ("an indifferent and unfocused student with little sense of direction or purpose") and said so in an evaluation report to Dr. Love. In addition, the research problem he wanted to pursue was not acceptable or appropriate to the work in genetics being conducted at Cornell. He returned to Missouri at the end of the summer and just prior to McClintock's coming to Cornell. In 1922, Stadler instead completed his Ph.D. with Prof. Etheridge, at the University of Missouri (Stadler 1938; Rhoades 1957; Redei 1971).[25] It was not until the spring of 1926 that Stadler and McClintock would become colleagues. By 1925, Stadler switched his focus to genetics of maize (Stadler 1925, 1926). He wrote a grant proposal that impressed Emerson and was awarded a National Research Council Fellowship to work at Harvard's Bussey Institution with E. M. East and at Cornell with R. A. Emerson.[26] Stadler first became interested in genetics after reading Morgan's (1919) monograph on *The Physical Basis of Heredity*, and by 1925 would use the X-ray techniques of the *Drosophila* geneticists to induce mutations in maize. Discussions with Randolph, McClintock, and others at Cornell would lead to cooperative efforts with these maize researchers. Stadler would continue to encourage and support McClintock's studies of maize cytogenetics. She worked in his laboratory on pioneering cytogenetics studies beginning in 1931, and she would later become a member of his research group as an Assistant Professor of Botany beginning in 1936 (see Chapters 5 and 6).

[24] R. A. Emerson to L. J. Stadler, January 14, 1930, Plant Breeding Records 21/28/889, Box 18, File L. J. Stadler, CU. Shortly thereafter, the Head of the Botany Department informed Randolph that it was not possible to furnish him an assistant for the coming year, L. F. Randolph to F. D. Richey, March 9, 1926, NARA, Box 121, Quisenberry to Randolph, Rg 54, BPI/SAE. Correspondence 1917–35. File Randolph, L. F. 1922–1926.

[25] L. J. Stadler's Graduate School Plant Breeding File 1919-1926, Department of Plant Breeding Archives, since transferred to CU; Stadler only spent one summer term at Cornell prior to obtaining his Ph.D. (not one year), and it was Eames (not Emerson) who wrote a poor assessment of Stadler, as reported for Emerson by Rhoades (1957). Stadler's entry in *American Men of Science* 1938 only records his 1926–1927 NRC Fellowship at Harvard and does not mention Cornell.

[26] Stadler, L. J. Graduate School File, see note 25; L. J. Stadler Correspondence with R. A. Emerson, Department of Plant Breeding Records, 21/28/889, Box 18, CU; Stadler (1938); *SCRB* 1920-1927. As a National Research Council Fellow, Stadler also attended many Plant Breeding Department Synapsis Club meetings during the spring (second) term of the 1925–1926 academic year (Murphy & Kass 2011, Figure 22, p. 152). McClintock had attended her first Synapsis Club meeting on November 7, 1923; LBK examined these *SCRBs* in the Plant Breeding Department's archives, before they were deposited in CU (Kass 2003).

By the summer of 1926, Cornell hosted the International Congress of Plant Sciences [Fourth International Botanical Congress (IBC) 1926] in Ithaca (August 16–23, 1926), where Barbara McClintock was in her last year as a doctoral student (Kass 1999). Stadler presented work on experimental error in field plot tests in the Section of Agronomy (IBC 1926a,c; Stadler 1929), but gave no presentation on his mutation studies with corn to the Genetics section. Randolph gave a presentation to the Section of Genetics on studies he had published with McClintock (IBC 1926a,c; Randolph & McClintock 1926), but did not submit a paper for publication in their *Proceedings* (published 1929). H. J. Muller gave a presentation in the Section of Genetics titled "The gene as the basis of life" (IBC 1926a,c; Muller 1929). He cited work of the maize geneticists and referenced a 1921-circular of mimeographed notes authored by E. G. Anderson (unpublished). Muller emphasized that genes were on chromosomes and the difficulty of measuring the size of a gene. He believed that a gene in a chromosome does not contain more than one molecule of a given kind, but more work would be necessary to decide whether or not the "typical" gene may at times become rearranged as suggested by work of Anderson, Eyster, and Demerec with studies of mosaic mutant organisms. The latter authors reported on mutable genes that arose spontaneously and noted that gene interactions might be responsible for such mutations. No X-rays were involved in those studies. At this time, Muller had shown that temperature affected the rate of mutation and made no reference to work with X-rays. An abstract of Blakeslee's IBC-1926 presentation was published in these same IBC *Proceedings* (Blakeslee 1929). Blakeslee's abstract mentioned the induction of mutations, "(2n+1) types," in *Datura* using "radium rays," and he notes that forthcoming papers are in press. He also notes a co-authored paper published in 1927 regarding "chromosomes and gene mutations following exposure to radium rays" (Gager & Blakeslee 1927, cited in Blakeslee 1929). It seems that Blakeslee's IBC-1926 presentation was the only one published that mentions induction of mutants by radiation.

Although Stadler and Muller had both started using X-rays to try to induce mutations in plants and insects, it would be about a year after the International Botanical Congress (IBC) that results would be announced. Muller's Nobel Award–winning work on induction of mutants in *Drosophila* by X-rays began around 1925 and was published in July 1927. Concurrently, Stadler, too, was applying X-rays to plants and obtained similar results.[27] Stadler and H. J. Muller simultaneously and independently reported their X-ray studies at the 1927 AAAS meetings (Kass 2018; see Chapter 4), but Muller's work had previously been hailed around the world when he presented it at the International Congress of Genetics in Berlin, in September 1927 (Nachtsheim 1928). It is not known why Stadler did not present his own research on the same topic, which had also been scheduled for the Berlin Congress.[28]

The 1926 IBC was intended to "bring together … leaders for the presentation and consideration of fundamental contributions to the research and educational aspect of plant sciences …" (Duggar 1929). The first Congress met in Paris in 1900. Congresses were held in successive five-year intervals in Vienna, Austria, and Bruxelles, Belgium. The 1915 Congress, which was scheduled for London, was precluded by the "Great War" [1914–1918] (Duggar 1929). L. H. Bailey, former dean of the College of Agriculture, was President of the 1926 Congress. Faculty in Cornell's College of Agriculture held positions on the Executive Committee and Program Committee, and as Secretaries of Congressional Sections (IBC 1926a). By 1926, the Allied and Axis powers had had eight years to recover from the devastation of war. This recovery "favored a renewal of these established Congresses." Every effort was made to "secure the outstanding active leaders in fundamental research and teaching in the respective fields." Sessions were held in the recently dedicated Baker Laboratory, and buildings of the New York State College of Agriculture. Headquarters were housed in Willard Straight Hall, the newly built Student Union Building. "Dining facilities of Willard Straight Hall [were] opened for the duration of the Congress. This allow[ed] men to eat in the tea

[27] According to Redei (1971), July 30, 1924, was the first record of Stadler's application of X-rays to plants.
[28] See note 16.

room and women as well as men to use the cafeteria, which privileges [were] not usually extended to the patrons of the Hall" (IBC 1926b).

As was the custom of the day, wives of Cornell botanical faculty were hostesses at tea for members of the Congress each afternoon in the library of Willard Straight Hall. They were assisted by *Sigma Delta Epsilon*, graduate women's scientific fraternity (McClintock was not a member). The teas were instituted as a "social medium" to foster "acquaintanceship" among members of the Congress. "Since young men desire to meet older and more illustrious men, and since older and more illustrious men desire to meet others, they will find a very pleasant opportunity to do so at the tea" [U.S. Prohibition of Alcohol 1919–1933]. "Introducers," who were identified by wearing yellow badges, were the "official catalyzers" of the Congress. They were present to "introduce strangers to each other" and especially to assist "younger scientists to meet the more eminent members of their fields" (IBC 1926b).

The Plant Congress facilitated the exchange of ideas between students and experienced researchers and catalyzed McClintock's scientific career. Program Committee Secretary, Lester W. Sharp, wrote to invited participants that "the promotion of personal acquaintance and mutual understanding by informal means [round-table discussions, exhibits, field excursions], is regarded as one of the most important functions of the Congress." Barbara McClintock, then a graduate student of Cytology and Genetics in Cornell's Department of Botany, was in charge of the "Exhibition and informal discussions of cytological preparations" for the Congress Sections of Cytology and Genetics (IBC 1926a). Thus, she was highly visible to leading researchers in her field. McClintock was appointed to this important responsibility by her Major Professor Lester Sharp, a member of the Congress Executive Committee and Secretary of the Section of Cytology. Randolph, Richey, and McClintock seem to have temporarily resolved their differences in a spirit of "mutual understanding" as reflected in a group photo taken at the 1926 Plant Sciences Congress (Figure 3.1).

"The [Congress'] sectional secretaries were in charge of selecting those who [were] invited to present papers." Participation in the formal sectional programs was by "invitation only." Although 156 professional women in science were registered members of the Congress, representing approximately 17% attending, approximately 3% are listed in the Congress program's scientific sessions, and fewer than 2% presented "invitation papers." McClintock was one of nine female scientists who contributed to the program. Four women presented invitation papers in three sections (Cytology, Pathology, and Mycology) out of 13 scheduled sections. Five women participated in demonstrations and exhibits throughout these three sections, and in the Section of Morphology, Histology, and Paleobotany. Gemma Jackson, a colleague of McClintock and a graduate student of Professor A. J. Eames in Anatomy and Morphology in Cornell's Department of Botany, was in charge of exhibits and discussions of anatomical and palaeobotanical materials for the Congress Section of Morphology, Histology, and Paleobotany (IBC 1926a,c). She and Dr. Winifred Goldring presented exhibits in the same session. As mentioned above, during 1924–1926, McClintock was a research assistant to L. F. Randolph, an employee of the U.S.D.A and a cooperative researcher in Cornell's Departments of Botany and Plant Breeding. As part of their research investigations on corn, she independently applied Belling's (1921a,b, 1923) aceto-carmine smear technique to corn microsporocytes. This led to their short note, Polyploidy in *Zea mays* L., published in *American Naturalist* in January/February 1926. Randolph presented part of their work as an "Invitation paper" to a joint session of the International Congress of Plant Sciences Sections of Genetics and Cytology, and as an Exhibit (IBC 1926a). Unfortunately, Randolph's more extensive presentation was not published in the Congress *Proceedings*.

Contacts with influential delegates led to recognition of McClintock's work and resulted in future recommendations for National Research Council [NRC] Fellowships for 1931–1932 and 1932–1933 (see Chapter 5). Among those present who encouraged and supported her early achievements were senior Cornell faculty in Botany & Cytology (Karl M. Wiegand, Lester Sharp), and Plant Breeding

FIGURE 3.1 Frederick D. Richey, Barbara McClintock, Lowell F. Randolph at the 1926 International Botanical Congress, Ithaca, New York. Venue is alcove of the Agronomy Building (later named Stone Hall). (Courtesy of W. B. Provine estate.)

& Genetics (R. A. Emerson, A. C. Fraser). Distinguished American cytologists and geneticists, who presented Invitation papers, Symposium talks, or Exhibits, and who had interests in McClintock's research, were C. E. Allen (University of Wisconsin), E. G. Anderson (University of Michigan), A. F. Blakeslee (Cold Spring Harbor), R. E. Cleland (Goucher College), E. M. East (Harvard University), D. F. Jones (Connecticut Agr. Exp. Sta.), H. J. Muller (The University of Texas), K. Sax (Maine Agr. Exp. Sta.), G. H. Shull (Princeton University), and L. J. Stadler (University of Missouri) (IBC 1926a,c; Figure 3.2).

Following this Congress, Randolph left for Europe on an International Education Board Fellowship, sponsored by the Rockefeller Foundation.[29] McClintock continued to work toward completing her doctoral thesis and would soon meet future Nobel Laureate George Beadle, who became a student of Emerson. Beadle arrived at Cornell in 1926 from the University of Nebraska. He was recommended by his Professor, Franklin D. Keim, who was a doctoral student of Dr. Love in the Department of Plant Breeding (Keim, Cornell Ph.D. 1927;

[29] *SCRB* 1927–1934, p. 7, Randolph returned from Europe; Randolph Faculty File, CU. Letter R. A. Emerson to Division of Biology & Agriculture, National Research Council, April 30, 1930, Randolph held an International Education Board Fellowship. Plant Breeding Records #21/28/889, Folder: National Research Council, Box 15.

FIGURE 3.2 1926 International Botanical Congress Cytology Group Photo; persons are identified from left to right. Front row: Lester W. Sharp (seventh). Second row: Irene Dobroscky (fourth), Barbara McClintock (fifth), A. F. Blakeslee (seventh, above Sharp). Fourth row: Lowell F. Randolph (first). Fifth row: R. E. Cleland (first). (Reprinted from *Photographs of the International Congress of Plant Scientists 1926*, public domain, courtesy of Cornell University Library.)

Murphy & Kass 2007, 2011) and an active member of the Synapsis Club.[30] Beadle was at first in the Department of Agronomy with a minor in Plant Breeding, but after a year as an Assistant in Agronomy, he soon transferred to Emerson's department in 1927 (Beadle 1930a,b; Beadle 1974; Kass & Bonneuil 2004).[31] Beadle attended the Plant Breeding Department's Synapsis Club, lectures, and dinners, where he would become acquainted with faculty and students majoring and minoring in Plant Breeding and Genetics (Murphy & Kass 2011, Figure 23; "1927 Synapsis Club," p. 153).[32] After transferring to the Plant Breeding Department, Beadle enrolled in Sharp's Cytology course, offered during the summer of 1927.[33] Barbara McClintock had previously assisted Sharp with his summer course, but in the summer of 1927, she was enrolled in a Botany class at the Marine Biological Station at Woods Hole (see Chapter 4). She and Beadle, however, soon became research collaborators (Beadle & McClintock 1928) and lifelong friends (Berg & Singer 2003).

[30] See *SCRB 1920–1927*, 1923–1924 academic year.

[31] Beadle's (1974) recollections include compressed memories; he misrepresents McClintock as a student of R. A. Emerson. See File of Beadle, G. W. in Cornell Department of Plant Breeding Records 1906-1970: LBK examined these original files when they were housed in Bradfield/Emerson Hall and in the possession of Cornell's Department of Plant Breeding; they were later deposited in the Cornell Archives. We thank R. P. Murphy for bringing these files to our attention, which we later brought to the attention of Maxine Singer for her research on George Beadle (Berg & Singer 2003); see also Murphy and Kass (2007, 2011). Beadle arrived at Cornell's Department of Agronomy in 1926; he transferred to Department of Plant Breeding in 1927. Beadle's Ph.D. was conferred on February 1930. His thesis was published the same year (Beadle 1930a,b).

[32] *SCRB* 1920–1927, Beadle attended his first meeting of the Cornell Synapsis Club during the first term of the 1926–1927 academic year, on October 4, 1926.

[33] L.W. Sharp *Roll Book* "S-6 Cytology summer 1927."

FIGURE 3.3 Barbara McClintock (fourth from the left, second from ground) with women from the class of 1923 at a four-year reunion with mascot, 1927. (Permission Division of Rare and Manuscript Collections, Cornell University Library.)

CORNELL ALUMNA AND INSTRUCTOR IN BOTANY

A devoted alumna, McClintock regularly attended Cornell reunions.[34] At a joint dinner for the "class of 1923" fourth reunion on Friday evening, June 10, 1927, their sister class of 1922 presented "1923 with a cow anatomically correct in every detail" (*Cornell Alumni News* 1927, p. 460). McClintock and women from the class of 1923 proudly posed for a photo with their new mascot (Figure 3.3).[35]

McClintock was awarded her doctorate the following Monday, June 13, 1927, at Cornell's 59th annual commencement (*Cornell Alumni News* 1927, p. 463). Professor Sharp recommended her for an Instructorship in Botany. She taught classes in Botany along with her botany colleague Gemma Jackson, who was also promoted to Instructor (Figure 3.4).[36]

This appointment held faculty rank toward tenure during those early years (Kass 2003). This employment afforded her the opportunity to write and prepare her doctoral thesis for publication, and to continue to pursue her studies on the triploid corn plant and its offspring. As will be detailed in Chapter 4, these investigations led her to devise a technique for distinguishing the plant's ten individual chromosomes (McClintock 1929c) and to correlate linkage groups with their chromosomes.

---

[34] Barbara McClintock Alumni File, CU.

[35] All persons are identified on the back of the photograph. See also reprint of this photo in *Cornell Chronicle* Vol. 26 (No. 20, February 9, 1995), p. 2; McClintock was mistakenly given a middle initial in the caption for this picture taken at the fourth reunion (1927) of the class of 1923.

[36] This photo was misidentified as McClintock with Harriet Creighton on p. 6 of *The Dynamic Genome* (Fedoroff & Botstein 1992).

FIGURE 3.4 Drs. Gemma Jackson and Barbara McClintock, Instructors at Cornell University, 1931. (Courtesy of Cold Spring Harbor Laboratory Archives.)

LITERATURE CITED

Announcement of the New York State College Agriculture 1923–24. (July 1, 1923). Cornell University, Ithaca, NY. https://ecommons.cornell.edu/handle/1813/41934

Beadle, G. W. 1930a. *Genetic and cytological studies of Mendelian asynapsis in Zea mays.* Ph.D. Cornell Univ. 1930 [February 1930]. https://newcatalog.library.cornell.edu/catalog/302057

Beadle, G. W. 1930b. Genetical and cytological studies of Mendelian asynapsis in Zea mays. *Cornell University Agricultural Experimentation Station Memoir No. 129.* Ithaca, NY. [May 1930] [Publication of Beadles Ph.D. Dissertation]. https://digital.library.cornell.edu/catalog/chla7251474_8569_002

Beadle, G. W. 1974. Recollections. *Annual Review of Biochemistry* Vol. 43, pp. 1–13.

Beadle, G. W. and B. McClintock. 1928. A genetic disturbance of meiosis in *Zea mays. Science* Vol. 68 (No. 1766, 2 November 1928), p. 433. [This became George Beadle's dissertation research project.]

Belling, J. 1921a. The behavior of homologous chromosomes in a triploid *Canna. Proceedings of the National Academy of Sciences* Vol. 7 (No. 7, 15 July 2021), pp. 197–201.

Belling, J. 1921b. On counting chromosomes in pollen-mother-cells. *American Naturalist* Vol. LV [55] (No. 641, November/December), pp. 573–574.

Belling, J. 1923. Microscopical methods used in examining chromosomes in iron-acetocarmine. *American Naturalist* Vol. LVII [57] (No. 648, January/February), pp. 92–96.

Berg, P. and M. Singer. 2003. *George Beadle, An Uncommon Farmer*, Cold Spring Harbor Laboratory Press, Cold Spring Harbor, NY.

Blakeslee, A. F. 1929. An attempt to analyze the composition of Nubbin, A compound (2n+1) chromosomal type in *Datura* (Abstract). *Proceedings of the International Congress of Plant Sciences*, Ithaca, NY, August 16–23, 1926. Vol. II, pp. 831–832. George Banta Publishing Company, Menasha, WI.

Creighton, H. B. 1992. Recollections of Barbara McClintock's Cornell years, pp. 13–18, *In*, N. Fedoroff and D. Botstein (Eds.), *The Dynamic Genome: Barbara McClintock's Ideas in the Century of Genetics*, Cold Spring Harbor Laboratory Press, Plainview, NY.

Duggar, B. M. (Ed.). 1929. *Proceedings of the International Congress of Plant Sciences*, Ithaca, NY, August 16–23, 1926. Vols. I & II. George Banta Publishing Company, Menasha, WI.

Emerson, R.A., Beadle, G. W., and Fraser, A. C. 1935. Summary of Linkage Studies in maize, *Cornell University Agricultural Experiment Station Memoir No. 180.*

Fedoroff, N. V. and D. Botstein (Eds.). 1992. *The Dynamic Genome: Barbara McClintock's Ideas in the Century of Genetics*, Cold Spring Harbor Laboratory Press, Plainview, NY.

Hook, E. B. 1986. James Watt Mavor (1883–1963): A forgotten discoverer of radiation effects on heredity. *Perspectives in Biology and Medicine* Vol. 29 (No. 2, Winter), pp. 278–291. https://doi.org/10.1353/pbm.1986.0053

International Botanical Congress [IBC] (Fourth 1926). 1926a. *Official Program of the International Congress of Plant Sciences* (Fourth International Botanical Congress), Cornell University, Ithaca, NY, August 16–23, 1926. [The Congress, Ithaca, NY]

International Botanical Congress [IBC] (Fourth 1926). 1926b. *Daily News Leaflet.* No. 1–6, August 16–21, 1926. Ithaca, NY.

International Botanical Congress [IBC] (Fourth 1926). 1926c. *Photographs of the International Congress of Plant Scientists* [*sic*] *1926* [held at Cornell University, Ithaca, NY, photos by John P. Troy; Ithaca, NY]

Kass, L. B. 1999. Barbara McClintock and the 1926 International Botanical Congress. *XVI International Botanical Congress, Abstracts*, p. 474. [Poster available in the Cornell Archives. https://newcatalog.library.cornell.edu/catalog/4031434]

Kass, L. B. 2003. Records and recollections: A new look at Barbara McClintock, Nobel Prize-winning geneticist. *Genetics* Vol. 164 (No. 4, 1 August 2003), pp. 1251–1260. https://www.genetics.org/content/164/4/1251

Kass, L. B. (Ed.). 2013ff. *Perspectives on Nobel Laureate Barbara McClintock's Publications (1926–1984): A Companion Volume*. Volumes I & II The Internet-First University Press. https://hdl.handle.net/1813/34897 [Online December 30, 2013, updates March 2014; Vol. III, Increments No. 1–3, May 2016; Vol. IV, Increment No. 4, April 2020]

Kass, L. B. 2018. Questions and reflections on H. J. Muller's Nobel Prize winning research. *Drosophila Information Service* Vol. 101 (December 2018), pp. 93–106. https://www.ou.edu/journals/dis/DIS101/DIS101.html

Kass, L. B. and C. Bonneuil. 2004. Mapping and seeing: Barbara McClintock and the linking of genetics and cytology in maize genetics, 1928–1935, pp. 91–118, *In*, J.-P. Gaudilliere and H.-J. Rheinberger (Eds.), *Classical Genetic Research and Its Legacy: The Mapping Cultures of 20th Century Genetics*, Routledge, London. [Routledge Studies in the History of Science, Technology and Medicine]

Kass, L. B. and E. Cobb. 2007. Landmarks and milestones in American plant biology, the Cornell connection. *Plant Science Bulletin* Vol. 53 (No. 3, fall), pp. 90–101. Available online September 2007: https://botany.org/PlantScienceBulletin/psb-2007-53-3.pdf

Kass, L. B and W. B. Provine. 1998, 1999. Formerly restricted interview with Barbara McClintock, now available at Cornell University Archives. *Maize Genetics Cooperation Newsletter* Vol. 73, p. 41. [Online 1998: https://mnl.maizegdb.org/mnl/73/11kass.html]

Keller, E. F. 1983. *A Feeling for the Organism. The Life and Work of Barbara McClintock*, 1st edition, W. H. Freeman, San Francisco, CA.

Longley, A. E. 1924. Chromosomes in maize and maize relatives. *Journal of Agricultural Research* Vol. 28 (No. 7, 17 May), pp. 673–682, plus 3 plates.

Mavor, J. W. 1921. On the elimination of the X-Chromosome from the Egg of Drosophila melanogaster by X-rays. *Science* Vol. 54 (Issue 1395), pp. 277–279. https://www.science.org/doi/10.1126/science.54.1395.277

Mavor, J. W. 1922. The production of non-disjunction by X-rays. *Science* Vol. 55 (Issue 1420), pp. 295–297. https://www.science.org/doi/10.1126/science.55.1420.295

Mavor, J. W. 1947. *General Biology*, 3rd edition, 738 pp., The Macmillan Co., New York.

Mayr, E. 1980. Protologue, *In*, E. Mayr and W. B. Provine (Eds.), *The Evolutionary Synthesis; Perspectives on the Unification of Biology,* Harvard University Press, Cambridge, MA and London, England. [Transcribed interviews based on a 1974 conference on the synthesis]. https://newcatalog.library.cornell.edu/catalog/794654

McClintock, B. 1925. *A Resume of Cytological Investigations of the Cereals with Particular Reference to Wheat.* Ithaca, NY, 52 pp. plus 25 unnumbered pp. of tables and bibliographies. Thesis, M. A. Cornell Univ. A literature review; no original research. Acknowledges Prof. L.W. Sharp.

McClintock, B. 1927. *A Cytological and Genetical Study of Triploid Maize*. Ithaca, NY, 104 pp. plus 39 unnumbered pp. of tables, plates, and bibliographies. Thesis, Ph.D. Cornell Univ. Acknowledges L. W. Sharp and A. C. Fraser.

McClintock, B. 1929a. A cytological and genetical study of triploid maize. *Genetics* Vol. 14 (No. 2) [11 March 1929, received 11 July 1928], pp. 180–222. [Publication of 1927 Ph.D. Thesis. *Genetics* was issued bimonthly (every two months) at this time] [no pachytene chromosomes]

McClintock, B. 1929b. A method for making aceto-carmin[e] smears permanent. *Stain Technology* Vol. IV (No. 2, April), pp. 53–56.

McClintock, B. 1929c. Chromosome morphology in *Zea mays*. *Science* Vol. 69 (No. 1798, 14 June 1929), p. 629. [The first published ideogram of *Zea* chromosomes]

Morgan, T. H. 1919. *The Physical Basis of Heredity*, J. B. Lippincott Co. Philadelphia, PA and London.

Morgan, T. H., A. H. Sturtevant, H. J. Muller, and C. B. Bridges. 1915. *Mechanism of Mendelian Heredity*, Henry Holt and Company, New York.

Muller, H. J. 1927a. Artificial Transmutation of the Gene. *Science* Vol. 66 (No. 1699, July 22), pp. 84–87.

Muller, H. J. 1927b. Effects of x-radiation on genes and chromosomes. Abstract #134. Program and Abstracts for the twenty fifty annual meeting of the America Society of Zoologists, *The Anatomical Record*, Volume 37 (No. 2, December 1927), p. 174 (see also p. 115, List of Titles, Genetics, Read, "134. Effects of x-radiation on genes and chromosomes, H. J. Muller, University of Texas (also by demonstration)."

Muller, H. J. 1929. The gene as the basis of life. *Proceedings of the International Congress of Plant Sciences*, Ithaca, NY, August 16–23, 1926. Vol. II, pp. 897–921, George Banta Publishing Company, Menasha, WI

Murphy, R. P. and L. B. Kass. July 2007. *Evolution of Plant Breeding at Cornell University: A Centennial History, 1907–2006*, pp. 1–98, Department of Plant Breeding & Genetics, Cornell University, Ithaca, NY, Appendices A1–A98, Photo Section P1–P38.

Murphy, R. P. and L. B. Kass. June 27, 2011. *Evolution of Plant Breeding at Cornell University: A Centennial History, 1907–2006*. Rev. edition i–x, 178 pp., Appendices, Photo Section. e-book, The Internet-First University Press, Ithaca, NY. https://hdl.handle.net/1813/23087

Nachtsheim, H. (Ed.). 1928. *Proceedings of the 5th International Congress of Genetics, Berlin, 1927* [*Verhandlungen des v. Internationalen kongresses für vererbungswissenschaft, Berlin, 1927*, hrsg. von Hans Nachtsheim] *Zeitschrift für induktive abstammungs- und vererbungslehre*; suppl. 1–2. [March 1928]

National Research Council. 1932. *CATALOG of the National Research Fellows in the Biological Sciences*. 1923–1932. March 1, 1932. NAS-NRC Archives, Central Files, ADM: Fellowships: Board of National Research Fellowships in the Biological Sciences: Postdoctoral List of Fellows: 1923–1932.

Navashin, M. 1925. Polyploid mutations in *Crepis*. Triploid and pentaploid mutants of *Crepis capillaris*. *Genetics* Vol. 10 (No. 6, 1 November), pp. 583–592. https://doi.org/10.1093/genetics/10.6.583

Plant Breeding Records, Cornell University Archives: Cornell University Department of Plant Breeding records, 1906–1970. Collection Number: 21-28-889 Division of Rare and Manuscript Collections. Cornell University Library. [Boxes 56–65 (student files) are restricted to Plant Breeding and Genetics Section Chair or their designate]. https://rmc.library.cornell.edu/EAD/htmldocs/RMA00889.html

Provine, W. B. and P. Sisco. 1980. Interview with Barbara McClintock at Cold Spring Harbor Laboratory, Long Island, New York, August 28, 1980. Division of Rare and Manuscript Collections, Cornell University Library, Ithaca, NY.

Randolph, L. F. and B. McClintock. 1926. Polyploidy in *Zea mays* L. *American Naturalist* Vol. LX (No. 666) [January/February 1926, received, no date given], pp. 99–102.[first report of a triploid corn plant & the behavior of its chromosomes, reprinted in Kass 2013ff.]

Redei, G. P. 1971. A portrait of Lewis John Stadler 1896–1954, pp. 5–20, *In*, G. P. Redei and G. Kimber (Eds.), *Stadler Genetics Symposia* (1969 &1970)Vols. 1 & 2, University of Missouri, Agricultural Experiment Station, Columbia, MO.

Rhoades, M. M. 1957. Lewis John Stadler. Biographical Memoir, National Academy of Sciences, Washington, DC. https://nasonline.org/publications/biographical-memoirs/memoir-pdfs/stadler-lewis.pdf

Rhoades, M. M. and B. McClintock. 1935. The cytogenetics of maize. *Botanical Review* Vol. 1 (August), pp. 292–325

Rossiter, M. W. 1982. *Women Scientists in America. Struggles and Strategies to 1940*. Johns Hopkins University Press, Baltimore, MD.

Rousseau-Gueutin, M. and J. F. Wendel. 2020. Barbara McClintock, a Pioneer in the Study of Polyploidy, pp. 1.3.4–1.3.11, *In*, L. B. Kass (Ed.), 2013ff. *Perspectives on Nobel Laureate Barbara McClintock's Publications (1926–1984): A Companion* Volume. Volume IV The Internet-First University Press. URI: https://hdl.handle.net/1813/34897; Kass Vol IV, 30 April 20 FINAL.pdf (3.253Mb), INCREMENT #04.

Rudolfs, W. 1924. Annual report of the sewage substation. July 1 1923–June 30, 1924. *New Jersey Agricultural Experiment Stations, Bulletin 403*. New Brunswick, NJ, pp. 1–7.

Rudolfs, W. and H. A. Trajkovich. 1924a. Fungi and algae of the sprinkling filter bed and their distribution. *New Jersey Agricultural Experiment Stations, Bulletin No. 403*. New Brunswick, NJ, pp. 82–84.

Rudolfs, W. and H. A. Trajkovich. 1924b. Experiments on the physiology of fungus No. 2 (Penicillium). *New Jersey Agricultural Experiment Stations, Bulletin No. 403*. New Brunswick, NJ, pp. 84–85.

Science News-Letter. 1922. X-rays upset heredity, control flies' sex, *Science News-Letter*, Vol. 1, No. 71, Saturday, August 19, 1922, pp. 1–3, https://www.sciencenews.org/sn-magazine/august-19-1922

Sharp, L. W. 1921. *An Introduction to Cytology*, 1st edition, McGraw-Hill Book Company, Inc., New York.
Sharp, L. W. 1926. *An Introduction to Cytology*, 2nd edition, McGraw-Hill Book Company, Inc., New York.
Sharp, L. W. 1924–1931 *Roll Book*, Beadle, G. Cytology course, S 6, Summer 1927.
Sharp, L. W. 1934. *An Introduction to Cytology*, 3rd edition, McGraw-Hill Book Company, Inc., New York.
Smocovitis, V. B. 2009. The plant drosophila. *Historical Studies in the Natural Sciences* Vol. 39 (No. 3), pp. 300–355. ISSN 1939-1811, electronic ISSN 1939-182X. © 2009 by the Regents of the University of California. https://people.clas.ufl.edu/bsmocovi/files/Crepis-Smocovitis.pdf
Sorrells, M. E. 2013. Commentary on Barbara McClintock's 1929 cytogenetics analysis of triploid maize: A cytological and genetical study of triploid maize, pp. 1.15–1.68, *In*, L. B. Kass (Ed.), 2013ff. *Perspectives on Nobel Laureate Barbara McClintock's publications (1926–1984): A Companion Volume*. Volume II, The Internet-First University Press. https://hdl.handle.net/1813/34897
Stadler, L. J. 1925. Variation in the intensity of linkage in maize. *American Naturalist* Vol. 59 (No. 663, July–August 1925), pp. 366–370.
Stadler, L. J. 1926. The variability of crossing over in maize. *Genetics* Vol. 11 (No. 1, 1 January 1926), pp. 1–37.
Stadler, L. J. unpublished [1927]. Abstract. Genetic Effects of X-Rays in Maize and Barley, Presentation at Joint Genetics Sections AAAS December 1927.
Stadler, L. J. 1928a. Genetic effects of X-Rays in maize. *Proceedings of the National Academy of Sciences*, Vol. 14 (No. 1, 15 January; communicated 23 November 1927), pp. 69–75.
Stadler, L. J. 1928b. Mutations in barley induced by X-Rays and radium. Special Article. *Science* Vol. 68 (No. 1756, 24 August), pp. 186–187.
Stadler, L. J. 1929. Experimental error in field crop tests. *Proceedings of the International Congress of Plant Sciences*, Ithaca, NY, August 16–23, 1926. Vol. II, pp. 107–127, George Banta Publishing Company, Menasha, WI.
Stadler, L. J. 1938. Stadler, Prof. L(ewis) J(ohn), p. 1343. NRC Fellow 1926, 6th edition, *In*, J. M. Cattell and J. Cattell (Eds.), *American Men of Science, A Biographical Directory*, The Science Press, New York.
The Register 1922–1923. 1923 (September 1, 1923). Cornell University, Ithaca, NY. https://ecommons.cornell.edu/handle/1813/22316
The Register 1923–1924. 1924 (September 1, 1924). Cornell University, Ithaca, NY. https://ecommons.cornell.edu/handle/1813/22333
The Register 1924–1925. 1925 (September 1, 1925). Cornell University, Ithaca, NY. https://ecommons.cornell.edu/handle/1813/22354
Trajkovich, H. A. 1923. *Yellow-Virescent Seedling in Maize*, M.S. Cornell Univ. June 1923, [Master's] Thesis 1923 T768. https://newcatalog.library.cornell.edu/catalog/5018346
Trajkovich, H. 1924. Inheritance of *Xantha* seedlings in maize. *Cornell University Agricultural Experiment Station Memoir 82* [December], pp. 1–13. [Publication of Her Master's thesis]. Digital document, Cornell University: https://digital.library.cornell.edu/catalog/chla7251474_8565_002
Weatherwax, P. 1923. *The Story of the Maize Plant*, Chicago University Press, Chicago, IL. https://archive.org/details/storyofmaizeplan00weat/page/n9/mode/2up

4 The Golden Age of Corn Genetics
Cornfests, Cornfabs, and Cooperation

POSTGRADUATE INSTRUCTOR AT CORNELL UNIVERSITY

McClintock became an Instructor at Cornell after obtaining her Ph.D. in June 1927.[1] Instead of continuing to assist Dr. Sharp in his Cytology class, offered by *The Summer School of Biology*, Instructor Barbara McClintock attended the Botany course at the Woods Hole, Marine Biological Laboratory at Cape Cod, Massachusetts.[2] The six-week class (June 28–August 8), "Morphology and Physiology of the Algae," admitted only 18 students; selection was limited to persons who "appear likely, by intellectual gifts and intentions, to contribute to the advancement of science," and preference was "given to the better prepared student." No grades or exams were given, but the offerings were intended "to aid in the production and training of investigators."[3] The class was taught by Ivey F. Lewis, Professor of Biology, University of Virginia, and William Randolph Taylor, Assistant Professor of Botany, University of Pennsylvania. Professor Lewis had taught the course for 30 years and was to retire at the end of this term. Taylor went on to become a Professor at the University of Michigan and an expert in the marine algae of the Northeast (Taylor 1937, 1957). The course consisted of daily lectures and laboratory work with weekly collecting trips. The aim was to provide an understanding of morphology and taxonomy of the algae, with respect to its evolution within the group by means of field studies of habits and habitats, and methods of collections, preservation, and classification, and to learn fundamentals of physiology by laboratory studies of algae. Woods Hole is situated at a point where the northern and southern marine floras overlap, thus providing the advantage for a broader study of the algal flora. Emphasis was placed on a study of fresh material collected during field trips to various points within a 20-mile radius of the laboratory. The "Laboratory steamer" was available for trips, but unfortunately during the summer of 1927, the weather prevented traveling to many of their intended venues.[4]

The obvious question is why would McClintock wish to enroll in this course if her focus was on the cytogenetics of the maize plant. McClintock had the opportunity to teach classes besides Cytology,

[1] *The Announcement of Winter Courses for the College of Agriculture, 1927–1928*, issued June 15, 1927, included Barbara McClintock as Instructor, which seems to indicate that she had been made Instructor upon receiving her Ph.D. The winter course began November 1927, and Instructor McClintock was promoted in salary in October 1, 1927, *Proceedings of the Board of Trustees 1927–1928*, p. 7, "Barbara McClintock Ph.D. Instructor in Botany," pp. 1722–1723, Instructors, Appointments, "Promotions: Barbara McClintock, Instructor in Botany," effective October 1, 1927, from $1,200 to $1,500 on state funds. Faculty File, McClintock, Barbara: Appointments Instructor in Botany, July 1, 1927, to June 30, 1931, Division of Rare and Manuscript Collections, Cornell University Library (henceforth CU).

[2] *Announcement of Summer School of Biology* 1926, Cytology offered by Prof. Sharp & Assistant McClintock; *Announcement of Summer School of Biology* 1927 offered by Prof. Sharp 1927; Lester Sharp *Roll Book 1924–1931*.

[3] Annual Announcement *Marine Biological Laboratory*, Woods Hole, Massachusetts, Fortieth Year 1927, Lancaster Press, PA.

[4] *Annual Announcement Marine Biological Laboratory 1927*, pp. 29–30; *The Collecting Net* 1927 Vol. 2, No. 1, July 8, 1927.

DOI: 10.1201/9781003332527-5

and she would go on to teach or assist in classes taught by Professors Sharp and Eames.[5] More importantly, the Marine Biological Laboratory (MBL) was a venue for "Independent Investigators" to conduct research, and to present and attend lectures by leading workers in the fields of Botany, Zoology, Embryology, and Physiology. The lectures were open to all participants at the Laboratory, and were often published in their *Biological Bulletin.* Many principal investigators had been affiliated with the MBL for decades and returned year after year. During the summer of 1927, researchers in Genetics who were affiliated with Zoology and Botany Departments throughout the United States were at MBL. In their labs or during informal teas or after lectures, McClintock and others met to discuss their work. Several investigators whose cytogenetics research would have held the most interest for McClintock were in residence that summer. They included H. J. Muller and J. T. Patterson, from The University of Texas; the Columbia University Zoologists, T. H. Morgan (and his wife Lilian V. and student Albert Titlebaum [Tyler]), Calvin Bridges, A. H. Sturtevant, Jack Shultz, and Helen Redfield; and Department of Genetics, Cold Spring Harbor researchers John Belling and Charles Mets. She also would have had an interest in the work of other investigators who were previously in Botany at Cornell: J. R. Schramm with whom she learned Botany, now Professor at the University of Pennsylvania, and W. M. Robbins, Professor at the University of Missouri who would later be her Department Chair in their Department of Botany; and B. M. Duggar, Professor of Plant Physiology, Washington University (St. Louis), who was then editing the *Proceedings* of the International Congress of Plant Sciences where she had a prominent role the previous summer (Duggar 1929; see Chapter 3). Furthermore, Professor Taylor, Co-instructor of the Woods Hole Botany Class, held a special interest in chromosome structure. He had given an invited presentation on the subject at the 1926 International Botanical Congress in the "Section of Cytology" and led the round-table discussion on the structure of chromosomes (IBC 1926a; Taylor 1929). Taylor's expertise in chromosome structure (morphology) held much relevance for McClintock's research on the chromosomes of maize.

By July 1927, H. J. Muller briefly announced his pioneering research on "The transmutation of the gene" in the journal *Science* (Muller 1927a). At that time, and still the practice today, groundbreaking research was typically announced in *Science* as brief reports. Muller then gave his first full report on the discovery of X-ray–induced mutations in *Drosophila* at the International Congress of Genetics (ICG) held in Berlin, September 1927. While he was still in Europe, his work was immediately heralded around the world (Kass 2018), and lauded again on his return home (Haynes 1998). A full account was reported in the ICG *Proceedings,* which was published the following year (Muller 1928; Nachtsheim 1928). At the AAAS Christmas meeting, in December 1927, Muller again presented his research, emphasizing how the "general frequency of gene mutation may be raised at will" (Muller 1927b). At this same meeting, L. J. Stadler also presented for the first time similar results with oats and maize (Stadler unpublished;[6] Rhoades 1957; Kass 2018). Stadler and Muller began their X-ray experiments simultaneously but at different institutions. Stadler (1928a) published his result in the mid-January issue of *PNAS,* followed by a second "Special Article" in *Science* (Stadler 1928b). Yet, Muller's work received priority, and would be recognized almost 20 years later with a Nobel Prize in Physiology or Medicine (Muller 1946; Kass 2018). Stadler continued his studies of irradiated maize plants and formed collaborations with several Cornell maize geneticists and cytologists, including L. F. Randolph and Barbara McClintock.

---

[5] Botany 141: History of Botany, Plant Anatomy: Graduate course (with Eames and Petry); Methods in Histology and Cytology: Graduate course, *Announcements of the New York State College of Agriculture* 1927–1931.

[6] Stadler, L. J. unpublished [1927] abstract, Genetic effects of X-rays in maize and barley, was presented at the Joint Genetics Sections of the AAAS, December 1927, in Nashville, Tennessee. Stadler's 1927 AAAS presentation on X-rays is summarized in the "Report of the Joint Genetics Sections" (F&G, Zoology & Botany) of the AAAS, *Science* Vol. 67 (No. 1727, February 2, 1928), p. 125. Muller (1928c *PNAS*, September) misdated Stadler's 1927 (unpublished) AAAS abstract as "Stadler (1928a) *Anatomical Record* Vol. 37, p. 176." An examination of *Anatomical Record* Vol. 37 (December 1927) reveals no title or abstract by Stadler, but the volume does include Muller's (1927b) abstract. Stadler mentions his 1927 AAAS presentation in his August publication in *Science* (Stadler 1928b), but without a citation for the abstract. Rhoades 1957 mentions Stadler's 1927 AAAS presentation, without a citation; see details for both Muller's and Stadler's 1927 AAAS presentations in Kass 2018.

Back at Cornell, during the summer of 1927, Lester Sharp was teaching his course in Cytology. Enrolled were two of Emerson's graduate students from the Department of Plant Breeding, G. W. Beadle and T. Bregger. Beadle had switched departments, and Emerson was now his major professor. The approximately five week course (July 5–August 12, 1927) was attended by seven graduate students and one undergraduate. The latter, Alfred G. Marshak, began a Master's degree with Sharp in the Botany department (Marshak 1931), and completed his Master of Arts at Harvard.[7] He was an outstanding student, and by March 1930, he had been recommended by Fraser for an assistantship at Harvard. E. M. East was looking for a graduate student, and Emerson submitted Marshak's name. East asked Emerson if Marshak is "one of the type so often found among Jews, who are so pushing that in two weeks they own the place?" Emerson replied: "What you ask about Marshak is too large an order for me. Whether he would show the ordinary characteristic of his race and want to act as your new dean, I can't say."[8] Marshak was accepted at Harvard and was awarded a Ph.D. (1934) on mouse development, under the mentorship of W. J. Crozier. Additionally, he continued his interest in plant chromosomes (Marshak 1934), which he began at Cornell (Marshak 1931), and continued with studies on the induction of chromosome mutations by X-rays (Marshak 1937). Alfred Marshak was McClintock's friend, and his son's family would play a large part in her care during the last years of McClintock's life (Dan Marshak Pers. Comm. July 29, 2020).

RECOGNITION AND DISCRIMINATION, 1928

Upon returning to campus in fall of 1927, Instructor McClintock continued writing up her thesis for publication, and mentored graduate students in their studies of maize genetics and cytogenetics. The rank of Instructor afforded McClintock faculty status with responsibilities for teaching and guiding graduate students (Kass 2003). In January 1928, Professor Sharp and Emerson, as a dean of the graduate school, nominated her for an International Education Board (IEB) Fellowship. Records of this event revealed that gender played a part in the IEB denying her an award. They asked for Emerson's recommendation of McClintock keeping in mind that "the applicant is a woman and may leave the field of science at any time."[9] Emerson's reply was very supportive. He stressed her commitment to continue in cytological research and that Cornell wished to continue to employ her at her current rank (Instructor). He emphasized that a year abroad would be most beneficial, but he did not specify that she had made any outstanding contribution to her field. Neither did Sharp's letter of recommendation nor McClintock's IEB application indicate that she had any innovative plan for her studies abroad. There is no mention of McClintock identifying the chromosomes of maize, or a plan to connect chromosomes with linkage groups. McClintock's application indicated that she wished to work with Professor [Victor] Gregoire at the University of Louvain [Belgium] and with [Øjvind] Winge at the University of Copenhagen [Denmark], and with one or more of the group of geneticists at Berlin [Kaiser Wilhelm Institute]. She proposed that these investigators would (1) assist her with further studies on the polyploid studies published previously (Randolph & McClintock 1926) and (2) further her investigations on cytological and genetic research for understanding results of the

[7] Marshak studied "Structure & Physiology of Chromosomes" (*American Men of Science* 1938, p. 938), but not in maize. Although Marshak was an assistant to East at the Bussey Institution, he independently investigated the morphology of chromosomes in peas (Marshak 1931), acknowledging Professors L. W. Sharp, K. Sax, E. M. East. He began his chromosome investigations at Cornell with Dr. Sharp and completed the study at the Bussey Institution, where he was awarded an MA in 1932. He then transferred to another program at Harvard, working on growth in mice and continuing chromosome studies in plants. He received his Ph.D. in 1934; Marshak, Graduate School File, Harvard University Archives. McClintock (1934) cited Marshak's 1931 chromosome paper in her study of the nucleolar organizer of maize chromosome 6.

[8] Plant Breeding Records, 21/28/889, File E. M. East, Box 10, correspondence with Emerson and East, March 12, 1930, March 18, 1930, CU (henceforth PB/CU).

[9] PB/CU, 21/28/889 Box 13, File International Education Board, Correspondence: Emerson's correspondence with IEB for IEB Fellowship for Barbara McClintock, CU; Rockefeller Foundation Archives Center, International Education Board Archives (henceforth RF/IEB), (1922–1930)–1964, Box 74, B. McClintock 1928, Folder 1245.

crosses of triploid-by-diploid plants as described in her Ph.D. thesis (McClintock 1927). However, she provided no details of that work. She also proposed to consult with geneticists at the University of Berlin [affiliations at Kaiser Wilhelm Institute Division of Biology] and to conduct a "thorough investigation of the literature ... with the aid of the facilities offered by a large university library." At the completion of her fellowship, she planned to return to the position she "now held." Her application was signed by McClintock and "Proposed by Lester W. Sharp and Seconded by R. A. Emerson" (February 28, 1928). In May 1928, the IEB committee notified Sharp that they did not fund McClintock's proposal and simultaneously returned the reprint of Randolph and McClintock (1926) that she had submitted with her application form.[10] McClintock never reminisced about her IEB rejection. It is possible that her feelings about discrimination in Cornell's Plant Breeding Department were related to Emerson's association with this process.

SIGNIFICANT CONTRIBUTIONS, 1928–1931

McClintock remained at Cornell for several more years. Within the year, she began to make significant contributions to her field, while working cooperatively with students and postdocs in Botany and Plant Breeding.[11] McClintock's collaborations with students contributed to and transformed the cytogenetics of maize in the "Emerson School" at Cornell, the time period often called the "Golden Age of Corn Genetics."[12] Here, I first will provide a brief summary of these contributions followed by a section detailing the pioneering early contributions of McClintock and the institutional setting for those collaborations at Cornell.

EMERSON AND HIS SCHOOL AT CORNELL

Beginning with the year 1928, maize genetics reached a turning point (Kass & Bonneuil 2004). Barbara McClintock contributed to and transformed the genetic mapping of maize, driven by the "Emerson School" at Cornell, where maize genetics was in its golden age. The network of the students of Emerson and E. M. East constituted the nucleus of the maize genetics cooperative that was strengthened in 1928 and formalized in 1932. It was on maize, not on *Drosophila*, that a "Cooperative News Letter" was first initiated (Kass & Bonneuil 2004; Kass et al. 2019). By then, maize rivaled *Drosophila* as an experimental organism until the significance of the banded nature of the giant salivary chromosomes of *Drosophila* was recognized (Painter 1933; Bridges 1935).[13] Parallel to Painter and Muller's cytogenetic work on *Drosophila*, McClintock's cooperative efforts fostered a strong connection between genetic mapping and cytogenetics using maize. McClintock (1929c)[14] cytologically identified the ten chromosomes of corn, which permitted a correlation of individual chromosomes with each of the ten known linkage groups (McClintock & Hill 1929, 1931). The cytogenetic proof of genetic crossing-over in 1931 was the result of two years of collaborations with Emerson's postdoc Charles Burnham (Burnham 1930; McClintock 1930a,b) and Botany student Harriet Creighton (McClintock 1931a; Creighton & McClintock 1931; Coe & Kass 2005a; see Chapter 5). As a postdoc with Stadler at the University of Missouri, McClintock (1931b) would later determine the physical location of genes within the chromosomes by studying translocations, inversions, and deficiencies (McClintock 1931b).[15]

[10] RF/IEB (1922–1930)–1964, Box 74, B. McClintock 1928, Folder 1245.

[11] Beadle and McClintock (1928), McClintock and Hill (1929), Beadle (1930a,b), Hill (1930, 1933), Kass and Bonneuil (2004), and Kass and Murphy (2003).

[12] Kass and Bonneuil (2004), Kass et al. (2005), Murphy and Kass (2007, 2011), and Kass et al. (2019).

[13] For a history of the discovery of the salivary chromosomes and the attack on the gene, see Painter (1934).

[14] The first published ideogram of *Zea* chromosomes; the chromosomes were identified in the "first division in the microspore" (Mitosis) not at the pachytene stage of meiosis I, as mistakenly described by some textbook authors.

[15] See also Rhoades (1984), Murphy and Kass (2007, 2011), Kass and Birchler (2014/2016), and Kass (2013ff.).

R. A. EMERSON, MAIZE, AND GENETICS[16]

R. A. Emerson (1873–1947) received a B.S. degree in 1897, from the University of Nebraska's College of Agriculture. Emerson then worked for the United States Department of Agriculture (U.S.D.A.), Washington, DC, until 1899, when he returned to his alma mater as Horticulturist in the Nebraska Agriculture Experiment Station and Assistant Professor of Horticulture. Taking his part in the agricultural science community's promotion of Mendelism in America, Emerson used beans and corn in his classes to demonstrate and validate Mendel's laws (Kimmelman 1987). During an experiment that was conducted to demonstrate a 3:1 ratio of starchy to sugary kernels, he found aberrant ratios. Thus, the quest to determine the cause of his aberrant results led to his studies of maize genetics. In 1908, Emerson began a study of quantitative inheritance in maize. His experiments were designed to test whether or not these differences were due to numerous factors inherited in a strictly Mendelian manner. Three years later, Emerson took a year's leave of absence from Nebraska to pursue graduate studies at Harvard's Bussey Institution with the distinguished geneticist Edward M. East. Around this time (and independent of the studies of the Swedish geneticist Nilsson-Ehle), East (1910) and East and Hayes (1911) published papers that proposed multiple factor inheritance in corn. They followed this with another important contribution (Hayes & East 1911) to the genetics of Mendelian segregation, inbreeding, yields, etc. These were important contributions to the country's agriculture. Emerson was awarded his D.Sc. in 1913, and that same year, Emerson and East published a collaborative paper titled "The inheritance of quantitative characters in maize." This paper applied and corroborated the multiple factor hypotheses and constituted a very important milestone in quantitative genetics, because for the first time the theoretical aspects were strongly supported by adequate data.

In 1914, Emerson was called to Cornell University to head the Department of Plant Breeding that he did until 1942 (Murphy & Kass 2007, 2011). Joining him as graduate students were Ernest W. Lindstrom (B.A. University of Wisconsin), and Ernest Gustaf Anderson, a former student from Nebraska, who was awarded his Bachelor of Science in 1915 (*AMS* 1938, p. 28).[17] They continued work on corn problems. On recommendations from distinguished plant breeders and geneticists (e.g., E. M. East at Harvard, W. C. Etheridge at Missouri, and D. F. Jones at the Connecticut Agricultural Research Station), students and established researchers from around the country and throughout the world soon joined Emerson's group and studied corn breeding and genetics at Cornell. As mentioned previously, C. B. Hutchison, a former Cornell graduate (MA 1913), was appointed Professor of Plant Breeding in 1916 and taught the courses in Genetics. By 1920, he had decided to discontinue his work on flax and devote the time he had for research to corn. He continued Emerson's unpublished study of *C–Sh* linkage and established (including a linkage map) that *Sh* was part of the C–Sh–Wx (see definitions of abbreviations on p. 63 of this chapter) linkage group (Hutchison 1921, 1922).

Linkage groups were defined by genetic factors (genes, characters) that were usually inherited together (Morgan 1911). Such genes were also demonstrated to be in a linear order on chromosomes (Sturtevant 1913). Genes farther apart from one another on a chromosome are less likely to be inherited together. By measuring the frequency of recombination (gene pairs exchanging places from one chromosome to another, crossing over), the distance between genes on a chromosome can be determined. The higher the percentage of recombination means that the genes are farther apart on the chromosome. The lower the percentage of recombination means that the genes are closer together on the chromosomes. By analyzing the recombination frequency data, the distances between each gene can be determined. A genetic map (linkage map) therefore shows the spatial arrangement of genes on chromosomes and the relative distance between genes.

---

[16] Unless specifically noted, this section is based on Kimmelman (1992) and Morris (1969), an unabridged copy is in the Department of PB/CU, 21/28/889, Box 51, CU. See also Kimmelman (1987, pp. 172–237), Rhoades (1949), Murphy and Kass (2007, 2011), Coe (2001, 2009), and Kass and Bonneuil (2004).

[17] Murphy and Kass (2007, 2011, p. 33), mistakenly reported that Lindstrom got his undergraduate degree at Nebraska.

Hutchison's study of colored or colorless (*C, c*) aleurone color, full or shrunken (*Sh, sh*) kernels, and non-waxy or waxy (*Wx, wx*) starch in kernels of corn showed that crossing plants with dominant genes with plants having recessive genes were linked. His linkage map for the genes (C–Sh––– Wx) showed that genes for color and shrunken were very close together (recombination frequency 3.4%) and the linked waxy gene was separated by a crossover distance of 25.2%. Ten years later, McClintock (1931a) would demonstrate that these genes were on chromosome 9. Her chromosome map showed that the arrangement of the genes *C, Sh,* and *Wx* was consistent with the linkage map published by Hutchison (1921). Her crossover data (3.3% and 20.3%) for the sequence of genes on the chromosome were taken from unpublished linkage maps (Coe & Kass 2005b) and were very close to the percentages published by Hutchison (1921). The gene sequence was key to the proof for cytological crossing-over co-authored by Creighton and McClintock (1931; see below). When A. C. Fraser succeeded Hutchison, he also turned (from wheat) to maize (Fraser 1924). In addition to Anderson and Lindstrom, numerous other students pursued graduate or postdoctoral work with Emerson on corn genetics. These included Thomas Bregger, William H. Eyster, Milislav Demerec, Paul Kvakan, Helen Trajkovich, George Sprague, Marcus Rhoades, George Beadle, Hsien W. Li, Swarm Singh, among others (see Murphy & Kass 2007, 2011).

Since Emerson's early work on multiple factor inheritance (Emerson & East 1913), his maize genetics school contributed simultaneously to the progress of corn breeding and to general knowledge in genetics. In this respect, Emerson's program may be considered a corollary with Morgan's group, and Emerson's students had close scientific associations with the *Drosophila* geneticists. Concepts, methods, standard nomenclatures, and students (including E. G. Anderson, M. Demerec, G. Beadle, and M. Rhoades), who were trained in corn genetics and later also worked on *Drosophila*, circulated between the two communities. Nevertheless, Emerson had a different agenda from Sturtevant and Bridges' mapping enterprise (Kohler 1994). Until 1918, he still had not pursued a systematic study of linkage groups in corn (linkage studies were "stumbled into" in the progress of other work) and showed little interest in genetic mapping. He explained to D. F. Jones:

> This type of work does not interest me as much as working out some complex Mendelian Case for the reason that it is more or less mechanical & from the added fact that we cannot hope to do other than trail along behind Morgan and his co-workers. I am hoping that all the men in this country who are working on related problems with corn may cooperate to such an extent that we can cover the field more quickly.[18]

Emerson saw greater scientific value in more theoretical questions such as multiple factor inheritance, and in the more chemical and physiological studies involving biochemical processes. As Barbara Kimmelman has shown, in the years 1917–1925, Emerson developed and promoted research on the genetics and biochemistry of kernel pigment patterns and hoped that maize would be the "right tool" for such a job. He envisioned a physiological genetics study that would connect the genetic aspects of heredity with physiology. Furthermore, these studies (useful for the country's agriculture) would interest American agricultural scientists, whose initial leading position in genetics was challenged by the shift to research in *Drosophila* genetics after 1910 (Kimmelman 1992).

The achievements of this physiological genetics agenda on maize remained meager. Emerson's former student, George Beadle, made decisive contributions to such a research program in the late 1930s, but not using maize as an experimental organism. By 1928, few significant general contributions to genetics were achieved with corn, and maize linkage studies and genetic mapping stood nearly a decade behind *Drosophila*. The diploid number of chromosomes in *Zea mays* ($2n = 20$) was in question when L. F. Randolph, Emerson's collaborator, commenced his cytological investigations in 1923 (and was still under investigation in 1928), and – contrary to *Drosophila* (Dobzhansky 1929a)[19] – no linkage group had been assigned to any determined chromosome. By early 1929, the ten linkage groups in corn were still not all clearly identified and the mapping work in each

[18] Letter Emerson to Jones, November 8, 1918, PB/CU, 21/28/889, Box 4. File Jones, D. F., CU.

[19] Dobzhansky (1929a), "the larger pair of the V-shaped chromosomes carries the third-chromosome linkage group of genes, the smaller pair carries the second-chromosome linkage group," p. 418.

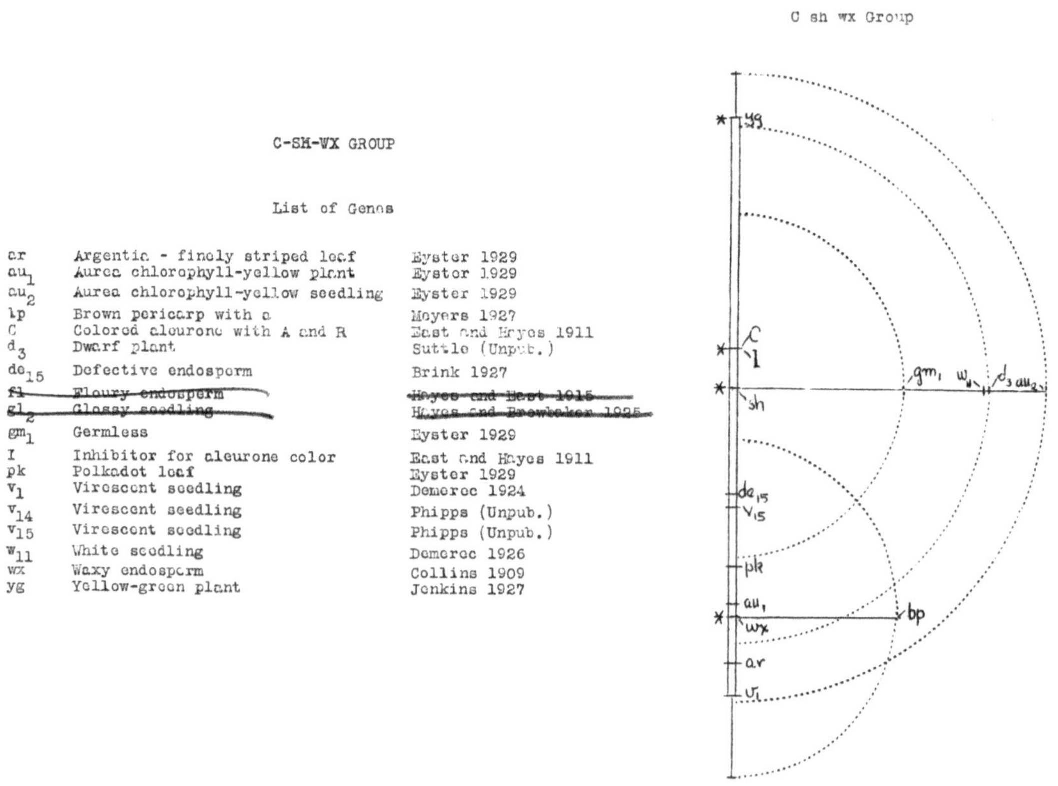
C-SH-WX GROUP

List of Genes

ar	Argentia - finely striped leaf	Eyster 1929
au1	Aurea chlorophyll-yellow plant	Eyster 1929
au2	Aurea chlorophyll-yellow seedling	Eyster 1929
bp	Brown pericarp with a	Meyers 1927
C	Colored aleurone with A and R	East and Hayes 1911
d3	Dwarf plant	Suttle (Unpub.)
de15	Defective endosperm	Brink 1927
~~fl~~	~~Floury endosperm~~	~~Hayes and East 1915~~
~~gl2~~	~~Glossy seedling~~	~~Hayes and Brewbaker 1925~~
gm1	Germless	Eyster 1929
I	Inhibitor for aleurone color	East and Hayes 1911
pk	Polkadot leaf	Eyster 1929
v1	Virescent seedling	Demerec 1924
v14	Virescent seedling	Phipps (Unpub.)
v15	Virescent seedling	Phipps (Unpub.)
w11	White seedling	Demerec 1926
wx	Waxy endosperm	Collins 1909
yg	Yellow-green plant	Jenkins 1927

FIGURE 4.1 Rainbow map as of April 12, 1929, "C–sh–wx group," maize linkage group 9 (after Emerson 1929, excerpted from E. G. Anderson's copy archived at University of Missouri; reprinted in *Maize Genetic Cooperation News Letter 53*, pp. 118–119, 1959; reprinted in Kass et al. (2019). (Courtesy of E. H. Coe.)

group was still very rough, as illustrated by the preliminary "rainbow maps" drawn by Beadle and Emerson in April 1929 (Emerson 1929; Figure 4.1).[20]

EMERSON'S MAIZE COOPERATION

During the period 1918–1920, Emerson realized that he could not avoid investigating the linkage of maize, which was crucial both to breaching the gap with *Drosophila* workers and to providing a deeper basis for the breeding work on corn (Kass & Bonneuil 2004). Whereas *Drosophila* linkage mapping remained the concern of a few laboratories, from 1913 to 1928, Emerson promoted the idea that maize genetic mapping should be a larger cooperative enterprise. It appears that four aspects can account for this approach (Kass & Bonneuil 2004):

1. Emerson's cooperative and enthusiastic, yet trustworthy, personality
2. His belief that collaborative work on the "more or less mechanical" job of genetic mapping would allow individuals to devote the best of their research time to more fundamental research projects

[20] See Emerson (1929) rainbow maps for the remaining maize linkage groups reproduced in *Maize Genetics Cooperation News Letter* 53, March 1, 1979, pp. 117–30, and reprinted in Kass et al. (2019); see also Kass and Bonneuil (2004): Figure 5.1, R-G group; Kass et al. (2005): Figure 4, *c-sh-wx* group.

3. His concern for building a network of former students so as to limit a brain drain to *Drosophila* genetics
4. His striving to benefit from opportunities opened by massive U.S.D.A. funding for corn research from 1920 onward (Fitzgerald 1990, pp. 56–74)

When Frederick D. Richey took charge of corn breeding at the U.S.D.A. in the early 1920s, a regular collaboration was solidified between Emerson and the U.S.D.A. (Richey 1927).[21] Thus, Emerson's school afforded training and scientific support for the U.S.D.A. programs. Emerson acted as an advisor, and trained many U.S.D.A. people at Cornell, including Randolph (who attended the Advanced Genetics course at Cornell in 1924). T. Bregger, Marcus Rhoades, and several others of Emerson's students found jobs at the U.S.D.A. (Randolph was appointed a U.S.D.A. Agent in 1923 to do corn cytology at Cornell, as described previously).

In 1921, Emerson convened the first "cornfest" (later "cornfab") to draw together maize geneticists at an informal gathering during the AAAS meetings in Toronto, Canada (see Chapter 2). From then on, such informal "cornfabs" were convened annually. Emerson circulated letters "to students of corn genetics"; the first communication suggesting ways to develop a standard notation for characters as adopted by *Drosophila* workers (Coe & Kass 2005b; Kass et al. 2005).[22] The Cooperation was reinforced at the 1928 "cornfab" when the mapping of the ten linkage groups was divided among ten groups of researchers with the aim of publishing a general linkage monograph (Kass et al. 2005). Beadle acted as the secretary, and he and Emerson prepared a state-of-the-art summary for each linkage group with provisional maps, called rainbow maps (see above, Emerson 1929, reprinted in Kass et al. 2019). A meeting of corn geneticists was convened during the 6th International Congress of Genetics at Ithaca in August 1932, which led to further development of the cooperation. The maize investigators elected a committee (Emerson, Brink, Mangelsdorf, Jones, and Stadler) to coordinate the work and seek funding support. The group's newsletter began to act as a stock exchange for tester strains. Cornell was chosen to host the repository for all genes and gene combinations (especially tester strains), with Marcus Rhoades as the custodian (Kass et al. 2005, 2019; see Chapter 5).[23]

The news circular that united the maize genetics group (later called the "Maize Genetics Cooperation") soon was not limited to offers and demands for strains but also disseminated unpublished results among the researchers. The rule was that any data appearing there could not be cited in publications without the direct consent of the contributor. So the *News Letter* became a vehicle for the exchange of non-published information, which accelerated the progress and the mapping work and allowed the diffusion of partial data not yet submitted or accepted in journals (Kass et al. 2005). By 1932, Emerson could proudly assert that no other group of researchers had yet so freely shared materials and unpublished results and praised this "unique, unselfishly cooperative spirit" (Emerson 1932, p. 142; Kass 2001; Coe 2001; also see Chapter 5).

THE GENETICS OF MAIZE

By 1930, maize and *Drosophila* were the two organisms whose genetic features were most thoroughly known to investigators (Emerson 1932; Rhoades 1984). In comparison with some other plants and with *Drosophila*, however, corn had the disadvantage of having a long life cycle, which in temperate regions permits only one generation a year or two generations within greenhouse facilities (instead of two weeks for *Drosophila*!). Because corn is a self-compatible plant, corn ears

[21] See Emerson & Richey communications in National Archives and Records Administration, U.S. (henceforth NARA 54-31 Box 43, NA-Emerson 1919–1935); see extensive literature cited in Richey (1927).

[22] See also Emerson to students of corn genetics (March 7, 1923), NARA 54-31 Box 4.

[23] "Materials supporting a request for funds to further in maize genetics studies," exhibit C, "Newsletter to maize geneticists," October 5, 1932, RF RG1.1 200D Box 136 f. 1679.

(silks) had to be covered to prevent unknown self or cross-pollination. The relatively large haploid number of chromosomes (n = 10) also complicated the cytological and linkage studies compared with *Drosophila* (n = 4). Nevertheless, corn also had many advantages for genetic research. It is easy to pollinate, many kernels result from a single pollination, and consequently up to 10,000 kernels can be collected from only "a single hour of work." Additional advantages are the wealth of variability in maize plants, providing much genetic material because of prevalent cross-fertilization. Many weak recessive mutations are perpetuated when they appear in a heterozygous condition with their dominant normal alleles (alternative forms). Yet, these are easily exposed by inbreeding, in contrast to naturally self-fertilized species. In addition, the phenomenon of xenia and double fertilization favors genetic investigation on endosperm and aleurone [3n tissues] characters, and maize lends itself readily to cytological examination because the chromosomes are much bigger than *Drosophila* chromosomes.[24]

The phenomenon of xenia and double fertilization in maize was known since the later part of the nineteenth century. Kernel characters, such as color, shape, size, endosperm (and its aleurone covering layer), exhibited the phenotypes of the male parent (xenia). This was known as "The immediate effect of pollen in maize." The explanation was provided by the discovery of "double fertilization," in 1899 (S. Nawashin 1899; Guignard 1899; summarized by Coe 2001). When plants are fertilized, one sperm (from the pollen) unites with the egg to form the embryo, and the other sperm unites with two nuclei in the embryo sac to form a triploid (3n tissues) endosperm and its aleurone covering. The endosperm provides food for the developing plant. Observations of the characters exhibited by triploid endosperm reveal the genotype of the embryo, without the need to germinate the plant.

As McClintock would soon show, maize haploid mitotic stage chromosomes are readily observed microscopically allowing their identification by size, form, and other morphological features (McClintock 1929c), and when they are long in the meiotic early prophase stage, the distinction of markers on the chromosomes (knobs, chromosomal alterations, etc.) can be observed with greater clarity (McClintock 1930a). Moreover, there were a large number of maize genetics investigators, and more was known about its genetics than any other crop plant, due to the economic importance of maize in the United States. Since the late nineteenth century, when the State and the Land Grant College's Experiment Stations and the U.S.D.A. started systematic corn breeding research, many mutants had been detected and studied (200 genes studied by 1932 and almost 400 by 1935). By 1930, several linkage tester strains were available for "three-point tests."[25] The maize cooperation rapidly distributed these strains (Kass et al. 2005, 2019). Rhoades (1984, p. 24) therefore hardly overstated the subject when he asserted that, "In depth of knowledge of its genetic constitution and in its cytological resolution, maize remained unsurpassed as an experimental organism until the significance of the banded nature of the giant salivary chromosome of *Drosophila* was recognized."

CYTOLOGY MEETS CORN GENETICS AT CORNELL

McClintock's work can be understood in the context of both of these opportunities and constraints. For an understanding of how McClintock became the connection between the geneticists and the cytologists, thus advancing the new field of maize cytogenetics at Cornell, one must look at the connection between the plant breeders and the cytologists. Until about 1931, Plant Breeding was housed in the Forestry Building [named Fernow Hall in 1922] east of the main buildings of the College of Agriculture. Botany was in the Agronomy Building (named Stone Hall in 1933), which was a wing to the west of the main building of the College of Agriculture (Figure 4.2).

[24] This was an advantage until the discovery of the giant salivary chromosome in *Drosophila* by Painter in 1933. Furthermore, compared with *Drosophila*, maize stocks can be stored for decades between investigations.

[25] When two genes are mapped by performing a cross, it is called a two-point cross. The three-point cross involves three pairs of linked genes, and is a valuable method for determining positions of genes in relation to each other and for mapping distances between genes; similarly, a four-point test cross involves four linked genes.

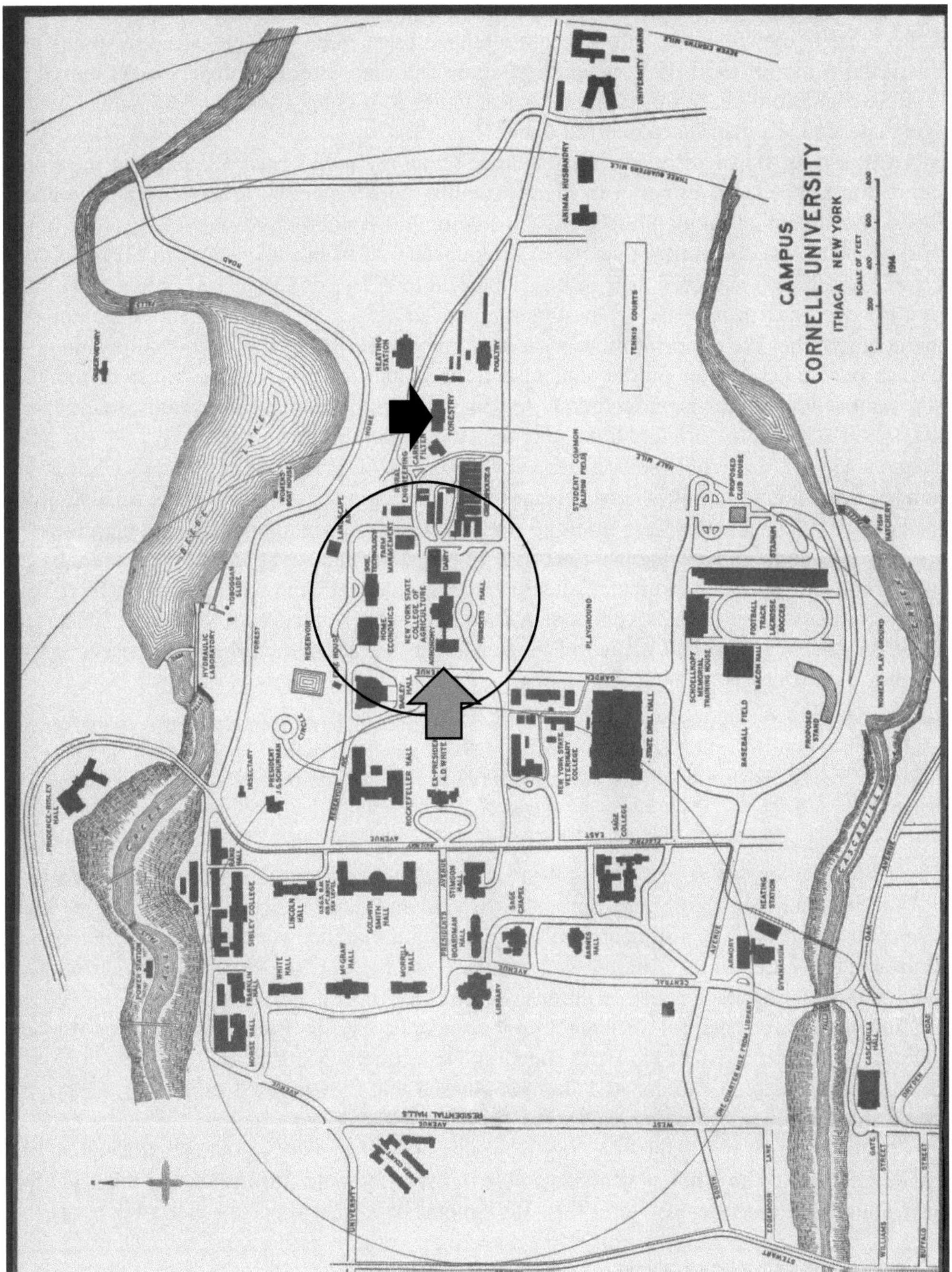

FIGURE 4.2 Map of Cornell University, 1919. Note the distance between the Forestry Building [Fernow Hall, black arrow] and the Agronomy Building [Stone Hall, grey arrow]. (Public domain.)

Thus, researchers in these disciplines were separated by some distance. To assist communications among the young investigators, Dr. Sharp established a space in the Agronomy Building where the corn geneticists could work and consult with the cytologists (Sharp, Randolph, and McClintock) on their studies of linkage groups. By this time and until Botany moved out of the Agronomy Building, McClintock worked in a laboratory that Dr. Sharp had established for graduate students and visiting researchers. There they undertook the preparation and observation of cytological specimens, from 1926 until the winter of 1930–1931, when the Botany and Plant Breeding departments moved into the Plant Science Building. In this cramped space, in the Agronomy Building, McClintock worked with graduate students that Sharp referred to her.[26]

As mentioned previously, Randolph had studied Cytology with Lester W. Sharp in the Botany Department at Cornell and minored in Plant Breeding with Emerson. Emerson recommended a U.S.D.A. appointment for Randolph at Cornell to investigate corn chromosome cytology. McClintock had worked with Randolph until she set out to work on her own after she discovered a triploid maize plant, and Randolph co-authored their results published in 1926 (see Chapter 3). When McClintock received her doctorate in June 1927, Sharp recommended her for an instructorship in Botany. This appointment gave her the opportunity to write and prepare her Ph.D. thesis for publication and to continue to pursue her studies on the triploid corn plant and its offspring. As we shall see, these investigations led her to devise a technique for distinguishing the plant's ten individual chromosomes (1928–1929) and to correlate linkage groups with their chromosomes.

Emerson, head of the Plant Breeding Department, understood the value of cytology. He therefore encouraged Randolph's pursuit of corn chromosome numbers to correlate with the linkage groups that students of maize genetics were finding.[27] In addition, Emerson knew that a cooperative effort with a cytologist should prove valuable because of the work that was coming from the laboratories of Morgan (at Columbia and then Caltech) and Blakeslee (at Cold Spring Harbor, New York). Emerson's former students, E. G. Anderson and M. Demerec, worked with Morgan's, Blakeslee's, and E. B. Babcock's groups and wrote to Emerson at Cornell about the exciting advances in those laboratories. For example, in 1918, Anderson (then at Columbia University) wrote:

> The drosophila [*sic*] work here is very interesting. Some notions have been changed very markedly during the last year or so. Among them deficiency. They now have several so called deficiencies, of which one at least shows a dominant effect in the wing ... Then too reduplication is an interesting phenomenon. A piece of one chromosome gets detached from its proper position and gets hooked on somewhere else. That sounds fishy and I have to bolt a little to get some of the stuff down. But maybe they are right.[28]

E. G. Anderson and Demerec often came back to visit and gave seminars at the Plant Breeding Department's Synapsis Club, a student/faculty organization that met weekly for social and intellectual get-togethers. Later while working with Stadler's plants, McClintock would use chromosome deficiencies in maize to place genes on chromosomes (McClintock 1931b: Chapter 5).

McClintock was very helpful to Sharp's minor students, George Beadle and Marcus Rhoades, although they were members of the Plant Breeding Department. McClintock taught them to make their own cytological preparations, and she was pleased that they learned so quickly – but not as quickly as she did – to interpret them (Provine & Sisco 1980).

After transferring to the Plant Breeding Department, as Emerson's graduate student in 1927, George Beadle worked on a maize genetics problem involving male sterility that he inherited from a former student of Emerson (Eyster 1921). He hypothesized that several genetic factors (genes)

---

[26] Many plant breeding graduate students minored in Cytology with Sharp.

[27] Letter L. F. Randolph to F. D. Richey, February 5, 1925: "A number of problems suggest themselves which seem to me to be of considerable importance ... not only to learn about the behavior of these extra chromosomes, but also of characters [genes, factors] which may be associated with them. Also the origin of the extra chromosome is still open for definite solution." NARA, Box 121, Quisenberry, K.S. to Randolph, L.F., RG 54, records of the Bureau of Plant Industry, Soils & Agricultural Engineering. Correspondence 1917–35. File Randolph, L.F. 1922–1926.

[28] Letter, E. G. Anderson to R. A. Emerson December 21–18, [1918] PB/CU, 21/28/889, Box 2, File: Anderson, CU.

caused sterility in many maize cultures (stocks). Another student of Emerson, I. F. Phipps, had maize stocks segregating for male sterility, which he shared with Beadle. He worked and consulted with the cytologists housed in the room that Professor Sharp had provided for cytological studies. The large but cramped room in the Agronomy Building afforded everyone access to each other's preparations and easily allowed them to discuss results. This atmosphere, combined with McClintock's communicative enthusiasm and the ingenuity of the young research community, created a very stimulating scientific atmosphere. This atmosphere was similar to the room where the *Drosophila* researchers worked at Columbia University (Sturtevant 1965). When Harriet Creighton arrived the following year, she recalled the "excitement of seeing what someone else had found ...," thus providing an "open exchange of observations and interpretations" (Creighton 1992, p. 16). By studying the cytology of microsporocytes (pollen mother cells) at meiosis, Beadle found a recessive Mendelian factor (gene) causing irregular meiosis. The character would later be identified by the gene symbol *as1*. By November 1928, he and McClintock published the new finding in *Science* (Beadle & McClintock 1928). The study would become the subject of Beadle's doctoral research. It seems, however, that Beadle preferred to interpret his own results after applying the cytological techniques he had learned in the Cytology course he had taken with Dr. Sharp. During an interview with history of science Professor William Provine, Beadle recalled that while working in the cytology lab along with McClintock, she looked at his material and interpreted it before him (W. B. Provine pers. com., 1993). Beadle continued his cytology studies and completed his thesis project in February 1930. He graciously acknowledged McClintock in his doctoral thesis (Beadle 1930a) regarding the cytological methods used in his study.

In September 1928, Marcus Rhoades, encouraged by E. G. Anderson and Sterling E. Emerson, came to Cornell with bachelor's and Master's degrees from the University of Michigan to begin his doctoral work with R. A. Emerson.[29] Rhoades became excited by McClintock's project to assign linkage groups to chromosomes. Rhoades (1984, p. 22) recalled that the maize chromosomes were not yet identified when he arrived at Cornell in the fall of 1928. Before he arrived, however, McClintock had been experimenting with different types of preparations under various conditions. She had discovered that late prophase or metaphase in the first microspore mitosis (male gametophyte) gave clearer chromosome observations than did root tip preparations. Sometime between the summer and the fall of 1928, McClintock thought she had recognized a few of the individual distinctive features of the 10 maize chromosomes.[30] In his second term of graduate school (spring of 1929), Rhoades enrolled in Sharp and McClintock's Cytology class. He was the only student in the class (of 14) to receive an A grade (Sharp *Roll Book 1929–1931*). By mid-June 1929, McClintock (1929c) had identified the morphology of the ten maize chromosomes in studies of the first mitotic division of the microspore (male gametophyte, pollen grain).

McClintock's reputation as the cytology expert who could see what others could not see and who was willing to cooperate with geneticists soon extended beyond Cornell and was established in the maize genetics community. In 1929, Harriet Creighton came to do graduate work in Cornell's Botany Department encouraged by Professor Margaret Ferguson of Wellesley College (Kass 2005). Ferguson had years earlier obtained her Ph.D. at Cornell (1901). Creighton would learn cytology from McClintock in the course she attended in the spring of 1930. She too would be only one of two students in the course to receive an A grade. Creighton and McClintock would provide the first cytological proof of crossing-over using maize (McClintock 1931a; Creighton & McClintock 1931; see Chapter 5).

Charles R. Burnham came to Cornell in the summer of 1929 as a postdoc with R. A. Emerson in the Plant Breeding Department. He hoped to corroborate cytologically that Belling's translocation

---

[29] McClintock had recalled that Rhoades arrived after working at Caltech (Provine & Sisco 1980); see also Letter from Barbara McClintock to Peter A. Peterson, September 27, 1972, Barbara McClintock Papers, Box 1. Folder 8. Spectre 1993: Series I.

[30] McClintock's culture cards no.'s MC 121, 8/8/[1928]; McC 137 [summer 1928]; McC 138, 8/3/[1928]. On these cards, she numbered three chromosomes counting from the smallest. McClintock MS Col 79, Series V, Box 68, Spectre 1993: Series V.

hypothesis (Belling 1925; Belling & Blakeslee 1926) was the underlying mechanism for the semisterile corn plants he and Brink had observed (see below). Burnham, a National Research Council (NRC) Fellow, had recently obtained his Ph.D. with R. A. Brink at the University of Wisconsin, and arrived during July 1929. He audited Sharp's summer session Cytology course (L.W. Sharp *Roll Book 1924–1931*), which permitted him to make his own cytological preparations with McClintock's help. His semisterile maize stocks would provide the material for McClintock's cytological observations of early prophase (pachytene stage) meiotic chromosomes (McClintock 1930a). Their fortuitous collaborations permitted them to describe the underlying cytology for semisterility, which indeed proved to be a reciprocal translocation between chromosomes (Burnham 1930). During these studies, McClintock (1930a,b) showed that chromosomes 8 and 9 were involved in the translocations (interchange) (Coe & Kass 2005a; Coe 2001, p. 903; Phillips 2013).[31] Employing semisterile plants and having the translocation that included chromosome 9 permitted them to establish the cytological proof of crossing-over (Creighton & McClintock 1931) and to locate the gene sequence on chromosome 9 (McClintock 1931a). The investigation combined the genetic study of three aleurone and endosperm genes, *C*, *Sh*, and *Wx*, which facilitated cytological observations because of an easily detected knob on the same chromosome (McClintock 1931a; Coe & Kass 2005a; see Chapter 5).

Beadle, Rhoades, and Burnham, who were members of the Plant Breeding Department, had a table and a microscope for cytological observations in the Botany Department lab in the Agronomy Building (Figure 4.3; Provine & Sisco 1980; Creighton 1992, p. 16).[32] Faculty members in Cornell's Botany Department often gave the most promising research problems to ambitious young graduate students. McClintock did not hesitate to give the cytological proof of crossing-over to Creighton and the trisomic tests for correlating linkage groups with chromosomes to Henry E. Hill. Unlike Plant Breeding Department faculty, who, as Rhoades (1949, p. 316) reported, would initially assign routine problems to students,

> When a new student appeared, he [Emerson] would usually assign him some routine problem ... if the student were good, he would soon find a more interesting and exciting problem for his doctoral dissertation, while if the student were mediocre, it didn't matter what kind of a problem he had.

Maize proved to be (until 1933) the best tool for cytogenetic studies. Although linkage studies in maize lagged behind *Drosophila*, until 1928, Emerson soon had the satisfaction of seeing maize cytogenetics rival *Drosophila*. The (10) linkage groups were cytologically assigned to the (10) *Zea* chromosomes only a few years after similar studies began in *Drosophila*. The cytological proof of crossing-over was demonstrated in corn and published by Harriet Creighton and Barbara McClintock only a few months before an equivalent study in *Drosophila* was reported by Curt Stern (1931; see Chapter 5). The union of cytology and genetics stimulated genetic mapping, and these accomplishments brought maize to the forefront of genetics research. Rhoades summarized the accomplishments of the early maize researchers in an unpublished memo that he handed out at the Synapsis Club reunion at Cornell in 1982 (Murphy & Kass 2007, 2011, p. 23).[33]

Their pioneering breakthroughs of 1928–1929 led Emerson to bring McClintock "back in the fold" of the Plant Breeding Department (Keller 1983, pp. 47–50; Provine & Sisco 1980). In a letter to L. J. Stadler in January 1930, Emerson wrote: "graduate students as Beadle and Rhoades and also Dr. Burnham went to McClintock who was willing to spend any amount of time in helping them with the cytological aspects of their problems. These three men with McClintock's help have turned

[31] For an updated list of McClintock's publications, see Kass 2013ff: Table 1. In this volume citations to McClintock's papers, follow chronology in Kass 2013ff: Table 1.

[32] Lee Kass interview with Harriet Creighton, August 17, 1996, Creighton made a sketch of the room reproduced here as Figure 4.3.

[33] A scan of the handout is on the NLM website, without context.

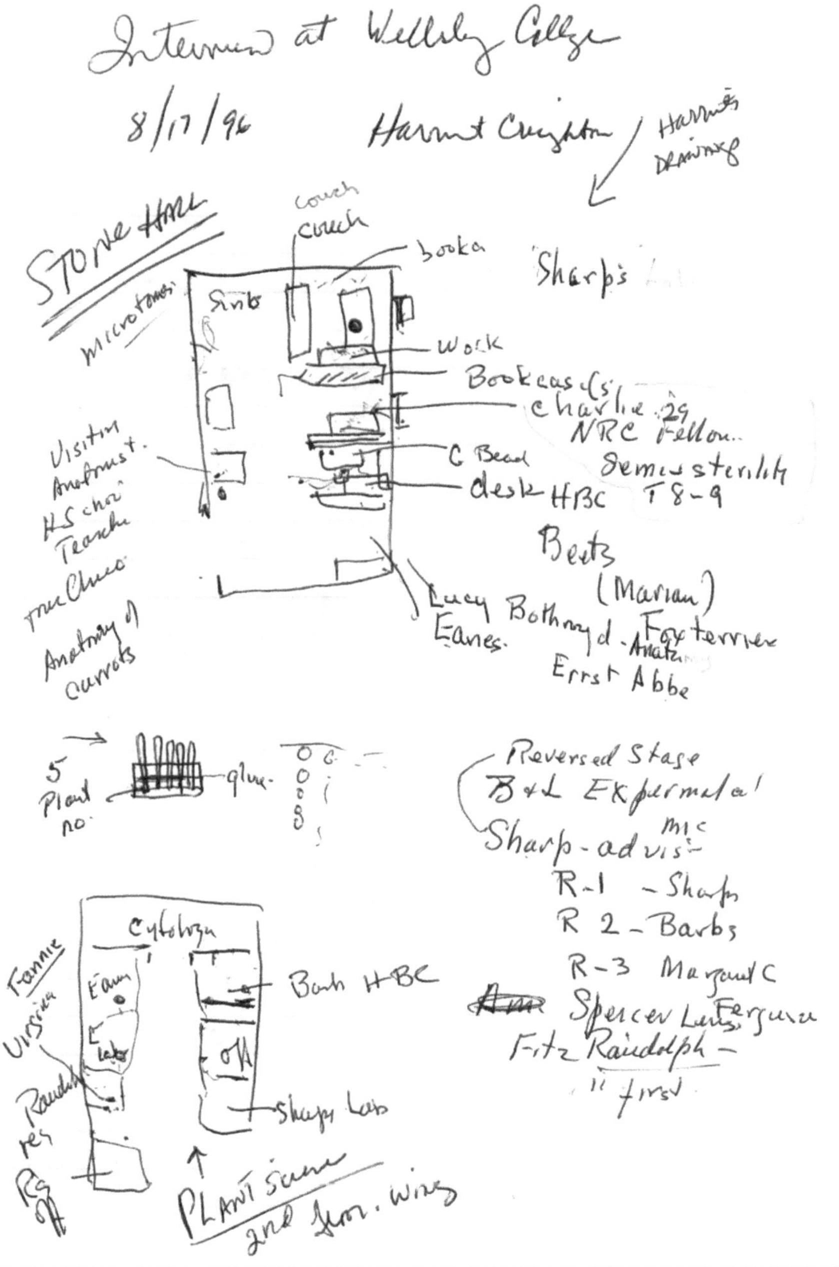

FIGURE 4.3 Professor Lester W. Sharp's Lab in the Agronomy Building [Stone Hall], Cornell University, College of Agriculture. Sketched by Harriet Creighton for Lee Kass, August 17, 1996. (L. B. Kass research notebook IV, 1996.)

up more interesting and important things the past season [corn growing season 1929] than usually come to light in several years" (Figures 4.4 and 4.5).[34]

[34] Letter R. A Emerson to L. J. Stadler, January 14, 1930, PB/CU 21/28/889, Box 18, File: L.J. Stadler, CU. See also acknowledgments to McClintock in the doctoral theses of M. M. Rhoades, G. W. Beadle, H. E. Hill, and the Master's theses of H. Besley, G. W. Scott, S. M. Allen, CU; see also Figure 5.4.

FIGURE 4.4 Professor Emerson, with students, postdocs, and Instructor Barbara McClintock in the back of the Plant Breeding Shed, Cornell University, 1929. Standing from left to right: Dr. Charles R. Burnham, National Research Council Fellow; Marcus M. Rhoades, graduate student; Professor Rollins A. Emerson, Head of the Plant Breeding Department; Dr. Barbara McClintock, Instructor in Botany; (kneeling) George W. Beadle, graduate student, and his dog Toto. (Permission Plant Breeding & Genetics, Cornell University [see Figure 4.5]).

ASSIGNING LINKAGE GROUPS TO CYTOLOGICALLY IDENTIFIABLE CHROMOSOMES

Assigning each linkage group to cytologically identifiable chromosomes was on Emerson's mind when he and Randolph began collaborative studies supported by the U.S.D.A. (see Chapter 3). While working with Randolph as a graduate student, McClintock found a triploid corn plant in the Cornell stocks. Applying John Belling's smear technique (1921a,b, 1923), she was able to describe the chromosomes in late meiotic prophase, but had not yet identified each chromosome individually (Randolph & McClintock 1926). Upon his return from studies abroad (1926–1927), Randolph (1928) published additional studies of maize chromosomes, but the individuality of each chromosome was not yet identifiable. Persistent and innovative McClintock modified staining techniques to enhance coloration in root tip preparations by warming (short of boiling) the sections after staining (McClintock 1927, 1929b). Her technique further modified the aceto-carmine smear technique previously published by Belling (1926). It provided a better technique to preserve microscope preparations for long-term study.[35]

[35] The slides she made were used for teaching in Sharp's cytology class and are deposited at the Hunt Institute for Botanical Documentation in Philadelphia, PA (Hunt Institute 2018). Kass deposited the Cornell Department of Botany cytology and cytogenetics microscope slide collection at Hunt Institute, and also the Department of Plant Breeding photograph collection; see *Bulletin of the Hunt Institute for Botanical Documentation*. Vol. 31 (No. 2) Fall 2019, News from the Archives, p. 5, 9 (recognition: a gift of Cornell's Department of Plant Breeding photographic slides glass lantern slides), https://www.soc-botanical-artists.org/wp-content/uploads/hibd-bulletin-31-2.pdf.

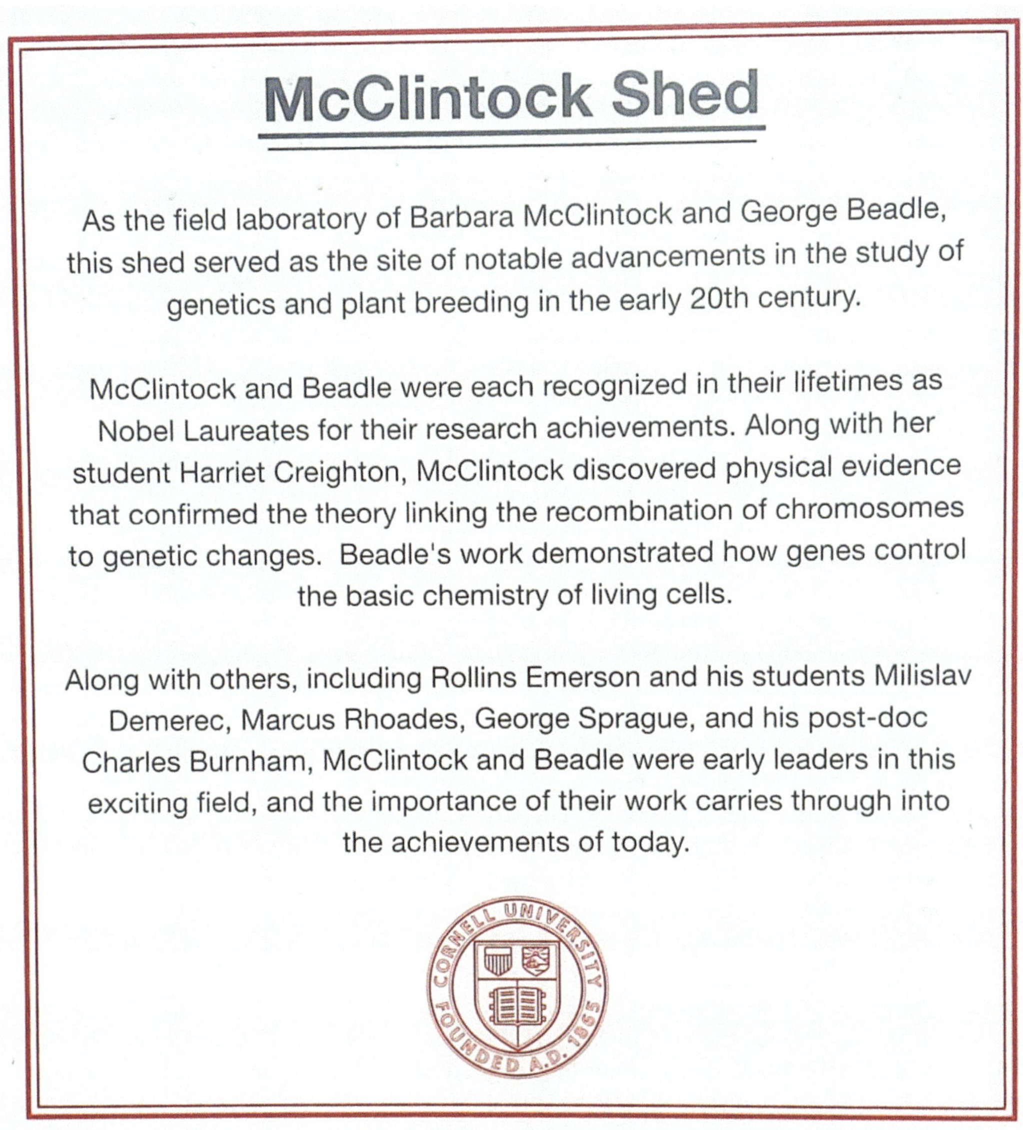

FIGURE 4.5 Plaque on the Plant Breeding Shed, now called "McClintock Shed," at Cornell University (see Figure 4.4). (Image courtesy of Leslie Larson, Plant Breeding & Genetics, Cornell University.

Two years after completing her doctoral thesis, she published the first ideogram (a schematic representation of the chromosomes) for maize (McClintock 1929c) in the June issue of *Science*.[36] The ideogram depicts the ten haploid chromosomes from left to right, smallest to largest, and they were not numbered (reprint in Kass 2013ff.; perspective by Wayne 2013). The chromosomes were identified in the "first division in the microspore" (Mitosis), not at the pachytene stage of meiosis. Identifying the chromosomes would permit the connection of linkage groups with chromosomes as she had proposed in her doctoral studies.

In the context of recalling the work she did with Randolph from September 1924 to February 1926 (Provine & Sisco 1980), McClintock explained how she identified the morphology of corn's

[36] See McClintock (1929c, Figure 1), in a reprint belonging to Professor Sharp, *In*, Kass 2013ff., p. 1.86.

ten chromosomes while a first-year graduate student (see also Keller 1983, pp. 39–40; Comfort 2001, pp. 49–51). McClintock did indeed elucidate the shape and form of these chromosomes. However, she worked out the technique between 1928 and 1929, after she had obtained her Ph.D. This was described in her doctoral dissertation (McClintock 1927, 1929a), revealed by her corn record cards, and is found among her papers at the American Philosophical Society (APS, Specter 1993), and with a careful reading of her primary papers (Kass & Bonneuil 2004; see details below; reprints in Kass 2013ff.). The records clearly show that it was not during her first year in graduate school that McClintock initially described the morphology of corn's chromosomes, nor did she use pachytene chromosomes to do so (Kass 2002a,b, 2003). In an unpublished book chapter, McClintock (unpublished results [1938], APS) explained that she had devised a technique for using late prophase and metaphase stage chromosomes in *mitosis* to describe the morphology of the corn chromosomes, as depicted in her articles published in 1929 and later (McClintock unpublished results, APS; McClintock 1929c; McClintock & Hill 1929, 1931). These documents show that between 1929 and 1930, she first observed pachytene stage chromosomes at *meiosis* while looking for translocations in Charles Burnham's semisterile corn strains (Burnham 1930; McClintock 1930a,b; Kass 2000, 2013ff.: Table I; see details below). Burnham's recollections of this event (Burnham 1982, p. 96; Burnham 1992) are strengthened by his primary papers and correspondence from the era.[37]

In her June 1929 *Science* paper on maize chromosome morphology, she reveals her project and experimental strategy: "it is desirable to determine what linkage group is represented by the extra chromosome in the several 2n + 1 individuals" (McClintock 1929c). As described earlier, McClintock had investigated triploid maize for her Ph.D. dissertation, and she planned to assign linkage groups to chromosomes using Belling's and Blakeslee's trisomic test (McClintock 1927, 1929a). During meiosis in trisomic plants (recovered from the mating of triploids and diploids), some spores and therefore gametes receive one of the triplicate chromosomes, and others receive two. Consequently, the genetic characters carried by this chromosome (and associated with this linkage group) appear in abnormal proportions (a "trisomic ratio"), while other characters, simultaneously, appear in normal Mendelian ratios. It then becomes possible to associate a linkage group with a chromosome by cytologically observing spores with extra chromosomes in the late mitotic prophase in the microspore if that extra chromosome is also present in triplicate in the corresponding strain (and if it can be recognized cytologically). McClintock prepared several crosses during the summer of 1929, put a graduate student, Henry Hill, on this important problem, and secured the help of Marcus Rhoades and George Beadle for the genetic part (McClintock & Hill 1929). They found that in a strain of maize with an extra chromosome, the smallest of the set [number 10] (by 1932, maize geneticists and cytologists had decided to number the chromosome from the largest to the smallest) was present in triplicate. At the same time, they observed that an aleurone color (purple or red aleurone), due to the gene *R*, showed trisomic ratios, contrary to characters in other linkage groups, which showed normal ("disomic") ratios. Therefore, by late 1929, they had identified the R–G linkage group (colored aleurone-yellowish plant) with chromosome 10, but simultaneously incorrectly identified the B–lg linkage group (plant-color intensifier booster-liguleless leaf) with chromosome 4 (McClintock & Hill 1929). Their results were presented at the December/January 1929/1930 winter meetings of the AAAS (in Iowa). They introduced data for two linkage groups; the smallest chromosome carries the r–g linkage group, and the fourth largest carries the b–lg linkage group. Her correction of the B–lg linkage group from chromosome 4 to chromosome 2 appeared in a footnote in a later publication (McClintock 1933). By 1931, McClintock and Hill reported that between March 1930 and March 1931, five additional linkage groups had been assigned to chromosomes

[37] There is much correspondence in Burnham's papers deposited at the University of Minnesota Archives (UMN), and Brink's papers at the University of Wisconsin Archives (UWI).

(chromosomes were not numbered in their paper, but instead only linkage groups were listed) (McClintock & Hill 1931)[38]:

P–br (kernel pericarp and cob color)
B–lg (plant-color intensifier booster-liguleless leaf)
A–d1–cr (plant aleurone and pericarp color-dwarf plant-crinkly leaf)
Y–P1 (yellow endosperm-purple plant color)
C–Sh–wx (colored aleurone-shrunken endosperm-waxy endosperm starch)

Summarizing these early studies, McClintock and Rhoades (1935) reported that by using trisomic methods, six of the ten linkage groups had been associated with the following chromosomes:

Chromosome 2 [B–lg linkage group] initially incorrectly assigned to 4
Chromosome 3 [a1–rg linkage group] [or A–d–cr linkage group (McClintock 1931b)]
Chromosome 5 [pr–v2 linkage group]
Chromosome 6 [Y–P1 linkage group]
Chromosome 7 [gl1–ra linkage group]
Chromosome 10 [r–g linkage group]

Other methods gave a definite check on previous trisomic determinations for linkage groups 1 [P–Br], 4 (su–Tu), and 9 [c–sh–wx]. Emerson's "Summary of Linkage Studies in Maize" reported that using trisomic ratios, McClintock identified eight linkage groups with chromosomes: 2, 3, 5, 6, 7, 8, 9, and 10 (Emerson et al. 1935). It is clear that by 1929, McClintock had first applied trisomic ratios and used mitotic chromosomes to correlate linkage groups with chromosomes and she was responsible for considerable progress in this area.

Before trisomic tests were completed for the remaining four linkage groups, other experimental strategies led to their correlation with chromosomes 1 (P–Br), 4 (su–Tu), 8 (j), and 9 (c–sh-wx). One strategy used semisterile strains, which were caused by a reciprocal translocation. While genetic studies could reveal the two linkage groups associated with a certain case of semisterility, cytological observation at meiotic prophase showed cross-shaped, synaptic (chromosome pairing) configurations involving all four chromosomes (resulting in a chromosome ring at diakinesis) and revealed therefore the identity of the chromosomes involved. McClintock first observed this interchange phenomenon in a translocation involving chromosomes 8 and 9 [T8–9] for which chromosome numbers had not yet been assigned (see McClintock 1930a: Figure 1, reprinted in Kass 2013ff., pp. 1.95–1.101). She first saw the translocation in a semisterile-2 stock that Burnham brought to Cornell, in the summer of 1929, where, as mentioned above, he specifically came to learn cytology from McClintock.

Burnham's semisterile 2 proved to be the most significant strain to demonstrate the cytological basis for semisterile maize plants. John Belling (1914, 1915) was the first to recognize semisterility in the velvet bean by observing 50% aborted pollen and 50% normal seed set. He called this phenomenon semi-sterility. He later observed in plants of *Datura* that non-homologous chromosomes could exchange parts (Belling & Blakeslee 1924). These observations led to his hypothesis (Belling 1925) that semi-sterility in the velvet bean was due to a "segmental interchange between non-homologues" (Burnham 1962, pp. 66–67). Blakeslee (1928) reported a strain of *Datura* having a ring of four chromosomes as observed in late prophase of meiosis. This observation accounted for the segmental interchange and was correlated with semi-sterile *Datura* plants. The first maize semisterile line was described in 1927, and in 1929, Brink and Burnham attributed the mechanism of semisterility to reciprocal translocation. Brink (1927) and Brink and Burnham (1929) had

[38] See McClintock and Hill (1931) Appendix, added after they submitted their paper in March 1930, in which they explain that five additional linkage groups had been correlated with the chromosomes.

observed semisterility in maize plants, similar to that found by Belling. They hypothesized that their observation of semisterility was also due to a segmental interchange. Burnham's cytological studies of semisterile-2 showed a ring of four chromosomes plus 8 pairs of chromosomes at diakinesis (late prophase). The ring of four was assumed to be due to an interchange chromosome. His semisterile-2 diploid (2*n*) maize plant, therefore, demonstrated the underlying cytology for semi-sterility, which was interpreted to be a reciprocal translocation between chromosomes (Burnham 1930).

McClintock (1930a) then pursued a study to determine which chromosomes belonged to the ring of four observed at diakinesis in Burnham's semisterile-2 plants (Phillips 2013). Crossing trisomic plants for the "smallest and fourth smallest" chromosomes [chromosomes 10 and 7], she eliminated these as part of the ring of four and concluded that, "The chromosomes involved in what proved to be a reciprocal translocation, or segmental interchange, were therefore, the second and third smallest chromosomes." The chromosomes were later numbered as maize chromosomes 8 and 9.

Once more, her attention to detail led to a cytological innovation. During this study, she also observed for the first time the mid-prophase [pachytene] stage of meiosis. She clearly saw the paired, "parasynapsed," interchange chromosomes before they had opened out into four rings at diakinesis (see McClintock 1930a: Figure 1, reprinted in Kass 2013ff.; and perspective by Phillips 2013). McClintock had noted that the pachytene (mid-prophase) stage of the first meiotic division of the microsporocyte (pollen mother cell) showed long chromosome threads (with translocation or interchange chromosomes staining quite darkly), and afforded much finer resolution than preparations of root tip sections or microspore smears of first division mitosis to demonstrate where translocation breaks occurred (McClintock 1930a; Phillips 2013). Her diagram of the interchange chromosomes showed a cross-shaped configuration (synaptic complex) where the position of the cross reflected the position of the breakpoints in the two non-homologous chromosomes [chromosomes 8 and 9]. She recognized for the first time that the knob at the end of chromosome 9 (the second smallest) and the centromere positions could be used to identify all four chromosomes (Phillips 2013). She showed for the first time that during meiosis, the interchange chromosomes might separate in a way that produced spores carrying duplicate or deficient genes. As a result when paired interchange chromosomes separate at meiosis, some of the interchange chromosomes carry deficient or extra chromosomes, thus resulting in 50% sterility.

Observing paired (parasynapsed) chromosomes at the mid-prophase [pachytene stage] of meiosis led McClintock to the idea that paired interchange chromosomes having different markers, such as chromosome knobs, could be used to demonstrate the cytological proof of crossing over. With the collaboration of Burnham and Harriet Creighton, she would continue an intense study to prove cytological crossing-over in the spring of 1931 (Coe & Kass 2005a; see Chapter 5).[39]

Following this breakthrough, the first use of an interchange to assign linkage groups to chromosomes was Burnham's and Brink's work on semisterile-1, which was shown, by late 1931, to be associated with linkage groups P–Br and B–lg and to involve the two largest chromosomes (1 and 2, respectively). Since the B–lg linkage group was assigned to chromosome 2, P–Br could be attributed to chromosome 1 (Rhoades & McClintock 1935).[40] In a similar way, a study of a strain carrying a translocation in chromosomes 8 and 4 [T8–4] led Anderson to assign linkage group *su–Tu* to chromosome 4 (Emerson et al. 1935; Rhoades & McClintock 1935).

[39] Letter McClintock at Cornell to Burnham at Caltech, March 12, 1931, Burnham collection # 4-3-91, Box 3, File McClintock, UMN.

[40] Burnham (1930) assumed that *Pbr* was associated with the largest chromosome; see Emerson (1932, p. 149) discussing reciprocal translocations, "By this means Burnham (1930) had shown that genes of the P–br linkage group (group I) are borne by chromosome 1." Note: Cooper and Brink (1931) reported that chromosomes in semi-sterile-1 are not easily identified at the open spireme [pachytene] stage; see also Brink and Cooper (1931); here, they used diakinesis, anaphase I, and metaphase 1 (not pachytene) chromosomes to confirm their association.

CYTOGENETICS HELPS GENETIC MAPPING

Cytogenetics did more than permit a correlation of chromosomes with linkage groups or provide cytological proof of genetic crossing over (summary in Chapter 5). It also accelerated and partly transformed the mapping work itself in the Maize Genetics Cooperation group whose organization was strengthened by Emerson in late 1928 at the AAAS meetings and again in 1932 at the International Congress of Genetics. Thanks to the localization of a breakage point on the chromosome seen in synaptic configurations, the interchange also allowed qualitative positioning of genes relative to the breakage point (McClintock 1930a, 1931a). McClintock's and Burnham's research stimulated interests in cytogenetics, and soon geneticists actively searched for translocation mutants. By 1935, more than 50 such strains were produced (mostly by Burnham and by Anderson at Caltech).[41] These strains were used to craft many tester lines that contained both usual (phenotypic) characters and a translocation. About one-third of the "three-point tests" and "four-point tests" whose results were reported in the Emerson et al. (1935) linkage monograph involved a translocation. Such three-point test combinations were extremely valuable, since they allowed confirmation of gene associations with specific chromosomes, and gave the order of genes and breakage points (Rhoades 1931). Another major cytogenetic approach to mapping used deletions. McClintock was the first to find and apply this technique in Stadler's plants, and the details will be described in Chapter 5.

Early biographies of McClintock portrayed her at Cornell as a lone female genius not understood by traditional plant geneticists. This account may have been due to strong reliance on interviews and memories necessarily blurred or compressed after 50 years, and often retrospectively biased by her later experiences on transposable ("controlling") elements (Keller 1983, repeated by Comfort 2001, pp. 32–36). These stories told about the incomprehension met by McClintock's cytogenetics research program were reconsidered by Kass and Bonneuil (2004), and Kass (2003, 2013ff.), and further detailed here.

It is clear from this chapter and the details provided elsewhere (Kass & Bonneuil 2004; Kass et al. 2005, 2019) that maize geneticists maintained strong relations with *Drosophila* geneticists during the 1920s. This connection was due primarily to Rollins Emerson, and to his students, who worked with both corn and *Drosophila* (see Murphy & Kass 2007, 2011). E. G. Anderson and M. Demerec studied with both Morgan's and Blakeslee's groups and kept Emerson informed about the exciting work that was progressing in these laboratories. Rhoades had worked with Anderson at the University of Michigan before arriving at Cornell in 1928, and subsequently collaborated with him at Caltech, while a teaching fellow, during the 1929–1930 academic year (Anderson & Rhoades 1931).[42] Consequently, Cornell maize geneticists were aware that the drosophilist's use of cytology had opened a fertile second front to tackle problems (e.g., see Painter, Dobzhansky).

Furthermore, the association between cytology and genetics was important in several research programs in plant genetics in the 1920s so that McClintock did not arrive with a *tabula rasa*. John Belling and Albert F. Blakeslee, working on *Datura* at Cold Spring Harbor, already had elegantly combined genetic-cross analysis and cytology. The use of trisomic strains and trisomic heredity to associate one specific chromosome (being present in three copies) with one specific linkage group was a strategy they used as early as 1923 (Blakeslee & Farnham 1923). They also had shared their results in seminars presented at Cornell (Chapter 3). This work, decisive for McClintock's

---

[41] See Anderson's impressive stocks *in Maize Genetics Cooperation Newsletter* Vol. 5, January 25, 1934, pp. 11–12, also reproduced in Kass et al. (2019).

[42] "Paper from the Department of Botany of the University of Michigan No. 273, issued 1930." Rhoads had been a student of Anderson's at University of Michigan and held Teaching Fellowship in the Division of Biology at Caltech during 1929–1930 (*Bulletin of the California Institute of Biology*, Vol. 38, December 1929, p. 36). Emerson had no funds to support Rhoades at Cornell during 1929–1930. Rhoades returned to Cornell in July 1930 to take Beadle's Assistantship after Beadle began his NRC Fellowship at Cornell. Beadle then transferred his fellowship to Caltech in November 1930; Letters R. A. Emerson to C. E. Allen, February 18, 1930; G. W. Beadle to E. E Conger, October 10, 1930; PB/CU 21/28/889, Folder: National Research Council; CU.

association of linkage groups with chromosomes, was known to Cornell's maize researchers since Randolph and Sharp visited Belling in 1924 and applied the aceto-carmine smear technique.

Additionally, even if there had been an early lack of understanding or acknowledgment of McClintock's work, this was reversed by her very productive summer of 1929. Emerson would report extensively on her cytogenetic achievements in his 1932 summary of the contributions of the Cornell maize workers (Chapter 5). It seems that after the relative failure of his physiological genetics program, in contrast to the Morgan group's achievements, Emerson was very gratified to see maize compete with *Drosophila* as a research organism for fundamental discoveries in the new field of cytogenetics.

Ultimately, not only Beadle and Rhoades, but also Anderson, Stadler, Burnham, and Brink were geneticists who immediately grasped the significance of McClintock's work and consequently modified their research projects accordingly. Anderson and Burnham devoted their life's work to translocations in corn and trained their students in this field. Beadle left corn but would years later invite McClintock to his laboratory at Stanford University in 1944 to carry out cytology of *Neurospora* chromosomes.[43] Rhoades devoted his career to corn cytogenetics and trained many students in this area. McClintock is considered the pioneer in all these areas of research.

LITERATURE CITED

American Men of Science, A Biographical Directory. 1938, 6th edition, J. M. Cattell and J. Cattell (Eds.). The Science Press, New York.

Anderson, E. G. and M. M. Rhoades. 1931. The distribution of interference in the X-chromosome of Drosophila. *Papers of the Michigan Academy of Science, Arts and Letters* Vol. XIII (1930), pp. 227-229. [Published 1931]

Beadle, G. W. 1930a. *Genetic and cytological studies of Mendelian asynapsis in Zea mays.* Ph.D. Cornell University 1930 [February 1930] https://newcatalog.library.cornell.edu/catalog/302057

Beadle, G. W. 1930b. Genetical and cytological studies of Mendelian asynapsis in *Zea mays*. *Cornell University Agricultural Experimentation Station Memoir No. 129.* Ithaca, NY. [May 1930] [Publication of Beadle's Ph.D. Dissertation] https://digital.library.cornell.edu/catalog/chla7251474_8569_002

Beadle, G. W. and B. McClintock. 1928. A genetic disturbance of meiosis in *Zea mays*. *Science* Vol. 68 (No. 1766, November 2, 1928), p. 433. [This became George Beadle's dissertation research project]

Belling, J. 1914. The mode of inheritance of semi-sterility in the offspring of certain hybrid plants. *Zeitschrift für Induktiv Abstammungs und Vererbungsleher* Vol. 12, pp. 303–342.

Belling, J. 1915. Inheritance of pod pubescence and partial sterility in *Stizolobium* crosses *University of Florida Agricultural Experiment Station Annual Reports for 1914*, pp. 81–105.

Belling, J. 1921a. The behavior of homologous chromosomes in a triploid *Canna*. *Proceedings of the National Academy of Sciences* Vol. 7 (No. 7, July), pp. 197–201.

Belling, J. 1921b. On counting chromosomes in pollen-mother-cells. *American Naturalist* Vol. LV [55] (No. 641, November/December), pp. 573–574.

Belling, J. 1923. Microscopical methods used in examining chromosomes in iron-acetocarmine. *American Naturalist* Vol. LVII [57] (No. 648, January/February), pp. 92–96.

Belling, J. 1926. The iron-acetocarmine method of fixing and staining chromosomes. *Biological Bulletin* Vol. 50 (No. 2, February 1926), pp. 160–162.

Belling, J. and A. F. Blakeslee. 1924. The configurations and sizes of the chromosomes in the trivalents of 25-chromosome *Daturas*. *Proceedings of the National Academy of Sciences* Vol. 10, pp. 116–120.

Blakeslee, A. F. 1928. Genetics of Datura. *Zeitschrift für Induktiv Abstammungs und Vererbungsleher*, suppl. 1, pp. 117–130. [Paper presented at the Fifth (V) International Congress of Genetics (Berlin), 1927]

Blakeslee, A. F. and M. E. Farnham. 1923. Trisomic inheritance in the poinsettia mutant of *Datura*. *The American Naturalist* Vol. 57 (November-December 1923), pp. 481–495.

Bridges, C. B. 1935. Salivary chromosome maps. *Journal of Heredity* Vol. 26, pp. 60–64.

Brink, R. A. 1927. The occurrence of semi-sterility in maize. *Journal of Heredity* Vol. 18, pp. 266–270.

Brink, R. A. and C. R. Burnham. 1929. Inheritance of semi-sterility in maize. *The American Naturalist* Vol. 63 (No. 687, July–August 1929), pp. 301–316.

[43] Beadle would later return to corn investigations (Horowitz 1990).

Brink, R. A. and D. C. Cooper. 1931. The association of semisterile-1 in maize with two linkage groups. *Genetics* Vol. 16 (No. 1, November 1, 1931), pp. 595–628.
Burnham, C. R. 1930. Genetical and cytological studies of semisterility and related phenomena in maize. *Proceedings of the National Academy of Sciences* Vol. 16 (No. 4, April 15, 1930), pp. 269–77.
Burnham, C. R. 1982. Personal recollections of events leading to a correlation of linkage maps and chromosomes in maize and barley, pp. 93–105, *In*, W. F. Sheridan (Ed.), *Maize for Biological Research*, Plant Molecular Biology Association, Charlottesville, VA.
Burnham, C. R. 1992. Barbara McClintock: Reminiscences, pp. 19–24, *In*, N. Fedoroff and D. Botstein (Eds.), *The Dynamic Genome: Barbara McClintock's Ideas in the Century of Genetics*, Cold Spring Harbor Laboratory Press, Plainview, NY.
Coe, E. 2009. East, Emerson and the birth of maize genetics, pp. 3–16, *In*, J. L. Bennetzen and S. C. Hake (Eds.), *Handbook of Maize: Genetics and Genomics*, Springer Science and Business Media LLC, New York.
Coe, E. Jr. 2001. The origins of maize genetics. *Nature Reviews Genetics* Vol. 2 (November 1, 2001), pp. 898–905.
Coe, E. and L. B. Kass. 2005a. Proof of physical exchange of genes on the chromosomes. *Proceedings of the National Academy of Sciences* Vol. 102 (No. 19, May 10), pp. 6641–6656; online May 2, 2005: https://www.pnas.org/content/102/19/6641.full
Coe, E. and L. B. Kass. 2005b. Maize genetics cooperation news letter files: Expanded chronological list of materials and related cooperation. *Maize Genetics Cooperation Newsletter* Vol. 79, pp. 72–76; available online April 2005: https://mnl.maizegdb.org/mnl/79/06CoeKass.htm; https://mnl.maizegdb.org/mnl/79/mnl79.pdf
Comfort, N. 2001. *The Tangled Field.* Harvard University Press, Cambridge, MA.
Cooper, D. C. and R. A. Brink. 1931. Cytological evidence for segmental interchange between non-homologous chromosomes in maize. *Proceedings of the National Academy of Sciences* Vol. 17 (No. 6, June 1, 1931), pp. 334–338.
Creighton, H. B. 1992. Recollections of Barbara McClintock's Cornell Years, pp. 13–18, *In*, N. Fedoroff and D. Botstein (Eds.), *The Dynamic Genome: Barbara McClintock's Ideas in the Century of Genetics*, Cold Spring Harbor Laboratory Press, Plainview, NY.
Creighton, H. B. and B. McClintock. 1931. A correlation of cytological and genetical crossing-over in *Zea mays*. *Proceedings of the National Academy of Sciences* Vol. 17 (No. 8, August 15, 1931) [Communicated July 7, 1931], pp. 492–497.
Dobzhansky, Th. 1929a. Genetical and cytological proof of translocations involving the third and fourth chromosomes of *Drosophila melanogaster*. *Biologisches Zentralblatt* Vol. 49, pp. 408–419.
Dobzhansky, Th. 1929b. A homozygous translocation in *Drosophila melanogaster*. *Proceedings of the National Academy of Sciences* Vol. 15 (No. 8, August 15, 1929), pp. 633–638.
Duggar, B. M. (Ed.). 1929. *Proceedings of the International Congress of Plant Science, Ithaca, New York, August 16-23, 1926*. Vols. I & II. George Banta Publishing Company, Menasha, WI.
East, E. M. and H. K. Hayes. 1911. Inheritance in maize. *The Connecticut Agricultural Experiment Station Bulletin No. 167* (April), pp. 1–142 (New Haven, CT). https://www.maizegdb.org/docs/EastHayes1911.pdf
Emerson, R. A. 1929. "To students of maize genetics," April 12, 1929; *Reprinted in Maize Genetics Cooperation News Letter 53*, March 1, 1979, pp. 117–130 and Kass et al. 2019. [See also Kass & Bonneuil 2004: Figure 5.1; Kass, Bonneuil & Coe 2005: Figure 4]
Emerson, R. A. 1932. The present status of maize genetics. *Proceedings of the Sixth International Congress of Genetics* Vol. 1, pp. 141–152. [Paper presented in August, published in December 1932]
Emerson, R. A., G. W. Beadle, and A. C. Fraser. 1935. Summary of linkage studies in maize. *Cornell University Agricultural Experiment Station Memoir No. 180.*
Emerson, R. A. and E. M. East. 1913. The inheritance of quantitative characters in maize, *Bulletin of the Agricultural Experiment Station of Nebraska, Research Bulletin No. 2.* https://core.ac.uk/reader/77948660 [Other R. A. Emerson papers are also available at this link]
Eyster, L. A. 1921. The linkage relations between the factors for tunicate ear and starchy-sugary endosperm in maize. *Genetics* Vol. 6 (No. 3, May 1, 1921), pp. 209–240.
Fraser, A. C. 1924. Heritable characters in maize XVII. Intensified red and purple aleurone color. *Journal of Heredity* Vol. 15, pp. 110–123.
Fitzgerald, D. 1990. *The Business of Breeding: Hybrid Corn in Illinois, 1890-1940*. Cornell Univ. Press, Ithaca, NY

Hayes, H. K. and E. M. East. 1911. Improvement in corn. *The Connecticut Agricultural Experiment Station Bulletin* Vol. 168 (June), pp. 3–21 (New Haven, CT). https://www.maizegdb.org/docs/HayesEast1911.pdf

Haynes, R. H. 1998. Heritable variation and mutagenesis at early International Congresses of Genetics. *Genetics* Vol. 148 (No. 4, April), pp. 1419–1431.

Hill, H. E. 1930. A *cytological and genetical study of certain trisomic types in Zea Mays L.* M.S. Cornell University, June 1930, Ithaca, NY.

Hill, H. E. 1933. *A cytological study of two tetrasomic types in Zea mays L.* Ph.D. Cornell University, June 1933, Ithaca, NY.

Horowitz, N. 1990. *George Wells Beadle 1903–1989.* National Academy of Sciences of the United States of America, Washington, DC. Biographical Memoir, https://nasonline.org/publications/biographical-memoirs/memoir-pdfs/beadle-g-w.pdf

Hunt Institute for Botanical Documentation. (August 15, 2018). Institute receives cytology microscope slide collection. News: https://www.huntbotanical.org/news/show.php?id=91

Hutchison, C. B. 1921. Heritable characters of maize. VII. Shrunken endosperm. *Journal of Heredity* Vol. 12 (No. 2, February), pp. 76–83.

Hutchison, C. B. 1922. The linkage of certain aleurone and endosperm factors in maize, and their relation to other linkage groups. *Cornell University Agricultural Experiment Station Memoir No. 60*, June 1922, pp. 1419–1473, https://digital.library.cornell.edu/catalog/chla7251474_8544_009

International Botanical Congress [IBC] (Fourth 1926) Program. 1926a. *Official Program of the International Congress of Plant Sciences* (Fourth International Botanical Congress), Cornell University, Ithaca, NY, August 16–23, 1926 [Ithaca, NY, The Congress].

Kass, L. B. 2000. Barbara McClintock, (1902–1992). American Botanical Geneticist, pp. 66–69, *In*, R. Robinson (Ed.), *Plant Sciences*, Macmillan Science Library, New York.

Kass, L. B. 2001. Ethics in science: preparing students for their career. *Plant Science Bulletin* Vol. 47 (No. 2, Summer 2001), pp. 42–48.

Kass, L. B. 2002a. Book review, *The Tangled Field*, by N. Comfort. *Isis* Vol. 93 (No. 4), pp. 729–730.

Kass, L. B. 2002b. Recollections vs. records: A new look at Barbara McClintock. *AAAS Annual Meeting Abstracts*, pp. A51–A52.

Kass, L. B. 2003. Records and recollections: A new look at Barbara McClintock, Nobel Prize-Winning geneticist. *Genetics* Vol. 164 (August), pp. 1251–1260, https://www.genetics.org/content/164/4/1251

Kass, L. B. 2005. Harriet B. Creighton: Proud botanist. *Plant Science Bulletin*. Vol. 51 (No. 4, Winter), pp. 118–125. Available online, December 2005: https://www.botany.org/psbarchive/issue/2005-v51-no-4.html

Kass, L. B. (Ed.). 2013ff. *Perspectives on Nobel Laureate Barbara McClintock's Publications (1926–1984): A Companion Volume.* Volumes I & II The Internet-First University Press. https://hdl.handle.net/1813/34897. [Online December 30, 2013, updates March 2014; Vol. III, Increments No. 1–3, May 2016; Vol. IV, Increment No. 4, April 2020]

Kass, L. B. 2018. Questions and reflections on H. J. Muller's Nobel Prize winning research. *Drosophila Information Service*, Vol. 101, pp. 93–106. December 2018. https://www.ou.edu/journals/dis/DIS101/DIS101.html

Kass, L. B. and J. A. Birchler. 2014/2016. Setting up a prepared mind: Summary and analysis of "Cytological observations of deficiencies involving known genes, translocations and an inversion in *Zea mays*," Vol. III, Increment #1, pp. 1.47.1–1.47.15, *In*, L. B. Kass (Ed.), *Perspectives on Nobel Laureate Barbara McClintock's Publications (1926–1984): A Companion Volume*, The Internet-First University Press. https://ecommons.cornell.edu/bitstreams/9711d284-f6c1-4f0a-88a0-68eab1c8d7c8/download (15.27 MB) [Online June 7, 2016]

Kass, L. B. and C. Bonneuil. 2004. Mapping and seeing: Barbara McClintock and the linking of genetics and cytology in maize genetics, 1928–1935, pp. 91–118, *In*, J.-P. Gaudilliere and H.-J. Rheinberger (Eds.), *Classical Genetic Research and Its Legacy: The Mapping Cultures of 20th Century Genetics*, Routledge, London, UK. [Routledge Studies in the History of Science, Technology and Medicine]

Kass, L. B., C. Bonneuil, and E. Coe. 2005. Cornfests, cornfabs and cooperation: The origins and beginnings of the Maize Genetics Cooperation News Letter. *Genetics* Vol. 169 (No. 4, April), pp. 1787–1797. https://www.genetics.org/content/169/4/1787 [Online May 6, 2005]

Kass, L. B., E. H. Coe, M. N. Cook, M. E. Smith, and J. L. Singer (Eds.). 2019. *Founding of the Maize Genetics Cooperation News Letter at Cornell University: A 90th Anniversary Tribute.* (Foreword by E. S. Buckler) [e-book] Internet-First University Press, Ithaca, NY. https://ecommons.cornell.edu/handle/1813/66550. [Online October 10, 2019]

Kass, L. B. and R. P. Murphy. 2003. Will the real maize genetics garden please stand up? *Maize Genetics Cooperation Newsletter* Vol. 77, pp. 41–43. https://mnl.maizegdb.org/mnl/77/79kass.html

Keller, E. F. 1983. *A Feeling for the Organism. The Life and Work of Barbara McClintock*, 1st edition, W. H. Freeman, San Francisco, CA.
Kimmelman, B. 1987. *A Progressive Era Discipline: Genetics at American Agricultural Colleges and Experiment Stations 1900–1920.* Ph.D. Thesis, University of Pennsylvania, Philadelphia, PA.
Kimmelman, B. A. 1992. Organism and Interests in Scientific Research: R. A. Emerson's Claims for the Unique Contributions of Agricultural Genetics, pp.198-232, *In,* A. E. Clark and J.H. Fujimura (Eds.), *The Right Tools for the Job*: Princeton University Press. Princeton, NJ.
Kohler, R. E. 1994. *Lords of the Fly. Drosophila Genetics and the Experimental Life*, University of Chicago Press, Chicago, IL.
Marshak, A. G. 1931. The morphology of the chromosomes of *Pisum sativum. Cytologia* Vol. 2 (No. 4, November 30, 1931), pp. 318–339.
Marshak, A. G. 1934. Chromosomes and compatibility in the aloinae. *American Journal of Botany* Vol. 21 (No. 9, November, 1934), pp. 592–597.
Marshak, A. G. 1937. The effect of X-rays on chromosomes in mitosis. *Proceedings of the National Academy of Sciences, U.S.A.* Vol. 23 (No. 7, July 15, 1937), pp. 362–369.
McClintock, B. 1927. *A cytological and genetical study of triploid maize.* Cornell University, Ithaca, NY, 104 pp. plus 39 unnumbered pp. of tables, plates, and bibliographies. [Thesis, Ph.D. Acknowledges L. W. Sharp and A. C. Fraser]
McClintock, B. 1929a. A cytological and genetical study of triploid maize. *Genetics* Vol. 14 (No. 2) [March 11, 1929, received July 11, 1928], pp. 180–222. [Publication of 1927 Ph.D. Thesis]. [No pachytene chromosomes.]
McClintock, B. 1929b. A method for making aceto-carmin[e] smears permanent. *Stain Technology* Vol. IV (No. 2, April), pp. 53–56.
McClintock, B. 1929c. Chromosome morphology in *Zea mays. Science* Vol. 69 (No. 1798, June 14, 1929), p. 629. [The first published ideogram of *Zea* chromosomes]
McClintock, B. 1930a. A cytological demonstration of the location of an interchange between two non-homologous chromosomes of *Zea mays. Proceedings of the National Academy of Sciences* Vol. 16 (No. 12, December 15, 1930) [Communicated November 6, 1930], pp. 791–796.
McClintock, B. 1930b. [Abstract #9] A cytological demonstration of the location of an interchange between two non-homologous chromosomes of *Zea mays. Anatomical Record* Vol. 47 (No. 3, December 25, 1930), p. 380.
McClintock, B. 1931a. The order of the genes *C*, *Sh*, and *Wx* in *Zea mays* with reference to a cytologically known point in the chromosome. *Proceedings of the National Academy of Sciences* Vol. 17 (No. 8, August 15, 1931) [Communicated July 7, 1931], pp. 485–491.
McClintock, B. 1931b. Cytological observations of deficiencies involving known genes, translocations and an inversion in *Zea mays. University of Missouri, College of Agriculture, Agricultural Experiment Station, Research Bulletin* 163 [December, authorized 23 December 1931], pp. 1–30.
McClintock, B. 1933. The association of non-homologous parts of chromosomes in the mid-prophase of meiosis in *Zea mays*, with 51 figures in the text and plates VII–XII. *Zeitschrift für Zellforschung und mikroskopische Anatomie* Vol. 19 (No. 2, September 22, 1933) [Received 21 April 1933], pp. 191–237.
McClintock, B. 1934. The relation of a particular chromosomal element to the development of nucleoli in *Zea mays* with 21 figures in the text and plates *VIII–XIV. Zeitschrift für Zellforschung und mikroskopische Anatomie* Vol. 21 (No. 2, June 23, 1934) [Received March 2, 1934], pp. 294–328.
McClintock, B. and H. E. Hill. 1929. [Abstract #32]. The cytological identification of the chromosomes associated with the '*R-golden*' and 'B-*liguleless*' linkage groups in *Zea mays. Anatomical Record* Vol. 44 (No. 3, December 25, 1929), p. 291.
McClintock, B. and H. E. Hill. 1931. The cytological identification of the chromosome associated with the R-G linkage group in *Zea mays. Genetics* Vol. 16 (No. 2, March 16, 1931) [Received March 1, 1930], pp. 175–190.
Morgan, T. H. 1911. An attempt to analyze the constitution of the chromosomes on the basis of sex-limited inheritance in *Drosophila. Journal of Experimental Zoology* Vol. 11 (No. 4, November 20, 1911), pp. 365–412.
Morris, R. 1969. Rollins Adams Emerson (1873–1947): Horticulturist, Pioneer in Plant Genetics, Administrator, Inspiring Advisor. *Proceedings of the Nebraska Academy of Sciences* Vol. 79, p. 37, Abstract. https://digitalcommons.unl.edu/agronomyfacpub/901/ [See full article here: https://digitalcommons.unl.edu/cgi/viewcontent.cgi?article=1901&context=agronomyfacpub]
Muller, H. J. 1927a. Artificial transmutation of the gene. *Science* Vol. 66 (No. 1699, July 22, 1927), pp. 84–87.

Muller, H. J. 1927b. Effects of X-radiation on genes and chromosomes. Abstract #134, *In*, Program and abstracts for the twenty fifty annual meeting of the America Society of Zoologists. *The Anatomical Record* Vol. 37 (No. 2, December 1927), p. 174 (see also p. 115, List of Titles, Genetics, Read, "134. Effects of x-radiation on genes and chromosomes, H. J. Muller, University of Texas (also by demonstration)."

Muller, H. J. 1928. The problem of genic modification, *In*, H. Nachtsheim (Ed.), 1928 [March]. *Proceedings of the 5th International Congress of Genetics, Berlin,* 1927, Vol. 1. [*Verhandlungen des v. Internationalen kongresses für vererbungswissenschaft, Berlin,* 1927, hrsg. von Hans Nachtsheim] *Zeitschrift für induktive abstammungs- und vererbungslehre*; suppl. 1, pp. 234–260. [Often cited only by supplement title]

Muller, H. J. 1946. The Nobel Prize in Physiology or Medicine 1946, "for the discovery of the production of mutations by means of X-ray irradiation," https://www.nobelprize.org/prizes/medicine/1946/summary/

Murphy, R. P. and L. B. Kass. 2007. *Evolution of Plant Breeding at Cornell University: A Centennial History, 1907–2006.* Department of Plant Breeding & Genetics, Cornell University, Ithaca, NY, pp. 1–98, Appendices A1–A98, Photo Section P1–P38. (July 2007).

Murphy, R. P. and L. B. Kass. 2011. *Evolution of Plant Breeding at Cornell University: A Centennial History, 1907-2006.* Rev. edition, https://hdl.handle.net/1813/23087. [Online June 27, 2011]

Nachtsheim, H. (Ed.). 1928. *Proceedings of the 5th International Congress of Genetics, Berlin, 1927* [*Verhandlungen des v. Internationalen kongresses für vererbungswissenschaft, Berlin, 1927*, hrsg. von Hans Nachtsheim] *Zeitschrift für Induktiv Abstammungs und Vererbungslehe*, suppl. 1–2. [March 1928]

Painter, T. S. 1933. A new method for the study of chromosome rearrangements and the plotting of chromosome maps. *Science* Vol. 78, pp. 585–586.

Painter, T. S. 1934. Salivary chromosomes and the attack on the gene. *Journal of Heredity* Vol. 25 (No. 12, December 1), pp. 465–476.

Philips, R. L. 2013. McClintock's presence of mind and forward vison as illustrated in the analysis of an interchange in maize, pp. 1.88–1.92, *In*, L. B. Kass (Ed.) *Perspectives on Nobel Laureate Barbara McClintock's Publications (1926–1984): A Companion Volume*, Volumes I & II, The Internet-First University Press. https://hdl.handle.net/1813/34897

Provine, W. B. and P. Sisco. 1980. Interview with Barbara McClintock at Cold Spring Harbor Laboratory, Long Island, New York, August 28, 1980. Division of Rare and Manuscript Collections, Cornell University Library, Ithaca, NY.

Randolph, L. F. and B. McClintock. 1926. Polyploidy in *Zea mays* L. *American Naturalist* Vol. LX (No. 666, January/February 1926), pp. 99–102. [First report of a triploid corn plant & the behavior of its chromosomes, reprinted in Kass 2013ff.]

Rhoades, M. M. 1931. Linkage values in an interchange complex in *Zea. Proceedings of the National Academy of Sciences* Vol. 17 (No. 12, December 1, 1931), pp. 694–698.

Rhoades, M. M. 1949. Biographical memoir of Rollins Adams Emerson 1873–1947. *National Academy of Sciences of the United States of America Biographical Memoirs* Vol. 25, pp. 313–323.

Rhoades, M. M. 1957. *Lewis John Stadler. Biographical Memoir*, National Academy of Sciences, Washington, DC. https://nasonline.org/publications/biographical-memoirs/memoir-pdfs/stadler-lewis.pdf

Rhoades, M. M. 1984. The early years of maize genetics. *Annual Review of Genetics*. Vol. 18, pp. 1–29.

Rhoades, M. M. and B. McClintock 1935. The cytogenetics of maize. *Botanical Review* Vol. 1 (August), pp. 292–325.

Richey, F. D. 1927. Corn Breeding, *Department Bulletin* No. 1489 (November 1927), U.S. Dept. of Agriculture, Washington, DC. https://archive.org/details/cornbreeding1489rich/page/n1/mode/2up

Spectre, M. B. 1993. *Barbara McClintock Papers. Guide to the Collection*, American Philosophical Society Library, Philadelphia, PA. [Online: Barbara McClintock Papers Mss.Ms.Coll.79]. https://search.amphilsoc.org/collections/view?docId=ead/Mss.Ms.Coll.79-ead.xml

Stadler, L. J. Unpublished [1927] Abstract. Genetic effects of X-rays in maize and barley, Presentation at Joint Genetics Sections AAAS, December 1927.

Stadler, L. J. 1928a. Genetic effects of X-Rays in maize. *Proceedings of the National Academy of Sciences* Vol. 14 (No. 1, January 15) [Communicated November 23, 1927], pp. 69–75.

Stadler, L. J. 1928b. Mutations in barley induced by X-rays and radium. Special Article. *Science* Vol. 68 (No. 1756, August 24), pp. 186–187.

Stern C. 1931. Zytologisch-genetische Untersuchungen als Beweise für die Morgansche Theorie des Faktorenaustauschs. *Biologisches Zentralblatt* Vol. 51, pp. 547–587.

Sturtevant, A. H. 1913. The linear arrangement of six sex-linked factors in Drosophila, as shown by their mode of association. *Journal of Experimental Zoology* Vol. 14 (No. 1, January 20, 1913), pp. 43–59.

Sturtevant, A. H. 1965. *A history of genetics*, Harper & Roe Publishers, New York. https://archive.org/details/historyofgenetic0000stur/page/n5/mode/2up

Taylor, W. R. 1929. Chromosome structure in mitosis and meiosis. *Proceedings of the International Congress of Plant Sciences, Ithaca, New York, August 16-23, 1926.* Vol. II, pp. 265–269. George Banta Publishing Company, Menasha, WI.

Taylor, W. R. 1937. *Marine Algae of the Northeastern Coast of North America* C. C. Jao (Illustrator), University of Michigan Studies. Scientific Series, Vol. 13, University of Michigan Press, Ann Arbor.

Taylor, W. R. 1957. *Marine Algae of the Northeastern Coast of North America*, C. C. Jao (Illustrator), [c1957] [2d] Rev. edition, University of Michigan Studies. Scientific Series, Vol. 13, University of Michigan Press, Ann Arbor.

Wayne, R. 2013. Identifying the individual chromosomes of maize, pp. 178–185, *In*, L. B. Kass (Ed.), *Perspectives on Nobel Laureate Barbara McClintock's Publications (1926–1984): A Companion Volume*, Volumes I & II, The Internet-First University Press. https://hdl.handle.net/1813/34897

5 Awards and Recognition
Chromosome Structure and Behavior

SEPARATION, COOPERATION, AND REUNION

By November 1930, the maize cytologists who had worked closely with McClintock in Sharp's cramped laboratory were dispersed. Beadle and Burnham, as National Research Council Fellows, continued their maize investigations with E. G. Anderson in the Division of Biology at the California Institute of Technology.[1] By January 1931, Cornell's Department of Botany had moved to new quarters on the second floor of the recently opened Plant Science Building. The Department of Plant Breeding had moved from Fernow Hall to the first floor of the same building. Professors, Instructors, and students had their own offices with adjacent research laboratories. McClintock's office was only doors down from Professor Sharp's, but his graduate student, Harriet Creighton, had been assigned a shared office on the fourth floor. She would later inherit McClintock's office. The cytology teaching laboratory was adjacent to McClintock's office with a pass-through door leading from one to the other (see Figure 4.3, lower left). Later, McClintock and Creighton would share the same office.

Cooperative efforts in maize cytogenetics continued despite the geographical separations. Encouraged by Emerson and Beadle, the maize cooperators continued to submit their unpublished results, which were shared in an updated "Mimeographed pamphlet on linkage in maize" (Coe & Kass 2005b; Kass et al. 2005). By January 1930, Randolph, who had been collaborating with Stadler since 1929, made the decision not to move to Missouri.[2] Stadler then invited McClintock to collaborate on his project to investigate cytologically the underlying mechanism for the mutations he had observed genetically with irradiated (X-rayed) plants.[3] Emerson expressed concerns about losing both Randolph and McClintock but was certain that if Randolph stayed at Cornell he must not continue to remove himself from the collaborative contributions being made by Beadle, Rhoades, Burnham, and McClintock[4]:

> … if he [Randolph] stays here he must overcome this handicap.
>
> If Randolph goes and McClintock stays, things will go about as they have this summer [Summer 1929]. If she goes and he stays, I believe he will make an earnest effort to work with my students who may need cytological help. If both stay … I honestly believe I can get them working together again. If both go, Lord help me!

McClintock accepted Stadler's invitation to join him in Missouri for the summer of 1931.[5] She wrote Burnham that she had decided to leave Cornell because under the new building conditions, their group work would be lost. She also expressed concern that Harriet would be "working on nearly alone." In January 1931, Burnham learned that McClintock had applied for a National Research Council [NRC]

---

1 Letters R. A. Emerson correspondence with the National Research Council, Plant Breeding Records, Box 15, Folder: National Research Council, Division of Rare & Manuscript Collections, Cornell University Library (henceforth CU).

2 Stadler Papers, collection 2429, File Folder 1: Stadler/Randolph correspondence 1927-1929, Western Historical Manuscript Collections, The State Historical Society of Missouri Research Center-Columbia (henceforth SHSMO-Columbia).

3 Letter L. J. Stadler to R. A. Emerson, January 7, 1930, Plant Breeding Records 21/28/229, Box 18, File Stadler L. J., CU.

4 Letter R. A. Emerson to L. Stadler, Jan 14, 1930, Plant Breeding Records, Box 18, L. J. Stadler, CU.

5 Letter B. McClintock to C. Burnham February 6, 1931, Charles R. Burnham papers, uarc1030, Box 1, Folder McClintock, Barbara, Correspondence, University of Minnesota Archives (henceforth UMN).

DOI: 10.1201/9781003332527-6

Fellowship to work at Caltech with E. G. Anderson.[6] Burnham suggested that she alert Morgan about her application because he believed the faculty would have to approve her appointment, since Caltech was not coeducational. Morgan, of course, was already apprised of her application, and within a week, he made a motion to the Executive Council for McClintock to spend her NRC Fellowship in his Department. The Council approved the appointment because she was not applying for a degree. It was not exceptional for Morgan to make this request because the Executive Council approved all appointments for staff members and visitors to work or attend classes at Caltech.[7]

It is interesting to note that the National Research Council had not yet voted on McClintock's application when the Caltech Executive Council approved her appointment to work at their institution. The NRC committee was planning to vote in May and asked Emerson for the "applicant's attainment and promise in research ..."[8] Emerson doubted that he had ever "taken more pleasure in recommending any person ... in urging a fellowship for Dr. McClintock. She is as brilliant a young woman as I have ever known." He explained that she had overcome difficulties that would have discouraged a less enthusiastic and persistent person. Those difficulties, including finding a job, he believed were due to the "fact that she is a woman." Despite a teaching schedule that would "leave many men no time for research," she accomplished "more in correlating maize cytology with maize genetics than has been done by anyone else with any organism." He praised her accomplishment in identifying the "ten maize chromosomes morphologically." Additionally, he mentioned that "she established a correspondence between eight of [the chromosomes] and eight of the ten known linkage groups of maize." He emphasized that she had "never been associated [with him] in any formal way ... and is a member of the staff of another department." He, however, was kept informed about her work because of his interest and her "enthusiasm." Her application was approved to begin in June. The NRC also renewed Beadle's NRC Fellowship at the same meeting.[9] Burnham was appointed a Caltech teaching fellow. The cooperators would soon be working together again at Caltech.[10]

Meanwhile, she and Burnham continued collaborating on the translocations he was finding in his maize studies at Caltech. McClintock and Creighton continued working together at Cornell on the proof for the cytological (physical) basis for "crossing-over" using Burnham's "semisterile-2" stock (Figure 5.1). One question they had not yet resolved was whether the knob belonging to the interchange chromosome was the carrier of the c-sh-wx linkage group in the semisterile-2 stock.[11] In a long letter to Burnham, in February 1931, McClintock reported that the knob interchange chromosome was found to be "C Sh wx-interchange."[12] She hoped to be able to tell which piece of the chromosome carries the "c-sh-wx genes" and asked Burnham if Beadle had any pollen counts for the "c-sh-wx trisome" that she had sent him. One month later she confirmed that the knobbed chromosome, the second smallest [chromosome 9], is the c-sh-wx chromosome. She believed that the c-sh-wx genes are arranged in the sequence "interchange wx-sh-C-knob."[13] She added that she would let him know whether the "third smallest (#8) comes thru in my other stock." Her next letter (March 12, 1931)

[6] Letter C. Burnham to B. McClintock, January 23, 1931, McClintock Collection 79, Box 8, Series I, Correspondence, A-Car, American Philosophical Society Library (henceforth APS).

[7] G. E. Hale Collection, Executive Council Minutes, January 31, 1931, p. 4, California Institute of Technology (Caltech) Archives.

[8] Letter F. R. Lilly to R. A Emerson, February 9, 1931, McClintock "indicates that you might be willing to address a letter to the Board in support of her application," Plant Breeding Records, Box 15, Folder: National Research Council, CU.

[9] Plant Breeding Records, Box 15, Folder: National Research Council, CU.

[10] *Bulletin Caltech 1931–1932.*

[11] Linkage groups as defined in Chapter 4. Letter C. Burnham to B. McClintock, January 23, 1931, McClintock Collection 79, Box 8, Series I, Correspondence, A-Car, APS.

[12] Letter B. McClintock to C. Burnham, February 6, 1931, Charles R. Burnham papers, uarc1030, Box 1, Folder McClintock, Barbara, Correspondence, UMN.

[13] Letter B. McClintock to C. Burnham, March 1, 1931, Charles R. Burnham papers, uarc1030, Box 1, Folder McClintock, Barbara, Correspondence, UMN.

FIGURE 5.1 Barbara McClintock and HBC [Harriet B. Creighton], Cornell University, June 1930. (Photo courtesy of H. B. Creighton, January 6, 1997, to L. Kass.)

provided details of how she reached her conclusions regarding "the order of the genes" on the knobbed chromosome [chromosome 9] using trisomic ratios and cytogenetic analysis.[14]

It is clear from this letter that McClintock now believed they had the perfect setup to study crossing over at the cytological level using the sequence of the known genes with respect to the knob at the end of the chromosome. She wrote:

> To study crossing over with a gene & the knob we should use genes in the short arm of the c-sh-wx chromosome [chromosome 9]. If 2 genes are used & the crossover types picked … there should be an almost perfect correlation between the crossovers & the knob-providing there is practically but 1 chiasma [crossover] in this arm. Harriet is working on some which are crossovers between c-sh. … The wx & granule [knob] cross-over evidence is still pretty meager to base so much on but I am cautiously aware of the gene map & chromosome size.

Burnham's response to this news was brief, indicating that he was glad to learn that the "knob chrom. = cshwx." He asked if the greenhouse crosses were successful before going on to report his news on new translocations he was finding and request seeds for trisomic plants.[15] McClintock then sent Burnham photos and sketches for "proof of the fact that the long-interchanged chromosome carries the c-sh-wx genes." She also "felt it is possible all genes are in the short arm

[14] Letter B. McClintock to C. Burnham, March 12, 1931, C. R. Burnham Papers, Charles R. Burnham papers, uarc1030, Box 1, Folder McClintock, Barbara, Correspondence, UMN.

[15] Letter C. Burnham to B. McClintock, April 2, 1931, McClintock Collection 79, Box 8, Series I, Correspondence, A-Car, APS.

of c-sh-wx chromosome."[16] McClintock's letter to Burnham included two important postscripts relevant to the legends told about McClintock and Creighton's publication for the cytological proof for crossing-over. The first is that Thomas Hunt Morgan was now at Cornell, but she had not yet been introduced. She hoped to see him during the week. The second is that "Harriet is getting great evidence on x-over & knob. *C* sticks with knob nearly always. *Wx* less & interchange probably less."

THE "PUSH" TO PUBLISH THE CYTOLOGICAL PROOF FOR THE THEORY OF CROSSING-OVER

The story of Thomas Hunt Morgan visiting the young women researchers at Cornell and encouraging them to publish their crossing-over data, although they believed it meager, was first reported by Keller (1983, 1993, p. 59) and based on Harriet Creighton's memories (Kass 2003; Coe & Kass 2005a). Creighton recalled that Morgan knew that Curt Stern was close to solving this problem using *Drosophila*, and pushed them to publish their data first. Rhoades (1992, p. 65) partly disagreed with Creighton's assessment. Garland Allan, Morgan's biographer, thought it would be uncharacteristic of Morgan to "encourage that kind of priority concerns" (Kass 2003). Stern had studied with Morgan's group at Columbia University, and *Drosophila* stocks were often shared among the researchers (Kohler 1994). Morgan was quite familiar with Stern's research, via letters from Stern to both Morgan and his lab colleagues Bridges and Sturtevant (Kohler 1994; see especially p. 142).[17] Stern (1971) recalled that he had received *Drosophila* translocation (interchange) stocks from H. J. Muller that allowed him to demonstrate crossing over at the cytological level (Stern 1931).[18] In early 1931, Stern received the *Drosophila* stock that had a piece of the X chromosome (No. 1) attached to a piece of chromosome IV (No. 4). It is known as "Translocation (1;4) Bar of Stone" and has variously been referred to as T(X-IV), T(1-4), T(1;4)A4, T(1;4)B^S, and T(1;4)Bar^S.[19]

Stern (1931) acknowledged Muller for his generosity, although the new mutant was actually induced using X-ray irradiation by Muller's student, Wilson Stone, who reported it first as "X-IV4" in his Master's thesis (Stone 1931, p. 12, Table III), which he had completed in August 1931. The new translocation was mentioned by Painter (1931) and cited briefly by Stone (1934) in the publication of his doctoral work. Clearly, cooperative efforts among researchers led to important contributions in the field of genetics. The founding of the *Maize Genetics Cooperation News Letter* and the *Drosophila Information Service* provided an additional venue for the sharing of unpublished data (Kass et al. 2005, 2019).

It was fortuitous that Morgan had been invited to give a series of lectures at Cornell during this time. He gave 12 Messenger Lectures beginning on April 7, 1931, through Friday, April 24.[20] The Cornell University Messenger Lectureship Series was established in 1924 by longtime mathematics instructor Dr. Hiram Messenger, an 1880 Cornell graduate. The visiting lectureship represents high recognition of scholarship by researchers across many disciplines. It continues today, and nominees are selected by a committee composed of faculty and students.[21]

---

[16] Letter B. McClintock to C. Burnham, April 11, 1931, Charles R. Burnham papers, uarc1030, Box 1, Folder McClintock, Barbara, Correspondence, UMN.

[17] Morgan's lab expected researchers, who used their *Drosophila* stocks, to inform them of their planned experiments (Kohler 1994; see especially Chapter 5, The *Drosophila* Exchange Network).

[18] Flybase, Stern 1931, https://flybase.org/reports/FBrf0002175. I thank Michael Gutensohn for help with German translations.

[19] Bridges in *Drosophila Information Service,* Volume 9, November 1938, p. 102. Flybase, https://flybase.org/reports/FBab0006529.html: T(1;4)Bar^S Stone 1931—no source for origin given; see Stone's (1931) Master's thesis for the origin of this translocation, which he called translocation "X-IV4."

[20] *Cornell Daily Sun,* April 7–24, 1931. Morgan's 12 Messenger Lecture series was titled, *The Experimental Study of Organic Evolution.* The publication of his lectures resulted in his book, *The Scientific Basis of Evolution* (Morgan 1932); Morgan's Preface (p. vii) states that his "... lectures have been somewhat expanded but stand essentially as delivered."

[21] Messenger Lectures history, https://theuniversityfaculty.cornell.edu/dean/messengeruniversity-lectures/.

After Morgan had presented his last lecture, he visited Emerson's and Sharp's students in their offices. As most students of genetics are aware, Morgan's work with *Drosophila* led him to propose an explanation of linkage based on the assumption that linked characters are located in the same chromosome and remain together in inheritance. Creighton, with McClintock's guidance (and Burnham's maize stock), now seemed to have the cytological proof for the exchange of genes on the chromosomes. The collaborations and correspondence between McClintock and Burnham described above show that she was working on this problem for some time. As mentioned previously (Chapter 4), Burnham's semisterile-2 maize stock provided the means to show that 50 percent sterility was due to interchange chromosomes. McClintock's observations of these reciprocal translocations at the pachytene stage of meiosis led to the cooperative efforts to demonstrate crossing-over using Burnham's semisterile-2 stock. She asked Burnham in early May if he wished to write up the order of the genes c–sh–wx linkage group with her.[22]

Their experiments were not yet completed, however, when McClintock began her NRC Fellowship at Missouri on June 1, 1931. Less than two weeks after arriving at Missouri, McClintock sent Burnham a draft manuscript for the order of the genes. She explained that the paper that goes with it is Harriet's, "– A correlation of cytological & genetical crossing over." She wrote "I am anxious to get Harriet's data in press as I have been advised to push it. Since this paper goes first in the same issue of the Proceedings of the National Academy of Science I am anxious to get it off as soon as possible." She invited him again to co-author the paper with her, explaining that she had used some of his unpublished crossing-over data in her analysis.

Coe and Kass (2005a, reprinted in Kass 2013ff.) provide background and details for Creighton and McClintock's experiment, which offered the first cytological proof for crossing over and which was not published until August 1931 (McClintock 1931a; Creighton & McClintock 1931; reprinted in Kass 2013ff.). Their proof was based first on knowing the sequence of the genes on the chromosomes, and second on a cytological demonstration of crossing-over between genes on two morphologically distinguishable chromosomes. The illustrations in McClintock's (1931a) paper demonstrating the sequence of the genes were required for an understanding of the crossing-over paper published collaboratively (Creighton & McClintock 1931). The two papers were expected to be read together (Coe & Kass 2005a). The illustration of interchange chromosomes at the mid-prophase of meiosis and the opening out at diakinesis (McClintock 1931a: Figure 1) was also essential for an understanding of the co-authored paper that followed in the journal (Creighton & McClintock 1931; Coe & Kass 2005a).

As detailed in Coe and Kass (2005a) and briefly summarized by Kass and Chomet (2009), the pioneering study used a semisterile maize plant found to be associated with the reciprocal translocation (segmental interchange) between the second and third smallest chromosomes (chromosomes 9 and 8), which McClintock had previously described (McClintock 1930a; see Chapter 4). By means of trisomic inheritance, McClintock (1931a) first observed that chromosome 9 of the haploid complement (n=10) was associated with genes of the c–sh–wx linkage group. Using Burnham's yet unpublished interchange data, and trisomic inheritance for the genes *wx*, *Sh*, and *c*, McClintock reported the order of the genes on chromosome 9 as *wx–sh–c,* beginning at the interchange point and running toward the end of the short arm. Knowing the order of the genes was critical to demonstrating the correlation of cytological and genetical crossing over (Creighton & McClintock 1931; Coe & Kass 2005a).

Burnham's semisterile-2 stock provided the diploid plant that showed 8 paired chromosomes and an interchange chromosome having a recognizable knob, observed at the mid-prophase of meiosis. Using a chromosome pair that was morphologically different in two regions, one chromosome of the homologue possessing an enlarged chromatic (heavily staining) knob on the end of the short arm of the chromosome and a piece of a non-homologous chromosome attached

[22] Letter B. McClintock to C. Burnham, May 12, 1931 (postscript), Charles R. Burnham papers, uarc1030, Box 1, Folder McClintock, Barbara, Correspondence, UMN.

to the same chromosome (translocation or interchange chromosome), both pairs of contrasting features (knob and translocation) can be observed in meiotic prophase. When a heterozygous plant (in which an interchange chromosome includes the knob, and the non-interchange chromosome is knobless) is crossed to a standard knobless plant, the combinations knobless-interchange and knobbed standard were found in the progeny. This indicated that chromosomal crossovers had occurred in the region between knob and point of interchange. Since the genes had been shown to be in the cytologically marked region of the chromosome (McClintock 1931a), genetic crossing-over was correlated with the occurrence of chromosomal crossing-over (Creighton & McClintock 1931; Figures 5.2 and 5.3). Professor Emerson submitted both manuscripts to *PNAS* on July 3, 1931. He specifically requested that both papers be published together, and if they could not, the paper by McClintock should be published first (Coe & Kass 2005a). Their publications appeared back to back in the August issue, and reprints were issued as one document (see reprint in Kass 2013ff.). The demonstration that genes on chromosomes had physically exchanged places was considered "one of the truly great experiments of modern biology" (Gabriel & Fogel 1955, p. 268).

In October, Stern's (1931) extensive 40-page paper demonstrating cytological crossing over in *Drosophila* was published in German.[23] His paper was dedicated to Morgan for his 65th birthday (September 25; Stern 1931, p. 547; 1971). Unfortunately, Morgan had suffered a debilitating car accident in the fall of 1931, and did not return to work until the following January (Shine & Wrobel

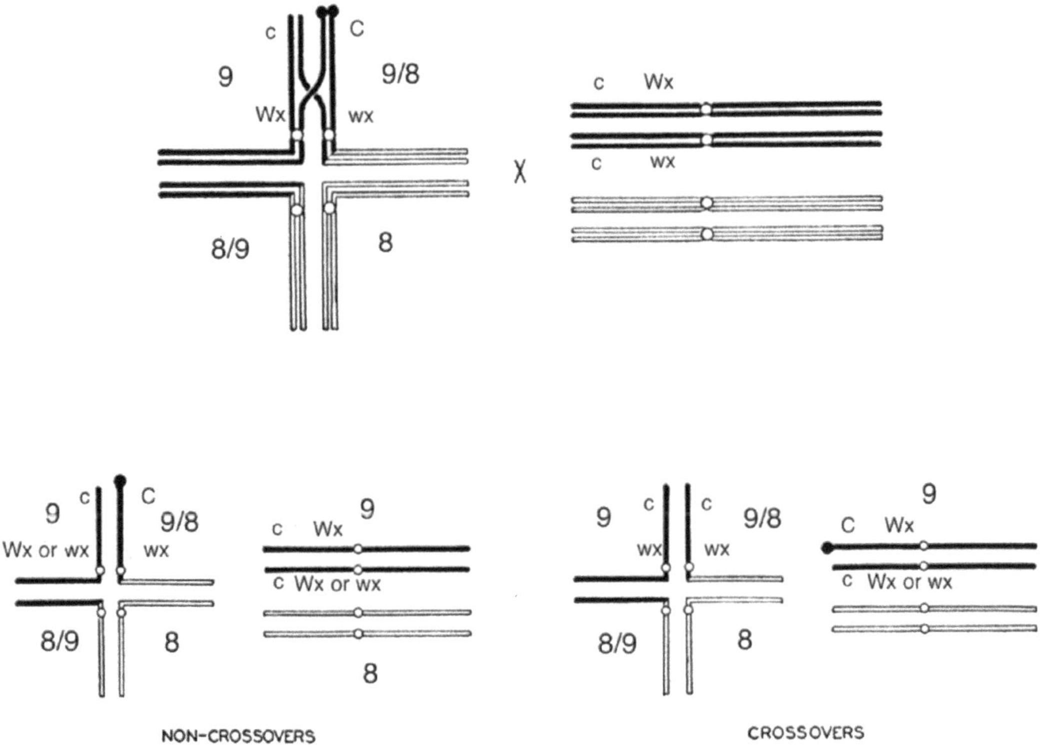

FIGURE 5.2 Experimental maize cross and progeny types designed by Creighton and McClintock (1931) to test the hypothesis that exchange between genes is accompanied by exchange of physical parts of chromosomes. (See Coe & Kass 2005a: Figure 4 for additional details; courtesy of E. H. Coe.)

[23] Flybase, https://flybase.org/reports/FBab0006529.html; https://flybase.org/reports/FBrf0002175

FIGURE 5.3 Maize ear carrying translocation 8-9 (T 8-9, Δ 8-9) that provided the data for Creighton and McClintock's (1931) experimental proof (cytological, physical) for crossing-over. They wrote, "The data are necessarily few since the ear contained but few kernels." (See Coe & Kass 2005a; photo by Ed Gerson©2023, used with permission.)

1976, p. 128). Only then was Morgan able to respond to Stern's gift with an interesting comment and a wish to see him soon (Stern 1971)[24]:

> I appreciate very much the dedication of your paper to me. The results you report are of importance for the crossover theory, and I am glad that at last we have some objective evidence upon which to rest the [cross-over] theory." Morgan also expressed hope to see Stern and his wife at the [International] congress [of Genetics] next summer [in Ithaca].

[24]Letter T. H. Morgan to C. Stern, January 8, 1932, Curt Stern Papers, APS Thomas Hunt Morgan—Correspondence, Folder 2, 1932-1941, https://diglib.amphilsoc.org/islandora/object/thomas-hunt-morgan-correspondence-folder-2-1932-1941#page/1/mode/1up

Morgan (1932) recognized the contributions of McClintock and of Stern in the publication of his book based on the Messenger Lectures that he had delivered at Cornell in April 1931. With respect to his "Theory," Morgan (1932, p. 50) first states that the major support for crossing over and exchange is that it fits the genetic evidence so well. He follows this statement with the proof: "the actual demonstration that such an exchange has occurred has only recently been made by McClintock (in corn) and by Stern (in Drosophila)." Neither Stern's (1931) nor Creighton and McClintock's (1931) publications are included in Morgan's "References to Literature," and the author's names are not listed in the index to the 1932 edition. Note that Morgan does not cite "Creighton & McClintock" but instead gives credit only to McClintock for having demonstrated crossing over. He must have assumed the idea was McClintock's, and of course, it was. Creighton's (1933, p. 53) doctoral thesis acknowledged McClintock for the study, "which she suggested." The fact that neither McClintock nor Stern's publications are listed in the "References to Literature" might indicate that Morgan was aware of their experiments before they were published. Morgan's biographer, G. Allen, suggested that Morgan probably wrote up his lectures before meeting with McClintock in April 1931, and may have incorporated their findings while working on the final manuscript for his book after returning to Caltech in May 1931.[25] Similarly, he may have been in touch with Stern via letters, although we have not located them. It must have pleased Morgan to have the "evidence upon which to rest the [cross-over] theory." The following year, he was awarded the Noble Prize in Physiology or Medicine "for his discoveries concerning the role played by the chromosome in heredity."[26]

SEARCHING FOR THE CAUSE OF MUTATIONS

Stadler had invited McClintock to Missouri for the summer of 1931 to learn if mutations caused by X-rays were due to "chromosomal irregularities" or due to a change in the "individual gene" (Stadler 1931; Kass & Birchler 2014/2016). The focus of her study was to analyze cytologically the nature of chromosomal alterations and their relation to phenotypic genetic changes induced by X-ray treatment of maize pollen and embryos. From her previous studies (McClintock 1930a,b; McClintock & Hill 1931), she knew which linkage groups had been associated with their respective chromosomes. She therefore believed it was possible to "correlate a genic loss with an alteration in the chromosome known to carry this gene" (McClintock 1931b, p. 3).

The publication that resulted from the summer's research at Missouri (McClintock 1931b) is a relatively unknown (and often misrepresented) series of investigations, which analyzed chromosomal changes in several plants from material subjected to irradiation and genetically characterized by Stadler. McClintock's publication, issued in December 1931, is reprinted in Kass (2013ff., pp. 1.149–1.180). Kass and Birchler (2014/2016) provided a detailed description of the methods and results of McClintock's (1931b) investigations, the results of which she presented in eight sections (I-VII A, B). Kass & Birchler also included a glossary for terms used in McClintock's paper, a *Corrigenda*, and a Table summarizing the results by chromosome number, chromosome name (linkage group name), chromosomal alteration, and associated phenotypes (Kass & Birchler 2014/2016: Table I). More importantly, they emphasized that McClintock's report clearly showed that (1) she first observed maize ring chromosomes during the summer of 1931, and she cited publications and unpublished work of others who had found them previously; (2) she published the first illustration of a pericentric inversion (inversion loop that includes the centromere) in maize, but it was not correlated with bridge formation at meiosis; and (3) she observed interchange chromosomes that produced single and double dicentric chromosome bridges at meiosis. It was during the latter study that she observed a centric [including a centromere] ring chromosome. Rhoades (1932) considered McClintock's report of a double dicentric bridge at anaphase I as "indirect cytological evidence,"

---

[25] Email G. Allen to L. Kass, December 1, 2000, Allen also added that it was also typical of Morgan to draft his books immediately following a lecture series.

[26] T. H. Morgan, Nobel Prize webpage, https://www.nobelprize.org/prizes/medicine/1933/morgan/facts/

along with his genetic evidence (Rhoades 1932, trisomic 5), to support *Zea* as the first organism in which double-strand crossing over had been demonstrated. Direct cytological evidence was later presented by Creighton and McClintock (1932), Creighton (1933, pp. 43–46), and McClintock (1933, pp. 203–204) using (1) heteromorphic synapsed chromosomes, (2) translocations, and (3) paracentric inversions showing bridges at anaphase.

Note that contrary to legends, McClintock was not the first person to find ring chromosomes in plants. As mentioned above, her initial finding of ring chromosomes in maize was reported in 1931 (McClintock 1931b; Kass & Birchler 2014/2016). Michael Navashin (1930), a student of Ernest Babcock at Berkeley, had first reported a ring chromosome in *Crepis* (Darlington 1932, p. 39; McClintock 1938, p. 368). Additionally, McClintock (1933, p. 230) interpreted variegations observed in tobacco plants by researchers at the University of California at Berkeley in the early 1930s as possibly due to ring chromosomes.[27] The Berkeley investigators later accepted that interpretation (Stino 1940; Clausen & Cameron 1944): "Occurrence of variegation types, presumably through spontaneous production of unstable ring chromosome fragments, has been recorded for the F chromosome, giving carmine-coral variegation (Clausen 1930)" (Clausen & Cameron 1944). While McClintock was not the first to describe ring chromosomes, she apparently was the first to suggest the relationship between unstable ring chromosomes and variegated plants.

A funny story has been circulated about McClintock falling off a chair that she had been standing on, while looking for the ring chromosome publication by M. Navashin (1930). She first cited his paper in 1931 (McClintock 1931b) and mentioned it again in her paper on ring-shaped chromosomes and variegation published the following year (McClintock 1932b). Apparently, she had forgotten she was on a chair while searching for Navashin's paper and stepped off into the air, falling directly into Harriet Creighton's lap.[28] The room where they worked together on the second floor of the Plant Science Building (Room 220) was directly adjacent to the Cytology Laboratory (Room 219; see Figures 4.3 and 9.1). It was a long narrow room with glass-fronted bookcases attached to the walls high above the desks and work benches. The bookcases were not easily accessible. It became Creighton's office when McClintock left for her National Research Fellowship in June 1931.[29] It was in this shared office where McClintock (1931b) composed the final draft for her pioneering paper.

A brief summary of her findings (after Kass & Birchler 2014/2016) is as follows:

Four of the ten maize chromosomes studied by McClintock were identified by their linkage group names and were correlated with the loss (deficiency) of four dominant gene loci (*Lg, A, Pl,* and *R*) in chromosomes 2, 3, 6, and 10, respectively. A deletion that McClintock found in chromosome 7 was responsible for the loss of genes Gl_1, V_5. Evidence for a hypothesized deletion in the chromosome 6 section of a T6-4 interchange chromosome provided a "more definite location" for gene *Pl* in chromosome 6. A deficiency in the chromosome 3 section of the T6-3 interchange chromosome was "almost identical in extent" to the chromosome 3 deficiency at the end of the long arm, and correlated with loss of gene *A*. The deletion in interchange chromosome T6-3 resulted in a ring chromosome having a centromere.[30] Two inversions, one in the centric section of chromosome 2 and one in the chromosome 4 centric section of the T6-4 interchange chromosome, were neither correlated with gene loss, nor associated with bridge formations at anaphase I.

[27] She visited the laboratory of R. E. Clausen at Berkeley in early March 1933, Letter B. McClintock to C. Stern, March 13, 1933; Stern Papers, APS.

[28] Creighton Synapsis Club Reunion, Golden Age of Corn Genetics, tape recording 1982; tape recording courtesy of Rosalind Morris, University of Nebraska-Lincoln to Lee Kass. In various interviews, McClintock recalled that the paper was sent to her by [H.] Olmo [a student of R. E. Clausen] and by [E.] Babcock, the mentor of Navashin. Creighton (1982) accurately recalled the paper as that of Navashin (1930). The reference to the Navashin paper in McClintock's 1931b and 1932b publications gives credence to Creighton's recollections.

[29] *Cornell University Directory, 1932–1933*, p. 12, Both rooms were transferred to Plant Pathology around 2010, and a new partition and door block the visibility of this area-LBK personal observations with George Hudler, June 2013.

[30] X-rays causing breaks in centromere regions would be observed in a later study of ring chromosomes (McClintock 1932b: Chapter 5; see Wayne 2013b).

Kass and Birchler (2014/2016) concluded:

> The experience of studying these materials no doubt prepared McClintock for subsequent studies of ring chromosomes (McClintock 1932b), especially the analysis of homozygous deficiencies using ring chromosomes (McClintock [1938], chr. 5), and with studies of paracentric inversions, a type not reported in this paper nor associated with ring chromosomes … All of these classic contributions led to her discovery of the breakage-fusion-bridge cycle initiated by anaphase bridge formation … In this paper, one can recognize foundational findings that would have prepared McClintock's mind for the seminal discoveries that followed.

NATIONAL RESEARCH FELLOW, CALIFORNIA INSTITUTE OF TECHNOLOGY

From 1931–1933, McClintock continued her National Research Council Fellowships with E. G. Anderson at the California Institute of Technology, Division of Biology, William G. Kerckhoff Laboratories. She arrived in November 1931 and collaborated with Anderson and Burnham. Excited by McClintock's work, Anderson wrote to Stadler encouraging him to "push your X-ray work hard with Miss McClintock working with you." He wrote that Stadler had a "fine set up" for getting very valuable information on "deficiencies of ends of chromosomes, deletions, inversions, etc." He concluded,[31]

> I am finding that I have to learn all of my cytology over again. The stuff Miss McClintock is handing out here doesn't sound at all like the stuff I was exposed to in school. It gives me quite a thrill.

McClintock studied the cytology of non-homologous pairing in *Zea* chromosomes in the mid-prophase (pachytene stage) of meiosis. She gave a Caltech "Genetics Seminar" on April 20, 1932, just before returning to Ithaca.[32] She prepared her study as a sectional paper (McClintock 1932a) for the International Congress of Genetics at Ithaca in August 1932.[33] In May 1932, she returned to Ithaca where she and Creighton also prepared an exhibit for the International Genetics Congress demonstrating four-strand crossing over in maize (Creighton & McClintock 1932; see below).

Just before the Genetics Congress, in July 1932, McClintock returned to Missouri to continue the research project on the cytology of X-rayed plants she began there the previous year. It was an investigation of ring-shaped chromosomes in corn, which was inspired by a study of similar chromosomes first reported by M. Navashin (1930) in *Crepis*. As previously mentioned, McClintock had observed "ring fragments" in maize (McClintock 1931b). She recalled (Provine & Sisco 1980) arriving in Missouri by train and was met there by Burnham and Stadler. The goal was to learn the cause of variegation that Stadler had observed in his maize plants. She employed the microsporocyte smear technique for the study of maize anthers. These would have been ready (approximately) in early July (before pollen shedding) through late July.[34] Since the ring chromosomes varied in size in different cells, and sometimes were totally eliminated, she suggested that the variegation was caused by the somatic elimination of that part of the ring chromosome that carried the dominant allele. Studies of the synapsis and morphology of the different ring chromosomes and their normal homologues at the pachytene stage of meiosis gave the approximate location of the *b* and *pl* gene loci. Gene *b* was found to be situated in the mid-portion of the short arm of the *B–lg* chromosome [chromosome 2], and *pl* was located toward the middle of the long arm of the satellite chromosome [chromosome 6] (McClintock 1932b; Kass & Chomet 2009; Wayne 2013). Burnham and Stadler reviewed the ring chromosome manuscript before it was submitted to *PNAS* by Emerson following the Genetics Congress.[35]

---

[31] Letter E. G. Anderson to L. Stadler, Stadler 1927-1955 f. 5 Corresp., October–December 1931, SHSMO-Columbia.
[32] *Cal Tech Weekly Calendars*, Caltech Archives.
[33] Her paper was submitted for publication in April 1933, and published the following September (McClintock 1933).
[34] I thank Jim Birchler for the timing when maize anthers would be ready for study in Missouri during summer.
[35] Letter C. Burnham to B. McClintock, October 5, 1932, APS. Barbara McClintock Papers, Box 1, Folder 6.

THE INTERNATIONAL CONGRESS OF GENETICS (ICG), 1932

McClintock returned to Cornell in time to register for the International Congress of Genetics. There, she presented her research on homologous pairing in maize in a 20-minute contribution to the first sectional papers on Cytology, and an exhibit with Harriet Creighton on cytological evidence for 4-strand crossing over in maize (McClintock 1932a; Creighton & McClintock 1932). McClintock was Vice Chairman and Secretary for the third set of sectional papers on Cytology.[36]

Thomas Hunt Morgan was the President of the Congress. He opened the Congress with a keynote address titled "The Rise of Genetics." Emerson was Chairman of the Local Committee. He praised the maize researchers for their cooperative efforts toward progress in maize genetics at the General Papers sessions on the first morning of the Congress (Emerson 1932):

> I cannot refrain from noting here a very real advantage experienced by students of maize genetics ... I am aware of no other group of investigators who have so freely shared with each other not only their materials but even their unpublished data. The present status of maize genetics, whatever of noteworthy significance it presents, is largely to be credited to this somewhat unique, unselfishly cooperative spirit of the considerable group of students of maize genetics. In this connection I want gratefully to acknowledge the help of many persons who have contributed directly or indirectly to this summary statement of the status of maize genetics.
>
> *[Thursday, August 25, 1932]*

As I noted in Chapter 4, McClintock's research featured prominently toward the status of maize genetics by this time. During the ICG, McClintock was introduced to international researchers, most notably Curt Stern. Morgan introduced Stern to McClintock and to Creighton at the ICG.[37] Morgan asked them to be friends (not competitors), knowing that Creighton and McClintock had published their cytological studies on crossing over in maize only three months before Stern had published similar findings in *Drosophila*. As mentioned above, their important results demonstrated the first physical [cytological] proof for crossing over that Morgan had hypothesized more than 20 years earlier, and their reports demonstrated the first cytological proof that contributed to "his discoveries concerning the role played by the chromosome in heredity," for which he was awarded the Nobel Prize in Physiology or Medicine in December 1933. Following the Congress, McClintock remained in Ithaca and completed her ring chromosome paper for publication. Her NRC Fellowship was renewed for a second year, and she also applied for a Guggenheim Fellowship to work with Stern in Germany. Stern began his Rockefeller Foundation Fellowship working first at the University of Illinois and then at Caltech (as detailed below).

Barbara McClintock did not wish to discuss her time in Germany (Provine & Sisco 1980).[38] However, archival documents can shed some light on (1) events that led her to propose a collaboration with Stern and the circumstances that prevented his return to Germany; (2) other persons she had anticipated consulting with upon arriving in Germany, but were fired or had resigned; and (3) why she only remained in Germany for fewer than five months – spending only about two months in Berlin-Dahlem (November 1933–January 1934), then going on to Freiburg before returning to Ithaca, via England.

BARBARA McCLINTOCK FROM CALTECH TO GERMANY

When Barbara McClintock was awarded a Guggenheim Fellowship in March 1933, she planned to work with Dr. Curt Stern at the Kaiser Wilhelm Institute for Biology in Berlin-Dahlem (KWIB/BD), beginning in the fall of 1933–1934. Stern, a *Privatdozent* at the University of Berlin (Friedrich

[36] *Sixth International Congress of Genetics Program*, August 24–31, 1932, Cornell University, Ithaca, New York.
[37] L. Kass: Creighton Interview August 17, 1996; see Stern in Jones, D. F. (1932).
[38] Letter B. McClintock to Evelyn and Curt Stern, December 11, 1934, Stern Papers, APS.

Wilhelm University, now Humboldt University), conducted his research at the KWIB/BD. The Institute was devoted to basic research, and scientists were permitted to pursue their own research without any teaching duties. The KWIB/BD was established in 1913–1914. Carl Correns became the first director of the newly founded Kaiser Wilhelm Institute for Biology in Berlin-Dahlem, and the director of Plant Biology within that unit. Unfortunately, his death in February 1933 would preclude McClintock from meeting with this notable scientist. He was one of three who had rediscovered Mendel's work in 1900. Following Correns' death, Richard B. Goldschmidt had been appointed an interim first director, until he stepped down in May 1933 (see below). He was succeeded in the fall of 1934 by another distinguished plant biologist Fritz von Wettstein, who was formerly Correns' assistant at KWIB (1919–1924) (Plarre 1987; Macrakis 1993, pp. 112, 115–116; Henning & Kazemi 2009, 2016).

Paul von Hindenburg, the second president of the German Weimar Republic, had appointed Adolph Hitler as Chancellor of Germany on January 30, 1933. Subsequently, many KWI researchers were rapidly forced to leave their positions; e.g., Charlotte Auerbach, Viktor Jollos, and Friedrich G. Brieger (Schnarrenberger 1987), among many others. Brieger had joined Correns as a scholarship holder and at his exit was a senior lecturer in "botany and the science of heredity." He found refuge at the John Innes Horticultural Institute (1933–1936; see below). Brieger then became Professor of Genetics at the University of Brazil in Sao Paulo (1936) and subsequently in Piracicaba. There, he mentored Almiro Blumenschein (Brieger et al. 1958). Years later, McClintock would train Blumenschein in cytogenetics at North Carolina State University, and, together with Kato Yamakake, publish a book entitled *Chromosome Constitution of Races of Maize* (McClintock et al. 1981; Kato 2013; see Chapter 9).

Beginning in 1933, the German government enacted many anti-Jewish laws restricting the rights of German Jews (or those with Jewish heritage) to earn a living, to enjoy full citizenship, and to gain an education. Probably the most damaging to the academic community was the "Law for the Restoration of the Professional Civil Service," decreed on April 7, 1933, which forbade Jews to work in the civil service. Employees of the KWI were considered civil servants (Harwood 1993). The subsequent 1935 Nuremberg Laws stripped German Jews of their citizenship and forbade Jews to marry non-Jewish Germans (Bradsher 2010). Max Planck, a Nobel Laureate in physics (1918), was the President of the Kaiser Wilhelm Society at the time, but he was not able to prevent the Nazi regime from removing Jewish scientists (Deichmann 1996; KWI-Biology).

Consequently, when McClintock arrived in Berlin-Dahlem in late 1933, she recalled that there were few researchers in the KWI für Biologie Laboratory, where Richard Goldschmidt was Head of the Department of Animal Heredity (genetics), and Second Director of the Institute. Upon the death of Carl Correns (February 14, 1933), Goldschmidt had assumed the [interim] first directorship (Henning & Kazemi, 2009, 2016; KWI-Biology). Upon receiving the directive from the Kaiser Wilhelm Society (KWG) administration in April 1933 to dismiss non-Aryan staff, Goldschmidt complied with the request to have his employees complete forms, which specifically requested ancestral heritage. To preclude dismissing his staff, he instead encouraged them to resign from their posts, which many did either immediately or soon after (KWI-Biology).

Richard Goldschmidt, the head of Stern's Department at KWIB, had become Acting Director on February 14, 1933, following Correns' death, yet he could not avert the ultimate outcome for Jewish Staff in his department. Goldschmidt, as did *all* KWI Directors, followed the directive to dismiss Jewish Staff. It was sent to them by the KW Society, dated April 27, 1933.[39] Goldschmidt did follow the directive he received from President Planck, in July 1933, to dismiss members of his department. In a letter dated 24 July 1933, he wrote[40]:

[39] U. Deichman 1996, p. 14, footnote 3 gives documentation. See also KWI-Biology.

[40] KWI-Biology, I thank Bretislav Friedrich for the translation; the employees are identified as Dr. Gross, D. Heimann, and D. Bodenstein; see also Macrakis 1993, p. 54, reports that Goldschmidt dismissed Jewish personnel at his institute.

> In response to your letter of 21 July [1933], I'm letting you know that I have carried out the task given me therein [in the said letter from Planck] to dismiss the three employees by September 1.

Even though Goldschmidt had been associated with the KW Institute for Biology from its inception (1914), he too would be forced to leave by 1936, after loss of his German citizenship on September 15, 1935, as decreed by the Nuremberg Laws.

CURT AND EVELYN STERN MEET BARBARA McCLINTOCK IN AMERICA IN 1932

The story of Stern's return to the United States in 1932 and his decision to remain there is more complicated than revealed in published reports (Neel 1987; Glass 1988). Records at the Rockefeller Foundation Archive Center (RF) and Stern's communications with KWI Directors provide a different outlook on the events under which Stern came to the U.S. in 1932, whom he worked with during his RF Fellowship, and circumstances that permitted his safe refuge in the U.S.[41]

Stern earned his Ph.D. (1923) from the University of Berlin (Kaiser Wilhelm University). His research was under the mentorship of Max Hartmann, Department Head of the KWIB, Section of Protozoology (Neal, 1987; Glass 1988, pp. 68–89). The following year (1924), Richard Goldschmidt recommended Stern for a fellowship funded by the International Education Board of the Rockefeller Foundation (RF) to study genetics with Thomas Hunt Morgan's research group at Columbia University. During that time, Stern made important contributions to the genetics and cytogenetics of *Drosophila*. The results were published upon his return to Germany (Stern 1926–1927; see the list of publications in Neal 1987). An extensive report of these early contributions, titled "Genetical and cytological proof of heredity in the Y chromosome of *Drosophila melanogaster*" (Stern 1927), reviewed by Richard Hesse and Carl Zimmer, qualified him for habilitation (Rürup & Schüring 2008, p. 326), and University faculty rank of *Privatdozent* (1928–1933), thus permitting him to teach as an unsalaried lecturer (Stern Papers APS; see also Harwood 1993). Stern's cytogenetic studies of *Drosophila* sex chromosome aberrations would also lead him to find a way to prove crossing over at the cytological level, using translocations between the X and Y chromosomes (Stern 1931). His study, as mentioned previously, was published only a few months after Creighton and McClintock's (Creighton & McClintock 1931; Coe & Kass 2005a).

By January 1932, Stern had been approved for a second Rockefeller Foundation Fellowship to work in America for 6 months, beginning September 1, 1932, immediately following the International Congress of Genetics at Ithaca (August 24–31, 1932).[42] He originally planned to work with Morgan's group, which had moved to the California Institute of Technology in 1928, and with H. J. Muller in Texas.[43] Prior to starting his RF Fellowship, Stern had arranged to teach summer classes at Western Reserve University in Cleveland, Ohio, during summer session, June 20–July 29, 1932.[44] He was scheduled to teach "Recent Advances in Genetics" and "Genetics of Drosophila."

By the end of April 1932, Stern had requested a change of plans,[45] and instead arranged to work "from September 1 to December 1, 1932 with Prof. of Zoology Charles Zeleny at the University of Illinois, and with Prof. T. H. Morgan, Calif. Inst. of Technology, from Dec 1, 1932 to February 28, 1933."[46] Stern had probably learned that Muller would be on a Guggenheim Fellowship in Germany beginning November 1932,[47] thus affecting Stern's original arrangements.

---

[41] RF Fellowship Record Cards: Stern. Stern papers, APS. I am grateful to archivists at APS and RF for access to Stern's files, and to M. Dietrich for translations of Stern's letters at APS.

[42] RF Fellowship Cards Stern.

[43] RF Fellowship Cards Stern.

[44] *Science* April 1932; RF Record Card entry, June 27, 1932.

[45] RF Fellowship Cards Stern: note, April 26, 1932.

[46] RF Fellowship Cards Stern.

[47] Muller, H. J. Guggenheim Foundation Archives published report, Fellowship begins November for one year 1932–1933, https://www.gf.org/fellows/all-fellows/hermann-j-muller/.

Stern and his wife, Evelyn Sommerfield Stern, an American who he had met while on his International Education Board Fellowship with the Morgan group at Columbia University (1924–1926), came to New York on the s/s Deutschland, arriving on May 27, 1932.[48] Curt and Evelyn had been married in New York City on October 29, 1931.[49] By June 1932, he was teaching at Western Reserve and his wife was planning to complete her Ph.D. at Columbia University in New York City that winter.[50]

While teaching in Cleveland, Ohio, during the summer of 1932, Stern was also preparing a monograph on linkage and crossing over and wrote to C. B. Hutchison at the University of California, Berkeley, asking for the photograph of linkage studies in corn, which Hutchison had previously published. Stern could not recall where he had seen the illustration. Hutchison reminded him that it had been published in 1922 and that he would ask Dr. Emerson at Cornell to send him a copy.[51] Stern asked Hutchison to send the photograph to him in New York City, where he would be after July 29, 1932.[52] Following the Ithaca Congress, Emerson sent the photograph to Stern.[53] Stern's monograph was published in German in *Handbuch der Vererbungswissenschaft* [*Handbook of Genetics*] (Stern 1933).

He presented an invited paper on the topic at the International Congress of Genetics at Ithaca in August 1932, which was published in the Congress *Proceedings* (Stern 1932). He referred readers to the references in his forthcoming monograph (Stern 1933), instead of listing them at the end of his *Proceedings* paper (Stern 1932, p. 303). Following the Ithaca Congress, which ended on August 31, 1932, Stern began his three months' Rockefeller Foundation Fellowship (September 1–November 30, 1932) at the University of Illinois with Charles Zeleny.[54] It is interesting to note that I have found no reports other than the record cards at the Rockefeller Foundation that accounts for Stern working with Zeleny during this time. There is no account of their collaboration or of letters between them reported in his papers at the APS Library (Glass 1988: Stern).

Why did Stern not mention that he had worked with Zeleny at the U of Illinois, before continuing his Fellowship with Morgan's group at Caltech? Is it because Zeleny, as a student of C. M. Child (University of Chicago), and T. H. Morgan had conflicting views on the nature of inheritance (Mitman & Fusto-Sterling 1992, p. 187)? Charles Zeleny was a developmental geneticist, who had been a graduate student of Morgan at Columbia University, but completed his Ph.D. at the University of Chicago in 1904 (*AMS 1938*, p. 1594). Developmental Genetics was a subject that interested Stern along with his colleagues at the Kaiser Wilhelm Institute for Biology (Harwood 1993). Both Morgan and Sturtevant also held an interest in development as is clear from studies initiated in the early 1920s in the Morgan Lab at Columbia University.[55] Letters between Morgan's lab colleagues and Stern make clear their common interests in genetics and development (Kohler 1994, p. 188). In December 1932, Zeleny gave the vice presidential address, titled Genetics and Embryology, at a

[48] RF Fellowship Cards Stern.

[49] Ship manifests for 1931 & 1932—Stern left Germany on October 14 and returned to Germany on October 30, 1931, Passenger Search, Ellis Island Website. Therefore, his marriage was in New York on October 29, 1931.

[50] RF Fellowship Cards Stern. 6.27.32 entry from *FBH Diary 1932*, Berlin, interview with R. Goldschmidt, p. 28 (stamped p. 353), Stern teaching in Cleveland, Ohio; see also Glass 1988, p. 68, Stern Vis. Prof. 1932, Western Reserve University obtained a temporary appointment at Western Reserve University (now Case Western Reserve University). Evelyn Stern did not complete a degree at Columbia, but years later received a Masters of Social Welfare at Berkeley in 1959. Evelyn Sommerfield Stern (A.B. U of Michigan 1925; living in Orinda, CA) received the Degree of Master of Social Welfare in the School of Social Welfare at U of California Berkeley on June 11, 1959; Ninety Sixth Commencement, University of California, Berkeley 1959, p. 117.

[51] Letter (copy) C. Stern, from Biological Laboratory, Western Reserve University, Adelbert College, Cleveland, Ohio, to C. B. Hutchison, July 15, 1932, at College of Agriculture, UC-Berkeley, and reply, Letter C. B. Hutchison to C. Stern, July 23, 1932, Plant Breeding Records, Box 18, File, Stern, CU.

[52] ... at the home of his father-in-law, A. W. Sommerfield.

[53] Letter C. B. Hutchison to R. A. Emerson, July 23, 1932, annotation: "Photo sent 8/31/32," Plant Breeding Records, Box 18, File, Stern, CU.

[54] RF Fellowship Cards Stern; Zeleny Papers, U of Illinois Archives, University Library (henceforth UI), Record Series No. 15/24/22, Box 13, Stern, Kurt [*sic*] 1932–1933.

[55] Kohler 1994: Chapter 6, improvisations pp. 176–177, 182–184, 186–188.

dinner for the American Society of Zoology, in cooperation with Section F (Zoology, Zeleny was Chair) of the American Association for the Advancement of Science, in Atlantic City, New Jersey. Published in *Science* the following February, Zeleny had made a plea for geneticist and embryologist to work together[56]: "Both genetics and embryology are profiting by functional contact. ... It will serve no good purpose to brand the offspring as illegitimate."

Drosophila geneticists had a keen interest in development in the mid-1920s, as Kohler (1994) clearly demonstrates. Stern, a visiting scientist with Morgan's group at Columbia University (1924–1926), played an active part in their investigations. Mosaic mutants loomed large in the Columbia researcher's studies of developmental genetics, and Stern found such mosaic spots in his studies.[57] Mosaics at the time seemed only to resolve questions of chromosome segregation (non-disjunction) and sex determination (Kohler 1994, p. 189). Later work by McClintock (1932b), Rhoades (1936), and Jones (1936, 1937, 1941) provided other interpretations for mosaic phenomena. Subsequent studies demonstrated the connection is even more profound (Blair 2003; Gehring 1978).

After completing his work at Illinois in November 1932, Stern requested permission to visit two or three scientific institutions *en* route to Pasadena, California, which was approved by the Rockefeller Foundation. He arrived at Caltech on December 1, 1932.[58] As an "International Research Fellow in Genetics of the Rockefeller Foundation," Stern worked in the Division of Biology.[59] Morgan's vision for his Department was a unified Biology, with no typical separation of Botany and Zoology.[60] Morgan was then no longer pursuing *Drosophila* studies, but had returned to Embryology, teaching the subject with his former student and now Assistant Professor of Embryology, Albert Tyler.[61]

Professor Alfred H. Sturtevant was on Sabbatical (as was Tyler), when Stern arrived at Caltech in December 1932, but Sturtevant's collaborator, Assistant Professor T. Dobzhansky, was available for consulting. Stern and his wife shared a house with the Dobzhansky's during this time, which permitted even closer discussion on research problems.[62] Other visiting fellows in the Division of Biology, who held an interest in genetics and development during this time, were National Research Fellows in Biology, Barbara McClintock, Berwind Kaufmann, and Alfred Ernest Clarke; International Research Fellow in Genetics of the Rockefeller Foundation, Cyril Dean Darlington; and Research Fellow in Biology, George W. Beadle.

McCLINTOCK'S GUGGENHEIM RESEARCH PROPOSAL

McClintock and Stern became friends at Caltech, where she continued her National Research Council Fellowship (1932–1933). While her decision to work with Stern had been decided months prior to his arrival at Caltech, it is not clear when they had first made plans to collaborate at KWIB, which she had proposed for her 1933–1934 Guggenheim Fellowship. They may have discussed the possibility of collaborating during the Genetics Congress at Ithaca.

In October 1932, Ralph Cleland, who had completed a Guggenheim Fellowship in Germany with Frederick Oehlkers, recommended McClintock for a Guggenheim Fellowship. Letters between the Foundation Secretary, Henry Allen Moe, and McClintock demonstrate that her application was almost delayed due to miscommunications of her whereabouts. Assuming she was at Caltech in Pasadena, the application was first mailed there. Not receiving an application, she then requested

[56] Zeleny 1933, *Science* Vol. 77 (No. 1990), pp. 177–181.
[57] Stern. 1927b. *Naturwissenschaften*, Vol. 15, pp. 740–746, cited in Stern 1936.
[58] RF Fellowship Cards Stern: December 1, 1932–March 3, 1933, Fellowship Termination date February 28, 1933.
[59] *Bulletin of the California Institute of Technology Catalogue 1933*, Published December 1932, Stern, p. 35 (henceforth *Bulletin Caltech*).
[60] *Bulletin Caltech 1933*, published December 1932; Division of Biological Sciences, p. 96.
[61] *Bulletin Caltech* 1933; to avoid antisemetic (Jewish heritage) consequence, Morgan had suggested that Tyler change his name from Titlebaum, when he moved from Columbia to Pasadena (see Chapter 2).
[62] Letter C. Stern to C. Zeleny, December 1932, C. Zeleny Papers, UI.

one herself, which she finally obtained, completed, and mailed from Ithaca on November 4, 1932, before she returned to Pasadena.[63]

McClintock had proposed three projects for her Guggenheim Fellowship. The first project was "A determination of the types of translocations resulting from the association of non-homologous parts of chromosomes, and if possible, something as to the mechanism responsible for these translocations." The second project was "A determination of the mechanism responsible for the inconstancy in size of ring-shaped chromosomes through nuclear cycles." And the final project was "A determination of the cause of constructions at the points of interchange in chromosomes of *Drosophila* and the possible relation of change in pycnotic regions to the genetically inert parts of the chromosomes."

The first project was a continuation of a study she presented at the International Congress of Genetics on the "intimate association of non-homologous parts of chromosomes in the mid-prophase of meiosis" (McClintock 1932a). She had found that in a "trisomic plant [three of one type of chromosome] associations between the three homologous chromosomes frequently bring non-homologous parts of the three chromosomes into intimate contact. Progeny of these plants occasionally produce individuals with altered chromosomes." She proposed to study trisomic plants to determine the kinds of progeny they produce and if the "mechanism is involved in the crossover process." The second project continued studies of ring-shaped chromosomes and variegation (McClintock 1932b). Her goal here was to determine the cause of "diminution or increase in size of ring-shaped chromosomes during the mitotic cycle." The third project's aim was to study the internal structure of chromosomes and their relation to metaphase morphology using *Drosophila* chromosomes. These were known to have constrictions at interchange points, which might be due to pycnotic regions [heavily staining thickened regions of a chromosome]. She hoped to learn "whether or not the presence of constrictions at places of interchange can be correlated with the internal morphology of normal chromosomes." Ultimately, McClintock went to Germany, but as Adolph Hitler rose to power, Stern was advised to remain in the U.S.

McCLINTOCK, STERN, AND THE RISE OF NATIONAL SOCIALISM IN GERMANY, 1932–1933

At Caltech during the winter of 1932–1933, Stern and McClintock worked on their respective research problems. On January 3, 1933, McClintock gave a "Genetics Seminar" titled "Translocations resulting from the association of non-homologous parts of chromosomes."[64] Stern spoke to a "General Biology Seminar" on "Recent Biological Work at the KWI for Biology," at the end of January.[65] Within the week (January 30, 1933), Hitler had been appointed Chancellor, and would decree, on April 7, 1933, that Jews be removed from their posts. Stern would soon find himself without a job, and McClintock would need to consider alternatives for research projects and collaborations in Germany.[66]

During the winter and spring of 1932–1933, McClintock collaborated with E. G. Anderson and C. R. Burnham on a maize translocation (T 6-9) chromosome problem that involved the nucleolus. Anderson (1934) reported on their experiments; cytological examination at pachytene showed the interchange was within the reticulate region of chromosome 6 and about one third of the way out on the long arm of maize chromosome 9. Anderson also reported that "Because of its interest in connection with the problem of the origin and function of the nucleolus, detailed cytological study of this interchange [was] made by Dr. Barbara McClintock," and would be published in a cytological journal. McClintock recalled (Provine & Sisco 1980) that Anderson permitted her to take the

[63] Guggenheim Foundation Archives (henceforth GFA).

[64] January 3, 1933, *Cal Tech Weekly Calendar*, Dr. Barbara McClintock, National Research Fellow.

[65] January 24, 1933, *Cal Tech Weekly Calendar*, Dr. Curt Stern Int. Res. Fellow.

[66] Letter B. McClintock to C. Stern, April 19, 1933, and August 12, 1933, Stern Papers APS.

problem with her to Germany. She lamented that the published report was not well presented. It is one of her most cited, however, and was the first to demonstrate the exact location of the deep staining region on chromosome 6 that organizes the nucleolus (Sharp 1934, pp. 119–121; Bloom 2013). McClintock would complete and publish the study during her Guggenheim Fellowship in Germany (McClintock 1934).

Meanwhile, as Stern's Fellowship at Caltech was coming to a close (end date February 28, 1933), he was offered an opportunity to work with future Nobel Laureate (1973) Carl von Fritch, at Munich, upon his return to Germany. Fritch had funding to establish a new institute and asked Stern to be curator of his collections. Goldschmidt encouraged him,[67]

> Frisch had naturally gotten in touch with me first, and I said to him that I would advise you to accept the position. It is clear that I must advise you thus, even though I am most reluctant to lose you; you cannot refuse a tenured position which is roughly the equivalent of an Extraordinarius and especially since this position is, at the moment, the best opportunity in Germany.

In March 1933, Stern left for New York, stopping in Columbia, Missouri, to visit with Stadler's genetics research group.[68] From there, he visited McGill University (Montreal, Canada) for a few days, and at the same time informed the Rockefeller Foundation that "because of his race" he had no position to which to return in Germany and he "Will remain in U.S. for a while."[69] At the end of March, Stern wrote to Stadler that he had postponed his departure [to Germany] due to "sickness of Mrs. Stern" and that he hoped to leave next week. He asked, "If impossible to work in Germany, any jobs?"[70]

Hitler had not yet made his decree on April 7, 1933, denying employment to persons of Jewish ancestry, when Stern made this decision to remain in America for a while. What precipitated Stern's decision? Was it because he had learned that in March 1933, the German SS had established the Dachau concentration camp, which opened on March 22? Or that on March 28, 1933, Einstein had resigned from the Prussian Academy of Sciences, Berlin, due to the destruction of German Jews through economic means.[71] Was it this news and more that prompted Stern's decision to remain in the U.S.? The *Jewish Telegraphic Agency's* published and unpublished reports document persecution against persons of Jewish ancestry, beginning in 1923 in Germany, Poland, and Austria, and by 1933, their publication offices, originally in the Hague, had moved to New York City.[72]

Contrary to published recollections, it seems that neither Stern nor his wife returned to Germany before the summer of 1933.[73] There is no record in Stern's Rockefeller Record Cards of a visit to Germany until one was recorded for the summer of 1933, when Stern and his wife went abroad. On April 4, 1933, Stern sent a postcard to Stadler updating his address for the present: Dr. M. Demerec, Carnegie Institution, CSH, L.I., N.Y.[74] On April 7, 1933, the Law for the Restoration of the Professional Civil Service was passed in Germany.[75] Shortly thereafter, McClintock learned that Stern would be at Cold Spring Harbor, where he held a temporary research appointment at the Department of Genetics.[76]

---

[67] Letter R. Goldschmidt to C. Stern, February 14, 1933, I thank Michael Dietrich for the translation, Stern papers APS.
[68] Letter B. McClintock, Caltech to C. Stern, Columbia [MO], March 13, 1933, Stern Papers, APS.
[69] March 1933, RF Stern Record Cards.
[70] Letter C. Stern to L. Stadler, March 28, 1933, Stadler file SHSMO-Columbia.
[71] For example, Hitler's announced boycott of Jewish businesses, March 28, 1933, scheduled for April 1 and onward, http://www.academia.edu/22059639/Albert_Einstein_Letters_to_the_Prussian_Academy_of_Sciences_and_the_Academy_s_Response_March_28-April_5_1933
[72] Archives of the *Jewish Telegraphic Agency* (*JTA*) are online at https://www.jta.org/archive
[73] See Neel (1987), NAS Memoir and Glass APS reports, which were based on Stern's autobiography on file at APS.
[74] Stadler file, SHSMO-Columbia. Stern at Carnegie Institution, Department of Genetics, Davenport Director, Department of Genetics—see his report in 1932–1933, *Carnegie Year Book No. 32*, p. 39.
[75] Timeline Holocaust Museum website, https://www.ushmm.org/learn/timeline-of-events/1933-1938
[76] Letter B. McClintock, Caltech, to C. Stern, Cold Spring Harbor, April 19, 1933, Stern Papers, APS; Letter B. McClintock, Botany Dept. Cornell, to C. Stern, Cold Spring Harbor, August 12, 1933, Stern Papers, APS.

On April 28, 1933, the day Goldschmidt received the directive from the KW Society, he informed Stern that he would be removed from civil service, since none of the articles of exclusion applied to him. He added that all Jewish lecturers had been ordered to suspend their activity as lecturers at the university. With some hope, he added that the law was not compulsory, but then stated that Stern would have "no future prospects," and personally believed that nothing would change this. Goldschmidt told Stern that he could not extend his leave beyond June 15; that his colleagues believed he "should return and see what happens," but if he finds a position "over there," then it made no sense to return. He continued that if Stern did not find a position by the end of his leave, he had no choice but to return to Germany. He repeated that he had no authority to extend his leave. He added a footnote that he had filled out his form, which accompanied the Society directive, "the main point being race," and added a postscript that Haldane was interested in him.[77] Perhaps knowing that he would have a place to work in England or that there might be opportunity for him at the University of Rochester (see below), Stern would soon resign from KWIB.

As Assistant Director of the John Innes Horticulture Institute in Merton, England, J.B.S. Haldane was aiding scientists who had been displaced by the rise of National Socialism. Haldane, among well-known academics, worked with the Academic Assistance Council, which the economist William Beveridge had set up, with the aim of rescuing Jewish and politically vulnerable academics. The organization helped 1,500 academics escape Germany and continued their research work in safety in Great Britain. The council's "rescue operation" arranged for academics to come to Great Britain and provided grants, accommodation, and most importantly jobs (Coughlan 2013; Rall 2017). Haldane gave them appointments at his institute, or arranged for appointments at Cambridge University or at the University of London.[78]

By subsequently relinquishing his appointment as Acting Director of KWIB, Goldschmidt showed his displeasure, but not as effectively as KWI colleague Fritz Haber. As Director of the KWI for Physics and Electrochemistry in Berlin-Dahlem, Haber, himself of Jewish ancestry, resigned on April 30, 1933, the day after he had sent letters of dismissal to his Jewish Staff.[79]

Detailed letters from the Society on how to proceed with staff dismissal soon followed, as did requests from many KWI Directors and Department Heads asking for special exemptions for their Jewish staff. Most requests were denied, and those that were postponed eventually were overturned (KWI-Biology). On May 5, 1933, at Goldschmidt's request, President of the KWG, Max Planck, transferred Acting Directorship of the KWI for Biology to Max Hartmann and Otto Mangold. The reason that Goldschmidt gave for his request was that he felt "impaired" to implement the April 7, 1933, Law for the Restoration of the Professional Civil Service, directive that he had received from MPG on April 27, 1933 (Henning & Kazemi 2016).

Five days later, students hurled "un-German" books into bonfires in cities across Germany, accompanied by singing, music, and vicious addresses against Jewish intellectualism. On May 10, 1933, Joseph Goebbels, Reich Minister of Public Enlightenment and Propaganda of Nazi Germany, spoke to students outside the State Opera in Berlin "… you do well in this midnight hour to commit to the flames the evil spirit of the past."[80]

H. Bentley Glass, a former student of *Drosophila* geneticist H. J. Muller at The University of Texas, recalled seeing the first book burning in Berlin in May 1933, while a National Research Council postdoctoral fellow at KWIB in Berlin-Dahlem and KWI for Brain Research (KWIBR) in Berlin-Buch, Germany (Glass 1990). As a National Research Council Fellow in the Biological Sciences (1932–1934), Glass worked with Otto Mohr, at the Anatomical Institute, The University, Oslo, Norway, then went to KWIB, intending to work with Stern. When Stern did not return to

[77] April 28, 1933, R. Goldschmidt to C. Stern, at Cold Spring Harbor, I thank Michael Dietrich for the translation, Stern Papers APS.

[78] John Innes Horticultural Institute Archives, *JIHI Annual Reports 1933–1936.*

[79] Stoltzenberg (1994, 2004) English translation; Friedrich (2019); April 29, 1933 Letter Haber to Farkas, copy, MPG Archives, original in National Library Israel. See also Szöllösi-Janze (1998).

[80] See Glass (1990).

Germany, Muller suggested Glass transfer to KWIBR, where Muller was then on a Guggenheim Fellowship with N. V. Timofeef-Ressovsky (Muller left by September 1933).[81] Disturbed by the political situation in Germany, Glass completed his fellowship at the University of Missouri with L. J. Stadler. As a faculty member at Stevens College (1934–1938), not far from the University of Missouri, Glass was affiliated with Stadler's research group, when McClintock arrived there in 1936 (Kass 2003; see Chapter 6: Figure 6.1).

Within the week of the book burnings, Stern told Max Hartmann, KWIB Acting Director who had succeeded Goldschmidt, that he had requested his own "dismissal" from the Kaiser Wilhelm Society via a communication with Goldschmidt, the head of his department.[82] A copy of Stern's letter was sent to Alfred Kühn at Gottingen, and Fritz von Wettstein in Munich, among other zoologists and geneticists in Germany. He mentioned that his appointment with Frisch had been voided and expressed disapproval of exclusion of "non-Aryans" from participating in scientific work without regard for their accomplishments and efforts. By early June 1933, Stern informed the Rockefeller Foundation that Professor Willier, formerly of the University of Chicago, was to head the Department of Zoology at the University of Rochester, Rochester, New York, and had offered him a fellowship for one year.[83] Stern eventually obtained a permanent appointment at Rochester, remaining there until 1947.[84]

Meanwhile, at Caltech, on her renewed National Research Council Fellowship (1932–1933), McClintock seems to have partly resolved the first question she had proposed for her Guggenheim Fellowship, by April 1933 – "the association of non-homologous parts of chromosomes." After conducting research at the University of Missouri with L. J. Stadler (summer 1932), and Caltech with E. G. Anderson (winter of 1932–1933), McClintock's results were published in September in a prestigious German journal (McClintock 1933, reprinted in Kass 2013ff.). In addition, a hypothesis for understanding the cause of change in ring chromosome size, as well as the function of pycnotic chromosome regions, was also addressed in this publication (McClintock 1933, pp. 194, 229–230). The details would be expanded in her review article, co-authored with Marcus Rhoades, upon her return from abroad (Rhoades & McClintock 1935).

As documented here, Richard Goldschmidt, Head of Stern's Department in Berlin, appears to have advised Stern to remain in the U.S., not during the 1932 International Congress of Genetics in Ithaca, but only after receiving the directive to dismiss Jewish staff in April 1933. It is doubtful that McClintock would have proposed to work with Stern at KWIB during her 1933–1934 Guggenheim Fellowship, had Stern not intended to return to Germany.

In response to McClintock's request for advice on who to contact upon arriving in Germany,[85] Stern offered McClintock detailed guidance before she left for her Guggenheim Fellowship in Berlin.[86] He provided names of persons working at the KWIB, and especially those working in Goldschmidt's and Max Hartmann's Departments at the Kaiser Wilhelm Institute for Biology. Unfortunately, Stern did not seem to be aware that Goldschmidt had stepped down as Acting Director of KWIB by May 1933 (Henning & Kazemi 2016). He also had not yet been informed that many of his former colleagues and students had already left Germany because they had resigned or were dismissed as required by the Nazi Civil Service Law for non-Aryans. They, like Stern, were encouraged to resign to preclude the Director from dismissing them, as required by the law enacted on April 7, 1933 (Friedrich 2019; Szöllösi-Janze 1998). By the time McClintock arrived at KWIB in

---

[81] Bentley Glass Papers Mss. Col. 105, Series I Correspondence: Otto Lous Mohr Correspondence, 1932–1956, APS.

[82] Letter C. Stern to M. Hartmann, May 16, 1933, APS Ms. Coll. 5, C. Stern. M. Hartman file; see the partial translation of Stern's letter in U. Deichmann (1996, p. 21).

[83] RF Stern Record Cards, June 1933; Director's Diary, *Frank Blair Hanson Diary* (1933), The Rockefeller Archive Center (henceforth RFA).

[84] Note that Neel (1987) misrepresents the sequence of events of Stern's life during this time period.

[85] Letter B. McClintock to C. Stern at CSH August 12, 1933, Stern Papers, APS.

[86] Letter C. Stern, University of Rochester to McClintock, Cornell University, October 4, 1933, Stern Papers APS.

November 1933, almost all persons identified by Jewish heritage were gone.[87] She recalled that the Institute had few people with whom she could discuss her work (Provine & Sisco 1980).

On October 29, 1933, McClintock sailed from NYC on the *SS Europa* for Europe, arriving in early November. On December 23, she sent a report to the Guggenheim Foundation indicating that she had been in Berlin for seven weeks and was in a "better position to evaluate" her situation.[88] "The scientific situation has so materially changed under the new government," she wrote, "that I find many of the (Jewish) people in my line [area of research] have left or making plans to leave very soon." Several of the people who she had planned to work with in Berlin had gone. She therefore had to change her research plans. She asked permission to remain in Germany until March, and visit the John Innes Institute in England, before returning to the states in early April. She wished to complete her fellowship in Ithaca, where she could "get her spring and summer plantings in order." She received an invitation to the laboratory of Dr. Friedrich Oelkers at Freiburg, I. Br. [im Breisgau, in Baden-Württemberg], and hoped to go in January. She made suggestions on how to handle the financial situation, all money being handled through the "American Express offices in New York and Berlin." McClintock arrived in Freiburg, even before receiving a reply from the Guggenheim Foundation.[89] She reported that the drastic change in conditions at Dahlem makes it profitless to remain there longer, and she sent a quick telegram requesting they cancel some of the money previously requested. She also mentioned that members of the Kaiser Wilhelm Institutes at Dahlem were "a skeleton of their former selves. Foreign investigators, formerly numerous, are almost entirely absent."

Therefore, McClintock remained in Berlin only a few months (November, December, and part of January) and it may have been Goldschmidt who suggested that she leave for Freiburg, Germany, as early as possible.[90] She decided not to return to Berlin and to remain at the Botanical Institute in Freiburg, which "agreed with her."[91] She told Stern that she was planning in early April to visit the John Innes Institute in England and hoped to see Chester Bliss there for a "California reunion." John Innes was also home to German emigres, F. G. Brieger and Emil Heitz, who were also there for a while in the mid-1930s. McClintock did meet Heitz during her visit,[92] but I have no record of her meeting Brieger at that time. She may have met Brieger when he visited Emerson at Cornell in 1926.[93] Of course, she would also have had a reunion with Darlington, who had been at Caltech and returned to his duties at the John Innes Horticultural Institute in 1933. She added that she had planned to be back in Ithaca by the middle of April 1934 and she was "job hunting." The Guggenheim Foundation was allowing their fellows to work in the U.S. because of foreign exchange, and she planned to "work in Ithaca" that summer. Before leaving Freiburg, she submitted her manuscript on the development of nucleoli in maize, which was published in June (McClintock 1934).

It seems that she left for home earlier than planned, as recorded by the manifest for the Sailing Ship *Europa*. She left Southampton, England, on March 31, 1934, sailing on the *SS Europa*, arriving on April 5, 1934.[94] *Europa* and her slightly larger sister ship *Bremen* were designed to have a cruising speed of 27.5 knots, allowing an Atlantic crossing time of 5 days.

---

[87] Including Ursula Philips, a student of Curt Stern (Vogt 2009).

[88] Letter B. McClintock in Berlin to H. A. Moe, December 23, 1933, GFA.

[89] January 10, 1934, Letter B. McClintock in Freiburg to H. A. Moe, GFA.

[90] Interview Provine and Sisco (1980), Letter B. McClintock, Freiburg, to C. Stern [Rochester], March 4, 1934, Stern Papers APS.

[91] The botanical Institute in Freiburg was part of the University of Freiburg, officially the Albert Ludwig University of Freiburg; Philosopher Martin Heidegger joined the Nazi Party (NSDAP) on May 1, 1933, ten days after being elected Rector [Chancellor] of the University of Freiburg. A year later, in April 1934, he resigned from the Rectorship and stopped taking part in Nazi Party meetings, but remained a member of the Nazi Party until its dismantling at the end of World War II.

[92] Letter B. McClintock to L. Stadler, SHSMO-Columbia. Brieger emigrated to Brazil, and years later, his student A. Blumenschein became a research collaborator of McClintock; see Chapter 9.

[93] Letter L. J. Stadler to R. A. Emerson about coming for a visit to Cornell with Brieger, February 1926, Plant Breeding Records, CU.

[94] SS Europa, Ship's manifest departing March 31, 1934, arrival April 5, 1934; owned by Norddeutscher Lloyd (NDL) (North German Lloyd).

Upon returning to the states in 1934, McClintock worked at Cornell through 1936, and often visited the Stern's at Rochester, which is only a two-hour drive west of Ithaca, NY. Stern was later appointed Professor of Zoology (1947), then Professor of Genetics (1958) at the University of California, Berkeley, following the death of Goldschmidt (April 1958), who had been there since 1936.

CORNELL UNIVERSITY, 1934–1936

McClintock completed her Guggenheim Foundation Fellowship at Cornell in the summer of 1934. When Emerson learned that she was considering a new career in meteorology because she had not yet found employment in her chosen field, he contacted the Rockefeller Foundation for financial assistance. The Foundation had recently awarded Emerson a grant to support the Maize Genetics Cooperation at Cornell for five years (Kass 2003; Kass et al. 2005). Emerson therefore requested a separate grant-in-aid to hire her as his research assistant to work on the cytogenetics of maize. She was free to work on any problem she wished.

Emerson justified his request in a meeting at Cornell with Warren Weaver, Director of the Division of Natural Sciences at the Rockefeller Foundation. He explained that the Botany Department did not wish to reappoint her because her interest was purely in research and partly because she was not "entirely successful [as] a teacher of undergraduate work."[95] Weaver noted that, "The Botany Department obviously prefers a less gifted person who will be content to accept a large amount of routine duty." Emerson told him that McClintock was the "best trained and most able person in this country on the cytology of maize genetics." Weaver suggested that Emerson make a request for a research aid grant in genetics. The grant would be an integral part of Emerson's own general program, but the money would be used to pay McClintock's salary for next year. He promised that Emerson's request would receive prompt attention. Weaver acknowledged that the process was "admittedly a mild form of skullduggery," but he also believed it was "thoroughly justified." He noted that T. H. Morgan and Curt Stern had both spoken of McClintock in terms just as enthusiastic as Emerson, and that both agreed that "it would be scientific tragedy if her work did not go forward without serious handicap." He made no mention that McClintock had been recognized for her scientific achievements in the most recent edition of *American Men of Science* (1933).

Weaver immediately wrote a note to Frank Blair Hanson, Assistant Director of the Rockefeller Foundation, alerting him of the forthcoming grant that Emerson would be submitting.[96] It would be for an assistant to Emerson, and they agreed that Miss McClintock "is admirably qualified to be the assistant." The salary was for $1800, for 12 months beginning July 1 [1934]. He believed it was an "important opportunity, & that it should be put through at once." The following day, Emerson submitted his request to Hanson for the grant to begin on October 1, 1934. It was approved on July 5, 1934, "to provide salary of Dr. Barbara McClintock, research assistant to R. A. Emerson through September 30, 1935."[97]

McClintock had proposed a study of "Variegation correlated with chromosome behavior." She wished to follow up on previous studies [McClintock 1931b, 1932b] that had shown that ring-shaped chromosomes become continually reduced in size through loss of sections of chromatin in the ring. From studies so far, she had observed that ring-shaped chromosomes decrease (or increase) in size by loss of sections (or duplications of sections) within the chromosome ring, and the chromosome always remains a ring. She believed that this was due to "two separate breakages within the

[95] Warren Weaver Diary, June 24, 1934, RF 1.1, Box 136, Folder 1679, RFA.

[96] Handwritten note, W. Weaver to F. B. Hanson, Sunday afternoon [June 24, 1934], RF 1.1, Box 136, Folder 1679, RFA.

[97] Letter R. A. Emerson to F. B. Hanson, June 25, 1934, approved by Dean C. E. Ladd, RF. To Cornell University, Grant-in-Aid-New York, July 5, 1934, RF 34042, RFA; also in Box 25, Folder: Rockefeller Foundation, Plant Breeding Records, CU, with an attached proposal by B. McClintock.

ring-shaped chromosome, which occur in some of the somatic anaphases and which are followed later by recombinations [crossing-over] of the broken ends." The aim of her study was to "prove that broken ends of chromosomes seek to unite."

Other studies, she said, had strongly suggested that "unions between chromosomes only takes place at broken ends of chromosomes." She described these broken ends as "unsaturated" and that the term implied that such broken ends seek to be saturated by 2-by-2 unions of their "broken ends." She then proposed a method to produce free broken ends, using an inversion in one arm of a chromosome not involving the insertion region [centromere], in a heterozygous plant. During meiosis, homologous associations followed by crossing-over would give rise to a chromatid with two spindle fiber attachment regions. At anaphase I in *Zea*, the separation of centromeres to opposite poles, would break the two-centromere chromatid. Consequently, a free ("unsaturated") broken end would enter each telophase I nucleus. She hoped to produce such inversions in two different chromosomes of the set. As a result, two broken "unsaturated" ends would be introduced into the same telophase nucleus. "If union occurred between these two broken ends, the following division (Anaphase II) should give indication of this fusion." She already had one inversion "of the right type" and believed she could obtain the second inversion "shortly." Actually, it would be some time before she would obtain the second "inversion of the right type."

Her second proposal was to study anaphase I disjunction in plants heterozygous for reciprocal interchanges. Previous studies in *Zea* had shown that such plants produce abortive pollen grains (McClintock 1930a; Burnham 1930), based on whether four or six types of spores are produced. Using the satellite chromosome [chromosome 6], she hoped to show that interchanges involving this chromosome, which contained a "certain body" responsible for the orderly development of the nucleolus (McClintock 1934), could be used to determine spore types. If the nucleolar body was absent, the nucleolus was filled with many small nuclei, instead of a well-organized nucleolus. She expected that a correlation would be found between the position of the interchanges and the types of disjunctions producing the different spore types.

That December, at the AAAS meetings in Pittsburgh, McClintock told Hanson, Assistant Director of the Rockefeller Foundation, that she had been in conversations with Professor L. J. Stadler at Missouri and she was looking forward to transferring her work there at a later date. Hanson noted that she had a very fine demonstration of her cytological preparations at the meetings.[98]

It was at this meeting, in December 1934, that Marcus Rhoades presented a paper titled "A new aleurone color in maize" (Rhoades 1935). It was a preliminary report of his finding of the dotted (*Dt*) gene that appeared to control the expression of aleurone color in maize plants in the presence of other aleurone genes *a1*, *A2*, *C*, and *R* (Emerson et al. 1935). At the time, the chromosome locus of the *dt* allele was "unknown." Following the Pittsburgh meeting, Curt Stern wrote to Rhoades about his paper and tried to explain it by somatic crossing over. He also mentioned Jones' (1935) studies that had just been published and that he thought might be of a similar occurrence.[99] As mentioned previously, by 1941 Jones no longer believed his results were due to somatic crossing over. McClintock was cognizant of Rhoades' and Jones' studies as preliminary to similar observations of mutable genes in the maize stocks that she went on to study in the mid-1940s and that ultimately led her to discover transposable elements (McClintock 1987; Kass 2013ff.).

Rhoades had continued his research at Cornell after receiving his doctorate in February 1932, and finally found work with the U.S.D.A. at Ames, Iowa, in early 1935. Just before leaving for Iowa, Emerson et al. (1935) submitted the long-awaited linkage study monograph. Rhoades had prepared the linkage map representing the genes on the ten chromosomes of maize. The map showed the loci of "those genes whose position can be determined with reasonable certainty." Since the

[98] *F. B. Hanson diary 1933–1934*, p. 302, December 28 and 29, 1934, & Folder 1679, Box 136, RG 1.1, RFA.
[99] Letter C. Stern to M. Rhoades, March 12, 1935, Stern Papers Mss Coll 5, APS.

chromosome locus of *Dt* could not be mapped at that time, it does not appear on Rhoades' linkage map. McClintock is credited for "her help in the formulation of statements concerning the association of the several linkage groups with particular chromosomes." Rhoades and McClintock would continue collaborations for a review paper on the cytogenetics of maize (Rhoades & McClintock 1935).

McClintock also continued collaborating with Creighton, who had successfully found a teaching job at Connecticut College for Women by September 1934 (Kass 2005). They submitted a manuscript to *PNAS* in February 1935 that presented further proof for their previous publication on crossing-over in maize (Creighton & McClintock 1935). Their paper had been criticized in the literature, and the manuscript hoped to alleviate the arguments. McClintock recalled that it was a mistake to try to argue their case, and she planned never to do so again (Provine & Sisco 1980).

Burnham was successful in finding a faculty appointment and a U.S.D.A. affiliation at West Virginia University (1934–1938; Nesius 1988), and Beadle was still affiliated with Caltech. McClintock wrote to Burnham in early April about the results she was getting with her ring-chromosome studies.[100] She wished to talk it over with him. She reiterated her proposal "to get the experiment to show that broken ends will certainly unite." She was also concerned that she was "still on the unemployed list although [she] was getting a decent salary just now." She was awaiting some translocations from Anderson for her disjunction studies but had not yet heard from him. She said Beadle had written to her that "Anderson was not doing much corn work since the *Drosophila* stuff came out. It rather took the wind away from his sails." The recent publications by Painter (1934a,b) and by Bridges (1935) on using *Drosophila* salivary gland chromosomes for studies of associating genes and chromosomes apparently had discouraged Anderson.

In early August, Emerson was successful in renewing the grant to pay McClintock's salary beginning in October 1935 through September 20, 1936.[101] Emerson's initial request for renewed funding included an update by McClintock on research that had been accomplished.[102] She was still trying to find the appropriate stock to prove her theory of fusion of broken chromosomes. She had also shown that variegation with small rings was due to the actual loss of the ring-shaped chromosome carrying the dominant gene in question, but how the break and reunion of broken chromosomes occurred was still in question. She was also an active member of the Plant Breeding Synapsis Club during this time (Figure 5.4).

McClintock's father was concerned about her future in science. On the same day (August 7, 1935) that the Rockefeller Foundation renewed Emerson's funding for McClintock's salary, her father called on Hanson in his Rockefeller Foundation office.[103] His interest was in seeking a permanent post for his daughter, and he hoped the Foundation might be able to assist in that placement. They assured him that her salary grant had been renewed and that there might be a permanent post for her with the proposed genetics institute at the University of Missouri.

A week later, Barbara McClintock attended the genetics meetings at Woods Hole (August 23–24) and approached Hanson to say how embarrassed she was that her father had interceded without her knowledge. Hanson assured her that he had no part in their decision. She also said that she wished to return to Caltech to do some work. Hanson told her that since she was working for Emerson she would need his approval for the trip. At the genetics meeting, McClintock took part in a round-table

[100] Letter B. McClintock, Cornell to C. Burnham, West Virginia University, April 2, 1935, uarc1030, Box 1, Folder McClintock, Barbara, Correspondence, UMN.

[101] August 2, 1935, Letter A.R. Mann to F. B. Hanson, Request for Renewal, Folder 1680, Box 136, Series 200 RG 1.1, RF. Letter F. B. Hanson to L. Farrand, August 7, 1935, funding renewed, cc to Emerson at Cornell University, Folder 1680, Box 136, Series 200 RG 1.1, RFA. August 7, 1935, GRANT-IN-AID-NEW YORK [Folder 1684, Box 136, Series 200 RG 1.1, RFA].

[102] Letter R. A. Emerson to F. B. Hanson, July 19, 1935, PB # 114, note by RF on Emerson letter "RF grant 1800 ends Sept. 30, 1935," Folder 1680, Box 136, Series 200 RG 1.1, RFA. Copy also in Plant Breeding Records, CU, Box 25 File Rockefeller Foundation.

[103] Copy from *F. B. Hanson Diary, August 7, 1935*, [Interview with] Dr. McClintock-(with WW present), Folder 1680, Box 136, Series 200 RG 1.1, RFA.

FIGURE 5.4 Dr. Barbara McClintock was an active member of the Cornell Plant Breeding Synapsis Club during 1934–1936. Group photo taken May 11, 1936, outside the Plant Science Building, Cornell University. Front row, from left to right: R. G. Wiggins, H. H. Love, M. J. Murry A. C. Fraser, M. Dawson, B. McClintock, S. M. Allen, C. Y. Tsao, J. R. Livermore, R. A. Emerson, H. E. Fisher, S. E. Tai. (Courtesy of Plant Breeding & Genetics, Cornell University.)

conference on "Chromosomes and their relation to genes" with E. M. East as a leader and C. B. Bridges and Barbara McClintock as introducers. She and many prominent colleagues took part in the discussion: C. E. McClung, M. Demerec, C. B. Bridges, Edgar Anderson, A. H. Sturtevant, C. Stern, R. E. Cleland, A. Marshak, among others (Demerec 1935).

McClintock spent the winter of 1935–1936 at Caltech (December to April). On her way there, she stopped off at Missouri to talk with Stadler about the possibilities of an appointment with his research group. Stadler and his Missouri colleagues were successful in obtaining Rockefeller Foundation funding for a genetics institute in Missouri. McClintock would soon play a prominent role in his staff as Assistant Professor of Botany, beginning in September 1936.

LITERATURE CITED

American Men of Science, A Biographical Directory. 1933, 5th edition, J. M. Cattell and J. Cattell (Eds.). The Science Press, New York.

American Men of Science, A Biographical Directory. 1938, 6th edition, J. M. Cattell and J. Cattell (Eds.). The Science Press, New York.

Anderson, E. G. 1934. A chromosomal interchange in maize involving the attachment to the nucleolus. *American Naturalist* Vol. 68 (July/August), pp. 345–350.

Blair, S. S. 2003. Genetic mosaic techniques for studying *Drosophila* development. *Development* Vol. 13. (No. 21, November 1, 2003), pp. 5065–5072.

Bloom, S. E. 2013. McClintock's pioneering studies of the nucleolar organizer region in maize: Exact chromosomal localization and its function, pp. 1.261–1.268, *In*, L. B. Kass (Ed.), *Perspectives on Nobel Laureate Barbara McClintock's Publications (1926–1984): A Companion Volume*, Volumes I & II. The Internet-First University Press. https://hdl.handle.net/1813/34897

Bradsher, G. 2010. The Nuremberg Laws. Archives receives original Nazi documents that "legalized" persecution of Jews. *Prologue Magazine* Vol. 42 (No. 4, Winter 2010), U.S. National Archives and Records Administration. https://www.archives.gov/publications/prologue/2010/winter/nuremberg.html

Brieger, F. G., J. T. A. Gurgel, E. Paterniani, A. Blumenschein, and M. R. Alleoni. 1958. *Races of Maize in Brazil and other Eastern South American Countries*. Publication 593, National Academy of Sciences-National Research Council, Washington, DC.

Bridges, C. B. 1935. Salivary chromosome maps. *Journal of Heredity* Vol. 26 (No. 2, February 1935), pp. 60–64.

Burnham, C. R. 1930. Genetical and cytological studies of semisterility and related phenomena in maize. *Proceedings of the National Academy of Sciences* Vol. 16 (No. 4, April 15, 1930), pp. 269–77.

Clausen, R. E. and D. R. Cameron. 1944. Inheritance in *Nicotiana tabacum.* XVIII. Monosomic analysis. *Genetics* Vol. 29 (No. 5, September 1944), pp. 447–477.

Coe, E. and L. B. Kass. 2005a. Proof of physical exchange of genes on the chromosomes. *Proceedings of the National Academy of Sciences* Vol. 102 (No. 19, May 10), pp. 6641–6656. [Online May 2, 2005]. https://www.pnas.org/content/102/19/6641.full

Coe, E. and L. B. Kass. 2005b. Maize Genetics Cooperation News Letter files: Expanded chronological list of materials and related cooperation. *Maize Genetics Cooperation Newsletter* 79, pp. 72–76. [Online April 2005]. https://mnl.maizegdb.org/mnl/79/06CoeKass.htm; https://mnl.maizegdb.org/mnl/79/mnl79.pdf

Coughlan, S. 2013. The scientists who escaped the Nazis. *BBC News*, July 17, 2013. https://www.bbc.com/news/business-23261289

Creighton, H. B. 1933. *A cytological study of crossing-over in Zea mays*. Ph.D. Thesis, February 1933, Cornell University, Ithaca, NY.

Creighton, H. B. 1982. The golden age of corn genetics. Panel Discussion, July 1, 1982. Recording, two sound cassettes. Cornell University Library. https://catalog.library.cornell.edu/catalog/3601898

Creighton, H. B. and B. McClintock. 1931. A correlation of cytological and genetical crossing-over in *Zea mays*. *Proceedings of the National Academy of Sciences* Vol. 17 (No. 8, August 15, 1931) [Communicated July 7, 1931], pp. 492–497.

Creighton, H. B. and B. McClintock. 1932. Cytological evidence for 4-strand crossing over in *Zea mays*. *Proceedings of the International Congress of Genetics* Vol. II [August 24–31, 1932], p. 392. [Exhibit/Abstract]

Creighton, H. B. and B. McClintock. 1935. A corroboration of cytological and genetical crossing-over in *Zea mays*. *Proceedings of the National Academy of Sciences* Vol. 21 (No. 8, March 15, 1935), pp. 148–150.

Darlington, C. D. 1932. *Recent Advances in Cytology* (Foreword by J. B. S. Haldane), J. & A. Churchill, London.

Deichmann, U. 1996. (Translation by Thomas Dunlap) *Biologists under Hitler*, Harvard University Press, Cambridge, MA; London.

Demerec, M. 1935. Woods hole summer meeting of the genetics society of America. *The Collecting Net* Vol. X (No. 9, August 31, 1935), pp. 261, 264–265.

Emerson, R. A. 1932. The present status of maize genetics. *Proceedings of the Sixth International Congress of Genetics* Vol. 1, pp. 141–152.

Emerson, R. A., G. W. Beadle, and A. C. Fraser. 1935. Summary of Linkage Studies in Maize *Cornell University Agricultural Experiment Station Memoir No. 180*. [Issued June 1935]

Frank Blair Hanson Diary 1932. Rockefeller Foundation records, Officer's Diaries RG12, and RG12.F-L., The Rockefeller Archive Center.

Frank Blair Hanson Diary 4 September – 26 December 1933. Rockefeller Foundation records, Officer's Diaries RG12, and RG12.F-L., The Rockefeller Archive Center.

Friedrich, B. 2019. Fritz Haber at one hundred fifty: Evolving views of and on a German Jewish Patriot. *Bunsen-Magazine* Vol. 21 (No. 3, March), pp. 130–144. https://hdl.handle.net/21.11116/0000-0003-553C-B

Gabriel, M. and S. Fogel. 1955. *Great Experiments in Biology*, Prentice-Hall, Englewood Cliffs, NJ. [Creighton & McClintock reprinted in part on pp. 267-272.]

Gehring, W. J. (Ed.). 1978. *Genetic Mosaics and Cell Differentiation*, Springer-Verlag, New York.

Glass, B. 1988. *A Guide to the Genetics Collections of the American Philosophical Society*, American Philosophical Society Library Pub. No. 13, Philadelphia, PA.

Glass, B. 1990. The grim heritage of Lysenkoism: Four personal accounts. I Foreword by Bentley Glass. *The Quarterly Review of Biology* Vol. 65 (No. 4, December 1990), pp. 413–421. [See also part IV Difficult Years in Soviet Genetics by S. M. Gershenson, pp. 447–456]

Harwood, J. 1993. *Styles of Scientific Thought; The German Genetics Community, 1900–1933*, University of Chicago Press, Chicago, IL.

Henning, E. and M. Kazemi. 2009. *Dahlem, Domain of Science 2009*, 4th expanded and updated edition [Translated from the German]. https://www.archiv-berlin.mpg.de/49075/hausreihe_16_eng.pdf [Walking tour of Berlin-Dahlem; see section on KWI Biology]

Henning, E. und M. Kazemi. 2016. *Handbuch zur Institutsgeschichte der Kaiser-Wilhelm-/Max-Planck-Gesellschaft zur Förderung der Wissenschaften 1911-2011, Daten und Quellen II*, Archiv der Max-Planck-Gesellschaft, Berlin 2016, Erster Band, S. 245.

Jewish Telegraphic Agency [JTA] Archive. 1923–present. https://www.jta.org/archive

Jones, D. F. (Ed.) 1932. *Proceedings of the Sixth International Congress of Genetics, Ithaca, New York, 1932*. Vol. 1, *Transactions and General Addresses*. Vol. 2, *Condensed articles* [Abstracts] *and Descriptions of Exhibits*. Published at Menasha, WI, by the Brooklyn Botanic Garden.

Jones, D. F. 1935. Somatic segregation due to hemizygous and missing genes and its bearing on the problem of atypical growth. *Proceedings of the National Academy of Sciences* Vol. 21 (No. 2, February 1, 1935), pp. 90–96.

Jones, D. F. 1936. Atypical growth. *The American Naturalist* Vol. 70 (No. 70, January–February), pp. 86–92.

Jones, D. F. 1937. Somatic segregation and its relation to atypical growth. *Genetics* Vol. 22 (No. 5, September 1, 1937), pp. 484–522.

Jones, D. F. 1941. Somatic segregation. *Botanical Review* Vol. 7 (No. 6, June 1941), pp. 291–307.

Kass, L. B. 2003. Records and recollections: A new look at Barbara McClintock, Nobel Prize-Winning geneticist. *Genetics* Vol. 164 (August), pp. 1251–1260. https://www.genetics.org/content/164/4/1251

Kass, L. B. 2005. Harriet B. Creighton: Proud botanist. *Plant Science Bulletin.* 51(No. 4, Winter), pp. 118–125. [Online December 2005]. https://www.botany.org/psbarchive/issue/2005-v51-no-4.html

Kass, L. B. (Ed.). 2013ff. *Perspectives on Nobel Laureate Barbara McClintock's Publications (1926–1984): A Companion Volume.* Volumes I & II The Internet-First University Press. URI: https://hdl.handle.net/1813/34897 [Online December 30, 2013, updates March 2014; Vol. III, Increments No. 1–3, May 2016; Vol. IV, Increment No. 4, April 2020]

Kass, L. B. and J. A. Birchler. 2014/2016. Setting up a prepared mind: Summary and analysis of "Cytological observations of deficiencies involving known genes, translocations and an inversion in *Zea mays*," Vol. III, Increment #1, pp. 1.47.1–1.47.15, *In*, L. B. Kass (Ed.), *Perspectives on Nobel Laureate Barbara McClintock's Publications (1926–1984): A Companion Volume*, The Internet-First University Press. https://ecommons.cornell.edu/bitstreams/9711d284-f6c1-4f0a-88a0-68eab1c8d7c8/download (15.27 MB) [Online 7 June 2016]

Kass, L. B. and C. Bonneuil. 2004. Mapping and seeing: Barbara McClintock and the linking of genetics and cytology in maize genetics, 1928–1935, pp. 91–118, *In*, J.-P. Gaudilliere and H.-J. Rheinberger (Eds.), *Classical Genetic Research and Its Legacy: The Mapping Cultures of 20th Century Genetics*, Routledge, London. [Routledge Studies in the History of Science, Technology and Medicine]

Kass, L. B., C. Bonneuil, and E. Coe. 2005. Cornfests, cornfabs and cooperation: The origins and beginnings of the Maize Genetics Cooperation News Letter. *Genetics* Vol. 169 (No. 4, April), pp. 1787–1797. https://www.genetics.org/content/169/4/1787 [Online May 6, 2005]

Kass, L. B., E. H. Coe, M. N. Cook, M. E. Smith, and J. L. Singer (Eds.). 2019. *Founding of the Maize Genetics Cooperation News Letter at Cornell University: A 90th Anniversary Tribute.* (Foreword by E. S. Buckler) [e-book] Internet-First University Press, Ithaca, NY. https://ecommons.cornell.edu/handle/1813/66550. [Online October 10, 2019]

Kass, L. B. and P. Chomet. 2009. Barbara McClintock, pp. 17–52, *In*, J. L. Bennetzen and S. C. Hake (Eds.), *Handbook of Maize: Genetics and Genomics*, Springer Science, New York.

Kato Y., T. A. [T. Angel Kato Y]. 2013. Significance of McClintock's pattern concept in the analysis of chromosome knob distributions in the races of maize, pp. 4.39–4.49, *In*, L. B. Kass (Ed.), *Perspectives on Nobel Laureate Barbara McClintock's Publications (1926–1984): A Companion Volume.* https://hdl.handle.net/1813/34897

Keller, E. F. 1983. *A Feeling for the Organism. The Life and Work of Barbara McClintock*, 1st edition, W. H. Freeman, San Francisco, CA.

Keller, E. F. 1993. *A Feeling for the Organism. The Life and Work of Barbara McClintock*, 10th Anniversary edition, W. H. Freeman, New York.

Kohler, R. E. 1994. *Lords of the Fly. Drosophila Genetics and the Experimental Life*, University of Chicago Press, Chicago, IL.

KWI-Biology. "File I. Abt., Rep. 1A, Nr. 534-4 KWI Biology." Archives of the Max Planck Society. https://www.archiv-berlin.mpg.de/aboutArchive

Macrakis, K. 1993. *Surviving the Swastika: Scientific Research in Nazi Germany*, Oxford University Press, New York.

McClintock, B. 1930a. A cytological demonstration of the location of an interchange between two non-homologous chromosomes of *Zea mays. Proceedings of the National Academy of Sciences* Vol. 16 (No. 12, December 15, 1930), pp. 791–796.

McClintock, B. 1930b. [Abstract #9] A cytological demonstration of the location of an interchange between two non-homologous chromosomes of *Zea mays. Anatomical Record* Vol. 47 (No. 3, December 25, 1930), p. 380.

McClintock, B. 1931a. The order of the genes *C*, *Sh*, and *Wx* in *Zea mays* with reference to a cytologically known point in the chromosome. *Proceedings of the National Academy of Sciences* Vol. 17 (No. 8, August 15, 1931), pp. 485–491.

McClintock, B. 1931b. Cytological observations of deficiencies involving known genes, translocations and an inversion in *Zea mays*. *University of Missouri, College of Agriculture, Agricultural Experiment Station, Research Bulletin* 163 [December, authorized December 23, 1931], pp. 1–30.

McClintock, B. 1932a [Abstract]. Cytological observations in Zea on the intimate association of non-homologous parts of chromosomes in the mid-prophase of meiosis and its relation to diakinesis configurations. *Proceedings of the Sixth International Congress of Genetics* Vol. II [August 24–31, 1932], pp. 126–128.

McClintock, B. 1932b. A correlation of ring-shaped chromosomes with variegation in *Zea mays*. *Proceedings of the National Academy of Sciences* Vol. 18 (No. 12, December 15, 1932), pp. 677–681.

McClintock, B. 1933. The association of non-homologous parts of chromosomes in the mid-prophase of meiosis in *Zea mays*, with 51 figures in the text and plates VII-XII. *Zeitschrift für Zellforschung und mikroskopische Anatomie* Vol. 19 (No. 2, September 22, 1933), pp. 191–237.

McClintock, B. 1934. The relation of a particular chromosomal element to the development of nucleoli in *Zea mays* with 21 figures in the text and plates VIII-XIV. *Zeitschrift für Zellforschung und mikroskopische Anatomie* Vol. 21 (No. 2, June 23, 1934), pp. 294–328.

McClintock, B. 1938. The production of homozygous deficient tissues with mutant characteristics by means of the aberrant mitotic behavior of ring-shaped chromosomes. *Genetics* Vol. 23 (No. 4, July), pp. 315–376.

McClintock, B. 1987. *The Discovery and Characterization of Transposable Elements*. The Collected Papers of Barbara McClintock. Genes Cells and Organisms; Great Books in Experimental Biology, J. A. Moore, series editor, Garland Publishing Co., New York.

McClintock, B. and H. E. Hill. 1931. The cytological identification of the chromosome associated with the R-G linkage group in *Zea mays*. *Genetics* Vol. 16 (No. 2, March 16, 1931), pp. 175–190.

McClintock, B., T. A. Kato Y., and A. Blumenschein. 1981. *Chromosome Constitution of Races of Maize. Its Significance in the Interpretation of Relationships Between Races and Varieties in the Americas*. Colegio de Postgraduados, Escuela Nacional de Agricultura, Chapingo, Edo. Mexico, Mexico, Vol. xxxi, 517 pp.

Mitman, G. and A. Fusto-Sterling. 1992. What ever happened to *Planaria*?. C. M. Child and the physiology of inheritance, pp. 172–197, *In*, A. E. Clarke and J. H. Fujimura (Eds.), *The Right Tools for the Job*. Princeton University Press, Princeton, NJ.

Morgan, T. H. 1932. *The Scientific Basis of Evolution* (Based on the Messenger Lectures, Cornell University, 1931), W. W. Norton, New York, ix, 13–286 pp.

Navashin, M. 1930. Unbalanced, somatic chromosomal variations in *Crepis*. *University of California Publications in Agricultural Science* Vol. 6, pp. 95–106.

Neel, J. V. 1987. *Curt Stern 1902–1981. A Biographical Memoir*, National Academy of Sciences, Washington, DC.

Nesius, E. J. 1988. *The First 100 Years. A History of the West Virginia Agricultural and Forestry Experiment Station*, West Virginia University, The Agricultural and Forestry Experiment Station, Morgantown.

Painter, T. S. 1931. A cytological map of the X chromosome. *Science* Vol. 73 (No. 1902, 12 June), pp. 647–648.

Painter, T. S. 1934a. A new method for the study of chromosome aberrations and the plotting of chromosome maps in *Drosophila Melanogaster*. *Genetics*, 19 (No. 3, May 1), pp. 175–188.

Painter, T. S. 1934b. Salivary chromosomes and the attack on the gene. *Journal of Heredity* Vol. 25 (No. 12, December 1, 1934), pp. 465–476.

Passenger Search, The Statue of Liberty-Ellis Island Foundation, Inc. https://heritage.statueofliberty.org/passenger

Plant Breeding Records, Cornell University Archives: Cornell University Department of Plant Breeding Records, 1906–1970. Collection Number: 21-28-889 Division of Rare and Manuscript Collections. Cornell University Library. [Boxes 56-65 (student files) are restricted to Plant Breeding and Genetics Section Chair or their designate]. https://rmc.library.cornell.edu/EAD/htmldocs/RMA00889.html

Plarre, W. 1987. A contribution to the history of science of heredity in Berlin, pp. 147–218, *In*, H. Scholz (Ed.), *Botany in Berlin, Englera 7*. Botanischen Garten und Botanischen Museum, Berlin-Dahlem.

Provine, W. B. and P. Sisco. 1980. Interview with Barbara McClintock at Cold Spring Harbor Laboratory, Long Island, New York, August 28, 1980. Division of Rare and Manuscript Collections, Cornell University Library, Ithaca, NY.

Rall, J. A. 2017. Nobel Laureate A. V. Hill and the refugee scholars, 1933–1945. *Advances in Physiology Education* Vol. 41 (No. 2, June 1, 2017), pp. 248–259.

Rhoades, M. M. 1932. The genetic demonstration of double strand crossing over in *Zea*. *Proceeding of the National Academy of Sciences* Vol. 18 (No. 7, July 15, 1932), pp. 481–484.

Rhoades, M. M. 1935. A new aleurone color in Maize. *American Naturalist* Vol. 69 (No. 720, January/February 1935), pp. 74–75.

Rhoades, M. M. 1936. The effect of varying gene dosage on aleurone colour in maize. *Journal of Genetics* Vol. 33 (No. 3, December 1936), pp. 347–354.

Rhoades, M. M. 1992. The early years of maize genetics, pp. 45–69, *In*, N. Fedoroff and D. Botstein (Eds.), *The Dynamic Genome: Barbara McClintock's Ideas in the Century of Genetics*, Cold Spring Harbor Laboratory Press, Plainview, NY.

Rhoades, M. M. and B. McClintock. 1935. The cytogenetics of maize. *Botanical Review* Vol. 1 (No. 8, August 1935), pp. 292–325.

Rockefeller Foundation Archive Center [RF] RG: 10.2 (Fellowship Recorder Cards), Series: MNS (Medical & Natural Sciences) Germany. Box: 11, File "Stern, (Dr.) Curt" [3 cards].

Rürup, R. and M. Schüring. 2008. *Schicksale und Karrieren. Gedenkbuch für die von den Nationalsozialisten aus der Kaiser-Wilhelm-Gesellschaft vertriebenen Forscherinnen und Forscher*, Wallstein Verlag, Göttingen.

Schnarrenberger, C. 1987. Botany at the Kaiser Wilhelm Institutes, pp. 105–146, *In*, H. Scholz (Ed.), *Botany in Berlin, Englera 7*. Botanischen Garten und Botanischen Museum, Berlin-Dahlem.

Sharp, L. W. 1934. *Introduction to Cytology*, 3rd edition, McGraw-Hill Book Company, New York.

Shine, I. and S. Wrobel. 1976. *Thomas Hunt Morgan: Pioneer of Genetics* [foreword by George W. Beadle; on the occasion of the Kentucky Bicentennial], University Press of Kentucky, Lexington.

Stadler, L. J. 1931. The experimental modification of heredity in crop plants. I. Induced chromosomal irregularities. *Scientific Agriculture* XI (No. 9, May 1931), pp. 557–572.

Stern, C. 1931. Zytologisch-genetische Untersuchungen als Beweise für die Morgansche Theorie des Faktorenaustauschs. *Biologisches Zentralblatt* Vol. 51 [Heft 10, October], pp. 547–587.

Stern, C. 1932. Neuere Ergebnisse über die Genetik und Zytologie des Crossing over, *In*, *Proceeding of the Sixth International Congress of Genetics* Vol. 1, pp. 295–303. [Recent results on the genetics and cytology of crossing over.]

Stern, C. 1933. Faktorenkoppelung und Faktorenaustausch [linkage and crossing over] Berlin, Verlag von Gebrüder Borntraeger, [Monograph] 331 pp. *In*, E. Baur and M. Hartmann (Eds.), the German Series *Handbuch der Vererbungswissenschaft*, Vol. 1.

Stern, C. 1936. Somatic crossing-over and segregation in *Drosophila melanogaster*. *Proceedings of the National Academy of Sciences* Vol. 21 (No. 6, November 1, 1936), pp. 625–630.

Stern, C. 1971. From crossing-over to developmental genetics, pp. 21–28, *In*, G. Kimbar and G. Redei (Eds.), *Stadler Genetics Symposia 1-2*, University of Missouri Press, Columbia.

Stino, K. R. 1940. Inheritance in *Nicotiana tabacum*. XV. Carmine-white variegation. *Journal of Heredity* Vol. 31, pp. 19–24.

Stoltzenberg, D. 1994. *Fritz Haber: Chemiker, Nobelpreisträger, Deutscher, Jude, eine Biographie*, VCH, Weinheim.

Stoltzenberg, D. 2004. *Fritz Haber: Chemist, Nobel Laureate, German, Jew, A Biography* [English Translation], Chemical Heritage Foundation, Philadelphia, PA.

Stone, W. S. 1931 An analysis of some induced X-chromosomal translocations in Drosophila melanogaster. Master's thesis, Master of Arts, August 1931, The University of Texas, Austin.

Stone, W. S. 1934. Linkage between the X and IV chromosomes in *Drosophila melanogaster*. *Genetica* Vol. 16 (September), pp. 506–520. [Publication of his Ph.D. Thesis]

Szöllösi-Janze, M. 1998. *Fritz Haber: 1868–1934. Eina Biographie*, Verlag C. H. Beck, Munich.

Vogt, A. B. 2009. Ursula Philips, 1908–1995. *The Shalvi/Hyman Encyclopedia of Jewish Women*. https://jwa.org/encyclopedia/article/philip-ursula [March 1, 2009, Jewish Women's Archive].

Wayne, R. 2013. An attempt at identifying the position of genes on the chromosomes of maize using X-ray induced chromosome deficiencies, pp. 1.184–1.187, *In*, L. B. Kass (Ed.), *Perspectives on Nobel Laureate Barbara McClintock's publications (1926-1984): A Companion Volume*, Volumes I & II. The Internet-First University Press. https://hdl.handle.net/1813/34897

6 Missouri Compromise, Tenure or Freedom? *Barbara McClintock Leaves Academe*

When The Legend Becomes Fact, Print The Legend

–*from The Man Who Shot Liberty Valance*

ASSISTANT PROFESSOR, UNIVERSITY OF MISSOURI

Lewis John Stadler invited Barbara McClintock to join his research group at the University of Missouri, Columbia, beginning in September 1936[1] (Table 6.1). Having few other job opportunities, she was excited to accept the offer. Yet within six years, McClintock chose to leave a promised tenured professorship at Missouri for a full-time research appointment in New York. The story of why she decided to forego a secure job opportunity is complex and has early roots, as I detail in this chapter. She and George Beadle were considered for two separate appointments as early as April 1935. Although Beadle initially planned to accept their offer, he instead joined the faculty at Harvard, while McClintock was hired on the very strong recommendation of a committee appointed by the President of the University of Missouri (Table 6.1).[2]

William Jacob Robbins, Head of the Department of Botany and a Cornell University alumnus, chaired the committee, which included Stadler and Winterton Conway Curtis, Chairman of the Department of Zoology, and two other members.[3] As discussed in previous chapters, McClintock had first met Stadler, then a National Research Council (NRC) Fellow, when he came to work with Rollins Adams Emerson, Head of the Plant Breeding Department at Cornell University in 1925–1926.[4] Upon completing her Ph.D. in June 1927, McClintock remained at Cornell as an Instructor. She later conducted part of her NRC Fellowship in the Biological Sciences (1931–1933) in Stadler's laboratory at Missouri. At that time, Stadler had suggested one of her initial research problems

1 Excerpt from Frank Blair Hanson Officer Diaries [hereafter FBH], Tuesday, April 30, 1935, August 14–27, 1935, January 6, 1936, Rockefeller Foundation, Rockefeller Archive Center (hereafter RF). Letter L. Stadler to Marcus Rhoades, February 28, [19]36, Marcus M. Rhoades Collection, Lilly Library (hereafter MMR and LL).

2 Excerpt from FBH, April 30, 1935, see also "Proposed Institute of Genetics for University of Missouri," [stamped] July 2, 1935 – B. McClintock's name is penciled in for cytogenetics position with a joint appointment in the "U.S.D.A."; G. W. Beadle's name is penciled in as Geneticist for Polyploid Studies in the "U.S.D.A.", all at RF; see also Berg and Singer (2003) for a life of George Beadle. Letter F. Middlebush to L. Stadler, January 17, 1936, President's Office Papers [University of Missouri], Western Historical Manuscript Collections, The State Historical Society of Missouri Research Center-Columbia (hereafter PO and SHSMO-Columbia). Letter L. Stadler to F. Middlebush, April 13, 1936; appointments were for "full time or approximately full time research," Lewis J. Stadler papers, SHSMO-Columbia (hereafter LJS).

3 L. J. Stadler (1896–1954), the Head of the genetics research group at the University of Missouri, Columbia, held a regular faculty appointment in the Department of Field Crops, College of Agriculture, and by 1937, he was Principal Geneticist, Bureau of Plant Industry, U.S.D.A., Missouri.

4 Letter L J. Stadler to R. A. Emerson, arrangements for L. Stadler to work with R. A. Emerson at Cornell, Plant Breeding Records, Division of Rare and Manuscript Collections, Cornell University Library, CU (hereafter PB/CU); L. Stadler's NRC Fellowship with R. A. Emerson is not listed in *American Men of Science* (hereafter *AMS*) 1938, p. 1343.

DOI: 10.1201/9781003332527-7

TABLE 6.1
Investigators in L. J. Stadler's Genetics Research Group at the University of Missouri, 1936, Appointed for "Full Time or Approximately Full Time Research" (see also Figure 6.1).

Name	Title
Barbara McClintock	Assistant Professor of Botany, College of Arts and Sciences
Joseph O'Mara	Assistant Geneticist, U.S. Department of Agriculture (hereafter U.S.D.A.)
Ernest Sears	Assistant Geneticist, U.S.D.A.
Luther Smith	Assistant Geneticist, U.S.D.A.
George F. Sprague	Associate Professor of Field Crops, College of Agriculture, and Associate Agronomist, U.S.D.A.
L[ewis] J. Stadler	Associate Professor of Field Crops, College of Agriculture, and Senior Geneticist, U.S.D.A.
Frank M. Uber	Assistant Professor of Physics, College of Arts and Sciences

involving mutations in irradiated maize plants, and he also supplied McClintock with the genetic corn stocks for her investigations (McClintock 1931; Kass & Birchler 2014/2016). They worked collaboratively, but published independently.[5] Stadler was one of several prominent geneticists who "very strongly" recommended McClintock for a Guggenheim Fellowship that supported her work at the Kaiser Wilhelm Institute in Berlin-Dahlem, Germany, and the University of Freiburg during the winter of 1933–1934.[6] By early April 1934, after a little over four months in Nazi Germany, McClintock returned to Cornell, where she completed her fellowship but worried about finding a job. She and other postdoctoral colleagues, including George Beadle, Charles Burnham, and Marcus Rhoades, found no permanent job opportunities owing to the worldwide depression that left countless people homeless and destitute.

In 1934, R. A. Emerson was awarded a five-year grant from the Rockefeller Foundation to support the Maize Genetics Cooperation at Cornell University. He recognized McClintock's abilities toward this enterprise and requested a separate Grant-in-Aid to hire her as his research assistant. The grant was renewed in 1935 (Chapter 5), and in September 1936, McClintock resigned from her assistantship for a faculty appointment at the University of Missouri. Emerson submitted the final summary of her accomplishments to the Rockefeller Foundation in November 1936.[7] Therein, McClintock reported an analysis of homozygous deficiencies in mosaic maize plants by means of ring chromosome behavior. She presented the work at the genetics society meetings the following December (McClintock 1937). She continued these studies of maize "ring-shaped chromosomes" while a faculty member at the University of Missouri (McClintock 1938a,b; Figure 6.1).

McClintock recalled that Stadler had created a job for her at the University of Missouri and believed it was not permanent (Provine & Sisco 1980). The story, however, is more complex. By 1936, Stadler and Curtis collaboratively procured funds from the Missouri Agricultural Experiment Station, the U.S. Department of Agriculture (U.S.D.A.), the University of Missouri, and from the Rockefeller Foundation to support "A Regional Laboratory of Plant Genetics" at Missouri.[8] They secured Rockefeller funding based partly on the rationale that McClintock could be hired as a member of their research team (Curtis 1949, p. 169). In summarizing the proposed use of the Rockefeller Foundation grant for support of work in genetics at Missouri, Stadler wrote: "Doctor McClintock is

[5] See B. McClintock's and L. Stadler's acknowledgments to each other in their publications between 1931 and 1941, and also McClintock's (1984) acknowledgments to L. Stadler.

[6] Unsigned draft Letter of Recommendation for a Guggenheim fellowship for B. McClintock, [Stadler, 1933], LJS. Provine and Sisco (1980) and *American Men of Science* (*AMS*) 1938, McClintock, p. 891.

[7] Letter R. A. Emerson to B. McClintock, and to F. B. Hanson the Rockefeller Foundation, November 19, 1936, attachment Report to the Rockefeller Foundation, PB/CU, Box 25, File: Rockefeller Foundation.

[8] Letter W. C. Curtis to L. Stadler, August 3, 1935, LJS; Letter L. Stadler to M. Rhoades, February 28, 1936, MMR. Letter L. Stadler to F. Middlebush, April 13, 1936, LJS; also see "Proposed Institute of Genetics, copy for F. B. Hanson, [stamped] July 2, 1935," RF.

FIGURE 6.1 L. J. Stadler's genetics group at University of Missouri, February 1938. Persons are identified from left to right. Front row: J. W. Cameron, K. O. Mills (DeBoer), F. J. Clark, M. J. Guthrie, B. McClintock, E. W. Landron. Second row: G. F. Sprague, F. M. Uber, H. B. Glass, L. Smith, G. Francis, H. Roman. Top row: L. J. Stadler, J. G. O'Mara, W. R. Singleton, R. Winters, E. R. Sears. (Negative provided by George Rèdei and used with permission. Originally printed in Rèdei (1976), reprinted in Kass (2003); see also Kass (2004).)

unquestionably the best cytologist who could have been appointed and is in my opinion unsurpassed in her field."[9] During earlier negotiations for the grant, Frank Blair Hanson, Assistant Director of the Rockefeller Foundation, was told that McClintock was reported to be the best plant cytologist in the world and could be hired at a relatively small salary.[10] The Rockefeller Foundation initially awarded the grant for three years to fund their plant genetics research and facilities at Missouri and renewed it in 1939 for an additional five years. The renewal included a separate $80,000 grant dedicated to the construction of a new building to house laboratories and greenhouses for the genetics projects, and the Public Works Administration (PWA 1933–1939) provided additional funds for the "Genetics Building" (Curtis 1949, pp. 169–170).[11]

When McClintock initially accepted her faculty appointment in April 1936, she was told that her salary of $2,700 was contingent on the University receiving Rockefeller Foundation funding. The contingency indicated temporary employment and is consistent with McClintock's recollections. Apparently, she was uncomfortable being placed in such a precarious position.[12] She may not have been aware, however, that during negotiations with the Rockefeller Foundation, she had been identified as the best plant cytologist in the world for their appointment, or that the grant renewal was also contingent upon the University of Missouri assuming all responsibility for fundamental research in genetics by 1944. The renewal also specified that the University make a commitment to

[9] Letter L. Stadler to F. Middlebush, April 13, 1936, LJS.

[10] Excerpt from FBH, Lunch with Prof. W. C. Curtis and Dr. [*sic*] F. C. Richey, Tuesday, April 30, 1935, RF.

[11] Letter L. Stadler to F. Middlebush April 13, 1936, LJS; Letter W.C. Curtis to E. Ellis, June 7, 1954, File 42, W. C. Curtis papers, SHSMO-Columbia [hereafter WCC]; *Missouri Alumnus*, May 1939, p. 5, SHSMO-Columbia; *Missouri Alumnus*, May 1940, p. 7, SHSMO-Columbia; *The Rockefeller Foundation Annual Report*, 1939; see photograph of Genetics Building on p. 216.

[12] McClintock (pers. com. 1973), Provine and Sisco (1980), Keller (1993, pp. 81–83). Contingent on … funding, Letter W. C. Curtis to B. McClintock, April 29, 1936, LJS.

offer McClintock a permanent research position within the genetics research project.[13] As a result, by 1939 McClintock held a regular faculty appointment at the University of Missouri.

Stadler and Curtis had at first proposed a position for her at Missouri with salary paid by the U.S.D.A. She expressed a preference, however, for a regular university faculty appointment "as compared with any place in the U.S. Department of Agriculture." They therefore arranged for her appointment as Assistant Professor of Botany, which resulted in her receiving a lower salary and added responsibilities.[14] Had she taken the U.S.D.A. appointment, she probably would not have been encumbered by teaching responsibilities but may have been less autonomous.

THE UNIVERSITY OF MISSOURI WITH STADLER'S GENETICS RESEARCH GROUP

Members of Stadler's genetics research group were supported by different funding agencies and belonged to various departments and colleges throughout the University (Table 6.1; Figure 6.1). The University of Missouri, founded in 1839, is considered the oldest public institution of higher learning west of the Mississippi. Throughout its history, the university was the victim of sectional disputes that characterized Missouri society, and state funding was not equal to the institution's needs or to its aspirations. In 1863, the Missouri State Legislature accepted the terms of the Morrill Act for the endowment of a College of Agriculture and Mechanic Arts, and after many political debates, the new College of Agriculture and Mechanic Arts was established (1907) in the University at Columbia.[15] McClintock was officially a faculty member in the Department of Botany, College of Arts and Sciences (Curtis 1949, p. 169), and as she wrote several years later to her friend Marcus Rhoades, she was "supposedly hired for research."[16] However, upon beginning her faculty position at Missouri, McClintock was expected to, and did, participate in departmental faculty committees, teach undergraduate and graduate courses (Table 6.2), advise and direct graduate students' projects (Table 6.3), and act as a consultant for both resident staff and visiting researchers (Table 6.4).[17]

Stadler was justified in believing that McClintock was the most qualified cytologist to join his research group. Her previous contributions and those she completed while working at Missouri gave recognition to her and to Stadler's Laboratory of Plant Genetics. Appreciation and acknowledgment of her pioneering work eventually led to a permanent research appointment elsewhere. As Assistant Professor of Botany at Missouri, McClintock's (1937, 1938a,b) first publications resulted from studies she later termed the "chromosomal type" breakage-fusion-bridge (BFB) cycle. This work on ring chromosomes was a continuation of previous investigations that she had conducted at Missouri and at Cornell.[18] She found that some plants harboring such chromosomes would produce dicentric [two centromeres] chromosome bridges at mitotic anaphase, thereby losing genetic material and uncovering recessive traits in some of the cell descendants. Sister mitotic cells would gain the genetic material that was lost in the breakage of the chromosome bridge. The broken chromosome ends would fuse again to reform a ring, ready to undergo breakage again. The repeated losses resulted in variegated plant tissues due to the continual uncovering of recessive traits (McClintock 1938b).

[13] Letter W. C. Curtis to F. B. Hanson, January 23, 1939; Letter F. A. Middlebush to F. B. Hanson, January 23, 1939, Attached is W.C. Curtis to F. Middlebush, January 3, 1939, Enclosure II, "In 1936 ... university created two new permanent research positions, ... filled by Drs. [B.] McClintock and [F. M.] Uber"; all in RF. See also budget 1939–[19]40; salary for B. McClintock is listed under general maintenance, not Rockefeller grant [June 2, 1940], PO.

[14] B. McClintock, May-16-36 [1936], Appointment Card, C-1, University of Missouri Archives [hereafter UMO]; Official Notice L. Cowan to F. M. Tisdel and W. Robbins, May 21, 1936, Botany Department Papers, SHSMO-Columbia [hereafter BD]. regular ... appointment, Letter W. C. Curtis to F. B. Hanson, January 25, 1936; Letter W. C. Curtis to F. B. Hanson, February 11, 1936; Letter W. C. Curtis to F. B. Hanson February 28, 1936, all in RF.

[15] Vilas (1939, pp. 104–137), Stephens (1962, pp. 174–175, 355), Olson and Olson (1988, pp. 3–28).

[16] "Faculty member," Letter L. Stadler to F. Middlebush, April 13, 1936, LJS. Official Notice L. Cowan to M. Tisdale and W. Robbins, May 21, 1936, BD. "College of Arts and Sciences," founded at University of Missouri 1907, Stephens (1962, p. 355). "Hired for research," Letter B. McClintock to M. Rhoades, January 29, 1939, MMR.

[17] Many memos and letters record McClintock's activities as a faculty member at the University of Missouri, BD.

[18] McClintock (1931, 1932, 1934–1936) unpublished, Rhoades and McClintock (1935).

TABLE 6.2
Barbara McClintock's Teaching Responsibilities at the University of Missouri, 1936–1943

Courses	Faculty Offering Classes, 1936–1943
SPECIAL PROBLEMS f, w	All faculty in B including McClintock
SEMINAR f, w	All faculty in B including McClintock
SPECIAL TOPICS f, w	All faculty in B including McClintock
ADVANCED GENETICS w	Guthrie, Stadler, McClintock (cross-listed in B, F, Z; alt. yrs. with Advanced Cytology) [McClintock on leave, 1941-42]
ADVANCED CYTOLOGY w	Guthrie, McClintock, Stadler (cross-listed in B, F, Z; alt. yrs. with Adv. Genetics) [replaced by Cytogenetics in 1940-41]
RESEARCH f, w	All faculty in B including McClintock
THE HISTORY OF BOTANY w	All faculty in B including McClintock (alt. yrs., first offered 1938–1939)
CYTOGENETICS w	McClintock (not cross-listed in F or Z, generally alt. yrs., prerequisite Botany 125 or equivalent, first offered 1940-41; not offered 1941-42 [McClintock on leave], scheduled for 1942-43.

Courses Barbara McClintock was scheduled to teach, as listed in the University of Missouri, College of Arts and Science, *Announcements* for 1936–1943 (w, winter term; f, fall term). Courses were offered in the Botany Department (B) and cross-listed in the Departments of Zoology (Z) and Field Crops (F). McClintock participated in courses offered by Botany Department faculty and co-taught classes with M. J. Guthrie (Z) and L. J. Stadler (F) (see Figure 6.1). She independently developed and offered Cytogenetics in the winter term, 1940–1941. McClintock was on leave of absence in 1941–1942 and was scheduled to teach in 1942–1943 [SHSMO-Columbia].

TABLE 6.3
Students Mentored and Advised by Barbara McClintock at the University of Missouri, between 1936 and 1942 ([a]directed research, [b]major advisor, [c]member of Ph.D. committee, [d]consultant) (Botany Department Papers, SHSMO-Columbia; *AMS* 1944); see also Figure 6.1.

[a,b,c]Edward Allen Weaver, Ph.D. 1942.

[a,b,c*]Spencer Brown, in residence for Ph.D. 1938–1941, completed degree 1942, U. California, Berkeley.

[a,b,c,*]Frances Janes Clark, Ph.D. candidate 1936–1938; completed degree 1941, Yale (Clark, 1940, 1941).

[a,c]Helen Crouse, Ph.D. 1942; advisor, M. J. Guthrie, Department of Zoology.

[c,*]Mary Amelia Bartley [Schmidt], Ph.D. candidate 1938; Advisor, W. J. Robbins, Department of Botany.

[c]James Wagner Cameron, Ph.D. candidate 1938, completed 1947, Harvard.

[c]Herschel Roman, Ph.D. 1942; major advisor, L. J. Stadler, Department of Field Crops.

[d]Jesse Singleton, M.A. 1941; advisor W. B. Drew, Department of Botany.

*Frances Janes Clark conducted research with McClintock at the University of Missouri and at Cornell University during summers of 1937–1939. She acknowledged Stadler and McClintock in the publications of her doctoral research (Clark 1940, 1941) (see Figure 6.1).
*Mary Amelia Bartley married C. R. Schmidt and was awarded her doctorate in 1938 under her married name.
*Spencer Warton Brown completed his degree from the Division of Genetics, U of California, Berkeley, working with Harold. P. Olmo at Davis, California (Brown 1942, 1943).

TABLE 6.4
Notable Visiting Researchers who Consulted with Barbara McClintock, between 1936 and 1941, at the University of Missouri, and Their Research Organisms (Botany Department Papers, SHSMO-Columbia; *AMS* 1944); see also Figure 6.1.

W. Ralph Singleton, Connecticut Agricultural Experiment Station, Bureau of Plant Industry, U.S.D.A., 1937–1938, genetics and cytology of *Zea mays* [maize] (see Figure 6.1).

Will Martin Myers, Northeast Regional Pasture Research Laboratory, State College, Pennsylvania, U.S.D.A., 1938, cytology and genetics of forage grasses.

Carl C. Lindegren and Gertrude Lindegren, U. Southern California, 1940, genetics and cytology of *Neurospora* [bread mold].

J. Gordon Carlson, Rockefeller Foundation Fellow, University of Alabama, 1940–1941, Orthopteran [Grasshopper] cytology.

Carl P. Swanson, Sheldon Traveling Fellow, Harvard University, 1941, plant cytology, *Tradescantia* [Spiderwort].

McClintock had previously performed extensive cytological examinations of plants carrying Stadler's X-ray-induced mutations.[19] Those studies revealed that many of the mutations were due to deficiencies or rearrangements of a chromosome segment carrying the dominant allele. This material was the major source of chromosome rearrangements that allowed McClintock to examine the behavior of broken chromosomes and later reveal the induction of transposable elements (Kass & Chomet 2009). After years of studying the behavior of mitotically unstable chromosomes in the form of ring chromosomes (McClintock 1931 through 1938b), McClintock then sought to experimentally determine the process of chromosome breakage and fusion in non-ring chromosomes. Her studies (McClintock 1938c) of an inverted segment of chromosome 4L clearly showed that "fusion of broken meiotic chromatids does occur" resulting in anaphase bridge configurations and the formation of an acentric fragment. McClintock asked: What happens to such chromosomes when they are passed into the next generation (McClintock 1938c)? The definitive answer came only a year later in a brief report published in *PNAS* (McClintock 1939). It was well known that chromosomes having missing sections (deficient chromosomes) would not transmit well through the haploid pollen or ovule, so she devised a method to transmit a newly broken chromosome that included a full complement of genes by using rearrangements on chromosome 9. Plants with these chromosome rearrangements were obtained from both Harriet Creighton and L. J. Stadler. One of the chromosomes was an inverted duplication of the short arm of chromosome 9, which carried a series of dominant color and starch conditioning genes that could be visualized in the kernel and plant when they were present or lost (including *Yg*, *C-I*, *Sh*, *Bz*, and *Wx*). She followed this study with an extensive and detailed report submitted to the journal *Genetics* in November 1940. It appeared the following March (McClintock 1941a), with special authority from the editorial board (as described below). She reported being surprised when the breakage-fusion-bridge cycle ceased in some of the sporophytic generation of the plants (McClintock 1939, 1941a). She interpreted the results as a "healing" of broken chromosome ends in this phase of plant development. McClintock later termed these curious but profound behaviors of newly replicated chromosomes the "*chromatid* type" and the "*chromosomal* type of breakage-fusion-bridge [BFB] cycle," depending upon where in plant development the cycles appeared (1942a, p. 459, 461). Recognition of McClintock's pioneering work led to invitations to present her research at meetings, conferences, and symposia. Ultimately, she obtained a research appointment commensurate with her qualification, as described below.

[19] McClintock (1931, 1932, 1933); see Chapter 5.

FIGURE 6.2 Barbara McClintock and geneticists who attended the gene conference, September 4–6, 1938, that followed the Genetics Society of America meeting held on September 1–3, 1938, at the Marine Biological Laboratory, Woods Hole, Massachusetts. Persons are identified from left to right: T. Dobzhansky, Curt Stern, B. P. Kaufmann, Barbara McClintock, C. H. Waddington, D. Raffel, and L. J. Stadler. (Used with permission of the American Philosophical Society Library; see also cover of Glass 1988.)

During her time at Missouri, McClintock continued to attend and present papers at national meetings of the Genetics Society of America (GSA). Her abstracts were preprinted in their *Records (1936, 1937)* and were subsequently published in *Genetics* (McClintock 1937, 1938a). In September 1938, she attended the GSA summer meeting (September 1–3) at the Marine Biological Laboratory, Woods Hole, MA, participated in a special gene conference following that meeting (September 4–6, Figure 6.2) organized by Milislav Demerec,[20] and played a role in selecting presentations for the GSA winter meetings held in Richmond, Virginia, at the end of the year[21] (Lindstrom 1939). She was a member of the Program Committee, along with Stadler, who as President of the Society had appointed her the representative to the Editorial Committee of the *American Journal of Botany* (*Records GSA* 1938, pp. 15, 19). That year, she and Milislav Demerec, formerly a student colleague trained in corn genetics at Cornell, were elected to the office of GSA Vice President and President, respectively, for 1939 (*Records GSA* 1939, pp. 18–19).

In 1939, GSA Vice President McClintock was an invited speaker at the AAAS Section (G) meeting on Botanical Sciences, held in Columbus, Ohio, December 27–30, along with Affiliated Societies. The Section Secretary (John T. Buchholz, University of Illinois) reported in *Science* that on Thursday, Dec 28, Dr. Barbara McClintock presented "Recent Investigations in the Cytogenetics of Maize," where over eight hundred were in attendance (Moulton 1940a, p. 112). She and other members of her department had been authorized travel expenses by C. M. Tucker, the University of Missouri, Botany Department Chair, to attend the meetings in Ohio.[22] Her new colleague in the Missouri Botany Department, Paul R. Burkholder, as secretary of the Botanical Society of America's Physiological Section, summarized his section's program for the Columbus meeting (Moulton 1940a, p. 114). Harriet Creighton, her former student and close friend, and Burkholder's former colleague at Connecticut College, gave a co-authored paper to the Botanical Society General Section (Moulton 1940a, p. 113; Kass 2005b). Presentations at meetings and contacts with fellow Cornellians and other colleagues held significant roles regarding recognition and future job prospects for many researchers. McClintock was no exception.

[20] Letter M. Demerec to M. Rhoades, August 12, 1938, MMR.
[21] Letter B. McClintock to M. Rhoades, November 13 [1938], MMR.
[22] Memo. C. M. Tucker, Botany Dept Papers 3453, University of Missouri, File 1266, SHSMO-Columbia.

Within a year, McClintock was elected a Fellow of the American Association for the Advancement of Science (AAAS) (Moulton 1940b, p. 654). At that time, she confided to several of her geneticist colleagues that these and other demands on her time had made it difficult to devote her efforts completely to research.[23] To some extent, these responsibilities, which may not have been a significant part of her earlier experiences, were unexpected limitations on her research time. In a 1980 interview (Provine & Sisco 1980), McClintock indicated that she had found herself restricted by the University of Missouri bureaucracy. She had become accustomed to the freedom she had as a student, Instructor, and Research Assistant at Cornell University, where graduate students in her college were free to choose their own thesis committees, and where both students and faculty members could choose their own research problems. Students in the Plant Sciences at Cornell were permitted, and often encouraged, to work in the labs at any hour of the day or night, when necessary.[24] At Missouri, deans and department chairs made such decisions, and both faculty and students had to secure permits if they wished to work in the buildings past 10 PM (Provine & Sisco 1980).[25] McClintock knew from experience that scientific research often required an irregular work schedule, but the restrictions at the University of Missouri created an atmosphere that was quite different from that which she had been accustomed. Within two years of receiving her faculty appointment at Missouri, McClintock was dissatisfied with her position. George Sprague Sr. recalled that McClintock argued with Stadler about many things and that she often left Stadler's office in tears.[26] Writing to her friend and collaborator Marcus Rhoades about other matters, she let him know that she was "becoming more serious about the possibility of looking for another job."[27] Rhoades and other colleagues understood her desire to move and recommended her for positions at other institutions when such opportunities arose.[28]

After returning from "a long vacation – an extensive trip through the north-west and the west" during the summer of 1940 – McClintock wrote to her friend and former postdoctoral colleague Charlie Burnham that she had decided she "must look for another job." In that letter, she concluded that there was "nothing more" for her at Missouri: "I am an assistant professor at $3,000 and I feel sure that is the limit for me." She added that the job was not "too secure" and that she could "not make [herself] needed in the University. If I am going to move I shall have to make it soon. ... I feel that I ought to get something lined up this year, if possible."[29] Her comments may have resulted from recent negotiations to attract Stadler to the California Institute of Technology (Caltech), where he had been on leave during winter and spring of 1940 and had only returned in mid-June. While Stadler was at Caltech, T. H. Morgan offered him an appointment in his Division of Biology.[30] During those negotiations, the President of the University of Missouri suggested that if Stadler should leave Missouri, his genetics group might be "liquidated."[31] Although these communications were not publicized, members of Stadler's genetics group heard rumors about the possibility of his not returning to Missouri and were concerned over their prospects for the future (Crouse 1992, p. 30).[32] At the same time,

[23] Letter B. McClintock to C. Stern, March 2, 1937, Curt Stern Papers, American Philosophical Society Library, [hereafter CS and APS]; Letter B. McClintock to M. Rhoades, January 29, 1939, MMR.

[24] Harriet Creighton taped interview with L. Kass, 17 August 1996; Natalie Uhl taped interview with L. Kass, July 13, 1995; Harlan Banks, Charles Uhl, John Kingsbury taped interview with L. Kass, June 28, 1995; Lee Kass personal experience 1971–1975, 1991–present.

[25] See BD. *University of Missouri Catalogue*, 1940, University of Missouri, Columbia, MO, p. 253, SHSMO-Columbia.

[26] Lee Kass and R. P. Murphy telephone interview with George Sprague Sr., August 7, 1998.

[27] Letter B. McClintock to M. Rhoades, December 12 [1938], MMR.

[28] E. G. Anderson recommended M. Rhoades, B. McClintock, and others in L. Stadler's group for a job at Minnesota; see Letter H. K. Hayes, Chief of the Division of Agronomy and Plant Genetics, University of Minnesota to M. M. Rhoades, March 15, 1938. Letter H. P. Rainey, University of Texas, to M. Rhoades, thanks Rhoades for suggesting McClintock, Letter H. Rainey to M. Rhoades, May 5, 1940, MMR.

[29] Letter B. McClintock to C. Burnham, September 16, 1940, Charles R. Burnham, University of Minnesota Archives [hereafter CRB]. Her only salary increase of $300, providing a salary of $3,000, is recorded in her May 1938 appointment contract, UMO.

[30] *The Missouri Alumnus,* Columbia Missouri, January 1940, Faculty Notes, Vol. 28, p. 5, SHSMO-Columbia; McClintock to Curt Stern, February 1, 1940, CS; Morgan to Weaver, May 23, 1940, PO.

[31] Memo, "Stadler Possibilities," PO.

[32] Helen Crouse (1992), McClintock's student at the time, described their fears when she recalled her experiences at MO.

possibly without Stadler's knowledge, he and other notable geneticists were being considered for the next Directorship of the Carnegie Institution of Washington's Department of Genetics at Cold Spring Harbor (CIW/CSH), Long Island, New York. Moreover, by October 1940, Vannevar Bush, Carnegie's recently appointed President, expressed doubts about Stadler's qualifications, and by April 1941, he was eliminated from consideration.[33] Milislav Demerec, then Assistant Director at CIW/CSH, Barbara McClintock, and many others, however, would soon be considered for the Director's position.

Stadler remained at Missouri, and by early November 1940, after his research group had moved into the new Genetics Building, McClintock confided to Rhoades that things were going all right and "being together in the new building has helped a lot."[34] Nevertheless, she expressed her concerns about "friction" over a loss of future funding, "trouble with students," and a problem with "Miss Guthrie," her colleague in the Zoology Department who intimidated a student desiring to work with McClintock instead of Stadler.[35] By the end of November, she wrote that she was experiencing "one interruption after another" caused by commitments to "seven graduate students" (Table 6.3) and visiting researchers (Table 6.4), and she even considered "not trying to do any research" the following term.[36]

INVITATIONS TO COLD SPRING HARBOR, 1941

By year's end of 1940, McClintock learned that Milislav Demerec (1895–1966; Figure 6.3) planned to invite her, and many other prominent geneticists, to give a presentation at the 1941 Cold Spring Harbor Symposia on Quantitative Biology. These symposia had been held at the Biological Laboratory [Bio-Lab] at Cold Spring Harbor (CSH), since 1933, and were sponsored by the Long Island Biological Association (LIBA). Demerec, Assistant Director, soon to be Director of the Bio-Lab, had discussed the upcoming symposia with McClintock, and other potential participants, during the 1940 Genetics Society meetings, held in Philadelphia with the AAAS.[37] He subsequently issued invitations for the June 1941 Symposia and invited many of the participants to work cooperatively at the Biological Laboratory during that summer. Demerec was concurrently Assistant Director of the Department of Genetics, CIW/CSH, which physically adjoined the Bio-Lab. As an officer of both institutions, he discussed with his director, Albert F. Blakeslee, how their two organizations might cooperate in "the advancement of genetic study."[38] They established a "closer collaboration" between these two organizations by offering McClintock and other guest investigators "opportunities for summer research" at their adjoining facilities along the inner harbor at Cold Spring Harbor (Blakeslee 1941, p. 211).[39] In his first annual report as Director of the Biological Laboratory, Demerec (1941, p. 23–24) described the cooperation between his institution and the CIW's Department of Genetics (Demerec would become director of both institutions by 1943).

On February 5, 1941, Albert F. Blakeslee (1874–1954), then Director of the Department of Genetics, CIW/CSH, invited Barbara McClintock to spend the summer of 1941 as a guest investigator in his department. Blakeslee encouraged McClintock to come to CIW/CSH by explaining that her good friend Marcus Rhoades would be planting his corn there that summer and he was sure

[33] FBH, Hanson discussion with W. Curtis at Woods Hole, July 11–August 31, 1940, RF. Letters Lewis Weed to V. Bush May 3, 1940, October 9, 1940, October 16, 1940; Letter W. C. Curtis to V. Bush, September 18, 1940; L. Spoehr to G. W. Corner, April 25, 1941; Report of G. W. Corner, Department of Genetics, June 1941, Possibilities for Directorship, all correspondence at Department of Genetics Collections, Carnegie Institution of Washington Archives, Washington, DC, "Genetics Director Selection 1941," File, May 1940–October 1941, Carnegie Institution of Washington, DC [hereafter CIWDC].

[34] Letter B. McClintock to M. Rhoades, November 1, 1940, MMR.

[35] "loss of future funding" – negotiations with the Rockefeller Foundation to fund the genetics research program at Missouri are at RF. "Miss Guthrie" – Mary Jane Guthrie (1895–1975), Professor of Zoology, and Chair of the Zoology Department; see *AMS*, 1938, p. 591.

[36] Letter B. McClintock to M. Rhoades, November 25, 1940, MMR

[37] Letter M. Demerec to A. F. Blakeslee, January 14, 1941, Carnegie Institution of Washington Archives, Cold Spring Harbor Laboratory [hereafter CSHA]. Demerec was appointed Bio-Lab Director in 1941.

[38] Letter A. F. Blakeslee to M. Demerec, January 16, 1941, CSHA.

[39] Letter M. Demerec to A. F. Blakeslee, January 20, 1941, CSHA.

FIGURE 6.3 Milislav Demerec, Director of the Long Island Biological Laboratory (1941) and Assistant Director of the Department of Genetics, Carnegie Institution of Washington, at Cold Spring Harbor, Long Island, New York, ca. 1940. (Used with permission of his daughter, Rada Dyson-Hudson.)

there would be plenty of land "in this plot for you also."[40] The previous summer (July 1, 1940), Marcus M. Rhoades (1903–1989), McClintock's trusted friend and former research colleague at Cornell University, began his position as Associate Professor of Botany at Columbia University in New York City, an hour's drive west of Cold Spring Harbor. When Rhoades was considering leaving his U.S.D.A. position at Arlington Farm in Washington, D.C., to accept a faculty appointment at Columbia, Blakeslee had assured him that there would be facilities for growing his corn at CIW/CSH.[41] Within the week of receiving Blakeslee's invitation to spend the summer at Cold Spring Harbor, McClintock replied that she would be pleased to have the opportunity to work in his department and that she would communicate with Rhoades about space for planting corn.[42] The *Maize Genetics Cooperation News Letter 15* of April 1, 1941 (p. 49. 3c; Kass et al. 2019), announced that M. M. Rhoades and B. McClintock, along with [H. J.] Muller, [S.] Wright, [B. R.] Nebel, and other geneticists, "will be at Cold Spring Harbor this summer."

40 Letter A. F. Blakeslee to B. McClintock, February 5, 1941, CSHA.
41 Letter A. F. Blakeslee to M. Rhoades, May 31, 1940, CSHA.
42 Letter B. McClintock to A. F. Blakeslee, February 10, 1941, CSHA.

Several months before receiving Blakeslee's invitation to spend her summer at CIW/CSH, McClintock learned that he would retire the following year (1941) and that Demerec was a contender for the Director's position.[43] She had been looking for an opportunity to find another job. Would Demerec's appointment as Director lead to an offer with his group? McClintock had known Demerec from their student years together at Cornell,[44] where he had been a graduate student of maize genetics under Rollins Adams Emerson (1873–1947) in the Department of Plant Breeding. McClintock had learned plant genetics in Emerson's department, while a student of Lester W. Sharp (1887–1961) in the Botany Department. When McClintock was consulted regarding Blakeslee's replacement, she "pumped for Demerec."[45] She thought he would make a good director because he seemed to be "honest and sincere," which she believed was "a great necessity for cooperation." These were the same qualities afforded to Emerson, the founder of the Maize Genetics Cooperation at Cornell, and who had supported McClintock on a Rockefeller Foundation grant, in his department from 1934 to 1936 (Chapter 5). During the discussion regarding a new director, McClintock would not have been aware that by June 1941, the Department of Genetics at Cold Spring Harbor might soon be terminated. Within a month of receiving Blakeslee's February 1941 invitation to spend the summer at CSH, McClintock must have given serious thought to her alternatives, because shortly thereafter, she told L. J. Stadler, the head of her research group at Missouri, that she had "decided to quit at the end of the term."[46]

McCLINTOCK CONSIDERS A LEAVE OF ABSENCE

Before making a final decision to leave her job, McClintock was determined to clarify the permanence and responsibilities of her position at the University of Missouri. If Stadler should leave, would his genetics group be eliminated as rumored? Had her department chair recommended her promotion and would the dean accept his recommendation?[47] Would she receive a salary raise commensurate with her qualifications? Her concerns were valid, because, as described above, only one year earlier (1940) Stadler was offered an appointment at California Institute of Technology [Caltech], and the President of the University of Missouri had recommended the liquidation of the "Genetics Project" if Stadler did not return. Missouri retained Stadler by offering him his highest priority – a reorganization of the administration of his program – which was tied to a large salary increase. The former would insure a strong commitment to the development of his genetics research projects.[48] Stadler did not accept the Caltech appointment, but McClintock was convinced that he was considering other offers and might leave in the future; W. C. Curtis, Dean of the College of Arts and Sciences, would shortly confirm her suspicions. Furthermore, McClintock anticipated that she would not be promoted because she believed she was not needed at Missouri.[49]

On Friday, April 18, 1941, McClintock met with Dean Curtis, to clarify the responsibilities of her position.[50] Curtis had been named Acting Dean in September 1939 and was permanently appointed

---

[43] At the AAAS meetings in December 1940, B. McClintock became aware that A. F. Blakeslee would retire the following year, Letter B. McClintock to M. Rhoades, May 16, 1941, MMR. FBH, September 25, 1941, RF.

[44] In 1923, M. Demerec and B. McClintock were concurrently members of student organizations such as Synapsis Club, Sigma XI, and Cosmopolitan Club, *Cornellian* (1923).

[45] Letter B. McClintock to M. Rhoades, May 16, 1941, MMR. After a long search for a Director that began in May 1940, one possible decision was to close the Department of Genetics at CSH; the decision to appoint an Acting Director postponed this possible outcome until the former had proved himself to be a successful Director, "Genetics Director Selection 1941," File, May 1940–October 1941, CIWDC.

[46] Letter L. Stadler to M. Rhoades, March 18, 1941, MMR.

[47] L. Stadler had written to M. Rhoades that C. M. Tucker, her department chair, had recommended her promotion. L. Stadler to M. Rhoades, March 18, 1941, MMR.

[48] Letter M. F. Miller to M. A. McCall, June 27, 1940, LJS, SHSMO-Columbia.

[49] Letter B. McClintock to C. Burnham, September 16, 1940, CRB; Letter B. McClintock to M. Rhoades, April 28, 1941, and May 16, 1941, MMR.

[50] McClintock to Rhoades, April 28, 1941, and McClintock to Rhoades, Monday night [April 21, 1941], MMR.

in April 1940 (Curtis 1949, p. 147).[51] She summarized the outcome of their meeting in a letter to her confidant Marcus Rhoades. She wrote that she had "applied for a leave of absence for next year," and continued:

> I had a long talk with the Dean (Curtis) ... on Friday ... I wanted my situation clarified officially. It is rather dark. He wouldn't make any statement about possibility for advancement nor would he state whether I would be retained if Stadler left. If I did go on permanent tenure and Stadler left, he would'nt [*sic*] state what they would do with me other than that my duties would be changed. I have concluded definitely I will have to get something else to do. This job, regardless of permanent tenure, would certainly kill my vitality because of the attitude the University has taken. Even permanent tenure is no inducement. When one is definitely not desired, it is a great handicap to one's enthusiasm.[52]

Learning that McClintock had requested a leave of absence for the coming year, Rhoades promptly arranged for her to spend it as a guest investigator in his Department of Botany, at Columbia University during the academic year 1941–1942, "with the privileges of the laboratory and of the University."[53] In responding to Rhoades' news of the forthcoming invitation to Columbia, and of a possible opportunity at Cold Spring Harbor [if Demerec were appointed Director of their Department of Genetics], she remarked that it sounded "very encouraging" but she hoped that "the female element won't be a barrier as it has been so much so in the past."[54]

McClintock's meeting with Curtis had also confirmed her suspicion that Stadler was again considering leaving, and she wrote to Rhoades on May 16, 1941, that Stadler had offered to take her to CIW/CSH if he got the job as Director of the Department of Genetics. "Stadler was anxious to get the job for himself," she confided to Rhoades. She felt certain that he would like "a much more influential position with more prestige." Nevertheless, McClintock expressed her misgivings about Stadler, asserting, "I shall never trust him again."[55]

During a conversation in May with Frank Blair Hanson, Associate Director of The Natural Sciences at the Rockefeller Foundation, Curtis told of his meeting with McClintock. Hanson noted this interpretation:

> McClintock ... has got the idea that the large increase in salary which it was necessary to give to Stadler in order to keep him at Missouri worked to prevent her from being promoted to an Assoc. Professorship, and she has requested a year's leave of absence, stating that there was a high probability she will not return to the University if she can find another post. ... [The administration recognizes] that they have

---

[51] *Missouri Alumnus*, July–August 1939, p. 5, and April 1940, p. 10, SHSMO-Columbia. Curtis, Dean of College of Arts and Sciences, *Official Directory*, 1940–1941, Columbia: U. of Missouri, SHSMO-Columbia.

[52] Letter B. McClintock to M. Rhoades, Monday night [April 21, 1941], MMR. McClintock may not have been aware that W. C. Curtis had earlier written a letter of recommendation to Carnegie Institution regarding L. Stadler's qualifications as an administrator for the Cold Spring Harbor Director's position, W. C. Curtis to V. Bush, September 18, 1940, File, "Genetics Director Selection 1941," CIWDC.

[53] Letter M. Rhoades to G. B. Pegram, May 23, 1941, MMR. Letter G. B. Pegram to B. McClintock, June 2, 1941, MMR.

[54] Letter B. McClintock to M. Rhoades, May 26, 1941, MMR. Diary entry, May 5, 1941, "If [M.] Demerec's appointed ... he has promised to work hard to have [B.] McClintock given an appointment," Diaries of Franz and Sally Hughes Schrader, Columbia University Archives [hereafter FS/SHS].

[55] Letter B. McClintock to M. Rhoades, May 16, 1941, MMR. Gerald Neuffer, a student of Stadler, recalled that many persons believed that L. Stadler was a manipulator and that B. McClintock had a love-hate relationship with Stadler. Neuffer believed that she was afraid of putting her trust in Stadler, G. Neuffer interview with L. Kass, May 28, 1997. Also see Poehlman's (1988, p. 64) portrayal of Stadler: "With his ... suave, persuasive manner," Stadler would invite faculty, and students to cooperate in "genetic experiments that, to their later dismay, utilized their time and resources at the expense of their own research." L. Stadler's associate and B. McClintock's colleague at Missouri, George Sprague said, "The Missouri corn group can't say anything bad about Stadler," but persons outside that group would disagree with their evaluation, L. Kass and R. P. Murphy, telephone interview with George Sprague Sr., August 7, 1998. The Director's search, organized by the Carnegie Institution of Washington's Trustees Standing Committee on Biological Sciences, had expressed doubts about Stadler as early as October 1940, due to lack of "generosity toward his staff," and by April 1941, he was eliminated from consideration as Director of the Dept. of Genetics, CSH; File, "Genetics Director Selection 1941," CIWDC.

> one of the outstanding cytologists of the world but ... they will probably also have a very difficult personality on their hands if she returns to the University.[56]

McClintock's request for a leave was acted upon quickly. Records in the University files, dated May 9, 1941, state that it was ordered, upon the recommendation of the Dean of the Faculty of the College of Arts and Science and the President of the University, that McClintock be "reappointed as Assistant Professor of Botany without salary for the year September 1, 1941 to August 31, 1942; and that she be granted leave of absence without salary" for that same period of time.[57] Although Curtis made no promise to McClintock for advancement, it is evident, from a note he made the following month, while negotiating her one-year replacement, that he was aware that she was eligible for promotion.[58]

REACTIONS TO McCLINTOCK'S REQUEST FOR A LEAVE

Papers deposited in University of Missouri Archives indicate that the University administration presumed that McClintock would return to her position following a leave of absence for 1941–1942, if she did not find another appointment. The University *Directory* showed that she was "on leave without salary 1941–1942 academic year," and the 1942–1943 University Course *Announcement* showed that she was scheduled to teach all of the courses she had offered before she began her leave (Table 6.2).[59] Additionally, the correspondence of her contemporaries, and communications with Stadler, indicates that she had "not finally decided against returning to Missouri," following her year's leave.[60]

Curtis' account of why McClintock requested a leave of absence, it seems, differed from Stadler's understanding that she wanted to leave because her job lacked permanence. Stadler did not wish to see McClintock leave permanently, but he appeared concerned about her expressed dissatisfaction with her University of Missouri appointment. When McClintock told Stadler she was planning to quit, he communicated with Rhoades, "in strict confidence," about finding a position for her at "Columbia, CSH, or New York Botanic Garden."[61] Stadler wrote that McClintock had never been happy at Missouri and was completely "fed up with the job." He believed, "There is nothing in the local situation except her own feeling to make it necessary for her to move." He told Rhoades that McClintock was "definitely slated for a promotion this spring, and [C. M.] Tucker [who had replaced Robbins as chair of the Botany Department] has told her so." He continued, remarking that McClintock had expressed concern that the job was "not permanent" ... and if "she can't get established in a research job that she considers permanent, she'd rather just quit now and go into something else." Stadler reiterated McClintock's comments that the Missouri "genetics project might be closed in time of stress" and that "research jobs are a luxury to a University." He further qualified that at the time she was hired, the University gave "official assurance that the research jobs would be just as permanent as teaching appointments." He conjectured: "Presumably the promotion this year would make her an associate professor, which is the grade here at which permanent tenure becomes automatic."[62] The "crux of the matter is that she is not happy here," he concluded.

---

[56] FBH, May 5, 1941, RF.

[57] Curators' files, May 9, 1941, UMO. McClintock's appointment cards reveal that she had no increase in salary but also there is no indication that her salary was cut, University of Missouri appointment cards, UMO.

[58] Letter W. C. Curtis to Alvin Johnson, May 24, 1941; a final comment at the bottom of Curtis' note reveals that he was aware that B. McClintock was eligible for tenure, LJS.

[59] Records, course *Announcements*, *Directories*, and State Records, SHSMO-Columbia.

[60] See correspondence in the following: W. C. Curtis in BD, PO, LJS; M. Demerec, CSHA; Goldschmidt, Richard Goldschmidt papers, Manuscript and Map Collections, The Bancroft Library, University of California, Berkeley, California; W. M. Myers, M. Rhoades, MMR; C. Stern, CS; L. Stadler, LJS. L. Stadler to M. Demerec, March 30, 1942, CSHA.

[61] Letter L. Stadler to M. Rhoades, March 18, 1941, MMR.

[62] L. Stadler's statement to M. Rhoades implies that the University of Missouri was cognizant of the American Association of University Professors (AAUP) guidelines and followed their guidance for promotion and tenure. A review of the *Bulletins of the AAUP* revealed that the University of Missouri had an AAUP chapter, and their records show that the University followed AAUP guidelines for Promotion and Tenure in 1938. AAUP Archives, Gelman Library, George Washington University, Washington, DC [hereafter AAUP].

Learning that McClintock might leave Missouri, Stadler immediately began negotiations to replace her for one year.[63] All negotiations were contingent on McClintock returning to Missouri for the 1942–1943 academic year. Neither Stadler nor Curtis believed they could find an American scientist to replace McClintock for one year at the $3,000 salary they were prepared to offer. They therefore considered scientists who had been driven from their native countries, by the war in Europe, to seek refuge in neutral territories. In March 1941, upon learning of McClintock's decision to quit her job at Missouri, Stadler invited Emil Heitz (1892–1965), to come to Missouri as a Visiting Professor. Heitz and his family had been forced to leave Germany because of their Jewish ancestry, and they were then living in Basel, Switzerland (Zacharias 1995). Heitz, also a world recognized cytologist, had worked with both plant and animal tissues, and was responsible for important contributions on salivary gland chromosomes in dipterans. Heitz and Bauer's (1933) morphological descriptions of salivary gland chromosomes in larvae of the March fly, *Bibio hortulanus* L., preceded, supported, and apparently inspired Painter's (1933, 1934a,b) cytogenetic investigations of these chromosomes in *Drosophila*. Heitz agreed to become a Visiting Professor of Botany at the University of Missouri for one year effective September 1941, at a salary of $3,000.[64] Dozens of communications between Heitz and the University administration indicate, however, that he postponed his acceptance of the appointment until the "end of hostilities" because of problems securing a visa, his fear of internment upon entering the United States, and lack of the University's commitment to a permanent appointment. He had also earlier expressed concerns to a friend of McClintock's about learning of "feelings against Jews in some American Universities."[65] Indeed, such a concern was justified. Prior to starting his job at Columbia University in 1940, Rhoades was advised not to accept Jewish graduate students, when McClintock had recommended a student from the University of Missouri.[66] All negotiations with Heitz were based on the assumption that McClintock would return to Missouri following her leave of absence.

CONSIDERATIONS FOR CHANGING JOBS

General Milieu

McClintock's belief that the "female element [had been] a barrier" in the past may have loomed large in her desire to leave her academic position.[67] McClintock remembered, more than 30 years later, that when she was hired at Missouri, she was warned that if she were to marry she would lose her appointment.[68] During an interview with Provine and Sisco (1980), she recalled not desiring a University appointment because she did not want to be part of the "constant exclusion" that she felt both at Caltech and then at Missouri. One letter she received and responded to implied that some members of the Missouri genetics group held condescending attitudes toward women.[69] She also recalled being excluded from University faculty and committee meetings (Provine & Sisco 1980; Keller 1993, p. 81), but apparently, this exclusion was not confined to her alone; as of 1939, University of Missouri Policy did not include Assistant Professors in membership in the University Faculty,[70] and evidently, they were not invited to faculty meetings. Although McClintock did participate in

[63] Letter L. Stadler to E. Heitz, March 20, 1941, LJS; L. Stadler to C. Stern, May 26, 1941, CS.

[64] Letter L. Cowan to E. Heitz, June 21, 1941, LJS.

[65] Letter E. Heitz to W. Curtis, January 31, 1942, 1941–1942, LJS. Letter C. Stern to B. McClintock, January 27, 1940, CS, APS.

[66] Letter M. Rhoades to B. McClintock, May 31, 1940, MMR.

[67] Letter B. McClintock to M. Rhoades, May 26, 1941, MMR.

[68] When I asked her directly who gave her that warning, she did not wish to identify the person [but it may have been W. Curtis or L. Stadler], McClintock (pers. com., 1973).

[69] Letter B. McClintock to M. and V. Rhoades, November 21 [1938], MMR.

[70] Letter W.C. Curtis to S. Woodson Canada, [Registrar] March 7, 1939, Committee on University Policy recommends "that membership in University Faculty include assistant professors who have been on the faculty as assistant professors for three years or more," BD. I found no record when or if the Board of Curators acted on this recommendation.

and contribute to departmental and University-wide professional activities, at the request of her Department Chair,[71] her feelings of exclusion, in part, may have been justified.

Dale R. Corson, one of McClintock's former colleagues at Missouri and a future President of Cornell University, recalled that female faculty members were unwelcome in the University of Missouri Faculty Club with their male colleagues. Male faculty shared meals and conversations in the atmosphere of this fraternity, but women faculty did not have a place of their own to socialize.[72] One campus organization that would have encouraged and supported women was Sigma Delta Epsilon (Graduate Women in Science, GWIS), whose Delta Chapter was founded at Missouri in 1924 (Robbins 1971, p. 7). For a variety of reasons, this was an organization in which McClintock may not have felt welcome. The Alpha Chapter was founded at Cornell in 1921 for the purpose of fellowship for women in science, although membership was by invitation only, and their records show that McClintock had not been invited to join. That organization was apparently antisemetic (excluded Jewish students) in those early days and had McClintock been asked, she may have declined. As a freshman, she had broken her Delta Zeta sorority pledge when she became aware of their discriminatory practices (antisemitic implied).[73] Similarly, she may not have been inclined to join the Delta Chapter of Sigma Delta Epsilon, at Missouri in 1941. Moreover, Professor Guthrie, who presumed McClintock as rival, was a charter member of that chapter and this alone may have precluded McClintock's association.[74]

The exclusion of women from male gatherings was not unique to the University of Missouri, but seems to have been part of accepted practice regarding women faculty in most universities during that era (Rossiter 1982, p. 214; 1995, pp. 141–144).[75] In contrast, McClintock and other women were welcome members of Cornell's Synapsis Club, the Plant Breeding Department's student/faculty, scientific/social organization (Murphy & Kass 2007, 2011). While a member of Synapsis Club, McClintock had been an active graduate student participant – attending weekly meetings and joining committees – and later, she was elected the treasurer of the organization.[76] Consequently, as is indicated in her letter to Rhoades, McClintock would have favored an opportunity to reside at a research institution, such as CIW/CSH, where she hoped her gender would not be a matter of concern.[77]

Another accepted practice in Missouri, and other formerly slave-holding states, which may have disturbed McClintock, was a university policy that African-American students were discriminated against solely because of their race. Although Black students filed lawsuits to gain entrance to the University of Missouri, they were denied admittance because of state claims that separate but equal education was available at Lincoln University for Negroes at Jefferson City, Missouri. In December

[71] There is much correspondence in BD to demonstrate B. McClintock's participation in department and University activities.

[72] Dale Corson, interview with L. Kass, April 30, 1998. At the University of Missouri, Columbia, Dale R. Corson was Assistant Professor of Physics (1940–1943), Associate Professor of Physics (1943–1945), on leave academic years 1941–1942, 1943–1945, resigned 1945, *University of Missouri Catalogs*, 1940–1941 through 1946–1947, SHSMO-Columbia. Dale Corson was President of Cornell University from 1969 to 1977.

[73] Keller 1993, p. 33; the Cornell University yearbooks confirm that B. McClintock broke her sorority pledge. Also see note regarding antisemitism for Sigma Delta Epsilon at Cornell in Chapter 2. Professor Alfred Kahn also perceived discrimination against Jews while a graduate student at Missouri during the academic year 1937–1938, L. Kass interview with Fred Kahn, June 10, 2004, Ithaca, NY. The Cornell Chapter of Sigma Delta Epsilon was open to everyone when I became a member in the early 1970s.

[74] The Delta Chapter of SDE was installed at Missouri on December 13, 1924; see Petition from University of Missouri to form a chapter signed by M. J. Guthrie, SDE. Guthrie is reputed to consider McClintock a rival, as described below.

[75] Cornell's Faculty Club excluded women until the 1970s; Lee B. Kass, Cornell University graduate student and teaching assistant, 1971–1975, personal observation; Professor Natalie Uhl, interview with L. Kass, July 13, 1995; President Emeritus Dale R. Corson, interview with L. Kass, April 30, 1998. See also Rossiter (1982), especially Chapter 7.

[76] Synapsis Club Record Books, PB; now deposited in the Cornell Archives; see Chapter 5.

[77] Letter B. McClintock to M. Rhoades, May 26, 1941, MMR. Note that A. F. Blakeslee had employed Sophie Satina in his Department of Genetics, at CSH/CIW, when she was having difficulty finding a job, PB; Sophie Satina *AMS* 1938, p. 1235; Sharp (1934, pp. 512–513) cites eight publications by S. Satina and A. F. Blakeslee – beginning with 1925.

1938, the Supreme Court of the United States ruled [*Gaines v. Canada*] that one of these students, who wished to study law, had to be admitted to the University "until a satisfactory law school is provided … for Negroes …."[78] It was customary and expected at that time that students attach photographs to their undergraduate and graduate school admission applications. McClintock may have been expressing her objections to having to play a part in this discrimination when, in early February 1940, she wrote to Stadler (at Caltech) that the applications from students had come in and she had "done the deeds to see their faces on the fellowship blanks," implying that none of the faces were black. Only months earlier, the case justifying admission of an African American to the Missouri Law School was dismissed because the plaintiff, Lloyd Gaines, had disappeared. It has been conjectured that he had been murdered in Chicago on March 19, 1939, or became disillusioned with the case. The Gaines standing helped the NAACP lay the foundation for the Supreme Court's landmark 1954 unanimous ruling in *Brown v. Board of Education*, holding that separate facilities for public schools [as previously determined in Plessy v. Ferguson (1896)] were inherently unequal (Brown v. Board 2004).[79]

Problems with University Administration

W. M. Robbins was Head of the Botany Department and Dean of the University of Missouri Graduate Faculty when McClintock joined his department in 1936. He was also a Cornell graduate and seemed to have a good working relationship with McClintock. He cordially arranged for her to obtain funds to attend meetings, recommended her publications to other researchers, and extended her summer leave at Cornell University to continue a research project.[80] His correspondence with colleagues shows that he recommended women for faculty and research appointments and he was disappointed with colleagues who would not consider them.[81] Within the year, however, Robbins accepted the post of Professor of Botany at Columbia University and assumed Directorship of the New York Botanical Garden. He left Missouri in February 1938, and was replaced as Chair by C. M. Tucker, a member of his department, who was recently promoted to Professor of Botany.[82]

Tucker's correspondence with McClintock was not as cordial as was Robbins'. His regard for women may be reflected in a letter of recommendation that he wrote for a recent male graduate:

---

[78] *Missouri Alumnus*, "U. S. Supreme Court Rules Missouri Must Admit Negro," December 1938, Vol. 27, p. 8; "State Supreme Court Denies Negro's Appeal," December 1937, Vol. 26, p. 8; "Negro Girl Again Seeks To Enter University," January 1942, Vol. 30, p. 9, SHSMO-Columbia. NOTE – Missouri had entered the Union in 1820 as a slave state, a compromise to Maine's entrance as a free soil state, and slavery was prohibited in the Louisiana Purchase north of latitude 36° 30′ N EXCEPT for Missouri [The Missouri Compromise]. The Kansas-Nebraska Act of 1854 annulled the Missouri Compromise and permitted those states' self-determination on the question of slavery. The U.S. Supreme Court's "Dred Scott decision" in 1857, denying freedom to Missouri slaves, Dred and Harriet Scott, on the grounds that slaves had no rights to sue, caused much dissension in Missouri and is thought to be one of the contributors to the American Civil War, 1861–1865. Although Missouri had entered that war on the side of the Union, they seceded from it in October 1861, but continued to send regiments to fight in both Union and Confederate armies. The controversy over equality for Negroes as reflected in the 1930s battles for equal rights, and continuing through the 1960s, ultimately led to the U. S. Civil Rights Act of 1964, but this struggle continues even today.

[79] Letter B. McClintock to L. Stadler, February 9, 1940, LJS; *Missouri Alumnus*, "Court Dismisses Gaines Case," January 1940, Vol. 28, p. 5, on the same page of January 1940; see "Faculty Notes," L. J. Stadler granted a leave, February 1 to May 31, 1940, SHSMO-Columbia; Garrison, Chad (April 4, 2007), "The Mystery of Lloyd Gaines." *The Riverfront Times*. St. Louis, Missouri. See Endersby and Horner (2016) on the subject of the Gaines story.

[80] There are many memos in BD files that record these negotiations between W. J. Robbins and B. McClintock; *e.g.*, W. Robbins granted B. McClintock time to complete her summer research project at Cornell and assured her that she would not have to rush back to begin the fall term, Letter W. Robbins to B. McClintock, September 13, 1937, BD.

[81] Letter W. Robbins to B. P. Kaufmann, Univ. of Alabama, January 20, 1937, Kaufmann was looking for a "man" (it's not known if Kaufmann's attitude changed when he was appointed Investigator at Dept. of Genetics, Cold Spring Harbor, beginning in 1937); Letter W. Robbins to A. S. Pearse, Duke University, January 15, 1937 (the female assistant was employed at Duke University), BD.

[82] W. Robbins to C. N. Frey, January 3, 1937 [*sic* 1938], BD. *Missouri Alumnus*: October 1937, p. 10; August 1937, p. 4, SHSMO-Columbia.

> I should mention that his wife ... secured a doctorate here. ... She is admirably suited by inclination and temperament to be a decided asset to her husband. I mention this because I have found that the attitude of a wife is an important factor in the success or failure of a man engaged in creative work of any sort.[83]

Dean W. C. Curtis, who initially supported, encouraged, and arranged McClintock's invitation to Missouri, did not seem to be prejudiced against women. He had female graduate students and recommended them highly for appointments at colleges and universities. Curtis respected the work of Dr. Mary Jane Guthrie and recommended her for Rockefeller Foundation research grants (RF). They co-authored and revised a zoology textbook (Curtis & Guthrie 1927–1947), and he recommended her promotion to Full Professor in 1937. Guthrie was a recognized leader in her field, having been awarded a star in zoology in the sixth edition of *American Men of Science* in 1938 (p. 591). McClintock's former colleagues at Missouri suspected that Guthrie's influence on Curtis played a role in his negative attitude toward McClintock. Although Guthrie succeeded Curtis as Head of the Department of Zoology in 1940, when Curtis became Dean of the College of Arts and Sciences Faculty (Curtis 1949, p. 147), she is reputed to have "viewed McClintock as a rival."[84] George Sprague Sr., who worked in Stadler's group with McClintock, recalled that Stadler and Mary Jane Guthrie were good friends (Figure 6.1). He thought the latter might consider McClintock encroaching on her territory – a view that can be supported by some evidence. In 1940–1941, McClintock's new Cytogenetics course replaced the Advanced Cytology class formerly taught collaboratively by Guthrie, McClintock, and Stadler (Table 6.2).

At Woods Hole Marine Biological Laboratory, during the summer of 1940, Curtis discussed his concerns regarding McClintock with F. B. Hanson. Curtis was convinced that she was a "trouble maker" and he attributed disparaging remarks about Stadler to her. Although this description may have been unfair, Curtis clearly hoped "an offer may come her way so that she can have her career elsewhere than at Missouri."[85] Almost a year later, during negotiations for McClintock's one-year replacement, Curtis communicated to President Middlebush that Tucker, Chair of Botany, preferred Heitz to McClintock because Heitz was "more distinguished" and the staff members would "find him helpful and interested not only in problems of genetics but also in other fields of botany."[86] Tucker's remark about Heitz's interest in botany implied that McClintock may have felt a closer connection to the genetics group than to the botanists. In 1937–1940, McClintock's lab and office were moved from Lefevre Hall, home of the Botany Department, to room 112 of the newly constructed Conservation Building (now Stephens Hall), just east of her previous location. This separation may have contributed to the perception that she was disinterested in Botany Department matters.[87] Graduate students (Table 6.3) and visiting researchers (Table 6.4) had little difficulty engaging McClintock in conversations (Crouse 1992, pp. 25–32; Clark 1940, 1941). McClintock frequently spent hours helping others with their work, although she often asserted that these interruptions distracted her from her research. By mid-1940, most of the members of the genetics group had moved into the newly completed Genetics Building (now Curtis Hall), where McClintock had

---

83 Letter C. M. Tucker to J. G. Horsfall, Conn. Agr. Exp. Station, October 12, 1939 (the "wife" is Mary Bartley, who married C. D. Schmidt, whom Tucker is recommending for a job), BD; compare letters of recommendation from W. J. Robbins and from C. M. Tucker in BD.

84 In 1989, E. R. Sears recalled Guthrie's influence on Curtis, Letter E. Sears to A. Chute, May 11, 1989, Ernest R. Sears papers, SHSMO-Columbia [hereafter ERS]; Sears joined Stadler's group in 1936; see Table 6.1 and Figure 6.1, and also *AMS* 1938, p. 1264. Letter B. McClintock to M. Rhoades, November 1, 1940, MMR. L. J. Stadler's daughter recalled stories that Guthrie viewed McClintock as a rival, Joan Stadler Martin, telephone interview with L. Kass, August 26, 2001. L. Kass and R. P. Murphy telephone interview with George Sprague, Sr., August 7, 1998.

85 FBH, July 11–August 31, 1940, RF.

86 Letter W. Curtis to F. Middlebush, May 17, 1941, LJS.

87 I thank Ed Coe for this insight. Office locations are documented by the *Official Directory* 1936–1939, Univ. of Missouri, SHSMO-Columbia. Two other members of L. Stadler's research group, previously housed in Lefevre Hall, were also moved to the Conservation Building at that time. B. McClintock was moved from "204 Lefevre" to "112 Conservation Lab." L. Stadler's office was with Field Crops in Waters Hall.

a large three-room laboratory on the top floor (Room 302), immediately above Stadler's office.[88] Her location, now farther removed from the Botany Department, may have precipitated Tucker's comment.

Teaching and Advising Responsibilities

McClintock had been told she was "hired for research," yet every year she worked at Missouri she was committed to lecturing in graduate classes, advising graduate students, and coordinating student and faculty seminars (Tables 6.2 and 6.3). During the summers of 1937–1939, McClintock conducted research at Cornell, while also guiding the work of University of Missouri graduate student Frances J. Clark (Figure 6.1, Table 6.3).[89] At the 1938 GSA meetings, Clark reported on work conducted at the University of Missouri and at the Connecticut Agricultural Experiment Station (CAES), New Haven, Connecticut. She eventually transferred her project to the latter (Clark 1939). She was awarded her doctorate at Yale in 1941 (*AMS* 1944). Part of Clark's thesis work was published in April 1940 in the *American Journal of Botany* on research that she had conducted at the University of Missouri, Cornell University, and CAES (Clark 1940, 1941). She acknowledged Stadler for use of facilities [at Missouri] and McClintock "for much valuable help" in that work.

As mentioned previously, during McClintock's last year in residence at Missouri (1940–1941), she developed and taught (winter term 1941) the first graduate course offered in "Cytogenetics" at the University. Stadler claimed it was the "best graduate course [he] had ever known."[90] Her meticulously prepared lecture notes for this class were completely written out and included full contemporaneous bibliographies and lists of slides for microscopic demonstrations and for visual aids.[91] McClintock herself processed all the demonstration slides for the course. Anyone who has prepared class lectures would know from examining her notes that writing these lectures consumed hours of her time. McClintock wanted to restrict this class to a few graduate students who needed the course. She declared, in a letter to Rhoades, that there was always a "row of visitors" in her classes and "it always made it so much harder for [her]." She reasoned that keeping out visitors would enable her "to talk more informally to the students" and they would "feel freer to express themselves."[92] Several weeks later, she reported that although she wanted to "hold the group down to six students," her "sympathies were worked upon" and she now had 11 students, and [Carl. P.] Swanson, on a traveling fellowship from Harvard, was "sitting in" (Table 6.4; Swanson 1988).[93]

Swanson had learned about McClintock's work while conducting his doctoral investigations on the genus *Tradescantia* (Swanson 1940). He had observed inversions, chromosome bridges, dicentric chromosomes, and acentric fragments during meiosis in his plants. He explained these results by comparing similar phenomena previously published by McClintock (1938c) in maize. In McClintock's Cytogenetics class, Swanson learned more about chromosome behavior first-hand, while completing his doctoral degree (Swanson 1941). Years later, Swanson would feature, in his widely acclaimed textbook *Cytology and Cytogenetics* (1957), McClintock's studies on the breakage-fusion-bridge cycle that she had conducted at Missouri.

---

[88] The Genetics Building was also called the Biochemistry Building and is currently named Curtis Hall (Poehlman 1988, p. 64; Curtis 1949, pp. 169–170). Staff members' offices are listed in the *Official Directory*, 1940, University of Missouri, SHSMO-Columbia.

[89] See correspondence in files at the University of Missouri Archives.

[90] Cytogenetics, C. M. Tucker to H. E. Bent, February 3, 1940, BD; U. of Missouri *Announcement*, 1940–1941, SHSMO-Columbia. "Best graduate course," FBH, June 24, 1941, RF.

[91] McClintock promised her cytogenetics lecture notes to Marcus Rhoades to assist him in preparing his course at Columbia University, Letter B. McClintock to M. Rhoades, January 30, 1941, MMR. They are in the Marcus Rhoades Papers at LL: Lectures 1941, MMR.

[92] Letter B. McClintock to M. Rhoades, January 30, 1941, MMR.

[93] Letter B. McClintock to M. Rhoades, February 17 [1941], MMR; [Carl P.] Swanson reported completing his doctoral investigations with Karl Sax at Harvard University; see Swanson (1988, p. 7). Swanson's (1940) research paper would be the foundation of this Ph.D. thesis, awarded in 1941.

Research and Publication Responsibilities

McClintock believed she was primarily expected to do research and to publish her results. If her duties were to be changed, as Curtis had warned during their meeting in April 1941, she certainly would not have welcomed a change to additional teaching responsibilities, which had clearly interfered with her research. Nevertheless, she acknowledged in correspondence with Rhoades that she had difficulty writing papers and became quite distressed whenever she was forced to write about her work. She agonized over manuscripts for months before sending them out for publication. McClintock expressed frustration while reviewing proofs of the manuscript she had submitted to *Genetics* in November 1940 (published March 1941), "At my tender age [38], I will never learn to write properly (nor spell!). It is always an ordeal that throws me into a depression from which it is hard to recover for I get an inferiority complex."[94] Although later in life she recalled that she did not show her papers to others before submitting them for publication (Provine & Sisco, 1980), her early correspondence, and acknowledgments in her published papers, indicates that she sent draft manuscripts to friends and colleagues for editing prior to submission.[95] If she remained in a University position, it would be essential that she present papers at scientific meetings and publish in the prominent scientific journals of the day, but it seems she considered that even limited teaching responsibilities would prevent her from fulfilling that primary goal.

McClintock is reputed to have shunned scientific meetings in her later years.[96] The facts speak otherwise as the University of Missouri Botany Department records show that during the 1930s, she regularly attended annual meetings of the Genetics Society of America (GSA), with travel expenses covered by the University. She often presented seminars at neighboring universities, and, as previously mentioned, she participated in the 1938 GSA summer meetings held at Woods Hole Marine Biological Laboratory.[97] During her first two years as Assistant Professor at Missouri, she attended the annual GSA Christmas meetings (1936, 1937), where she contributed demonstration papers and published abstracts of the research she had conducted at Cornell between 1934 and 1936 (McClintock 1937, 1938a).[98] Her cytological studies had been supported by the Rockefeller Foundation and were conducted while she was a research assistant in the Plant Breeding Department at Cornell. By that time, Cornell's Plant Breeding and Botany Departments were physically housed in the Plant Science Building, and McClintock had easy access to the facilities of both. Her research at Cornell was inspired by the investigations she did with Stadler (McClintock 1987, pp. vii–xi) during her NRC Fellowship at Missouri between 1931 and 1933.[99] McClintock returned to Cornell during the summers of 1937–1939 to continue this project. As mentioned above, she soon published a complete account of that work, and the research that grew from the initial studies (McClintock 1938a–c). During the next three years, at the University of Missouri, she wrote three papers that continued those investigations (McClintock 1939, 1941a,b). The second paper (McClintock 1941a),

---

[94] Letter B. McClintock to M. Rhoades, January 24, 1941, MMR. B. McClintock expressed this frustration while revising her paper submitted in November 1940. Perhaps her writing and spelling difficulties were the result of her dyslexia, which she told Provine and Sisco (1980) was first recognized while she was in graduate school; note that during this 1980 interview, she used the term dyslexic, which came into wide usage in the early 1960s. It is estimated that 5 percent to 15 percent of the population suffers from some degree of dyslexia, Dr. Guinevere Eden and Dr. Thomas Zeffiro, February 16, 2001, AAAS meeting San Francisco.

[95] See B. McClintock correspondence in papers of Charlie Burnham, CRB; L. J. Stadler, LJS; Marcus Rhoades, MMR; Ralph Cleland (Ralph E. Cleland Papers, RC), LL; and George Beadle (George Beadle/Biology Department Papers, GB), California Institute of Technology Archives (CIT).

[96] Letter E. Sears to A. Chute, May 11, 1989, ERS.

[97] See also *Records of the Genetics Society of America* and the *Anatomical Record* for documentation on B. McClintock's attendance at scientific meetings. "Neighboring universities," Letter B. McClintock to M. Rhoades, January 29, 1939, MMR.

[98] Preprints of her abstracts are published in *Records of the Genetics Society of America*, 1936, 1937. "Cornell," Letter R. A. Emerson to F. B. Hanson, July 19, 1935, PB. See full list of McClintock's publications in Kass 2013ff, Volume I, Table I.

[99] Her ring chromosome work was also influenced by the research of M. Navashin, whom she cites in her papers published in 1931 and 1932, and whose work she and Harriet Creighton mentioned at the Synapsis Club reunion, 1982, taped symposium used with permission of Rosalind Morris. A paper published in 1931 by S. Ohno could not have inspired McClintock's ring chromosome studies, as reported by Comfort (2001, p. 58), because Ohno was only two years of age at the time; see Kass and Birchler (2014/2016).

a study of stability of broken ends of chromosomes, was submitted for a special volume in honor of R. A. Emerson's retirement.[100] As editor of *Genetics*, Rhoades requested from the editorial board, "approval for setting aside the page limit," which would "run approximately fifteen pages over the limit."[101] The results, therein, reported on the stability of broken chromosome ends, elucidated mechanisms for the maize breakage-fusion-bridge cycle, and would lead to her later discovery of transposable elements. She submitted the third paper (McClintock 1941b) to *Genetics* just one month before leaving for the summer Symposia at Cold Spring Harbor. This third report was a continuation of her earlier studies of homozygous deficiencies and ring chromosomes (McClintock 1938b). It was a detailed study of the mitotic BFB cycle, which she later designated the "chromosome type" of BFB cycle. McClintock admitted that she had "struggled for four months on the paper" and that it was a "hell of a paper to write."[102] It appeared in print in September 1941, four months after she began her leave of absence from Missouri. McClintock's invited Cold Spring Harbor Symposium presentation provided an excellent summary of the investigations she had conducted at the University of Missouri and at Cornell University. The Cold Spring Harbor Symposium presentations were published in December 1941 and included McClintock's contribution (McClintock 1941c). Therein, she also reported previously unpublished studies on crossing-over in chromosome 6 (the satellite or nucleolus chromosome). She showed how a dicentric chromatid could be formed and how breakage at meiotic anaphases would produce altered genetic constitutions (McClintock 1941c, pp. 75–80, Figures 4 & 5).

Promotion, Tenure, and the AAUP at the University of Missouri

Consideration of other social and political conditions during this time reveals a lack of "equal opportunity" laws and a paucity of tenure review procedures for university faculty. University faculty members were traditionally appointed on recommendations from major professors or former employers because professors were expected to find jobs for their students. Not all universities followed the American Association of University Professors' (AAUP) recommended guidelines for Tenure and Promotion as formulated in 1915, revised in 1925 and 1938, and finalized in 1940 (Joughin 1967). Historically, tenure and/or promotions were made on recommendations of department chairs to deans and university presidents (as is commonly the practice at many colleges and universities today). There was (and still is) no guarantee that a positive recommendation for promotion would result in advancement or permanent appointment. Although Tucker, McClintock's Department Chair, may have told her that he would recommend her for promotion to Associate Professor, as Stadler confidentially wrote in his March 1941 letter to Rhoades, McClintock may have been aware that deans in general and Dean Curtis in particular did not have to accept such recommendations. Curtis was a founding member of the AAUP, and belonged to the University of Missouri's AAUP Chapter; yet, he did not consistently follow their guidelines, nor did he always accept recommendations from Department Chairs. He believed that it was his prerogative to make the final decision on hiring, firing, and promotion.[103]

---

[100] Letter B. McClintock to M. Rhoades, November 1, 1940, MMR.

[101] Letter M. Rhoades to *Genetics* Editorial Board, December 2, 1940, PB, CU.

[102] Letter B. McClintock to M. Rhoades, April 30, 1941, MMR.

[103] See letters regarding W. C. Curtis' refusal to accept the recommendation of a department chair for promotion; e.g., E. B. Branson to Dean W. C. Curtis, October 10, 1941, "your recent decisions seem unjust ..."; W. C. Curtis to E. B. Branson, October 20, 1941, "I have not always followed the recommendations made by department chairmen. ... I think the mat ter involves only a difference of opinion between Chairman and Dean regarding promotion ...," all in BD. The AAUP investigated the Univ. of Missouri in 1941, and their report was published in the *Bulletin of the American Association of University Professors,* 1945 (see also Joughin 1967 and AAUPGW). The AAUP chided W. C. Curtis for his role in awarding and then retracting a tenure appointment eight months later as a result of anonymous and unsupported statements concerning a teacher's conduct. Consequently, the University of Missouri was censured in 1946 for failure to follow AAUP guidelines. For an excellent history and perspective on tenure in American Universities, see Hofstadter and Metzger (1955).

A review of the *Bulletin of the AAUP* reveals that University of Missouri faculty members were not only cognizant of but were active participants in formulating AAUP statutes. W. C. Curtis had been a charter member of the AAUP, and served on its Committee A, Academic Freedom and Tenure, from 1932 through 1937. Although he had been a faithful member of the AAUP Missouri Chapter during his tenure as a faculty member at Columbia, he became an associate member, as required by their bylaws, upon his promotion to Dean of the Faculty of Arts and Sciences, beginning in 1940. L. J. Stadler was also a member of the University of Missouri's AAUP Columbia Chapter, and served as its President in 1934 and 1935. McClintock was not a member of this AAUP Chapter; however, she may have been familiar with the AAUP guidelines for promotion and tenure. Her Department Chairman, C. M. Tucker, had been elected to the AAUP Missouri Chapter in 1939, and he, along with Curtis and Stadler, presumably was well versed in the rules and regulations governing tenure at the University of Missouri. McClintock most probably would not have been comfortable discussing the guidelines for her promotion with Mary Jane Guthrie, who by January 1941 was University of Missouri AAUP Chapter President.[104]

The 1940 AAUP Statement of Principles for Academic Tenure and Freedom was not meant to apply retroactively, and an appended interpretation states, "that all tenure claims of teachers appointed prior to the endorsement should be determined in accordance with the principles set forth in the 1925 Conference Statement on Academic Freedom and Tenure" [the latter interpreted as having a ten-year probationary period] (Joughin 1967, pp. 33–39, 175). Records at the University of Missouri have not revealed which AAUP Statement of Principles (1940, 1938, or 1925) Stadler was considering when, in March 1941, he wrote to Rhoades about McClintock's pending promotion to Associate Professor, but published and unpublished documents show that (1) the University of Missouri's Board of Curators published and approved its policy on tenure for Associate Professors and Full Professors in 1938; (2) the University assumed full responsibility for McClintock's salary as Assistant Professor beginning in 1939; and (3) in May 1941, McClintock was reappointed Assistant Professor, for one year without salary, and simultaneously granted a leave of absence for the 1941–1942 academic year.[105] I have found neither Tucker's letter recommending McClintock's promotion, nor an official letter from Curtis accepting his recommendation. Although Curtis may have avoided a commitment to promote McClintock in April 1941, a memo he wrote on May 24, 1941, concerning finding a replacement during her one-year leave of absence reveals that Curtis obviously was aware that she was eligible for promotion to Associate Professor:

> Finally it should be noted that if McClintock were staying regularly next year [1941–1942] and this question [her replacement] had not arisen, we should very properly be making her an Associate Professor at some increase in salary if money were available.[106]

The available documents indicate that (1) McClintock may have been eligible for tenure had Tucker recommended her for promotion to Associate Professor; (2) when McClintock met with Dean Curtis, he did not promise possibilities for advancement or retention, but instead clarified that if she were to be tenured, and Stadler should leave, her duties would be changed – probably from a research

---

[104] *Bulletin of the AAUP*, February 1941, Vol. XXXVII (No. 1), p. 124.

[105] (1) Miscellaneous Rules and Regulations, The University of Missouri, Columbia, Approved by the Board of Curators, August 1, 1938; also see curators cards November 25, 1932, 1/1/33; (2) budget 1939–1940 salary for B. McClintock is listed under general maintenance not Rockefeller grant [June 2, 1940], PO; (3) Curators' files, May 9, 1941. B. McClintock's rank in 1941–1942 was Assistant Professor, and she is listed as "On leave of absence (Without salary 9/1/41-8/31/42), Asst. Prof. of Botany–Regular," *Official Manual State of Missouri: 1941–1942.* A number of male faculty members at Missouri were also on leave of absence during WWII. Dale R. Corson, President Emeritus, Cornell University, recalled that while he was on leave from University of Missouri during WWII, he was promoted from Assistant to Associate Professor, "in absentia," Dale Corson, email to Lee Kass 2/16/01. The *Official Manual State of Missouri* supports his memory – Dale Corson *Assistant* Professor 1941–1942 on leave, *Associate* Professor 1943–1944 on leave, *Official Manual State of Missouri: 1941–1942, 1943–1944* [issued in alternate years]. See also *University of Missouri Catalogs* for 1941–1946, mentioned previously.

[106] "Note re E. Heitz, as per letter W. C. Curtis to Alvin Johnson, May 24, 1941," LJS. See also W. Curtis to F. B. Hanson, May 30, 1941, LJS.

professorship to an appointment that might include extensive teaching; and (3) if McClintock had been offered and had accepted permanent tenure, she might have compromised her chances of remaining in a research appointment. McClintock believed that her Assistant Professor position was insecure, but an appointment with permanent tenure would not prevent future duties encumbered with teaching responsibilities and would most probably preclude time for research. These considerations, coupled with a possibility of a job opportunity at CIW/CSH, seemingly led to her decision to ask for a leave of absence to seek employment elsewhere.

Clearly, within two years of arriving at Missouri, McClintock seriously considered finding another position. Prior to accepting her Missouri appointment, she had confided to R. A. Emerson that she was reluctant to go there in the first place.[107] Soon after arriving at Missouri, she encountered problems she had not anticipated: teaching responsibilities (Table 6.2), limited secretarial assistance for preparing manuscripts, and a general atmosphere that she was not willing to accept. When she ultimately decided to take a one-year leave to seek employment elsewhere, she mused on alternative career options such as "going over into Meteorology." Since her undergraduate days, she had pursued an interest in that area, and often thought about switching to that field (Provine & Sisco 1980; Creighton, Pers. Com. May 2, 1998).[108] She also considered using her leave to complete "a book on cytogenetics," which she and Rhoades began in 1938 at the invitation of L. C. Dunn, editor of *Genetics*.[109] That book was to be third in the reestablished Columbia University monograph series (the first of which appeared in 1937, with Dobzhansky's *Genetics and the Origin of Species*). Dunn's request was a great honor and a measure of her esteem. She also proposed writing a "usable textbook in cytogenetics."[110] She was, however, pleased with the possibility of a research appointment at CIW/CSH and preferred it to her other options. McClintock had many reasons to leave Missouri, but lack of a firm promise of promotion with tenure did not seem to be her first consideration.

SUMMER 1941 – COLD SPRING HARBOR

In McClintock's later years, after the Carnegie Institution of Washington had honored her with the title of Distinguished Service Member, McClintock recalled that one day she got fed up with her position at Missouri and packed her bags and left (McClintock, pers. com., 1973; Provine & Sisco 1980; Keller 1993, p. 83; Crouse 1992, p. 32). Her recollection may have revealed her determination of 40 years earlier; however, it does not provide the picture of what led to her decision. Her letters to colleagues, particularly Rhoades and Burnham, show that she had been contemplating leaving Missouri for quite some time, and finally requested a leave of absence to consider finding an alternative position. Creighton had no recollection of McClintock mentioning leaving Missouri permanently or leaving because she had been denied tenure – "If she had been denied tenure we would have talked about it, she said" (Creighton, pers. com. May 2, 1998).[111] Had she not eventually been offered a permanent job in Carnegie's Department of Genetics, she might indeed have returned to Missouri. In May 1941, she prepared for her leave by carefully separating and packing her personal items from her lab materials and from her books. She wrote to Rhoades that if she needed the latter, they could easily be sent to her.[112] Although she was not happy at Missouri, Stadler's and Curtis' correspondence at the time indicates that she had not yet definitely decided to leave permanently.

---

[107] Letter R. A. Emerson to L. Stadler, November 15, 1935, PB.

[108] "Going over into Meteorology," Letter B. McClintock to M. Rhoades, May 11, 1941, MMR; B. McClintock, undergraduate Registrar's Card, 1919–1923, CU.

[109] Letter B. McClintock to M. Rhoades, March 7, 1938, MMR, draft chapters of their book are at APS. B. McClintock to M. Rhoades, November 13 [1938], "I have written the larger part of the chapter on chromosome morphology," MMR.

[110] Letter B. McClintock to M. Rhoades, May 11, 1941, MMR.

[111] H. Creighton also emphasized that B. McClintock needed a job "to put food on the table." Jesse Singleton, who replaced B. McClintock in a faculty research appointment at the University of Missouri, was terminated due to financial exigency in 1954, Letter L. Stadler to G. Beadle, March 17, 1954, GB, CIT.

[112] Letter B. McClintock to M. Rhoades, May 26, 1941, MMR.

McClintock left Missouri on June 1, 1941, and had planned to drive directly to her lodgings at the Bio-Lab at CSH, but she visited first with colleagues at Columbia University.[113] During that summer, she lived in the facilities provided by the Long Island Biological Association and roomed with her former colleague Harriet Creighton.[114] McClintock's summer research assistant, Edward A. Weaver, who had recently completed the investigations for his Ph.D. under McClintock's direction at Missouri, arrived shortly after.[115] A colleague from Columbia University described McClintock's June 21st presentation at the Cold Spring Harbor Symposium as "splendid ... with a wealth of evidence ... well given too"[116] – It was work she had completed before beginning her one-year leave from Missouri and would be published in the *Cold Spring Harbor Symposia on Quantitative Biology* in December 1941 (McClintock 1941c). In their annual reports, both Blakeslee (1941) and Demerec (1941) each summarized their success in bringing prominent geneticists to Cold Spring Harbor that summer. Blakeslee (1941, p. 211) reported that among other "outstanding geneticists," Dr. Barbara McClintock of the University of Missouri was a guest investigator in his department, and grew her maize plants on departmental grounds. Blakeslee's Annual Research Report (Blakeslee 1941, p. 224) also mentioned that a member of his research group was pursuing a technique whereby the chromosomes of *Datura* may be adequately stained for comparison of paired strands at early prophase divisions. Such a technique through chromomere similarity would "determine the extent of inverted sections as Dr. McClintock has been able to do by use of special technique with maize chromosomes." Demerec (1941, p. 23) wrote that she and other guest investigators, Rhoades, Edgar Anderson, and Harriet Creighton, among others, conducted research on maize and worked cooperatively at the Bio-Lab that summer.[117] Demerec stressed the advantages of his lab's research atmosphere – emphasizing that, "scientific research cannot be carried on effectively with a rigid schedule and set working hours (Demerec 1941, p. 25)." McClintock would have been most comfortable in his unrestricted environment. Each of these maize cooperators described the results of their summer's research in *The Annual Report of The Biological Laboratory* published in 1941 (pp. 35–45).[118] Harriet Creighton (pers. com. May 2, 1998) recalled that during the summer of 1941, she helped prepare slides for Edgar Anderson's research, and McClintock taught Anderson the corn sporocyte smear technique that she had perfected.[119] McClintock (1941d) reported that her summer investigation "was made to determine the types of mutant characters which would appear when two or more homozygous mutant deficiencies were present in a plant."

In July, McClintock participated in the Thursday "Evening Lecture" series sponsored jointly by the Journal Club of the Department of Genetics and the Bio-Lab. Dr. E. C. MacDowell chaired these sessions, which were "well attended by members of the two institutions and of the nearby hospitals."

---

113 Letter B. McClintock to M. Rhoades, May 26, 1941, MMR. Bruce Wallace, a student at Columbia University at the time, recalled that she stopped there to give a seminar prior to driving out to Long Island, taped interview with Lee Kass, October 28, 1995. Wallace's recollection may be confirmed by FS diary entry, June 4, 1941 – Marcus Rhoades and Barbara McClintock visited from 12 to 12:15, FS/SHS.

114 "Investigators live in the proximity of their Laboratories" (Demerec 1941, p. 25). H. Creighton recalled that she and B. McClintock roomed together in the "fire house," H. Creighton (pers. com., May 2, 1998).

115 E. Weaver to C. Tucker, June 17, 1941, BD.

116 Diary entry, Sat. June 21, 1941, FS/SHS. The symposium was held for two weeks, June 18–July 2, 1942, condensed from 5 weeks as in previous years (Demerec 1941, p. 20).

117 Arthur Galston recalled that B. McClintock was very friendly to him while he worked as [J. T] Buchholz's student assistant in a nearby field that summer at CSH, L. Kass interview with Arthur and Dale Galston, September 21, 2000; see also *Records of the GSA* 1941.

118 Reports of Investigators at the CSH Bio-Lab summer 1941, included H. Creighton, Edgar Anderson, B. McClintock, M. Rhoades, in addition to other plant researchers, O. J. Eigsti, Norman H. Giles, and B. R. Nebel, Mabel Ruttle.

119 Letter, H. Creighton to L. Kass, February 28, 1998. Letter H. Creighton to M. Rhoades, April 30, 1941, "will be at CSH to grow corn this summer, ... would be interested to work on Papago corn with [Edgar] Anderson," MMR. Edgar Anderson was Geneticist to the Missouri Botanical Garden and Professor of Botany, Washington University in St. Louis. It seems that H. Creighton along with B. McClintock may have taught Edgar Anderson the corn sporocyte smear technique – see H. Creighton's summary in *Annual Report of the Biological Laboratory*, 1941, pp. 35–36. In 1944, Edgar Anderson (1944, p. 319) published work on *Tripsacum* using acetocarmine smears, "following McClintock's directions as to ultra-clean slides and the gentle use of heat to spread the cells."

She shared the platform with other prominent geneticists featured on the evening programs. These included Myron Gordon, a friend from her student years at Cornell; Edgar Anderson, Missouri Botanical Garden; Sewall Wright, University of Chicago; Alfred Mirsky, Rockefeller Institute; Max Delbruck from Vanderbilt University, working collaboratively that summer with Salvador Luria [Rockefeller Foundation Fellow, Columbia University]; R. F. Kimball, Johns Hopkins University; and her former colleague, A. H. Sturtevant, from Caltech.[120]

That summer, Stadler also participated in the 1941 Cold Spring Harbor Symposium and, while in New York, he contacted Frank Blair Hanson to arrange for Heitz's visit. During these negotiations, he explained to Hanson that he was "attempting to secure a post" for McClintock. He believed she would accept a teaching position, "but her strength lies in research and the teaching of graduate students." Stadler praised the graduate Cytogenetics class that she had recently offered at Missouri. He also emphasized that he did not want her decision to leave Missouri to be held against her.[121]

By the end of September 1941, during a discussion with Hanson of the Rockefeller Foundation, Demerec mentioned that he expected to be appointed the next Director of the Department of Genetics, CIW/CSH, and "was working toward the appointment of McClintock" and *Drosophila* geneticist H. J. Muller.[122] The following month, McClintock continued her work at Schermerhorn Hall, Columbia University, hosted by her friends and colleagues Marcus and Virginia Rhoades (Provine & Sisco 1980).[123] Meanwhile, Rhoades, continuing to assist McClintock in her job search, and possibly unaware of her reservations about a U.S.D.A. appointment, made inquiries to M. A. McCall, his former supervisor at the U.S.D.A., regarding the prospect of an opening for McClintock. McCall replied that there was no opening at the present time, but if a possibility opened up, it "will be brought to her attention." McCall, who, as Assistant Chief of the Bureau of Plant Industry, was in regular contact with Stadler and his genetics group, may have been mindful of McClintock's rejection of a U.S.D.A. job offer when he recounted Stadler's assertion that her University "position is open to her if she does want to go back." McCall also appended his belief that "they [the Missouri genetics group] would be very regretful if she should find another opening."[124]

GUEST INVESTIGATOR, DEPARTMENT OF GENETICS, COLD SPRING HARBOR, 1941–1942

Blakeslee officially retired as Director of the Department of Genetics, CIW/CSH, at the end of November 1941 (Demerec 1942, p. 172). Following a long and laborious search to replace the Director, on November 10, 1941, the President of the Carnegie Institution notified Demerec that he would be appointed Acting Director of the Department of Genetics. True to his word, one of his first initiatives was to invite McClintock to "join [their] group as Guest Investigator during the month of December," and he explained to McClintock that he had "recommended that a provision be made in the budget for 1942 to extend this arrangement for the duration of your leave of absence from the University of Missouri … until September 1, 1942."[125] On December 1, 1941, Milislav Demerec was officially appointed Acting Director of the Department (Hartmann 1988;

[120] B. McClintock's lecture presented on July 17, 1941, was titled: "The relation of homozygous deficiencies to mutation in maize," her affiliation address is Univ. of Missouri (Demerec 1941, p. 25). M. Delbruck and S. Luria would later share with A. Hershey the 1969 Nobel Prize in Physiology or Medicine, "for their discoveries concerning the replication mechanism and the genetic structure of viruses."

[121] FBH, June 24, 1941, RF.

[122] FBH, September 25, 1941, RF. Muller would share the Nobel Prize in Medicine or Physiology in 1946.

[123] Recollection confirmed by Diary entry, October 23, 1941, McClintock took up residency in Schermerhorn Hall, FS/SHS. McClintock worked in the facilities at Columbia University and lived in the Rhoades' home, Lee Kass interview with Virginia Rhoades, May 18, 1997.

[124] Letter M. McCall to M. Rhoades, November 1, 1941, MMR.

[125] M. Demerec accepted the recommended Acting Director appointment in November 1941, Letter A. F. Blakeslee to V. Bush, 10 November 1941, CSHA. A. Blakeslee retired end of November 1941, Demerec 1942, p 172. Letter M. Demerec to B. McClintock, December 1, 1941, Guest Investigator, CSHA.

Glass 1971). Only days before the United States was drawn into World War II, McClintock officially accepted Demerec's offer and thanked him for the "generous provisions ... offered ... for continuation of my investigations."[126] Within the week, Vannevar Bush, President of the Carnegie Institution of Washington, sent Demerec a copy of the budget for his department. Barbara McClintock was listed under the heading of Scientific Staff as a "Guest Investigator" from January 1 to August 31 [1942].[127] During this second Guest Investigator appointment at CIW/CSH, McClintock gave seminars at neighboring institutions, Fordham, Columbia, and Yale Universities, and the New York Botanical Garden. The publication resulting from her 1941 CSH Symposium presentation appeared in December 1941 (McClintock 1941c).

As a Guest Investigator at Cold Spring Harbor (1941–1942), McClintock prepared her first report for *Carnegie Institution of Washington Year Book No. 41*, 1941–1942 (McClintock 1942b, pp. 181–186), issued in December 1942.[128] After proposing the question in 1934, she had finally found a way to demonstrate fusion of broken chromosomes in non-ring chromosomes. The simultaneous introduction of broken chromosomes from male and female gametes into a zygote had allowed her to study the behavior of two newly broken chromosomes in the embryo and endosperm of maize plants. Each of the chromosomes was marked by genes she could observe phenotypically in the endosperm. She observed fusion of broken chromosomes in the endosperm (3*n*), prior to embryo development, and after formation of the diploid sporophyte. She showed that two broken chromosomes would fuse prior to replication in the embryo and undergo what she termed the "chromosome type" of breakage-fusion-bridge (BFB) cycle. In the developing diploid sporophyte, replicated broken "unsaturated" chromatid ends would fuse and undergo a "chromatid type" of BFB for a short time before it ceased, because the broken chromosome ends had "healed."

Her contribution was highlighted in the Carnegie Institution of Washington President's report for 1942 (Bush 1942, pp. 8–9), "Dr. McClintock has obtained important evidence regarding the fusion of broken chromosome ends. ... broken ends retain their capacity for fusion for a certain period of time after breakage, but after that period has passed they 'heal' and are unable to fuse again." Demerec's first report as Acting Director of the Department of Genetics provided an extensive summary of the two research problems that McClintock had continued as a guest investigator in his unit during 1941–1942 (Demerec 1942, pp. 172, 174–175). He, too, emphasized McClintock's significant findings that answered the question of whether fusions occur immediately after a chromosome is broken or whether a broken chromosome end may remain "unsaturated." McClintock had provided a convincing answer to this question by demonstrating that "two broken ends, derived from separate nuclei, could fuse." She had also observed fusions between recently broken ends of chromosomes within a dividing nucleus. If three broken chromosomes enter a nucleus, "fusions may occur between two of the three unsaturated chromosome ends." The third broken chromosome "loses its unsaturated state," and the ends "heal." Such a healed chromosome can no longer fuse

[126] Letter B. McClintock to M. Demerec, December 2, 1941, CSHA. Diary entry, November 11, 1941, Rhoades reported that Demerec phoned McClintock asking her to come for an interview at CSH. Rhoades advised her to hold out for a good salary, and SHS enters details of the negotiations as reported by Rhoades, FS/SHS. Bruce & Miriam Wallace recalled that McClintock told them that she was insulted by Demerec's salary offer and argued with him until he gave her a more reasonable salary, Bruce & Miriam Wallace interviewed by L. Kass, October 28, 1995; their recollection is substantiated by the Schrader's diary entry, November 13, 1941 – McClintock talked with F. Schrader for about an hour regarding Demerec's offer, FS/SHS. Demerec was reluctant to offer a higher salary than was being paid to the current cytologist at CSH, FS/SHS. On November 24, 1941, Sally Shrader recorded that McClintock had "gone to Cold Spring Harbor – at $300 a month for 8 months," FS/SHS. Clearly, McClintock had pushed back to the insufficient salary first offered by Demerec, and finally accepted an annual salary of $3,600, which would subsidize her through her leave of absence without pay, and was $600 per year more than she was making as an Assistant Professor at the University of Missouri.

[127] V. Bush to M. Demerec, December 12, 1941, budget for 1942, CSHA.

[128] Report to Dr. Bush 1942, CSHA; McClintock (1942b) cited her recently published work as "McClintock 1941; see bibliography," but the reprints do not include the bibliography, which is found in the *Carnegie Year* Book *No. 41*, and which lists three of her publications – 1941a–c, published in March, September, and December 1941, respectively; all publications have her affiliated with the University of Missouri.

with any other chromosome having an "unsaturated broken end." McClintock had also observed that healing may occur before union of broken chromosome ends takes place and that subsequently "no fusions occur."[129]

Demerec then summarized McClintock's second investigation involving a study of homozygous deficiencies in the terminal segments of maize chromosome 9 in sporophyte tissue. She had found that pollen grains with small missing chromosome segments (deficiencies) were functional. Seedlings harboring these small deficiencies showed modified recessive mutant phenotypic expression: "pale-yellow seedlings occur when the homozygous deficiency is short, and white seedlings occur when [the deficiency] is slightly longer." Larger deficiencies were studied by "means of the 'variegation method,' utilizing the behavior of recently broken chromosomes ... As the homozygous deficiency becomes progressively longer, the rate of development, the color of the aleurone, and the starch formation are progressively reduced except for the layer of cells immediately adjacent to normal cells."

McClintock's important contribution, which Demerec and Bush had emphasized in their 1942 reports, was completed over the summer at Cold Spring Harbor. The manuscript was submitted to *PNAS* in September and was published in November 1942 (McClintock 1942a). Her second project, which continued studies of homozygous deficiencies, was published the following year (McClintock 1943), and reports of additional investigations were published in 1944 (McClintock 1944a–c). Both projects would lead to an understanding and discovery of transposable elements (McClintock 1987, pp. ix–x).

NEGOTIATING A PERMANENT APPOINTMENT AT DEPARTMENT OF GENETICS, CIW/CSH, 1943

The pioneering research project demonstrating fusion of broken chromosomes had been completed by March 1942, while McClintock was still a guest of Demerec's department. Demerec emphasized McClintock's still unpublished results to provide confirmation of why she should be appointed permanently to the Staff of Investigators at the Department of Genetics at Cold Spring Harbor. Knowing of McClintock's dissatisfaction with the University of Missouri, Demerec took the opportunity to persuade her to become a permanent member of his Department. He was well acquainted with her accomplishments in the field of genetics, and in early March 1942, he first submitted a Memorandum to Carnegie President Bush recommending McClintock for their staff:

> ... I am confident that if an opinion were asked of Dr. McClintock's colleagues they would rate her as a first-class scientist and would place her among the first ten leading cytogeneticists of the world. ... she is primarily responsible for establishing maize genetics in the eminent position it now occupies. ... She functions to greatest advantage in an institution where there are no teaching obligations ... It can be said without hesitation that Dr. McClintock would be a great asset to this Department, and that she would replace Blakeslee's work better than any other single person. ... She would be willing to leave the University of Missouri, I feel, and the staff members of this Department concur with me, that there is a good opportunity for the Institution to acquire a first-class scientist, thus strengthening the work of one of its major departments. This we unanimously recommend.[130]

The staff members who recommended that McClintock would ably fill Blakeslee's vacancy were established researchers at CIW/CSH's Department of Genetics.[131] Earlier that summer, Curt Stern, the internationally recognized geneticist, had sent Rhoades a copy of a letter to share confidentially

[129] Years later, the healing process was explained by the formation of a telomere on the broken end of the chromosome entering the zygote (McClintock 1984). Elizabeth Blackburn and her colleagues ascertained the mechanism and won a Nobel Prize in Physiology or Medicine for the work in 2009 (https://www.nobelprize.org/prizes/medicine/2009/summary/).

[130] M. Demerec to V. Bush, Memorandum Regarding Dr. Barbara McClintock, March 12, 1942, CSHA – same memo at CIWDC.

[131] Oscar Riddle, E. C. MacDowell, B. P. Kaufmann etc., Bush to Demerec, budget for 1942, December 12, 1941, CSHA.

with "Barbara and some other good friends." Stern's original letter, sent to the Carnegie Institution of Washington, during their search to replace Blakeslee as Director, listed "in order of excellency ... ten men whom [he regarded] to be among the best botanical geneticists in this country" –– McClintock's name appeared second only to Stadler's.[132] Furthermore, in early 1942, Curtis had visited McClintock at Columbia University and encouraged her not to consider Missouri "a thing of the past."[133] McClintock was indeed respected and in demand.

Demerec immediately supplemented his Memorandum to Bush with a description of McClintock's current research project, "... a really ingenious technique for determining the time of the fusion of broken chromosomes, as described above."[134] Although Demerec encouraged McClintock to remain at CIW/CSH, by March 1942, she had not yet decided to accept his offer of a permanent position, nor had she yet resolved in her mind if it would be more advantageous to return to Missouri. Meanwhile, the Carnegie Institution was delaying action on Demerec's recommendation. Carnegie's Executive Officer, W. M. Gilbert, hastened the process by suggesting to Demerec that McClintock spend a few minutes with Bush, "to bring about a prompt decision."[135] Demerec encouraged her to talk with Bush; so McClintock flew to Washington on March 26, 1942, and had a delightful meeting with the President of the Carnegie Institution of Washington.[136] The following day, Demerec received instructions to clarify negotiations with McClintock and then to send his official letter recommending McClintock's appointment.[137] Demerec immediately informed Stadler of McClintock's reservations, "Barbara wants me to wait until she hears from you before writing my letter." Demerec added that it would be unwise to delay the matter very long, and he knew that McClintock wanted to "notify the University by April first with regard to returning" in the fall.[138]

Upon receiving Demerec's communication, Stadler immediately dashed off a handwritten response:

> I am glad to learn that you were able to secure approval of the permanent appointment at CSH. I should think it would be an ideal position for Barb and a most desirable arrangement for the laboratory. However, from her letter to me I got the impression that she had not finally decided against returning to Missouri, and I have therefore discussed with Dean Curtis and President Middlebush the question of her reappointment and the remedying of conditions here which she found objectionable. ... there is nothing I would like better than to have Barb come back to Missouri if she is willing to come back, though I hope she will stay at Cold Spring Harbor if she is convinced that would be better for her work. I spoke to her by phone this morning and she will probably discuss with you the whole question of her permanent position. It is good to know that she will be welcome at either place, and I hope she chooses the one that will be best for her.[139]

[132] Letter C. Stern to [G. Corner], April 22, 1941, attached to C. Stern to M. Rhoades, July 7, 1941, MMR. C. Stern's original list is at CIWDC. It seems that Stern's list was sent first to Corner during the search for A. F. Blakeslee's replacement at CIW/CSH; see also Curt Stern's list in order of merit, [McClintock is listed second after L. Stadler], CIW/CSH.

[133] Diary entry, January 7, 1942, FS/SHS.

[134] M. Demerec to V. Bush, March 14, 1942, CSHA. B. McClintock submitted the manuscript describing this work to *PNAS* in September 1942; it appeared in the November issue – see McClintock 1942a; for an annotated list of McClintock's publications, see Kass 2013ff.

[135] Letter W. Gilbert to M. Demerec, March 20, 1942, CIWDC.

[136] M. Demerec wrote to L. Stadler about B. McClintock's meeting with V. Bush, Letter M. Demerec to L. Stadler, March 27, 1942, CSHA. B. McClintock fondly recalled her meeting with V. Bush (Provine & Sisco 1980). W. Gilbert to M. Demerec, "McClintock came in this morning," March 26, 1942, unposted letter, CIWDC.

[137] Letter W. Gilbert to M. Demerec, March 26, 1942 unposted letter, handwritten note "not sent, covered in telephone conversation March 27, Demerec will write again," CIWDC.

[138] Letter M. Demerec to L. Stadler, March 27, 1942, CSHA. B. McClintock required to notify the University by April 1, 1942, "whether or not she will return," L. Stadler to E. Heitz, July 24, 1941, LJS. Gilbert to Dr. Bush, Memorandum March 13, 1942, B. McClintock wants to notify the University of Missouri by April first with regard to her return next fall, CIWDC.

[139] Letter L. Stadler to M. Demerec, March 30, 1942, CSHA.

MISSOURI COMPROMISE

McClintock's associates at Columbia University were kept apprised of these negotiations, and her colleagues Franz and Sally Schrader recorded Rhoades' account in their diary. In summary, McClintock had written to Stadler at Missouri, who thereafter offered her an Associate Professorship and an increase of $1,000 above her current salary of $3,000, at which she was "much pleased." After talking with Stadler, she immediately consulted Rhoades about which offer to accept, but Rhoades, wisely, only reminded her how unhappy she had been at Missouri.[140] Within the day, she had decided to accept Demerec's offer.

On April 1, 1942, Demerec formally recommended extending McClintock's Guest Investigator appointment through December 31, 1942, and requested she be appointed a regular staff member in their department beginning January 1, 1943. Demerec had an expert sense of how to foster another "prompt decision," and he added that McClintock had received word from Missouri not to make a decision until she heard from them:

> Today a letter came from Dean Curtis offering an advancement to a research associate professorship at a salary of $3,990. This position has tenure. Research facilities at the University of Missouri would match facilities available here. Dr. McClintock has told me that she would prefer to work in this Department and would accept the appointment if received.[141]

Two days later (April 3), McClintock received a letter from Bush continuing her appointment as guest investigator and recommending she be included in the 1943 budget as a regular staff member at a salary of $4,000 per annum.[142] Upon hearing of McClintock's pending appointment, one of the staff investigators remarked, "We should mark today's date with red letters in the Department Calendar!"[143] Perhaps Demerec's comment regarding research facilities was exaggerated, but his statement about her preference to remain at CIW/CSH is confirmed by comments McClintock wrote to Edgar S. Anderson, on April 12, 1942, while replying to his questions regarding corn cytogenetics. Here again, the "female element" reappears as a strong contributing factor in her desire to leave Missouri:

> When Demerec became acting-Director he invited me for the rest of my leave of absence to be a Guest Investigator (with salary, thank goodness). In the meantime he has offered me a permanent position here which I have accepted. The University of Missouri came across in good fashion to meet the offer here but I have decided to remain here rather than return to Missouri. Missouri is very nice, and I thought my job there was exceptionally good. It was a hardship to turn down the excellent equipment and conditions for the meager conditions here. However, all-in-all, I believe remaining here is the wisest thing to do – being a woman![144]

McClintock waited until the end of May 1942 to officially resign from her faculty appointment at the University of Missouri. She did not complete the optional "Reasons" for resigning. Her Department Chair, C. M. Tucker, and Dean, W. C. Curtis, were required to approve her resignation, "date effective August 31, 1942," before forwarding it to the President. Their signatures were appended on June 1, 1942, and Curtis added the requisite "Comment":

> Dr. McClintock is resigning to accept a position on the staff of the Genetics Division [*sic*] of the Carnegie Institution at Cold Springs [*sic*] Harbor, New York. We regret exceedingly that she is leaving the University since her distinguished reputation as an investigator has brought so much added reputation to our program in Genetics. It is hoped that a satisfactory replacement can be made eventually.

---

[140] Diary entry, March 31, 1942, M. Rhoades reports on B. McClintock's counter offer from Missouri, FS/SHS.

[141] Letter M. Demerec to V. Bush, April 1, 1942, CSHA. I have not located an official letter from Missouri offering B. McClintock a promotion but see letter from L. Stadler to W. Curtis, regarding an offer made to B. McClintock, June 13, 1942, LJS.

[142] Letter V. Bush to B. McClintock, April 3, 1942, CSHA.

[143] Letter M. Demerec to W. M. Gilbert, April 6, 1942, CSHA.

[144] Letter B. McClintock to Edgar Anderson, April 12, 1942, Missouri Botanical Garden Archives.

> For the present I understand that Dr. Stadler is proposing a program of visiting appointees, each one on a purely temporary basis. WCC[145]

The University accepted her resignation on June 8, 1942 (effective August 31). Five days later, Stadler wrote to Dean Curtis that McClintock's resignation was a blow to the genetics research program, and he was still concerned that they would not be able to replace her "with a person of comparable attainments at the salary which she received [$3,000]." He added that they might find someone for a one-year appointment at $3,600, which is "considerably less than [McClintock's] salary would have been if she had accepted the appointment recently offered her" [$3,990 or $4,000].[146]

On July 1, 1942, McClintock succeeded MacDowell as Chairman of the Department of Genetics Journal Club, "the informal gatherings of the scientific group and friends." She arranged joint meetings with the Bio-Lab and with the Columbia University Genetics group. Although gasoline shortages limited travel and hence interactions with the Columbia group, she succeeded in persuading many of her friends and colleagues to present seminars on their work.[147] As recorded in a colleague's diary, in November she gave a "superb seminar" at Columbia University on her methods for "producing and marking genetically, chromosomes with one unhealed broken end. ... She carried you with her so completely that you are worn out and weary as well as excited when it's over."[148]

At the end of December, Bush confirmed McClintock's regular appointment for 1943, at the Department of Genetics, CIW/CSH. He added, "Notification of such annual appointment will be sent to you regularly as long as your name appears in the budget." The department's budget for that year specifically listed McClintock's salary at $4,000 and designated an additional $1,300 for her research assistant.[149]

Paradoxically, her appointment at CIW/CSH was even more precarious than she feared would be the case at Missouri. McClintock had told Stadler that she did not wish to remain in an insecure position, and Missouri's offer of tenure rectified that situation. The appointment she finally accepted at CIW/CSH, however, was contingent on her name appearing in the annual budget. She accepted responsibilities at CIW/CSH that she had considered distracting at the University of Missouri, such as taking time to arrange research seminars, and responding to queries from colleagues. McClintock was empowered to shape her future, and she wisely considered both opportunities. She rejected an anticipated offer of tenure for a full-time research investigator appointment with no interference and complete freedom. She exchanged tenure and security at Missouri for an uncertain future at Cold Spring Harbor with freedom to pursue research without mandatory teaching responsibilities, committee work, graduate student advising, or deadlines for publications. She felt uncomfortable with this compromise, but it seemed to be the best decision at the time. It was a turning point in her career.

Gender was a strong contributor in her desire to leave Missouri, but it was not the only factor. Other considerations that played a part in her decision to leave academe were a restrictive university atmosphere, teaching distractions, an uncertain future in research (caused by lack of trust in the head of her program), and the value she placed on her freedom (Kass 2003). McClintock was not denied tenure at the University of Missouri. On the contrary, she declined their offer of an Associate Professorship with tenure and a salary of approximately $4,000, in exchange for an

---

145 Letter of resignation, B. McClintock to President, May 27, 1942, UMO.

146 Resignation, McClintock, Barbara, appointment card, UMO. Letter L. Stadler to W. Curtis, June 13, 1942, LJS. I assume Stadler is referring to the offer of a promotion to a "research associate professorship" at $3,990 described by M. Demerec in his letter to V. Bush dated April 1, 1942, CSHA, or perhaps L. Stadler is recalling the telephone conversation he had with B. McClintock, in which he offered her $1000 above her current salary of $3000, as recorded in FS/SHS diary entry of March 31, 1942. If the latter, then "considerably less" means $400 or 10% less.

147 Report to Dr. Bush, 1940–1941, 1941–1942, CSHA.

148 Diary entry, November 17 [1942], FS/SH; a seminar on her work submitted to *PNAS* in September 1942, and published in the November issue (McClintock 1942a).

149 "Notification," V. Bush to B. McClintock, December 21, 1942, CSHA; department's budget, M. Demerec to B. McClintock, Office Memorandum, January 26, 1943, CSHA.

uncertain appointment at CIW/CSH for a similar amount of money but with more independence. Almost 40 years after she left Missouri, she still expressed strong feelings of rejection: "I knew I was going to be fired sooner or later, so I fired myself" (Provine & Sisco 1980). And indeed as McClintock had feared, "research jobs" at Missouri did not continue to receive "official assurance … as permanent as teaching appointments." Statler must have believed the permanence of McClintock's job when he confided such "assurance" to Rhoades in his letter of March 18, 1941. McClintock's prophecy materialized when in 1954 Stadler explained to George Beadle that Jessie Singleton (Table 6.3), Assistant Professor of Botany at University of Missouri, had been laid off due to financial exigency.[150] In 1946, as newly appointed Chair of Caltech's Division of Biology, Beadle had asked McClintock to mentor J. R. Singleton for his doctoral research (Singleton, J. R. 1953). Upon completing his degree in 1948, Singleton was hired as a replacement for McClintock's research appointment in the Missouri Department of Botany. Singleton's termination of a research appointment may have been also on her mind when McClintock reminisced in 1980 about being fired "sooner or later."

The investigations McClintock began as a postdoctoral research fellow at Missouri in 1931 and 1932, and which she continued at Cornell and Missouri between 1934 and 1941, had great bearing on the significant research she would carry out at CIW/CSH, and for which she would ultimately be awarded the Nobel Prize in 1983 (McClintock 1984). Years later, she returned to the University of Missouri and explained to her audience at the 10th Stadler Genetics Symposium how her early work at Missouri "exposed innate mechanisms that initiate various grades of genome reorganization" (McClintock 1978). She added, "The full extent of their effects was not appreciated until the summer of 1944 and thereafter."

McClintock's desire for autonomy is summed up concisely in her reply to an invitation requesting that she return to Missouri to head the genetics project following Stadler's death in 1954:

> My present situation with the Carnegie is unique … I feel it would be difficult to acquire anywhere else the degrees of freedom that this position offers. The new President will continue the policy of no interference and complete freedom. I just go my own pace here with no obligations other than that which my conscience dictates. This seems to fit my personality rather well.[151]

RECORDS AND RECOLLECTIONS

Previous biographies and reports of McClintock's early life and career relied almost exclusively on oral histories or undocumented recollections. This has led to the unsubstantiated legends about her career at the University of Missouri and her reasons for leaving. McClintock tried to dissuade Keller from writing her biography, because she believed she was "too much of a maverick for her story to be of interest to others" (Keller 1993, pp. xxii). McClintock told other interviewers that she was reluctant to grant interviews because she might not remember events accurately (Provine & Sisco 1980; see also Bronte 1993).[152] As McClintock had feared, her life became a legend. The stories she told to Keller and to others – though not as transparent as she recalled – were embellished and fictionalized, *e.g.*, the legend of the University of Missouri denying McClintock tenure is but one example of such exaggeration.

The story reported by Keller (1983, 1993) and often repeated (*e.g.*, Kittridge 1991; Buckner 1997) is that Stadler created a faculty position especially for McClintock at Missouri in 1936, because she needed a job. She eventually left that position because she saw no chance for promotion. When

150 Letter L. Stadler to G. Beadle, March 17, 1954. GB, CIT & see above; In 1954, J. R. Singleton was appointed Assistant Professor of Cytogenetics at Purdue University, where he remained until his early death due to cancer in 1962 (Smith 1994, pp. 41, 65–67, 76, 156, 164).

151 Letter B. McClintock to E. Sears, May 9, 1955, ERS. For a biography of L. J. Stadler, see Rhoades (1957), and Rèdei (1971).

152 Provine and Sisco's (1980) interview was restricted at McClintock's request, CU; Lydia Bronte interview with Barbara McClintock, January 8, 1990, CSHA, cited here with permission of Lydia Bronte.

McClintock learned that she would probably be fired if Stadler left Missouri, she requested a leave of absence, intending never to return. Within the year, she reluctantly accepted a position at the Department of Genetics, Carnegie Institution of Washington, at Cold Spring Harbor, New York, although she was ambivalent about losing her freedom if she committed herself to any position. It is an earnest but abridged story and in the context of women's battle for legitimacy (Hechinger 1985), it was misrepresented after McClintock received the Nobel Prize (Anonymous 1985; Nash 1999).

Exaggerated accounts of this story have resulted in a legend that McClintock left Missouri because she was denied tenure (Geyer 1983; Hitt 1983; Anonymous 1985; Bennett et al. 1993; Rossiter 1995) or was dismissed (Nash 1999) or had quit science (McGrayne 1993) and that she eventually accepted a job at the Department of Genetics at Cold Spring Harbor because she had no place else to go (Kittridge 1991). McClintock was not denied tenure at Missouri, she was not dismissed, and she did not quit science (Kass 2003, 2005a).

The abbreviated and inflated accounts of McClintock's experiences at the University of Missouri (*e.g.*, Kittridge 1991; Rossiter 1995; Buckner 1997; Nash 1999) have relied almost exclusively on the interviews that Keller (1993) conducted with McClintock and her colleagues and on interviews she gave elsewhere (Bronte 1993; McGrayne 1993). This remains the case even in a previous biography that interprets the tenure legend in light of current academic guidelines and suggests that McClintock was not eligible for tenure (Comfort 2001, p. 64; Kass 2002). The reasons McClintock left Missouri are more obscure than is known from the popular tales (Kass 2003, 2005a). The "contemporaneous evidence" as presented in this chapter gives support for the complex events that led McClintock to leave academe.

McClintock may not be a typical example for women because her colleagues and mentors had recognized her earlier scientific contributions, and they had supported her with recommendations for fellowships and academic appointments. In the *Dynamic Genome*, a gift to McClintock on her ninetieth birthday, Nina Fedoroff wrote, "The influence of her early work is greater than that of any of her peers … Had she done no more, McClintock would have become a major figure in the history of genetics" (Fedoroff & Botstein 1992). McClintock was not the somewhat pitiful figure she is often portrayed, but in retrospect, she did not receive the same opportunities or recognition as men who had made comparable (or less than comparable) contributions. She was wise enough to recognize that her male colleagues had been recommended for preferred career opportunities that only a man could attain – e.g., appointments at exclusively male colleges and universities. The documents from the time show that she was quite cognizant of academic politics and was her own agent in this arena. McClintock's network of influential friends and scientific colleagues provided the opportunities that would permit her to reject a tenured academic job and accept a research appointment at Cold Spring Harbor. It would afford her the freedom to work independently and would be influential in her career.

This account raises the question: Why do histories based on oral interviews differ from the contemporaneous primary documents? Research on memory has shown that stories people tell about their past are shaped by the beliefs they hold in the present and are often reexamined in terms of current experiences. Psychologists who study the nature of autobiographical memory conclude that current beliefs can shape and sometimes distort recollections of past events. This quest for precision can lead biographers to criticize the factual inconsistencies, exaggerations, falsehoods, or self-deceit often found in autobiographies. Bias clouds memories of past impressions and feelings, which are filtered and made consistent with current impressions and feelings (Schacter & Scarry 2000; Schacter, 2001, pp. 14, 163). McClintock's memories (Provine & Sisco 1980) led me to seek documents that clarify and supplement the reminiscences and recollections in many stories told by and about Barbara McClintock (Kass 2000a,b, 2003, 2013ff.). Archived documents and records are undoubtedly more reliable sources than interpretations founded on oral histories, memoirs, or autobiographical recollections. Papers and correspondence written at the time portray a more complete and intriguing picture of the environment at Missouri when McClintock was employed there between 1936 and 1942.

Clearly, a variety of evidence must be integrated when writing a complete biography of a contemporary individual. Methodologically, there is no way of discovering the "objective" truth about what actually happened in a person's life. We can come closer to precision by using personal recollections and oral histories to guide us to records and documents that clarify those memories. The outcome is of course subjective, but can provide a clearer view of what transpired than memories and recollections compressed and beclouded by years of life.

LITERATURE CITED

American Men of Science, A Biographical Directory. 1938, 6th edition. J. McKeen Cattell and J. Cattell (Eds.). The Science Press, New York.

American Men of Science, A Biographical Directory. 1944, 7th edition. J. McKeen Cattell and J. Cattell (Eds.). The Science Press, Lancaster, PA.

Anderson, E. 1944. Cytological observations on *Tripsacum dactyloides*. *Annals of the Missouri Botanical Garden* Vol. 31, pp. 317–323.

Annual Report of the Biological Laboratory. 1941. Long Island Biological Association, Cold Spring Harbor, NY, 57 pp.

Anonymous. 1985. Nobel Prize winner's tenure denial illustrates an academic problem. *Columbia Missourian*, August 11, p. 7A.

Bennett, M., C. Bunce, N. Comfort, S. Cooper, and L. Hyman. 1993. *McClintock Laboratory Dedication Ceremony,* Public Affairs Department, Cold Spring Harbor Laboratory, Cold Spring Harbor, NY.

Berg, P. and M. Singer. 2003. *George Beadle, An Uncommon Farmer*, Cold Spring Harbor Laboratory Press, Cold Spring Harbor, NY.

Blakeslee, A. F. 1941. Annual Report of the Director of the Department of Genetics. *Carnegie Institution of Washington Year Book No. 40,* for the year 1940–1941 [December], pp. 211–257.

Bronte, L. 1993. *The Longevity Factor: The New Reality of Long Careers and How It Can Lead to Richer Lives*, Harper Collins, New York.

Brown, S. W. 1942. *The origin and nature of variability in the California Blackberries* (*Rubus ursinus* Cham. and Schlecht, and *Rubus lemurum* sp. nov.). Thesis (Ph.D.), Univ. of California, October 1942, Berkeley, CA.

Brown, S. W. 1943. The origin and nature of variability in the Pacific coast blackberries (*Rubus ursinus* Cham. & Schlecht. and *R. lemurum* sp. nov.). *American Journal of Botany* Vol. 30 (No. 9, November 1943), pp. 686–697.

Brown V. B. 2004. Timeline of school integration in the U.S. Trace school integration from 1849 to 2007. *Learning for Justice* (Issue 25, Spring 2004). https://www.learningforjustice.org/magazine/spring-2004/brown-v-board-timeline-of-school-integration-in-the-us

Buckner, V. L. 1997. Barbara McClintock (1902–1992), pp. 310–318, *In*, L. S. Grinstein, C. A. Biermann, and R. K. Rose (Eds.), *Women in the Biological Sciences. A Biobibliographic Sourcebook*, Greenwood Press, Westport, CT.

Bulletin of the American Association of University Professors. 1945. Academic Freedom and Tenure, University of Missouri. Vol. 31 (2, Summer), pp. 278–315. Washington, DC.

Bush, V. 1942. Report of the President. *Carnegie Institution of Washington Year Book No. 41,* for the year 1941–1942 [through December 18, 1942], pp. 1–9.

Cameron, J. W. 1947. Chemico-genetic bases for the reserve carbohydrates in maize. *Genetics* Vol. 32 (No. 5, September), pp. 459–485. [See Table 6.3]

Clark, F. J. 1939. A gene for abnormal meiotic spindle formation in maize. *Genetics* Vol. 24 (No. 1, January), p. 68, Abstract). [Affiliation Connecticut Agricultural Experiment Station, New Haven, Connecticut, and University of Missouri, Columbia, MO, presented at December 1938, Meeting of the GSA, Richmond, VA]

Clark, F. J. 1940. Cytogenetic studies of divergent meiotic spindle formation in *Zea mays*. *American Journal of Botany* Vol. 27 (No. 7, July 1, 1940), pp. 547–559.

Clark, F. J. 1941. *Cytogenetic studies of divergent meiotic spindle formation in Zea mays.* Ph.D. Dissertation, Yale University. [Acknowledgments to D. F. Jones, Barbara McClintock & L. J Stadler]

Comfort, N. 2001. *The Tangled Field*, Harvard University Press, Cambridge, MA.

Crouse, H. 1992. Barbara McClintock: Recollections of a graduate student, pp. 25–32, *In*, N. Fedoroff and D. Botstein (Eds.), *The Dynamic Genome: Barbara McClintock's Ideas in the Century of Genetics*, Cold Spring Harbor Laboratory Press, Plainview, NY.

Curtis, W. C. 1949. History of the Department of Zoology, University of Missouri. *Bios* Vol. 20, pp. 147–172.

Curtis, W. C. and M. J. Guthrie. 1927–1947. *Textbook of General Zoology*, John Wiley & Sons, Inc., New York [1933, 2nd edition; 1938, 3rd edition; 1947, 4th edition, seventh printing, November 1954; the books include a glossary and index but no bibliography or literature cited]

Demerec, M. 1941. Report of the Director, pp. 19–31, *In, Annual Report of the Biological Laboratory*, Long Island Biological Association, Cold Spring Harbor, NY.

Demerec, M. 1942. Annual report of the Director of the Department of Genetics, pp. 169–218. *In, Carnegie Institution of Washington Year Book No. 41*, for the year 1941–1942 [through December 18, 1942].

Endersby, J. W. and W. T. Horner. 2016. *Lloyd Gaines and the Fight to End Segregation*, University of Missouri Press, Columbia, MO.

Fedoroff, N., and D. Botstein. 1992. Introduction, pp. 1–4, *In*, N. Fedoroff and D. Botstein (Eds.), *The Dynamic Genome: Barbara McClintock's Ideas in the Century of Genetics*, Cold Spring Harbor Laboratory Press, Plainview, NY.

Geyer, G. 1983. Barbara McClintock represents the best of American Spirit. *Columbia Missourian* Vol. 75 (No. 30, Tuesday, October 18), p. 4A.

Glass, B. 1971. *Milislav Demerec (1895–1966), Biographical Memoir*, National Academy of Sciences, Washington, DC.

Glass, B. 1988. *A Guide to the Genetics Collections of the American Philosophical Society*. American Philosophical Society Library Pub. No. 13, Philadelphia, PA. [Figure 6.2 on cover]

Hartmann, P. E. 1988. Between Novembers: Demerec, Cold Spring Harbor, and the Gene. *Genetics* Vol. 120 (No. 3, November 1, 1988), pp. 615–619.

Hechinger, F. M. 1985. Commentary: Women in academia: battle for legitimacy often difficult. *Columbia Missourian* (excerpted from *The New York Times*), August 11, p. 7A.

Heitz, E. and H. Bauer. 1933. Beweise für die Chromosomennatur der Kernschleifen in den Knäuelkernen von *Bibio hortulanus* L. (Cytologische Untersuchungen an Dipteren I). *Zeitschrift für Zellforschung und Mikroskopische Anatomie* Vol. 17, pp. 67–82. (Received July 31, 1931) [English translation: *Evidence for the chromosome nature of the nuclear loops in the tangled nuclei of Bibio hortulanus* L.; published in *Journal of Cell Research and Microscopic Anatomy*]

Hitt, G. 1983. UMC scientists recall Nobel winner as 'the best'. *Columbia Missourian* Vol. 75 (No. 24, Tuesday, October 11), p. 1A.

Hofstadter, R. and W. P. Metzger. 1955. *The Development of Academic Freedom in the United States*, Columbia University Press, New York.

Joughin, L. (Ed.). 1967. *Academic Freedom and Tenure: A Handbook of the American Association of University Professors*, University of Wisconsin Press, Madison.

Kass, L. B. 2000a. Barbara McClintock, *Botanist, cytologist, geneticist. *American Journal of Botany* Vol. 87 (6), p. 64; Symposium Botany in the Age of Mendel, Abstract #193, https://www.2000.botanyconference.org/sympos4/abstracts/1.shtml

Kass, L. B. 2000b. Barbara McClintock, (1902–1992). American Botanical Geneticist, pp. 66–69, *In*, R. Robinson (Ed.), *Plant Sciences*, Macmillan Science Library, New York.

Kass, L. B. 2002. Book review, *The Tangled Field*, by N. Comfort. *Isis* Vol. 93 (No. 4), pp. 729–730.

Kass, L. B. 2003. Records and recollections: A new look at Barbara McClintock, Nobel Prize-Winning geneticist. *Genetics* Vol. 164 (No. 4, August 1), pp. 1251–1260. https://www.genetics.org/content/164/4/1251

Kass, L. B. 2004. Identification of photographs for the Barbara McClintock papers on the National Library of Medicine Website. *Maize Genetics Cooperation Newsletter* 78, pp. 24–26; online 2003: https://mnl.maizegdb.org/mnl/78/04kass.html [See Figure 6.1]

Kass, L. B. 2005a. Missouri compromise: tenure or freedom. New evidence clarifies why Barbara McClintock left Academe. *Maize Genetics Cooperation Newsletter* 79, pp. 52–71.

Kass, L. B. 2005b. Harriet B. Creighton: Proud botanist. *Plant Science Bulletin*. Vol. 51 (No. 4, Winter), pp. 118–125. Available online, December 2005. https://www.botany.org/psbarchive/issue/2005-v51-no-4.html

Kass, L. B. (Ed.). 2013ff. *Perspectives on Nobel Laureate Barbara McClintock's Publications (1926–1984): A Companion Volume*. Volumes I & II The Internet-First University Press. URI: https://hdl.handle.net/1813/34897 [Online December 30, 2013, updates March 2014; Vol. III, Increments No. 1–3, May 2016; Vol. IV, Increment No. 4, April 2020]

Kass, L. B. and J. A. Birchler. 2014/2016. Setting up a prepared mind: Summary and analysis of "Cytological observations of deficiencies involving known genes, translocations and an inversion in *Zea mays*," Vol. III, Increment #1, pp. 1.47.1–1.47.15, *In*, L. B. Kass (Ed.), *Perspectives on Nobel Laureate Barbara McClintock's Publications (1926–1984): A Companion Volume*, The Internet-First University Press. https://ecommons.cornell.edu/bitstreams/9711d284-f6c1-4f0a-88a0-68eab1c8d7c8/download (15.27 MB) [Online June 7, 2016]

Kass, L. B., E. H. Coe, M. N. Cook, M. E. Smith, and J. L. Singer (Eds.). 2019. *Founding of the Maize Genetics Cooperation News Letter at Cornell University: A 90th Anniversary Tribute.* (Foreword by E. S. Buckler) [e-book] Internet-First University Press, Ithaca, NY. https://ecommons.cornell.edu/handle/1813/66550. [Online October 10, 2019]

Kass, L. B. and P. Chomet. 2009. Barbara McClintock, pp. 17–52, *In*, J. Bennetzen and S. Hake (Eds.), *Handbook of Maize: Genetics and Genomics*, Springer, New York.

Keller, E. F. 1983. *A Feeling for the Organism. The Life and Work of Barbara McClintock*, 1st edition, W. H. Freeman, San Francisco, CA.

Keller, E. F. 1993. *A Feeling for the Organism. The Life and Work of Barbara McClintock*, 10th Anniversary edition, W. H. Freeman, New York.

Kittridge, M. 1991. *Barbara McClintock: Biologist* (American Women of Achievement series), Chelsea House, New York.

Lindstrom, E. W. 1939. Abstracts of the Papers presented at the 1938 meetings of the Genetics Society of America 1938. *Genetics* Vol. 24 (No. 1, January), pp. 65–111.

McClintock, B. 1931. Cytological observations of deficiencies involving known genes, translocations and an inversion in *Zea mays. University of Missouri, College of Agriculture, Agricultural Experiment Station, Research Bulletin* 163 [December, authorized December 23, 1931], pp. 1–30.

McClintock, B. 1932. A correlation of ring-shaped chromosomes with variegation in *Zea mays. Proceedings of the National Academy of Sciences* Vol. 18 (No. 12, December 15, 1932), pp. 677–681.

McClintock, B. 1933. The association of non-homologous parts of chromosomes in the mid-prophase of meiosis in *Zea mays*, with 51 figures in the text and plates VII-XII. *Zeitschrift für Zellforschung und mikroskopische Anatomie* Vol. 19 (No. 2, September 22, 1933), pp. 191–237.

McClintock, B. 1934–1936 unpublished. Reports to the Rockefeller Foundation, November 1936. Plant Breeding Grant-in-Aid to R. A. Emerson. Plant Breeding Records, file Rockefeller Foundation, Division of Rare and Manuscript Collections, Cornell University Archives.

McClintock, B. 1937. The production of maize plants mosaic for homozygous deficiencies: Simulation of the bm1 phenotype through loss of the Bm1 locus. *Genetics* Vol. 22 (No. 1), p. 200; Abstract preprinted in *Records of the Genetics Society of America* (1936), Paper presented December 29, 1936. [Ring Chromosomes]

McClintock, B. 1938a. A method for detecting potential mutations of a specific chromosomal region. *Genetics* Vol. 23 (No. 1), p. 159; Abstract preprinted in *Records of the Genetics Society of America* (1937), Paper presented December 28, 1937. [Ring Chromosomes]

McClintock, B. 1938b. The production of homozygous deficient tissues with mutant characteristics by means of the aberrant mitotic behavior of ring-shaped chromosomes. *Genetics* Vol. 23 (No. 4, July), pp. 315–376.

McClintock, B. 1938c. The fusion of broken ends of sister half-chromatids following breakage at meiotic anaphase. *University of Missouri, College of Agriculture, Agricultural Experiment Station, Research Bulletin 290* [authorized July 12], pp. 1–48. [Inversion 4a, not a ring chromosome]

McClintock, B, 1939. The behavior in successive nuclear divisions of a chromosome broken at meiosis. *Proceedings of the National Academy of Sciences* Vol. 25 (No. 8, August) [Communicated July 7], pp. 405–416. [Inversion 9]

McClintock, B. 1941a. The stability of broken ends of chromosomes in *Zea mays*. *Genetics* Vol. 26 (No. 2, March) [received November 27, 1940], pp. 234–282 [Published 3 months prior to McClintock's academic leave]. [BFB p. 281]

McClintock, B. 1941b. The association of mutants with homozygous deficiencies in *Zea mays*. *Genetics* Vol. 26 (No. 5, September 1941) [received May 3], pp. 542–571 [note-both the journal article fascicle and reprints are incorrectly dated as September 1940]. [A study of ring chromosomes, which was not included in McClintock 1987; reprinted in Kass 2013ff.]

McClintock, B. 1941c. [Symposium held June 1941]. Spontaneous alterations in chromosome size and form in *Zea mays,* pp. 72–80. *In*, *Genes and Chromosomes – Structure and Organization. Cold Spring Harbor Symposia on Quantitative Biology,* IX [June 1941, Issued December]. [Not included in McClintock 1987, reprinted in Kass 2013ff.]

McClintock, B. 1941d. Reports of investigators, pp. 41–42, *In*, *Annual Report of the Biological Laboratory*, Long Island Biological Association, Cold Spring Harbor, NY.

McClintock, B. 1942a. The fusion of broken ends of chromosomes following nuclear fusion. *Proceedings of the National Academy of Sciences* Vol. 28 (No. 11, November) [Communicated September], pp. 458–463. [First defines the terms "chromatid type" and "chromosomal type" to describe the BFB cycles.

McClintock, B. 1942b. Maize Genetics: The Behavior of unsaturated broken ends of chromosomes; phenotypic effects of homozygous deficiencies of distal segments of the short arm of chromosome 9. *Carnegie Institution of Washington Year Book No. 41*, 1941–1942 [June, issued December 1942], pp. 181–186.

McClintock, B. 1943. Maize genetics: Studies with broken chromosomes. Tests of the amount of crossing over that may occur within small segments of a chromosome. Deficiency mutations: Progressive deficiency as a cause of allelic series. *Carnegie Institution of Washington Year Book No. 42*, 1942–1943 [Issued 7 December 1943, submitted June 1943], pp. 148–152.

McClintock, B. 1944a. Carnegie Institution of Washington, Department of Genetics, Cold Spring Harber [sic], Long Island, N.Y. [This report is untitled in the *MGCNL*. This is a report on deficiencies in Chromosome 9]. *Maize Genetics Cooperation News Letter* 18 [January 31, 1944, submitted 1943], pp. 24–26. [The report concludes, "... the chromosomal breakage mechanism is a 'mutation' inducing process which 'induces' the same mutant time and again." McClintock (1987) cites title as: "Breakage-fusion-bridge cycle induced deficiencies in the short arm of chromosome 9." However, the term "Breakage-fusion-bridge cycle" is not used in this report.]

McClintock, B. 1944b. The relation of homozygous deficiencies to mutations and allelic series in maize. *Genetics* Vol. 29 (No. 5, September 1944, received February 8, 1944), pp. 478–502.

McClintock, B. 1944c. Maize genetics: Completion of the study of the allelic relations of deficiency mutants. The chromosome-breakage mechanism as a means of producing directed mutations. Continuation of the chromatid type of breakage-fusion-bridge cycle in the sporophytic tissues. Homozygous deficiency as a cause of mutation in maize. *Carnegie Institution of Washington Year Book No. 43*, 1943–1944 [Issued December 15, 1944, submitted June 1944], pp. 127–135.

McClintock, B. 1978. Mechanisms that rapidly reorganize the genome. *Stadler Genetics Symposia* 10, pp. 25–47 [Symposium held April 7–8, 1978].

McClintock, B. 1984. The significance of responses of the genome to challenge. *Science* Vol. 226 (No 4676, November 16, 1984), pp. 792–801.

McClintock, B. 1987. Introduction, pp. vii–xi, *In*, J. A. Moore (Ed.), *The Discovery and Characterization of Transposable Elements. The Collected Papers of Barbara McClintock*, Garland Publishing Co., New York.

McGrayne, S. B. 1993. *Nobel Prize Women in Science: Their Lives, Struggles, and Momentous Discoveries*, Carol Publishing Group, New York.

Moulton, F. R. (Ed.). 1940a. The third Columbus meeting of the American Association for the Advancement of Science and Associated Societies. *Science* Vol. 91 (No. 2353, February 2, 1940), pp. 101–124. [Meeting held December 1939.]

Moulton, F. R. (Ed.). 1940b. *A Brief History of the Association from its Founding in 1848 to 1940, Its Present Organization and Operation, Summarized Proceedings for the Period from January, 1934, to January, 1940 and a Directory of Members as of July 1, 1940*, American Association for the Advancement of Science, Washington, DC; William Byrd Press, Inc. Richmond, VA.

Murphy, R. P. and L. B. Kass. 2007. *Evolution of Plant Breeding at Cornell University: A Centennial History, 1907–2006*, pp. 1–98, Department of Plant Breeding & Genetics, Cornell University, Ithaca, NY, Appendices A1–A98, Photo Section P1–P38. (July 2007).

Murphy, R. P. and L. B. Kass. (June 27, 2011). *Evolution of Plant Breeding at Cornell University: A Centennial History, 1907-2006.* Rev. edition i–x, 178 pp., Appendices, Photo Section. e-book, The Internet-First University Press, Ithaca, NY. https://hdl.handle.net/1813/23087

Nash, J. 1999. Freaks of nature: Images of Barbara McClintock. *Studies in the History and Philosophy of Biomedical Science* Vol. 30, pp. 21–43.

Olson, J. and V. Olson. 1988. *The University of Missouri, An Illustrated History*, University of Missouri Press, Columbia, MO.

Painter, T. S. 1933. A new method for the study of chromosome rearrangements and the plotting of chromosome maps. *Science* Vol. 78 (No. 2034, December 22, 1933), pp. 585–586.

Painter, T. S. 1934a. A new method for the study of chromosome aberrations and the plotting of chromosome maps in *Drosophila Melanogaster. Genetics* Vol. 19 (No. 3, May 1), pp. 175–188.

Painter, T. S. 1934b. Salivary chromosomes and the attack on the gene. *Journal of Heredity* Vol. 25 (No. 12, December 1), pp. 465–476.

Plant Breeding Records, Cornell University Archives: Cornell University Department of Plant Breeding Records, 1906–1970. Collection Number: 21-28-889 Division of Rare and Manuscript Collections. Cornell University Library. [Boxes 56-65 (student files) are restricted to Plant Breeding and Genetics Section Chair or their designate]. https://rmc.library.cornell.edu/EAD/htmldocs/RMA00889.html

Poehlman, J. M. 1988. *A History of Field Crops 1870–1967 in the University of Missouri.* College of Agriculture, Special Report 385, University of Missouri-Columbia.

Provine, W. B. and P. Sisco. 1980. Interview with Barbara McClintock at Cold Spring Harbor Laboratory, Long Island, New York, August 28, 1980. Division of Rare and Manuscript Collections, Cornell University Library, Ithaca, NY.
Records of the Genetics Society of America, 1936, 1937, 1938, 1939, 1941. Published by the Secretary-Treasurer, The Genetics Society of America.
Rèdei, G. P. 1971. A portrait of Lewis John Stadler 1896–1954, pp. 5–20, *In*, G. P. Rèdei (Ed.), *Stadler Genetics Symposia* (3rd) [Proceedings], Vols. 1 & 2. University of Missouri, Agricultural Experiment Station, Columbia, MO.
Rèdei, G. P. 1976. PROLEGOMENA. *Stadler Genetics Symposia* Vol. 8, p. 3.
Rhoades, M. M. 1957. Lewis John Stadler. Biographical Memoir, National Academy of Sciences, Washington, DC.
Rhoades, M. M. and B. McClintock 1935. The cytogenetics of maize. *Botanical Review* Vol. 1 (No. 8, August 1935), pp. 292–325.
Robbins, M. L. 1971. *Delta Chapter, University of Missouri, 1924. A History of Sigma Delta Epsilon, 1921–1971.* Sigma Delta Epsilon, Washington, DC.
Rossiter, M. W. 1982. *Women Scientists in America. Struggles and Strategies to 1940*, The Johns Hopkins University Press, Baltimore, MD.
Rossiter, M. W. 1995. *Women Scientists in America. Before Affirmative Action*, The Johns Hopkins University Press, Baltimore, MD.
Schacter, D. L. 2001. *The Seven Sins of Memory. How the Mind Forgets and Remembers*, Holton Mifflin Co., Boston, MA.
Schacter, D. L. and E. Scarry (Eds.). 2000. *Memory, Brain and Belief*, Harvard University Press, Cambridge, MA. [See especially part III, Memory and Belief in Autobiographical Recall and Autobiography]
Sharp, L. W. 1934. *Introduction to Cytology*, 3rd edition, McGraw-Hill Book Company, New York.
Singleton, J. R. 1953. Chromosome morphology and the chromosome cycle in the ascus of *Neurospora crassa. American Journal of Botany* Vol. 40 (No. 3, March), pp. 124–144.
Smith, R. W. 1994. *Biology at Purdue 1874–1994; The Evolution of a Department of Biological Sciences.* Abridged Version, December 1994. Published by the Department of Biological Sciences, Purdue University.
Stephens, F. F. 1962. *A History of the University of Missouri*, University of Missouri Press, Columbia, MO.
Swanson, C. P. 1957. *Cytology and Cytogenetics*, Prentice Hall, Inc., Englewood Cliffs, NJ. [Sixth printing 1964]
Swanson, C. P. 1988. Cytogenetics and Karl Sax. *Genetics* Vol. 119 (No. 1, May 1, 1988), pp. 5–7.
Swanson, C. R. [*sic*]. 1940. The distribution of inversions in Tradescantia. *Genetics* Vol. 25 (No. 5, September 1, 1940), pp, 438–465.
Swanson, C. R. [*sic*]. 1941. *The distribution of inversions in Tradescantia.* Thesis (Ph.D.), Harvard University. https://id.lib.harvard.edu/alma/990039628150203941/catalog, ["Reprinted from *Genetics* Vol. 25 (No. 5, September 1, 1940), pp, 438–465, September 1940."]
The Rockefeller Foundation Annual Report. 1939. Genetics, University of Missouri, pp. 214–218, The Foundation, New York.
Vilas, J. 1939. (With the collaboration of several of his colleagues). *The University of Missouri, A Centennial History*, E. W. Stephens Co. Press, University of Missouri, Columbia, MO.
Zacharias, H. 1995. Emil Heitz (1892–1965). Chloroplasts, heterochromatin, and polytene chromosomes. *Genetics* Vol. 141 (No. 1, September 1, 1995), pp. 7–14.

7 The Road to Transposition

We are often more comfortable with treasured legends than with hard facts

– Michael G. Berger, NYT

How transposable elements were discovered has been detailed by McClintock herself and by others who worked closely with her.[1] In addition, her pioneering breakthrough was described by researchers who summarized the investigations she reported in the *Year Books of the Carnegie Institution of Washington*, and in comprehensive data found in her unpublished manuscripts.[2] As we have seen, McClintock had been studying the genetics and cytology of maize since the 1920s. She believed that her undergraduate courses in Genetics and Cytology at Cornell and her cooperative studies with graduate students and colleagues, especially Charles Burnham, George Beadle, and Marcus Rhoades "were, by far, the most influential in directing [her] scientific life" (McClintock 1984b, p. 172). "The communal experience profoundly affected each one of us," she wrote (McClintock 1984b, p. 172; see Chapter 4). In this respect, she recalled that finding that genetic loci could move (transposition) had its beginnings with her earliest studies on linking genes with chromosomes in maize (McClintock 1931, 1984b, p. 177). Later, studies conducted while she was at the University of Missouri in 1932 led her to study broken chromosomes. Following up those studies at Caltech, Cornell, the University of Missouri, and the Department of Genetics, Carnegie Institution of Washington, at Cold Spring Harbor, New York, led her to discover mobile genetic elements in 1948 (McClintock 1984a), as shown below.

McClintock was a well-respected scientist who was extensively recognized for her accomplishments. During these years, she was elected to the National Academy of Sciences (1944), elected the first female President of the Genetics Society of America (1945), elected to the American Philosophical Society (1946), starred in *American Men of Science* (1944), given the Achievement Award of the American Association of University Women (1947), awarded honorary degrees from the University of Rochester (1948) and Western College for Women (1949), and invited to present her pioneering work on "Chromosome organization and genic expression" at the *Cold Spring Harbor Symposia on Genes and Mutations* (McClintock 1951b). In addition, geneticists who recognized the importance of her pioneering early investigations invited her to present her work at their institutions, teach her work to their students, and encourage their graduate students to pursue similar studies.

The first accurate and detailed summaries of the investigations that led McClintock to discover transposable elements, while she was employed as a Resident Investigator at the Department of Genetics between 1943 and 1951, were given by Nina Fedoroff (1992, 1998a,b, 2013a). Fedoroff learned maize genetics directly from McClintock, and she and her colleagues in maize genetics provided the molecular basis for transposition (Fedoroff 1992, pp. 389–391, 1989; see also Fedoroff 2013b,c). Therefore, in this chapter, I will only briefly highlight the major events that led to McClintock's discovery of transposable elements. Additionally, I will call attention to the study that seems to have launched her on the path to transposition. It was first submitted to her maize genetics colleagues with the 1943 call to report current research to the *Maize Genetics Cooperation News Letter* (Coe & Kass 2005; McClintock 1944a), during her first year as Resident Investigator at the Department of Genetics.

McClintock's discovery of transposition originated with her studies of the behavior of broken chromosomes (Chapter 6). In 1945, she reported "an interesting type" of broken chromosome

[1] McClintock (1984a,b); see Federoff (1983–2017), and also Brown (1995), Fedoroff and Brown (2004).

[2] R. N. Jones (2005), Kass and Chomet (2009), Comfort (2001), Shapiro (1995), Keller (1983, 1993); see link for *Carnegie Institution Year Books*. See Barbara McClintock Papers, Collection No. 79, American Philosophical Library, Philadelphia, PA (APS; Spectre 1993).

DOI: 10.1201/9781003332527-8

behavior that had appeared in her maize cultures. Broken chromosome 9 was continually being lost from cells during development. The loss was neither due to chromosome bridge formation nor due to ring chromosome behavior. "Several of the plants showed variegation characterized by a change from mutant to normal-appearing tissues. ... [Seedling leaf] sectors with changed rates of variegation appear in all plants, especially in later-appearing tissues. It is significant that twin sectors accompany many if not most of the alterations" (McClintock 1945a; Fedoroff 1998a). During the 1945–1946 growing season, McClintock discovered a phenomenon in maize that had long been known in the genetics literature by the terms spontaneous mutations, mutable loci, variegation, unstable loci, mutable genes, and position effect (McClintock 1946, p. 178). McClintock provides no citations to that "genetics literature," but it was in this context that she interpreted her findings.

Fedoroff (1992, 1998a,b, 2013a) provided the background for the previous studies of variegation, i.e., "mutable" or "unstable genes," to which McClintock and other geneticists were long familiar. Since the early twentieth century, Rollins Emerson and Marcus Rhoades had taken an interest in these phenomena and had "carried out systematic genetic studies on mutable genes in maize and contributed substantial insights into their nature and genetic behavior" (Fedoroff 1998a, p. 90, 1998b). Rhoades had identified a genetic locus, which he named dotted (*Dt*), in whose presence a formerly stable recessive gene (*a*, colorless aleurone kernel) became unstable [a mutable gene].[3] He observed plants and kernels that were homozygous recessive (*a,a*, colorless aleurone), which should not have the ability to produce color, and had exhibited sectors of pigmented tissue when the dominant *Dt* gene was present. He also found that Emerson's *a* tester allele, used for years as a stable recessive allele, responded to the *Dt* gene the same way as his own *a* allele. Rhoades' genetic data clearly showed that the "presence or absence of a second unlinked gene could make the difference between a gene's stability and instability and ... that a mutable gene could mutate to a stable normal [non-mutable] form" (Fedoroff 1992, p. 391). Fedoroff considers that, in retrospect, Rhoades was gathering the first clues ... that one mobile element could promote transposition of a second unlinked element (Fedoroff 1992, p. 391).

McClintock had these earlier studies in mind as she continued investigations of the induction of maize mutants. She observed maize plants expecting to have a particular phenotype that instead had unexpected observable characteristics, such as variegated mutant sectors on maize seedlings. The "unstable mutants" appeared as variegated white, light-green, or luteus [yellow] seedlings (McClintock 1946, pp. 178–182). The variegation was observed as sectors on leaves that seemed to have spontaneously reverted from the recessive to the dominant phenotype. Her investigations revealed that the unexpected mutations were the result of a genetic locus that caused chromosome breaks and could be observed at the phenotypic level in the maize kernel and plants. The name for these new genetic loci would much later be termed "controlling elements" (McClintock 1956 [Brookhaven Symposium 1955]), transposable elements, mobile genetic elements, or transposons (Hedges & Jacob 1974). The mutation phenomenon McClintock observed was initially interpreted as caused by chromatin breakage, which she termed "the chromosome-breakage variegation" (McClintock 1946, p. 182). She believed that the new mutations she observed were correlated with the losses in the short arm of chromosome 9, which included the genetic loci for phenotypes she had also observed in variegated maize kernels. During plant development, the chromosome pieces were lost while undergoing the "chromosome type of breakage cycle" (McClintock 1946, p. 182).

The detailed report of these studies appears in the *Carnegie Institution of Washington Year Book No. 45*, for 1945–1946. It concludes with the possibility that the new unstable mutants (variegation types) that appeared in her cultures are associated with the chromosome breakage cycle:

> Although the factors responsible for this breakage phenomenon are not understood, nevertheless the factors that control the frequency and the timing of such occurrences are similar to those that control the frequency and timing of dominantly directed mutations in the variegation types previously described. In the case of the chromosome-breakage variegation just described, however, the breakage

[3] Fedoroff (1992, p. 391) and Fedoroff (1998a), reporting on studies of Rhoades (1936, 1938); see also Fedoroff (1998b).

itself corresponds to the "gene" mutations observed in the other variegations. Possibly the resemblance is more than coincidental, in that the underlying phenomena are basically similar.

(McClintock 1946, p. 186)

During the following year (1946–1947), "a common underlying phenomenon" associated with unstable loci was quickly identified. It was attributed to a locus that broke chromosomes, which McClintock named *Ds* for dissociation, "because the most readily recognizable consequence of its actions is this dissociation" (McClintock 1947, p. 147). The *Ds* locus would only undergo "dissociation mutations" in the presence of another chromosomal locus which she designated Activator (*Ac*) because it activated *Ds* (McClintock 1947, p. 147). In October 1946, she gave a report on the new phenomenon, titled "Variegation in maize-an interpretation of the unstable-gene problem," at a Department Staff Meeting.[4] In December 1946, at the Caltech "Biology Seminar," she presented a "lecture" on "The Genetic Effects of the Removal of Blocks of Genes in Maize." Thus, she reported her new findings to a wider audience, and to those in attendance who were familiar with similar findings in *Drosophila*.[5]

By 1948, McClintock observed that the phenotype of kernels harboring modified chromosomes had changed. The genetic locus causing chromosome breaks, and resulting in mutations, did not remain in one place on the chromosome but seemed to move to other locations. McClintock's years of mapping chromosomes led her to quickly reach the conclusion that the locus was in a place it had not been found previously. She reported, "It is now known that the [chromosome breaking] *Ds* locus may change its position in the chromosome …" (McClintock 1948, p. 158). A very brief summary of how she came to this conclusion follows. She likewise reported that "… the *Ac* locus may have been removed from its former position in chromosome 9 in a manner similar to that observed for the transposition of the *Ds* locus …" (McClintock 1948, p. 159). In two unpublished manuscripts, dated January 1949 and April 1949, McClintock provided extensive details of how she came to the conclusion that transposition of a chromosome locus had occurred in her stocks.

Fedoroff (1992, pp. 392–397, 1998a, 2013a) summarizes the details, including excerpts from the unpublished manuscripts. The first, titled "Memorandum to Marcus Rhoades" (January 1949), describes the relationship of the first mutable *c* locus, designated ($c^{m\text{-}1}$), to the position of the *Ds* locus in the chromosome.[6] The second unpublished manuscript titled "The origin of transposed *Ds*, Case I" (April 1949), was also intended for Rhoades, and explains in additional detail the origin of the transposition events.[7] McClintock's observations clarified how her study of variegation of the *c* locus is related to the mutable genes (variegations, mosaicism) "previously reported in the genetics literature" (McClintock 1946, p. 176). Specifically, Fedoroff (1992, p. 397) explained McClintock's January 1949 unpublished report:

the *c-m1* [$c^{m\text{-}1}$] locus is a mutable gene, like Rhoades' *a* allele. It is a stable recessive allele except when the *Ac* locus is present. *Ac*, like Rhoades' *Dt* locus, induces genetic instability, causing the recessive allele to mutate to the dominant, wild-type allele. McClintock explains this by transposition. The *C* locus became mutable when the *Ds* element transposed into it and inhibited its expression, giving the recessive, colorless *c* allele. Mutations to the dominant allele, whether somatic or germinal, remove the inserted transposable element from the locus. Thus, the temporary inhibitor that Emerson [1929] had postulated at the *P* [pericarp] locus is a transposable element. (We now know that Emerson's element was *Ac*, but Rhoades' *Dt* belongs to a different element family.) In addition, although [McClintock's] *Ds* is the causative agent of both mutability and chromosome breakage, it can do neither on its own, requiring something that the *Ac* element provides, what we today [1992] would call the transposase.

4 "Report to Dr. Bush 1941 [–1953]," November 1947, staff meeting, October 28, 1946, Cold Spring Harbor Archives (henceforth CSHA).

5 *California Institute of Technology Weekly Calendar*, November 29th, 1946, Biology Seminar, Tuesday, December 3, 1946; Collection Weekly Calendar, Vol. 22, No. 9, Caltech Archives.

6 See McClintock Unpublished, January 1949.

7 See McClintock Unpublished, April 1949.

> It is evident from the quoted passages [McClintock unpublished January 1949] that the ability of McClintock's *Ds* element to both break chromosomes and transpose was extremely important to explicating the relationship between transposition and instability.

In April 1950, McClintock submitted a manuscript titled "The origin and behavior of mutable loci in maize" to the *Proceedings of the National Academy of Sciences*, published in June (McClintock 1950a). It reported on the study of "insertions of a *Ds* into known gene loci." She recalled that "It was clear from responses to this report that the present thesis, and evidence for it could not be accepted by the majority of geneticists or by other biologists. Genetics was in an unformed state compared to the rapid changes in concepts that occurred subsequently in the 1950s and 1960s and there was no clear notion of the nature of the gene. It remained a hypothetical unit until proven otherwise" (McClintock 1987, p. x).

In a perspective on McClintock's 1950 paper, Weil (2013) notes that the paper, written for a more generic scientific audience, "has stood for decades at the center of transposable element biology." It would be another 30 years before the maize transposable elements would be isolated, thus allowing an understanding of their mechanism at the molecular level. Nearly every one of the mechanisms McClintock describes in this paper "using just her keen powers of observation, deductive reasoning and talents as a geneticist" would be validated after "a dozen years of work by a dozen labs around the world."

Additional details were described in her annual reports for the Department of Genetics (McClintock 1950b, 1951a). She presented this work at the Cold Spring Harbor Symposium in the summer of 1951, and recalled that "the response to it was puzzlement and in some instances hostility" (McClintock 1951b, 1987, p. x). She submitted a more detailed paper to *Genetics* in April 1953, which was published in November (McClintock 1953). The volume was not issued until January 1954. She recalled receiving only three reprint requests for that paper, which convinced her that "no amount of published evidence would be effective" (McClintock 1987, p. x).

In a perspective for McClintock's 1950 *PNAS* paper, Fedoroff (2012b) reminds us that McClintock's discovery of transposition was approximately contemporaneous with Watson and Crick's "landmark," but very brief report, published in *Nature* in April 1953. Within a decade, their work was recognized with the award of a Nobel Prize in Physiology or Medicine (1962). It provided the structure of the DNA molecule and "immediately explained the mechanics of gene inheritance that had by then been under intense study for half a century since the rediscovery of Mendel's work" (Fedoroff 2012b, p. 20201). The recognition of the importance of McClintock's work began when transposons were found to be ubiquitous throughout all eukaryotic organisms and when the underlying mechanisms of their transposition events were revealed at the molecular level.

One geneticist who was excited by McClintock's 1950 hypothesis was emigre Richard Goldschmidt, then on the faculty at the University of California, Berkeley, since 1936. He had hosted McClintock during her brief stay in Berlin (1933–1934) as a Guggenheim Fellow. He was familiar with her Carnegie Institution *Year Book* reports (1946–1950) and at the end of 1950 published a paper in the *American Naturalist* (Goldschmidt 1950) describing how the results presented in her most recent paper (McClintock 1950) might provide an understanding of his own thoughts on the action of genes in development for organisms other than maize[8]:

> From the point of view of basic genetic conceptions the most remarkable of McClintock's results is that mutation to a so-called mutable locus is the consequence of an insertion of the material composing *Ds* at or near the locus which becomes mutable. This means that the locus, called mutable, is not changed at all, but that the presence of the transposed *Ds* impedes its actions so that it behaves like its recessive. … if the break near a given locus brings heterochromatin adjacent to the locus the position effect is of a mosaic nature … In the light of McClintock's work, it is possible that in the course of development the break is again removed by return inversion. … thus restoring the normal function in some cells and their derivatives leading to a mosaic. … I had always a feeling that a position effect of a hitherto

[8] Comfort (2001, p. 160) presents a different view of Goldschmidt's emphasis.

> unknown kind should be involved [for *Drosophila* mosaics reported in the literature]. McClintock's work suggests an explanation in such terms: an invisible transposition into this region after the manner of the *Ds* locus might inhibit the action of neighboring loci just as *C* is inhibited in the maize case by transposition of *Ds* adjacent to it.
>
> *(Goldschmidt 1950, pp. 453, 454)*

McClintock presented her pioneering work on "mutable loci" at the Cold Spring Harbor Symposia in 1951 (McClintock 1951b). Although she believed that her symposium presentation was not well received (Keller 1983, 1993; Kittredge 1991), records of her colleagues offer a different picture of her remarkable 1951 presentation.

Goldschmidt opened the 1951 Cold Spring Harbor Symposia "with a confident indictment of the gene as particulate entity," as Columbia University Professor Franz Schrader carefully recorded in his *Diary*.[9] "He pulled no punches and pointed out that this view of his, simulating position effect vs. particulate gene had been indicated for 20 or more years …" Goldschmidt "proclaimed" that he had now accepted "new ideas and changed his old ones-something that many of his opponents had never been able to do." After Goldschmidt's (1951) talk, Edgar Altenburg, from the Rice Institute [University, 1960], Houston, Texas, "expressed displeasure over all this revolutionary stuff which is so distinctly and definitely to be deplored 'when we were getting along so well before' …" Goldschmidt told Schrader afterward that the Columbia group, where Altenburg had received his doctorate (1916), had been pioneers "in genetic development but that they had never gone beyond the limit of the early days and are always opposed to new ideas."

This was not the first time that Goldschmidt had publicly expressed his thoughts regarding the particulate nature of the gene. At a Symposium at Stanford University in 1939, he "boldly declared that there is no such thing as a separate gene particle. The chromosome, not the gene, is the hereditary unit and the classical theory of the gene corresponds to the conception of the indivisible atom of the old physics" (Stanford University 1939, p. 10).[10] The symposium on "The cell and Protoplasm" included biologists, biochemists, and chemists, who engaged in discussions of "all problems of growth," including "heredity and variation" (Mouton 1940, Foreword).

Goldschmidt's (1940) published essay on the subject introduced many examples for his arguments that the chromosome is the important unit for heredity. He argued that "the action of X-rays upon the hereditary material is to produce breaks in the chromosomes, thus resulting in rearrangements of the chromosomal material." This he believed was the cause of observed mutations. How the X-ray produces the break, and how the two breaks necessary for rearrangement are produced were important unanswered questions at that time (Goldschmidt 1940, p. 59). Types of gene action are enhancing or inhibiting effects of so-called modifiers, which he believed were parallel to position effects. The latter example was the bar-eye mutants in *Drosophila*, which were found to be caused by the movement of a piece of a chromosome near a dominant "gene" that caused a change in the chromosomal pattern without a change within an individual gene. If parts of the chromosome containing heterochromatin (inert regions having no known mutant loci) are transferred to other regions of the chromosome, "the result is a developmental mosaicism for the effects of loci brought in contact with the inert material." He concluded that "Any change of serial pattern within … sections [of a chromosome] produces … a definite abnormality of development [a mutation], with a possible variability of the effect, according to the type of the new pattern." He suggested that the "changed action of the [chromosome] segments (the mutant and/or position effect) is in some respects also controlled by the pattern of the whole chromosome." For Goldschmidt, the most important consideration in rejecting the theory of the particulate gene as derived from [point] mutation

[9] Franz & Sally Shrader's Diary, entry June 8, 1951, Columbia University Archives.

[10] Pamphlet in Folder 3680/7, Box Bio. Sci. 3680/6-3680/9, Stanford University C. H. Green Library, Special Collections and University Archives; see Moulton 1940 for Goldschmidt's published essay on the subject of "Chromosomes and Genes."

was "the serial pattern of the chromosome and its blocks [of chromatin material] of different sizes (in some respects independent), containing a smaller or larger number of loci." Years later, many of these ideas would be applicable toward an understanding of the production of mutable loci in maize. The underlying molecular biology would later clarify how the chromosomal pieces moved along the chromosome [DNA molecule] to control the action of genes.

Schrader also praised McClintock's 1951 presentation, which directly succeeded Goldschmidt's on the morning program. He recorded his impression immediately upon returning home. "She came right out handsomely with 'Nobody took any stock in Professor Goldschmidt's idea in 1938 or earlier. Neither did I, until my own work led me to the same conclusions.' These 2 morning talks were definitely the cream of the day and almost certainly of the whole shape of the symposium, so that Stadler's first talk of the afternoon came as an anticlimax, an unusual position for a lecture from him."

Years later, others offered additional recollections or interpretations to the reactions of McClintock's 1951 Cold Spring Harbor Symposium presentation.[11] Zinder (1992), then a graduate student at the University of Wisconsin, recalled that Goldschmidt had denied the existence of the gene and instead believed that the "chromosome was the unit of function."[12] He added that this was also the meeting that "spawned the myth" that McClintock's description of transposition was not understood. He recalled there being "a separate meeting with Barbara to discuss details of her experiments." He reminded readers that in attendance were corn geneticists, R. A. Brink, and his students also from the University of Wisconsin, who had obtained similar results and undoubtedly understood the phenomenon of transposition. He added that he and most of Brink's colleagues at Wisconsin were certainly aware of transposition since they were "all drafted to plant and pollinate his corn." He also mentioned that Horowitz and Leupold had also provided some "substantial evidence" for the "one-gene, one-enzyme hypothesis," but there were reservations about the name. They hypothesized that genes act through the productions of enzymes, with each gene responsible for producing a single enzyme that in turn affects a single step in a metabolic pathway, which had first been proposed in 1941 by George Beadle and Edward Tatum.

Horowitz (1995) recalled that his 1951 symposium presentation was a defense of their hypothesis against their most influential critic, as opposed to the interpretation that Beadle's hypothesis was already part of the entrenched genetic canon (Comfort 1995). Horowitz recalled Beadle's reaction to the published 1951 symposium volume: ... supporters of their hypotheses "could be counted on the fingers of one hand with a couple of fingers left over." Horowitz's recollection of McClintock's presentation differs from Zinder's and Schrader's and adds another dimension to the question of the reliability of memory or documentation. Horowitz attended McClintock's lecture, which held much interest (Horowitz 1995, p. 519). She gave it twice, he recalled, because the small lecture room could not hold all the attendees. Demerec, the program organizer, announced that McClintock would give her lecture again after dinner for those who could not find seats in the room. Unfortunately, no document for his recollections has been located. Horowitz recalled attending McClintock's evening presentation, and "like many others was mystified."

Metzenberg (2005) further emphasized Horowitz's displeasure with the interpretations of his work as given by historians of science. At the time, their hypothesis was not the "canonical view among genetics," and neither was it being challenged by "fascinating mavericks and intellectual pioneers" as interpreted by science historians (e.g., Comfort 1995). Metzenberg (2005) reported that Horowitz was "deeply disturbed by post-modern historians of science who, with varying degrees of certitude, continue to argue that scientific findings are cultural artifacts." Horowitz's (1996) critical analysis of Comfort's (1995) Perspectives article "exemplifies his rejection of what he saw as an attempt to rewrite the 'history of science.'"

Another interpretation of the 1951 reaction to McClintock's presentation was provided by Bruce Wallace, a *Drosophila* geneticist and a colleague of McClintock at Cold Spring Harbor from 1947 to

[11] Zinder (1992), Comfort (1995), Horowitz (1996), and Metzenberg (2005).

[12] At the time, Zinder was a graduate student of bacterial genetics at the University of Wisconsin with J. Lederberg.

1958.[13] Many symposium attendees were not maize geneticists, he said. Thus, they were not familiar with the complex maize life cycle, and especially the cytogenetics of maize, which was a prerequisite if one were to follow McClintock's innovative thoughts on "nuclear organization and gene function." As a staff member at Cold Spring Harbor during the years of her pioneering discovery, Wallace attended the joint weekly staff meetings of the Biological Laboratory and the Department of Genetics, which often included colleagues from Columbia University. As a Resident Investigator at the Department of Genetics beginning in 1943, McClintock had presented no fewer than 16 staff reports on new discoveries of mutable genes in maize, 13 of which were given between 1947 and 1951.[14] At these meetings, Wallace heard about her pioneering discoveries and the discussions that followed. Years later, he offered a model for understanding control regions of gene loci and how they might relate to mutable loci (Wallace & T. L. Kass 1974).

I asked Bruce Wallace about the reports by Keller (1983) and Kittredge (1991) that McClintock's talk was greeted with "stony silence." He said that McClintock's published presentation had no resemblance to the talk that she presented at the 1951 meeting.[15] She had to explain the life history of the corn plant first, he told me, before she could get to the meat of the talk. She was running out of time and rushed through the data. He did not recall any specifics regarding discussions that followed her presentation. There is no record in the symposium volume to demonstrate whether participants asked questions following her talk. Previously (since 1933), a recorder was responsible for compiling the questions and answers that followed speaker's talks. A transcript was sent to speakers, who then edited it for accuracy. However, in 1942 the practice ended and speakers were asked to submit "manuscripts of their discussions if they felt that their statements contributed information not otherwise covered" (Demerec 1942, p. v).

The fact that there is no discussion following McClintock's 1951 presentation cannot be an indication that no questions were asked or that any discussion ensued. The only assumption we can make is that if any discussions took place, she did not submit them for publication in the meeting's proceedings (McClintock 1951b). One discussion relevant to McClintock's presentation did follow a talk on "Gene-enzyme relationships in *Neurospora*" (Comfort 1995). That paper by David Bonner (1951) included three pages of discussion. Many other papers provided no discussions following the published text.

In September, developmental geneticist Ernst Caspari (1951) provided a review of the Sixteenth Symposium on Quantitative Biology in the pages of *Science*. Each of the 34 papers presented was followed by "an extended and lively discussion." Research is accumulating, which is "leading toward a revision of the older concept of the gene as a particulate, well-defined unit," he wrote. Although no agreement was reached on the "kind of revision necessary," recent work seems to "favor a more flexible and functional description of heredity units." In light of these developments, McClintock's work "aroused particular interest." It certainly would have interest for Caspari, whose studies beginning in the early 1930s were on developmental genetics. He investigated the action of genes in development, especially on the physiological action of eye color mutants in the moth *Ephestia*. As a Research Associate in the Department of Genetics at Cold Spring Harbor (1946–1948), he observed first-hand the investigations that led McClintock to discover transposition, and would have understood its implications to an understanding of genetic control and development.[16]

McClintock (1987, p. xi) believed that the difficulty in communicating her paper stemmed from her "emphasis on the regulatory aspects" of the movable elements. "In the mid-1940s there was little if any awareness for the need for genes to be regulated during development." Yet, it was "just that

[13] Bruce Wallace interview with Lee B. Kass, October 28, 1995.

[14] "Report to Dr. Bush 1941 [–1953]," Reports of Staff meetings, Demerec File, CSHA.

[15] Bruce Wallace interview with Lee Kass, October 28, 1995.

[16] Caspari was completing his Ph.D. in developmental biology in Germany just as the Nazis were coming to power. He had to flee the country because of his Jewish heritage. He emigrated to the U.S. in 1938 and became a naturalized citizen in 1944 (Grossbach 2009); see his publications and affiliation with the Department of Genetics, Cold Spring Harbor in *Carnegie Year Books Nos. 44, 45, 46.*

aspect" that caught her attention. She believed that "... some genetic mechanism was controlling the patterns of gene expression, as ... illustrated by the twin sectors in leaves and kernels." It was her intention "in the summer of 1945 to attempt to find out what this regulatory event might be." As described above, the path to solving the underlying cause of these events resulted in the publications of the early 1950s that she believed were not well received.

In spite of Goldschmidt's comments regarding the Columbia genetics group not accepting new ideas, as mentioned above, Altenburg had a definite interest in McClintock's pioneering study. He seems to have been the first textbook author to have incorporated her new findings into a chapter on "Mutable Genes" for his revised *Genetics* text for teaching genetics (Altenburg 1957: Chapter 19). Therein, he provided extensive details of her studies on the cytology and genetics of mutable genes in maize, devoting 14 of the chapter's 19 pages to a discussion of McClintock's published work. "Genes are very stable as a rule," he wrote, "But there is a class of genes which mutate rather frequently in the course of development." Throughout the chapter, his discussion of mutable genes is in the context of plant development.

In the same year, Swanson (1957, pp. 442–444), who had previously been with McClintock at University of Missouri (see Chapter 6), had a critical view of the subject. He also included a discussion of McClintock's "*Ac-Ds* position effect" in maize in his "reference work" *Cytology and Cytogenetics*. He considered her recent work to be of "obvious importance to the problem of differentiation" (p. 443). Yet, he concludes, "The variegations arising in the maize endosperm under the influence of *Ac-Ds* and other similar systems are of course abnormal; it is not expected that such are to be found in well-regulated developmental patterns. ... It may be that McClintock's *Ac-Ds* system is a mechanism of differentiation which ... had gotten out of normal control" (p. 444).

Even though it could be argued that McClintock's new findings were acknowledged and accepted with reservations, they were later cryptically criticized as I will describe in later chapters.

STUDIES WITH BROKEN CHROMOSOMES

As briefly mentioned previously, McClintock received much recognition during the years she was working toward understanding the mechanism of broken chromosomes and gene expression. As the third woman elected to the National Academy of Sciences (1944), she clearly held the respect of her peers. Colleagues requested her assistance in furthering their research investigations and encouraged their graduate students to pursue similar studies. Before describing these other "outside activities" in which McClintock was engaged during this time, I would like to return to an experiment that she reported that provided "first insights about a correlation of mutation and chromosome breaks," which she conducted during 1943 (McClintock 1944a).

McClintock's discovery of transposable elements can be traced back to extensive research on mutable loci studied by members of the maize genetics community. Maize researchers maintained contact at national and regional meetings, by personal letters and by submitting annual reports to the *Maize Genetics Cooperation News Letter* beginning in 1929 (Kass et al. 2019). Furthermore, traveling to national and international universities and colleges, maize investigators gave invited presentations on their current, and yet unpublished work. McClintock was no exception. It would be a mistake to say that McClintock worked in isolation as is often a story repeated and sometimes published (e.g., Gardner 1997).

The path to understanding transposition began with studies of broken chromosomes. The discovery unfolds in late 1943, when McClintock responded to the call to submit new findings for the forthcoming annual *Maize Genetics Cooperation News Letter 18* (*MNL*). Thus, by the end of January 1944, she had shared with her maize genetics colleagues the first insights about a correlation of mutation and chromosome breaks (McClintock 1944a, pp. 24–26).[17] That report on

[17] McClintock (1944a), *MNL*, pp. 24–26, is reprinted in Kass (2013ff.), and available at MaizeGDB; it was not reprinted in McClintock (1987).

mutations and deficiencies in Chromosome 9 was an untitled two-page account that provided the first understanding that a

> … chromosomal breakage mechanism is a 'mutation' inducing process which 'induces' the same mutant time and again …

Years later, she cited the investigation as "Breakage-fusion-bridge cycle induced deficiencies in the short arm of chromosome 9" (McClintock 1987, p. xiv). However, the term " Breakage-fusion-bridge cycle" is not used in this seminal report.

In the same issue of the *MNL*, where McClintock first announced that a chromosome breakage mechanism can induce mutants in maize, L. J. Stadler (1944, pp. 21–24) reported his results with X-rays and ultra-violet mutations. He announced that UV radiation could be considered a mutation-causing agent. He also correlated his UV radiation finding of mutations in maize kernels with the phenomena that McClintock was finding on broken chromosomes. He reasoned that although the cause of the differences he had found using X-rays vs. UV radiation was unknown, it might be somehow connected with the difference in breakage-fusion phenomena in endosperm and embryo, which might appropriately be termed the "McClintock effect." However, he also concluded that the slight increase in the frequency of mutations that he had observed as colorless sectors of endosperm on maize kernels might not be limited to deficiencies but also mutations – defined as, "The transformation of the gene to the recessive allele," which he had observed in UV-treated material over previous years.

As mentioned in Chapter 4, both Stadler and *Drosophila* geneticist H. J. Muller (among others) had been searching for the underlying cause of mutation. These researchers simultaneously and independently reported evidence, in 1927, that X-rays increase mutation rates. About 15 years later, Stadler's studies of UV radiation as a cause of increased mutations hypothesized that "deficiencies," as well as "mutations" – by then defined as a change within the gene – might also be responsible for observed abnormal phenotypes. Beginning around 1938, he and physicist Fred Uber began investigations on the various effects of the wavelengths of light causing induced chromosomal deficiencies in the maize endosperm or seen as embryo abortions. In 1942, they reported that UV light produced chromosomal deficiencies, but they did not rule out that some loss of their dominant marker genes could be the result of "mutation" rather than deficiency (Stadler & Uber 1942). They found that UV wavelengths between 2652 and 2483 Angstroms (Å) were most effective at inducing mutations, and they conjectured,

> Considering the various approximations involved, the agreement is surprisingly good and lends further support to the frequently advanced hypothesis that nucleic acid is intimately associated with the functioning of the germinal material.

Their study was probably the first to correlate the absorption of UV light with nucleic acid modifications. Only years later would DNA definitively be shown to be the heredity material, having an absorption peak of 254 nanometers (nm) [equal to 2540 Å] and a correlation would be made between the UV absorption spectrum and the DNA molecule – providing evidence for possible effects on DNA mutation rates.

After submitting her results to the maize community in late 1943, McClintock sent a manuscript to the journal *Genetics* detailing her studies of mutation and its relation to homozygous deficiencies in February 1944. It was published months later in the September issue (McClintock 1944b). This paper provided extensive details for the brief summary reported in the *Maize News Letter*, and it concludes with the statement that "The described chromosomal breakage mechanism is, then, a 'mutation inducing' process which 'induces' the same mutation time and again" (McClintock 1944b, p. 496).[18] Her discussion compares the mutants she found in maize with those observed in *Drosophila* having homozygous deficiencies, and other chromosomal anomalies including position effects. Those mutants were attributed to having a "genic change." She states, if a mutation is not

[18] McClintock (1944b) is reprinted in Kass (2013ff.); it is not included in McClintock (1987).

associated with a visible change in a chromosome, it is not possible with present methods to know what the underlying cause might be – "whether inactivation or a molecular change in a so called gene has occurred" or whether a "position effect" has occurred. She suggested that such mutants might not be caused by "genic change," if that meant a "molecular change in an isolated unit." It appeared to her that the interchangeable use of the terms "mutant" and "gene" should be avoided in order "not to prejudice ones thinking of genic action." She goes on to say that the induction of mutations had been caused by various means including X-rays, neutrons, "U.V. light," heat, age, moisture content of seeds, etc. With these methods it has not been possible "to control the particular mutation which will appear." To this list she now added "the method described in this paper which involves the repeated occurrence of breaks that are confined within the limits of a single arm of a particular chromosome." She had demonstrated this by repeatedly finding that the same mutants were associated with broken chromosomes. She correlated her findings with the similar behavior seen in the *Dt* mutant in maize, "which repeatedly induces mutations at a particular locus in another chromosome" as was found by Rhoades. The mutation process she concludes is "entirely different."

That year, her Carnegie report for 1943–1944 summarized the completion of her study of allelic relations of deficiency mutants and more importantly emphasized the "chromosome-breakage mechanism" as a means of producing directed mutations (McClintock 1944c, p. 135). She also reported collaborating with Rhoades on a project to investigate homozygous deficiencies at the *Bz* locus.[19] She recalled that it was the expansion of the experiment that she began in the summer of 1944 on the chromosome breakage mechanism and the progeny from those experiments that revealed the presence of transposable elements (McClintock 1987, p. x–xi). By the following year, she reported that several new mutants showed variegation characterized by a single change from mutant to normal-appearing tissue (McClintock 1945a, p. 110). Searching for the underlying causes led to her discovery of transposable elements as described above.

PRESENTATIONS AND COLLABORATIONS

In addition to submitting research reports for the *Carnegie Institution of Washington Year Books*, the Director of the Department of Genetics, Milislav Demerec, provided yearly reports on the "outside activities" of the members of his unit to Vannevar Bush, the President of the Carnegie Institution of Washington. These reports show the additional responsibilities McClintock assumed while engaged in studies of "Maize Genetics" and "Mutable Loci" during her early years as a Resident Investigator (1943–1951) in the Department of Genetics at Cold Spring Harbor.

McClintock shared her results at weekly staff meetings, and gave presentations at their Journal Club, and at the Noon Seminar Series.[20] In addition, she was invited to give lectures and seminars at local and distant universities, at scientific societies, and at symposia at the annual meeting of the Genetics Society of America. She gave no fewer than 15 such presentations on her maize research at venues outside of Cold Spring Harbor.

Furthermore, she gave almost as many lectures on her studies of *Neurospora* chromosomes, which she had conducted at Stanford University at the invitation of George Beadle during the fall of 1944 (McClintock 1945a,b). Years later, McClintock recalled her experience studying *Neurospora* chromosomes at Stanford as initially difficult. Then, as the story goes, after taking a walk outdoors she suddenly knew how to solve the problem, which was then completed in only a few days (Keller 1983; Comfort 2001). Of course, there is more to this story, which is revealed in her Carnegie annual reports and in the published manuscript resulting from her studies. In August 1945, she sent a manuscript to Beadle on the results of her *Neurospora* investigation. He submitted it for publication to the *American Journal of Botany*.

[19] Rhoades (1943) had first reported his project in *Maize News Letter 17*, item 2, The interaction of bronze (*bz*) with factors determining anthocyanin colors.

[20] "Report to Dr. Bush 1941 [–1953]," File Genetics Reports, CSHA.

Her study of *Neurospora* chromosomes was conducted over a period of ten weeks at Stanford during late October through December 1944. After one week, she wrote to Demerec that she "felt considerable fright" at first, but after working "all day and burning the midnight oil," her confidence improved.[21] They were trying some new methods for staining the prophase chromosomes. She was "roped in for some talks on chromosomes in general," which was not "too much of a task." She gave several seminars on the cytology of *Neurospora* to Beadle's group at Stanford.[22]

As her published paper indicated, this was not the first time she had observed *Neurospora* chromosomes. She and one of her former students at the University of Missouri had first observed a different number of chromosomes in *Neurospora* than had been reported in the literature. And another former graduate student at Missouri had prepared excellent aceto-carmine preparations and also observed seven chromosomes (McClintock 1945, p. 671), which she confirmed to be the haploid number for species *Neurospora crassa*. The investigations at Stanford were done in collaboration with Mrs. Mary B. Houlahan, Dr. Herschel K. Mitchell, and Dr. Lotti Steinitz, who supplied the mutant and wild-type strains and offered helpful suggestions regarding techniques (McClintock 1945, p. 671).[23] It took some time to determine that aceto-orcein was the best chromosome stain followed by aceto-carmine if nucleoli were to be viewed. "The observations of *Neurospora* included determinations of chromosome number, absolute and relative sizes of chromosomes, centromere positions, internal organization of the chromosomes, zygote formation, and chromosome behavior in the two meiotic mitoses and the equational mitosis that follows, as well as scattered observations of several chromosomal translocations" (Demerec 1945, pp. 106–107).

During the ten weeks that she was at Stanford, she also presented a lecture on "The production of specific mutation in Maize" to the Biology Seminar. In December, she was invited to the University of California, Berkeley, to give a seminar on "Deficiency as a cause of Mutation."[24] The latter seminars covered her recent investigations on deficiencies and the mutation-inducing process that was published just before she arrived in California (McClintock 1944b). Plant geneticist and evolutionary botanist, G. Ledyard Stebbins, a professor in the Division of Genetics at Berkeley by that time, recognized the importance of McClintock's 1944 studies of homozygous minute deficiencies. He had known McClintock's work since his graduate school days at Harvard and as a faculty member in a nearby college in the 1930s (Smocovitis & Ayala 2004). Yet by 1946, Stebbins had "no reason to believe that these regions [of deficiency] are more mutable than other parts of the maize complement [genome] or that the maize plant is unusual in the type of mutations it can produce" (Stebbins 1950, p. 79).[25]

Such deficiencies were also found in *Drosophila*, Stebbins wrote, and "may play a role in evolution similar to that of point mutations" (Stebbins 1950, p 79). Fedoroff (2000) noted that Stebbins' book reflected the prevailing view that deficiencies and other chromosomal aberrations were not the materials "to fashion … the products of evolution" and he concluded that "the majority of evolutionarily important changes … are attributable to classical genetic 'point' mutations." In his discussion of the effects of point mutations on evolution, Stebbins emphasized that it must be kept in mind that "the most direct, immediate action of genes is on the process of development, and genes produce effects on the visible, morphological characters of the mature organism only through their influence on development" (Stebbins 1950, p. 85). With respect to chromosome deficiencies, McClintock (1944b) had demonstrated that a piece of a chromosome disappeared from a genetic locus causing mutation. A few years later, she inferred that somatic variegation (mutation) reflects the frequent

[21] Letter B. McClintock to M. Demerec, November 3. 1944, File. B. McClintock, CSHA.

[22] "Report to Dr. Bush 1941 [–1953]"; several seminars on the cytology of *Neurospora* fall 1944; CSHA.

[23] McClintock (1945a): "Progress was greatly accelerated … by the generous and cooperative support of the members of the department."

[24] "Report to Dr. Bush 1941 [–1953]"; Biology Seminar Stanford, November 22, 1944; Berkeley Seminar, December 14, 1944; CSHA.

[25] Stebbins' (1950) *Variation and Evolution in Plants* was based on a series of lectures given at Columbia University in 1946; see Preface dated April 1949.

movement (transposition) of the chromosome-breaking locus (*Ds*) during development (McClintock 1948, 1949; Fedoroff 2000).

In 1946, McClintock was again invited by Beadle to work on the cytology of *Neurospora*. He had become Chairman of the Division of Biology, William G. Kerckhoff Laboratories of the Biological Sciences at the California Institute of Technology that summer. Thomas Hunt Morgan, who had established the Division in 1928, had retired, and Alfred H. Sturtevant then assumed the Chairman's role when Morgan died in 1945. Beadle was recruited for the Chairman's job, and he arrived in Pasadena in July 1946, along with many of his former Stanford associates. He had initially urged her to return to Stanford to continue her studies of the chromosomes of *Neurospora*, and now he suggested she spend the winter of 1946 at Caltech.[26] McClintock planned to leave for Caltech in November, and the results of the summer harvest had given her some new and interesting results that required some "thought and planning." She also mentioned receiving a note from Jesse Singleton, a former University of Missouri graduate student, who recently enrolled at Caltech. He had "stage fright," and she suggested that he would need encouragement to get over his fears.[27]

Beadle told McClintock that Singleton wished to work on the cytology of *Neurospora* for his doctoral work. He and Sterling [Emerson] explained how difficult this would be for someone who had virtually no experience. However, Beadle said Singleton could pursue this problem under the condition that she would come out to Caltech and "get him started." It was an option that she could accept or decline since he "did not know how good this fellow is."[28] McClintock quickly replied that she would be very glad to help Singleton get started and suggested the type of phase-contrast microscopic lenses needed for the project. She regretted not having taken photographs of *Neurospora* chromosomes during her time at Stanford, and a photographic series of all stages were needed. Singleton, she said, had been a graduate student in Botany at the University of Missouri during her last year there. He had come to her for help with his project when he received no direction. While in Europe, he wrote to her asking about continuing his graduate career, and of the many places she recommended, he chose Caltech.[29]

McClintock drove to California in the company of her father in November 1946.[30] At Caltech, she was appointed a Visiting Professor of Biology in the Biology Division for the academic year 1946–1947.[31] Residing at the home of relatives in town, she spent the winter working in cooperation with Singleton on the cytology of *Neurospora* (McClintock 1947, pp. 151–152).[32] Together, they accomplished her goals (1) to improve the techniques so that *Neurospora* chromosome conditions in the important stages might be more readily revealed, and (2) to obtain a set of photographic illustrations of the behavior of the chromosomes during the development of the fungal ascospore (ascosporogenesis). McClintock later applied their modified *Neurospora* staining procedures to study maize chromosomes (McClintock 1947, p. 148).

Indeed, Beadle and Sturtevant had nothing to fear regarding Singleton's cytology experience. At the University of Missouri, McClintock had helped him with his Master's thesis problem on "A study of colchicine-induced tetraploidy and other colchicine effects in *Xanthium chinense* Miller" (Singleton 1941; see also Chapter 6). While serving in the European theater during World War II, he made the decision to return to graduate work after his discharge from the Army in 1945 (*Proceedings ... Indiana* 1962). Under McClintock's guidance, and with new microscopic

---

[26] Letter B. McClintock to G. Beadle, November 27, [1945], time for *Neurospora* chromosomes; Letter G. Beadle to B. McClintock, August 13, 1946, arrived in Pasadena; spend winter of 1946; McClintock File, Beadle Collection, Caltech Archives (henceforth BC/CA).

[27] Letters B. McClintock to G. Beadle, September 4, 1946; October 6, 1946, note from Singleton; McClintock File, BC/CA.; see also Chapter 6.

[28] Letter G. Beadle to B. McClintock, October 9, 1946, McClintock File, BC/CA.

[29] Letter B. McClintock to G. Beadle, October 11, 1946, McClintock File, BC/CA.

[30] Letter B. McClintock to G. Beadle, November 8, 1946, Leaving for California on November 10, McClintock File, BC/CA.

[31] Letter President L. A. DuBridge to B. McClintock, December 12, 1946, McClintock File, BC/CA.

[32] Letter B. McClintock to G. Beadle, November 8, 1946, residing with relatives, McClintock File, BC/CA.

equipment, Singleton took dozens of photographs of the developmental stages of *Neurospora* asci using different staining techniques, during the winter of 1946–1947.[33] Many of these were depicted in his doctoral thesis, which he completed in 1948 (Singleton 1948).

At the annual meeting of the Genetics Society of America, McClintock and Singleton gave an invited presentation on their findings to the joint Genetics Society and Society of Naturalists Symposium, in Chicago, Illinois (December 29, 1947; *Records* 1947, p. 17). Over the next two years, McClintock also presented their *Neurospora* chromosome findings at the New York Microscopical Society in November 1947, and at lectures and seminars that included illustrated lantern slides from their collections.[34] After graduating, Singleton became a faculty member in the Department of Botany at the University of Missouri (1948–1954; see Chapter 6). His 1953 publication based in part on his Ph.D. thesis provided a full account of *Neurospora* chromosome morphology and behavior (Singleton 1953). He used many of McClintock's photographs, along with his own, to document the reproductive cycle of *Neurospora*.

Zlata Demerec-Dayton, the oldest daughter of Milislav Demerec, was completing her bachelor's degree in Botany at Cornell when she took an interest in the cytogenetics of *Neurospora*. In October 1949, she approached McClintock about the possibility of doing a problem on the subject for a Master's degree in the Department of Plant Breeding at Cornell. McClintock "was very encouraging and offered to show [her] the technics."[35] In November 1949, she was recommended as a Master's candidate to work with Professor [Adrian] Srb, in "cooperative work with Doctor McClintock." Srb had obtained his doctorate in biochemical genetics with George Beadle at Stanford University. He and fellow graduate student Norman Horowitz had studied the genetic control of the ornithine cycle in *Neurospora*. Their work, along with those of others in Beadle's laboratory, led to the formulation of the "one gene-one enzyme" hypothesis of gene action, for which Beadle, and his research associate Edward Tatum, shared the Nobel Prize in 1958 (Nasrallah 2012).

Upon completing her undergraduate program in February 1950, Mrs. Dayton was accepted into the Ph.D. program.[36] In January/February 1950, she spent time with McClintock at Cold Spring Harbor, where she learned the "aceto-orcein smear" technique "to stain *Neurospora* chromosomes." Unfortunately, she "could not get consistent results so the project never materialized."[37] Zlata left the program due to ill health and returned to Cold Spring Harbor, where she was born and raised. She later completed a Master's degree at Cornell on a study of mutability in *Salmonella* (Hartman 1956). Her project was relevant to the collaborative research she had worked on with her father at Cold Spring Harbor (Demerec et al. 1955). McClintock, however, would continue to study and inspire others to investigate the chromosomes of *Neurospora* (McClintock 1954; Perkins 1992).

McClintock was also asked to review doctoral dissertations for students at Columbia University and to participate in their thesis defense. It was the custom at the time for the Major Professor to appoint outside reviewers for such occasions. Franz Schrader asked Kenneth Cooper of Princeton University and McClintock to serve on the committee to review and examine his student, Hans Ris, on the cytological and experimental analysis of meiotic behavior in the bearberry aphid. Ris recalled that the Columbia Zoology Department required that a thesis be accepted for publication in a reputable journal before the Ph.D. would be awarded. He had no memory of McClintock making any comments or recommendations on his thesis.[38]

[33] *Neurospora crassa* photographs, Winter 1946–1947, Folder 1, McClintock Collection, Caltech Archives.

[34] "Report to Dr. Bush 1941 [–1953]," *Neurospora* seminars, November 19, 1947, Yale; December 7, 1947, Princeton; April 6, 1948, Torrey Botanical Club; April 29, 1948, Academy of Natural Sciences, PA; CSHA. See also Perkins (1992).

[35] Letter Mrs. Zlata Demerec-Dayton to Dr. [S. F.] Atwood [Dept. chair], October 25, 1949; Graduate School File, Department of Plant Breeding, Cornell University.

[36] Letter Dr. Sanford S. Atwood [Dept. chair], to Mrs. Zlata D. Dayton, November 8, 1949; B.S., February 1950, Ph.D. Candidate, February 1950, Major Genetics, A. M. Srb; Cytology L. F. Randolph; Biochemistry, H. H. Williams, Graduate School File, Department of Plant Breeding, Cornell University.

[37] Letter B. McClintock to G. Beadle, January 31, 1950, BC/CA; Letter B. McClintock to M. Rhoades, February 9, 1950, Rhoades Collection, Lilly Library (henceforth RC/LL). Letter Zlata [Demerec] Hartman to Lee B. Kass, May 19, 1998.

[38] email Hans Ris to Lee B. Kass, June 9, 1998; Ris' exam was May 22, 1942, Letter B. McClintock to M. Rhoades, May 15, 1942, RC/LL. The thesis was published in the *Journal of Experimental Biology* in July 1942.

McClintock was on the thesis committee of Rhoades' graduate student Ruth Sager. She entered Columbia Graduate School in 1945 and began a research project on "The mutability of the waxy locus" in maize with Marcus Rhoades (Biermann 1997). By this time, Rhoades had tenure and he accepted Sager, a Jewish student, despite the warnings he had received from his Department Chair and other colleagues in the Columbia Department of Botany (see Chapter 6). The following year, Drew Schwartz, another Jewish student, would also join Rhoades' laboratory as a graduate student of maize genetics. Sager consulted with McClintock on her thesis problem and received maize stocks from McClintock for the work. Sager submitted annual reports to the *Maize Genetics Cooperation News Letter*, and completed the writing of her thesis in 1948 (Sager 1948, cited in Sager 1951). At this time, Rhoades left Columbia for a new faculty position at the University of Illinois, Champaign-Urbana. Sager remained in New York, then visited Stanford University, and ultimately defended her thesis in December 1949 (Schmitt 2018). It is not clear who appointed McClintock to her committee, but McClintock was a reader and examiner for the thesis defense. McClintock initially was not pleased with the analysis of Sager's thesis data. Apparently, it did not agree with the results she had found in her own maize stocks for the cause of mutability of the waxy locus.[39] After receiving further data on biochemical analysis of maize starch, which Sager had performed at Columbia and at Stanford University, McClintock seems to have resolved her differences with part of Sager's thesis work.[40] Sager (1951) published her thesis results in *Genetics*.[41]

The results Sager reported for the mechanism underlying the mutability of the waxy locus were indeed compatible with, but in some ways different from, those found by McClintock. Sager concluded that "the time and frequency of mutation of wx^m [mutable waxy] are genetically controlled." Yet, in contrast to what McClintock had found, her data supported a hypothesis that the mosaic-waxy phenomenon "appeared to be a case of 'point mutation': mutation with which no other genic or cytological event has been correlated" (Sager 1951, p. 530). Furthermore, Sager concluded that the instability of the mutant waxy allele [wx^m] "could not be ascribed to the influence of another gene" and in addition "could not be accounted for by any gross chromosomal aberration or correlated with meiotic crossing over." Sager soon changed the focus of her research and made her reputation in the field of cytoplasmic inheritance, and cancer genetics (Schmitt 2018). It is notable that Sager's biographical memoir published for the National Academy of Sciences makes only a passing mention of a "1948 Ph.D." in maize genetics with Marcus Rhoades (Pardee 2001).

SCIENTIFIC SOCIETIES, COMMITTEE MEMBERSHIPS, AND RECOGNITION

In April 1944, at age 41, McClintock was the third woman elected to the U.S. National Academy of Sciences in the section of Botany. That year, she also received a star in botany next to her entry in the seventh edition of *American Men of Science*. Although the U.S. entry into World War II precluded holding annual meetings of the Genetics Society of America from 1942 through 1945, the mail ballot of late 1944 resulted in the election of Barbara McClintock as President for 1945. McClintock had been a member of the Joint Genetics Sections of the American Society of Zoologists and the Botanical Society of America since 1927 and became a member of the newly formed Genetics Society of America in 1932.[42] At the end of 1945, McClintock was appointed to the Panel of the Division of Biology of the Committee on Growth of the National Research Council (NRC, a unit of the National Academy of Sciences).

In January 1946, McClintock also served on the Panel on Virus and the Panel on Mutation with the Division of Chemistry of the Committee on Growth for the NRC. That same month, McClintock, Milislav Demerec, and L. J. Stadler organized and participated in the three-day Conference on

39 Letter B. McClintock to M. Rhoades, May 6, 1949, RC/LL.

40 Letter B. McClintock to M. Rhoades, August 4, 1949, Letter Ruth Sager to Barbara McClintock, July 25, 1949, RC/LL.

41 See Columbia University Catalog, which lists R. Sager (1951) as date of her Doctoral Dissertation, in contrast to the citation as Ph.D. 1948, in Sager's (1951) *Genetics* article. See also Pardee 2001 reporting the Ph.D. as 1948.

42 See *Records of the Genetics Society of America*, 1944–1951.

Mutations (January 26–28, 1946), sponsored by the Committee on Growth of the NRC. The report of the Conference was prepared by McClintock and gave "an excellent picture of the current trend of research in that branch of genetics."[43] The conference was organized "to bring to focus the present concepts of the process of mutation and genic action and their bearing on various types of cellular activity." To answer these questions requires the cooperative endeavors of the "biologist, the biophysicist and the biochemist." The Conference members, therefore, included those with interests in genetics, biochemistry, and biophysics. The sessions were devoted to Spontaneous Mutations, Induced Mutations, Extranuclear Factors in Heredity, The Nature of Genic Action, Genetic Aspects of Atypical Growth, and Present Concepts of Physical Nature of Genic Substances. Many of the attendees at this conference would answer the questions raised, and receive recognition with election to the National Academy of Science, the Albert Lasker Basic Medical Research Award, or a Nobel Prize in Physiology or Medicine.

At the end of WWII, and after a delay of four years, the 1945 annual meeting of the Genetics Society was held in March 1946. As Society President, McClintock was responsible for appointing all society committees. Although she and members of the executive committee were unable to attend the meeting in March 1946, she appointed a committee to consider methods for preparing its nominations (*Records* 1946, No. 15). The two immediate past presidents along with the current President, Vice President, and the Secretary of the Genetics Society constituted the executive committee. McClintock therefore continued active membership in society affairs.

In 1946, McClintock was elected to the American Philosophical Society (APS), founded in Philadelphia by Benjamin Franklin and others in the mid-1700s, "for promoting useful knowledge." Membership of the APS "honors extraordinary accomplishments in all fields." The following year, McClintock was recognized with the American Association of University Professors (AAUW) Achievement Award. She accepted the award in April 1947 at their annual meeting held in Dallas, Texas. The award was given in recognition of her years of work in maize genetics research. She acknowledged the importance of her former fellowships, which allowed her to be free of teaching and other commitments. Such comments were no doubt inspired by the committee she was working with to promote Science Foundation Legislation, an idea stemming from the recommendation of Carnegie President Vannevar Bush. In accepting the award, she appealed for more government encouragement to science via a National Science Foundation (University Women 1947).

While McClintock and Past President George Beadle (elected 1946) were serving on the Genetics Society executive committee, in 1947, she and Curt Stern were asked to represent the Genetics Society on the Inter-Society Committee on Science Foundation Legislation. They submitted a report to Society President H. J. Muller with the disappointing result that President Truman had vetoed the bill that had been passed by both houses of the U.S. Congress. Three years later, a revised bill was put forward. On May 10, 1950, Truman signed into law the National Science Foundation Act of 1950.[44]

Additional Genetics Society Committee responsibilities included Chairman of the Genetics Society of America Nominating Committee (1949) and Program Committee Chairman for 1951. During these years, she also held memberships in the Botanical Society of America, the American Society of Naturalists, the American Association for the Advancement of Science (AAAS), and the Society of Sigma Xi.

In summary, during McClintock's first years as a Resident Investigator at the Department of Genetics (1943–1951), she conducted pioneering research that led to her discovery of transposition, and the genetic loci, *Ds* and *Ac* (transposable elements). Her breakthrough was based on years of

[43] "Report to Dr. Bush 1941 [–1953]," November 1946, CSHA. Letter M. Demerec to W. M. Gilbert, Executive Officer, Carnegie Institution of Washington, February 16, 1946, includes the Program and Summary of the meeting, File Genetics Reports, Carnegie Institution of Washington, DC, Archives.

[44] For a detailed history of the founding of the National Science Foundation, see Blanpied (1998).

maize cytogenetics research, which she began as a graduate student in the 1920s (Chapters 3–5). When her conclusions began to question the prevailing dogma of the theory of the gene, she felt that many of her colleagues could not, at that time, accept the new paradigm toward an understanding of the term "gene" or its regulation during development.

McClintock shared her groundbreaking results in lectures and seminars, while concurrently mentoring and collaborating with others who were working on related problems in cytogenetics. She received awards and recognition from her peers and held influential committee roles with the Genetics Society of America and the National Academy of Sciences.

LITERATURE CITED

Altenburg, E. 1957. *Genetics, Revised Edition*, Henry Holt and Company, New York.

Bonner, D. 1951. Gene enzyme relationships in *Neurospora*, pp. 143–158, *In*, K. B. Warren (Ed.), *Genes and Mutations, Cold Spring Harbor Symposia on Quantitative Biology, Volume XVI*. The Biological Laboratory, Cold Spring Harbor, NY. [©1952]

Biermann, C. A. 1997. Ruth Sager, pp. 463–476, *In*, L. S. Grinstein, C. A. Biermann, and R. K. Rose (Eds.), *Women in the Biological Sciences: A Biobibliographic Sourcebook*, Greenwood Press, Westport, CT.

Blanpied, W. A. 1998. Inventing U.S. Science Policy, *Physics Today* Vol. 51 (No. 2, February 1998), pp. 34–40. American Institute of Physics. See reprint at NSF https://www.nsf.gov/about/history/nsf50/science_policy.jsp

Brown, N. M. 1995. Life's Jumps [An interview with Nina V. Fedoroff]. *Research at Penn State*. Vol. 16 (No. 3. September, 1995) 11 p. https://www.rps.edu/sep95/jumps.html [Accessed July 5, 2006, no longer available online.]

Carnegie Institution Year Books 1 through 99, years 1902 through 2000. https://carnegiescience.edu/carnegie-year-books

Caspari, E. 1951. Quantitative Biology. *Science*. Vol. 114 (No. 2958, September 1, 1951), p. 3. [A review of the Sixteenth Symposium on Quantitative Biology, sponsored by the Biological Laboratory of the Long Island Biological Association, Cold Spring Harbor, June 7–15.]

Coe, E. H. and L. B. Kass. 2005. Maize Genetics Cooperation News Letter files: Expanded chronological list of materials and related cooperation. *Maize Genetics Cooperation Newsletter* 79 (October 31), pp. 72–76, http://mnl.maizegdb.org/mnl/79/06CoeKass.htm (accessed April 2005).

Comfort, N. 1995. Two genes, no enzyme: a second look at Barbara McClintock and the 1951 Cold Spring Harbor Symposium, Perspective. *Genetics* Vol. 140 (No. 4, August 1995), pp. 1161–6. https://doi.org/10.1093/genetics/140.4.1161.

Comfort, N. 2001. *The Tangled Field*, Harvard University Press, Cambridge, MA.

Demerec, M. 1942. Foreword, *In*, K. S. Brehme (Ed.), *Cold Spring Harbor Symposia on Quantitative Biology* Vol. X, 1942. *The Relation of Hormones to Development*, The Biological Laboratory, Long Island Biological Association. Inc. https://symposium.cshlp.org/content/10/local/front-matter.pdf

Demerec, M. 1945. Department of Genetics, *Carnegie Institution of Washington Year Book No. 44* (1944–1945) (July 1, 1944–June 1, 1945). [Issued December 1945]

Demerec, M., I. Blomstrand and Z. E. Demerec. 1955. Evidence of complex loci in *Salmonella. Proceedings of the National Academy of Sciences* Vol. 41 (No. 6, June 15), pp. 359–364.

Fedoroff, N. V. 1983. Controlling elements in maize, Chapter 1, pp. 1–63, *In*, J. A. Shapiro (Ed.), *Mobile Genetic Elements*, Academic Press, New York.

Fedoroff, N. V. 1984. Transposable genetic elements in maize. *Scientific American* Vol. 250 (No. 6, June), pp. 84–90, 95–98.

Fedoroff, N. V. 1989. Maize transposable elements, pp. 375–411, *In*, M. Howe and D. Berg (Eds.), *Mobile DNA*, American Society for Microbiology, Washington, DC.

Fedoroff, N. V. 1991. THE RESTLESS GENE, how the colors of Indian corn have led to an understanding of wandering DNA. *The Sciences* Vol. 31 (No. 1, January–February 1991), pp. 22–28. [No literature cited; a publication of The New York Academy of Sciences]

Fedoroff, N. V. 1992. Maize transposable elements: A story in four parts, pp. 389–415, *In*, N. Fedoroff and D. Botstein (Eds.), *The Dynamic Genome: Barbara McClintock's Ideas in the Century of Genetics*, Cold Spring Harbor Laboratory Press, Plainview, NY. [See especially "Part II: How Transposition was discovered," pp. 392–393; For references see Fedoroff 1983 and 1989]

Fedoroff, N. V. 1998a. The discovery of transposable elements, pp. 89–104, *In*, S.-D. Kung and S.-F. Yang (Eds.), *Discoveries in Plant Biology*, Vol. 1, World Scientific Publishing Co. Pte. Ltd., Singapore. [February 18, 1998]

Fedoroff, N. V. 1998b. Marcus Rhoades and transposition. *Genetics* Vol. 150 (No. 3, November 1998), pp. 957–961. https://academic.oup.com/genetics/issue/150/3?login=false

Fedoroff, N. V. 2000. Transposons and genome evolution in plants. *Proceedings of the National Academy of Sciences* Vol. 97 (No. 13, June 20, 2000), pp. 7002–7007.

Fedoroff, N. V. 2001. How jumping genes were discovered. *Nature Structural Biology* Vol. 8 (No. 4, April), pp. 300–301. https://www.nature.com/nsmb/journal/v8/n4/pdf/nsb0401_300.pdf

Fedoroff, N. V. 2012a. Transposable elements, epigenetics, and genome evolution. *Science* Vol. 338 (No. 6108, November 9, 2012), pp. 758–767. [AAAS Presidential Address, given February 2012]

Fedoroff, N. V. 2012b. McClintock's challenge in the 21st century. *Proceedings of the National Academy of Sciences* Vol. 109 (No. 50, December 11, 2012), pp. 20200–20203. [Reviews the history of the discovery of transposable elements and includes excellent illustrations of maize kernels that demonstrate the phenotypic characters that McClintock had observed.]

Fedoroff, N. V. 2013a. The discovery of transposition, pp. 3–13, *In*, N. V. Fedoroff (Ed.), *Plant Transposons and Genome Dynamics in Evolution*, John Wiley & Sons, Inc., New York. [First published February 4, 2013; Modified from Fedoroff 1998a]

Fedoroff, N. V. (Ed.) 2013b. *Plant Transposons and Genome Dynamics in Evolution*, John Wiley & Sons, Inc., New York. [First published: February 4, 2013]

Fedoroff, N. V. 2013c. McClintock and epigenetics: Insights from changes in phase of transposition activity, pp. 3.34–3.39, *In*, L. B. Kass (Ed.), 2013ff. *Perspectives on Nobel Laureate Barbara McClintock's Publications (1926–1984): A Companion Volume*, Volumes I & II, The Internet-First University Press. https://hdl.handle.net/1813/34897

Fedoroff, N. 2017. From maize transposons to the GMO Wars, pp. 39–58, *In*, L. S. Privalle (Ed.), *Women in Sustainable Agriculture and Food Biotechnology*. Part of the Women in Engineering and Science, Springer, Cham. https://doi.org/10.1007/978-3-319-52201-2_3

Fedoroff, N. and D. Botstein. 1992. Introduction, p. 1–4, *In*, N. Fedoroff and D. Botstein (Eds.), *The Dynamic Genome: Barbara McClintock's Ideas in the Century of Genetics*, Cold Spring Harbor Laboratory Press, Plainview, NY.

Fedoroff, N. V. and N. M. Brown. 2004. Tinkering with evolution, Chapter 5, pp. 85–105. *Mendel in the Kitchen : A Scientist's View of Genetically Modified Foods*, Joseph Henry Press, Washington, DC. [Barbara McClintock and transposable elements mentioned on pp. 83, 97–105, 164, 323.]

Gardner, A. L. 1997. Barbara McClintock, geneticist, 1902–1992, pp. 307–324, *In*, M. L. Matyas and A. E. Haley-Oliphant (Eds.), *Women Life Scientists: Past, Present, and Future – Connecting Role Models to the Classroom Curriculum*, American Physiological Society, Bethesda, MD.

Goldschmidt, R. B. 1940. Chromosomes and genes, pp. 56–66, *In*, F. R. Moulton (Ed.), *The Cell and Protoplasm*, American Association for the Advancement of Science Publication No. 14, The Science Press, Lancaster, PA.

Goldschmidt, R. B. 1950. Marginalia to McClintock's work on mutable loci in maize. *The American Naturalist* Vol. 84 (No. 819, November–December, 1950), pp. 437–455.

Goldschmidt, R. B. 1951. Chromosomes and genes, pp. 1–11. *In*, K. B. Warren (Ed.), *Genes and Mutations, Cold Spring Harbor Symposia on Quantitative Biology, Volume XVI*, The Biological Laboratory, Cold Spring Harbor, NY. [©1952]

Grossbach, U. 2009. Seventy-five years of development genetics: Ernst Caspari's early experiments on insect eye pigmentation, performed in an academic environment of political suppression. *Genetics* Vol. 181 (No. 4, April 1, 2009), pp. 1175–1182. https://doi.org/10.1534/genetics.109.102723

Hartman, Z. 1956. *Induced mutability in Salmonella typhimurium*, M. S. Cornell University, September 1956, 118 pp. https://newcatalog.library.cornell.edu/catalog/304440

Hedges, R. W. and A. E. Jacob. 1974. Transposition of ampicillin resistance from RP4 to other replicons. *Molecular and General Genetics* Vol. 132 (No. 1, 30 August 1974), pp. 31–40. https://link.springer.com/article/10.1007/BF00268228

Jones, R. N. 2005. McClintock's controlling elements: the full story. *Cytogenetic and Genome Research* Vol. 109 (No. 1–3, March), pp. 90–103. [References are listed for McClintock's annual reports in the *Carnegie Institution of Washington Year Books*, but not cited in the text.]

Kass, L. B. (Ed.). 2013ff. *Perspectives on Nobel Laureate Barbara McClintock's Publications (1926–1984): A Companion Volume*. Volumes I & II, The Internet-First University Press. https://hdl.handle.net/1813/34897 [Online December 30, 2013, updates March 2014; Vol. III, Increments No. 1–3, May 2016; Vol. IV, Increment No. 4, April 2020]

Kass, L. B., E. H. Coe, M. N. Cook, M. E. Smith, and J. L. Singer (Eds.). 2019. *Founding of the Maize Genetics Cooperation News Letter at Cornell University: A 90th Anniversary Tribute*. (Foreword by E. S. Buckler) [e-book] Internet-First University Press, Ithaca, NY. https://ecommons.cornell.edu/handle/1813/66550. [Online October 10, 2019]

Kass, L. B. and P. Chomet. 2009. Barbara McClintock, pp. 17–52, *In*, J. Bennetzen and S. Hake (Eds.), *Handbook of Maize: Genetics and Genomics*, Springer, New York.

Keller, E. F. 1983. *A Feeling for the Organism. The Life and Work of Barbara McClintock*, 1st edition, W. H. Freeman, San Francisco, CA.

Keller, E. F. 1993. *A Feeling for the Organism. The Life and Work of Barbara McClintock*, 10th Anniversary edition, W. H. Freeman, New York.

Kittridge, M. 1991. *Barbara McClintock: Biologist* (American Women of Achievement series), Chelsea House, New York.

McClintock, B. 1931. Cytological observations of deficiencies involving known genes, translocations and an inversion in *Zea mays*. *University of Missouri, College of Agriculture, Agricultural Experiment Station, Research Bulletin* 163 [December, authorized December 23, 1931], pp. 1–30.

McClintock, B. 1943. Maize genetics: Studies with broken chromosomes. Tests of the amount of crossing over that may occur within small segments of a chromosome. Deficiency mutations: Progressive deficiency as a cause of allelic series. *Carnegie Institution of Washington Year Book No. 42*, 1942–1943 [Issued December 7, 1943, submitted June 1943], pp. 148–152.

McClintock, B. 1944a. Carnegie Institution of Washington, Department of Genetics, Cold Spring Harber [*sic*], Long Island, N.Y. [This report is untitled in the *MGCNL*. This is a report on deficiencies in Chromosome 9]. *Maize Genetics Cooperation News Letter*. 18 [January 31, 1944, submitted 1943], pp. 24–26. [p. 26. "In contrast to most mutation inducing agents, the chromosomal breakage mechanism is a "mutational inducing process which 'induces' the same mutant time and again."]

McClintock, B. 1944b. The relation of homozygous deficiencies to mutations and allelic series in maize. *Genetics* Vol. 29 (No. 5, September 1944), received February 8, 1944), pp. 478–502. [Reprinted in Kass 2013ff., https://hdl.handle.net/1813/34897]

McClintock, B. 1944c [July 1, 1943–30 June, 1944]. Maize genetics: Completion of the study of the allelic relations of deficiency mutants. The chromosome-breakage mechanism as a means of producing directed mutations. Continuation of the chromatid type of breakage-fusion-bridge cycle in the sporophytic tissues. Homozygous deficiency as a cause of mutation in maize. *Carnegie Institution of Washington Year Book No. 43*, 1943–1944 [Issued December 15, 1944, submitted June 1944], pp. 127–135.

McClintock, B. 1945a [July 1, 1944–30 June 1945]. Cytogenetic studies of maize and Neurospora: Induction of mutations in the short arm of chromosome 9 in maize. Preliminary studies of the chromosomes of the fungus *Neurospora crassa*. *Carnegie Institution of Washington Year Book No. 44*, 1944–1945 [Issued December 14, 1945], pp. 108–112.

McClintock, B. 1945b. *Neurospora*. I. Preliminary observations of the chromosomes of *Neurospora crassa*. *American Journal of Botany* Vol. 32 (No. 10, December 1945, issued January 14, 1946) [Received August 28, 1945], pp. 671–678.

McClintock, B. 1946 [July 1, 1945–30 June, 1946]. Maize genetics: Continuation of the study of the induction of new mutants in chromosome 9. Modification of mutant expression following chromosomal translocation. The unexpected appearance of a number of unstable mutants. *Carnegie Institution of Washington Year Book No. 45*, 1945–1946 [Issued December 13, 1946, submitted June 1946], pp. 176–186.

McClintock, B. 1947 [July 1, 1946–30 June, 1947]. Cytogenetic studies of maize and Neurospora: The mutable Ds locus in maize. Continuation of studies of the chromosomes of *Neurospora crassa*. *Carnegie Institution of Washington Year Book No. 46*, 1946–1947 [Issued December 12, 1947, submitted June 1947], pp. 146–152.

McClintock, B. 1948 [July 1, 1947–30 June, 1948]. Mutable loci in maize: Nature of the Ac action. The mutable c loci. The mutable wx loci. Conclusions. *Carnegie Institution of Washington Year Book No. 47*, 1947–1948 [Issued December 10, 1948, submitted June 1948], pp. 155–169.

McClintock, B. Unpublished, January 1949. Memorandum to Marcus Rhoades 54 pages plus, 6 tables; Available at APS, McClintock Coll. No 79, series III, Works by, and online; This link provides access to the full January 1949 54 page Memo, plus Tables (Totaling 64 pages) to Marcus Rhoades, page 17 is missing; https://collections.nlm.nih.gov/catalog/nlm:nlmuid-101584613X139-doc

McClintock, B. Unpublished, April 1949. "The origin of transposed *Ds*, Case I," 50 pages plus 40 tables and diagrams. [Available at APS, McClintock Coll. No 79, series III, Works by, filed with "The Ac locus".]

McClintock, B. 1949 [July 1, 1948–30 June, 1949]. Mutable loci in maize: The mechanism of transposition of the *Ds* Locus. The origin of *Ac*-controlled mutable loci. Transposition of the *Ac* locus. The action of *Ac* on the mutable loci it controls. Mutable loci *c m-2* and *wx m-1*. Conclusions. *Carnegie Institution of Washington Year Book No. 48*, 1948–1949 [Issued December 9, 1949, submitted June 1949], pp. 142–154.

McClintock, B. 1950a. The origin and behavior of mutable loci in maize. *Proceedings of the National Academy of Sciences* Vol. 36 (No. 6, June 15, 1950), pp. 344–355.

McClintock, B. 1950b [July 1, 1949–June 30, 1950]. Mutable loci in maize: Mode of detection of transpositions of Ds. Events occurring at the Ds locus. The mechanism of transposition of Ds. Transposition and change in action of Ac. Consideration of the chromosome materials responsible for the origin and behavior of mutable loci. *Carnegie Institution of Washington Year Book No. 49*, 1949–1950 [Issued December 15, 1950, submitted June 1950], pp. 157–167.

McClintock, B. 1951a [July 1, 1950–June 30, 1951]. Mutable loci in maize. *Carnegie Institution of Washington Year Book No. 50*, 1950–1951 [Issued December 14, 1951, submitted June 1951], pp. 174–181.

McClintock, B. 1951b [©1952, Symposium held June 1951]. Chromosome organization and genic expression, pp. 13–47, *In*, K. B. Warren (Ed.), *Genes and Mutations, Cold Spring Harbor Symposia on Quantitative Biology*, Volume XVI [June 7–15, 1951], The Biological Laboratory, Cold Spring Harbor, Long Island, NY. [©1952. [This reference has been cited as 1951 or 1952 – see *Science Citation Index*; McClintock (1987) lists it as 1951]

McClintock, B. 1953. Induction of instability at selected loci in maize. *Genetics* Vol. 38 (No. 6, November 1953, issued 20 January 1954, received April 14, 1953), pp. 579–599.

McClintock, B. 1954 [July 1, 1953–June 30, 1954]. Mutations in maize and chromosomal aberrations in Neurospora: Mutations in maize. Chromosome aberrations in *Neurospora*. *Carnegie Institution of Washington Year Book No. 53*, 1953–1954 [Issued December 10, 1954, submitted June 1954], pp. 254–260. [*Neurospora* study in collaboration with Mary B. *Mitchell.]*

McClintock, B. 1956 [Issued February 1956, Symposium held June 15–17, 1955]. Intranuclear systems controlling gene action and mutation, pp. 58–74, *In, Mutation, Brookhaven Symposia in Biology,* No. 8. Biology Department, Brookhaven National Laboratory, Upton, NY. [No editor listed for this volume. R. C. King, Symposium Chairman. Cited in McClintock (1987) as "Issued 1956" but listed chronologically with the 1955 publications. Listed in *Carnegie Year Book No. 55* Bibliography as a 1956 publication]

McClintock, B. 1984a. The significance of responses of the genome to challenge. *Science* Vol. 226 (No. 4676, November 16, 1984, pp. 792–801. [Nobel lecture, presented December 8, 1983; Footnotes indicate that "Minor corrections have been made by the author," and that this lecture "will be included in the complete volume of *Les Prix Nobel* en 1984 as well as in the series Nobel Lectures (in English) published by the Elsevier Publishing Company, Amsterdam and New York."

McClintock, B. 1984b. The significance of responses of the genome to challenge, pp. 174–193. *Les Prix Nobel 1983*. Almqvist & Wiksell International, Stockholm-Sweden. © by the Nobel Foundation. [Nobel lecture presented, December 8, 1983]. [In addition to McClintock's Nobel Prize lecture, this volume includes McClintock's photograph (p. 170), biographical sketch (pp. 171–173), and the short speech she delivered at the Nobel Banquet in Stockholm (p. 42).]

McClintock, B. 1987. *The Discovery and Characterization of Transposable Elements*. The Collected Papers of Barbara McClintock. Genes Cells and Organisms; Great Books in Experimental Biology, J. A. Moore, series editor, Garland Publishing Co. New York. [Published December 1, 1987; Currently in the series Genes Cells and Organisms, Routledge, Taylor and Frances Group]

Metzenberg, R. L. 2005. Norman Harold Horowitz, 1915–2005. *Genetics*. Vol. 171(No. 4, December), pp. 1445–1448. https://doi.org/10.1093/genetics/171.4.1445

Moulton, F. R. (Ed.). 1940. *The Cell and Protoplasm*, American Association for the Advancement of Science Publication No. 14, The Science Press, Lancaster, PA. ["Chromosomes and Genes" by Richard B. Goldschmidt, pp. 56–66.]

Nasrallah, J. B. 2012. Adrian M. Srb 1917–1997. *Biographical Memoirs*, National Academy of Sciences, Washington, DC.

Proceedings of the Indiana Academy of Sciences. 1962. Jesse Robertson Singleton [May 25, 1918–July 5, 1962]. Vol. 72, 1962, published 1963, p. 45.

Pardee, A. B. 2001. *"Ruth Sager, 1918–1997." Biographical Memoirs*, Vol. 80, National Academy of Sciences, Washington, DC.

Perkins, D. D. 1992. Neurospora chromosomes, pp. 33–44, *In*, N. Fedoroff and D. Botstein (Eds.), *The Dynamic Genome: Barbara McClintock's Ideas in the Century of Genetics*, Cold Spring Harbor Laboratory Press, Plainview, NY.
Records of the Genetics Society of America. 1944–1951. Published by the Secretary of the Genetics Society of America.
Records of the Genetics Society of America. 1947. Program of the Fifteenth Annual Meeting, Chicago, IL, December 29–31, 1947, pp. 17–22; Symposium: Genetics of microorganisms, Dr. B. McClintock and J. R. Singleton: Chromosome behavior in Neurospora during ascosporogenesis.
Sager, R. 1951. On the mutability of the waxy locus in maize. *Genetics* Vol. 36 (No. 5, September 1, 1951), pp. 510–540. [Submitted November 24, 1950]. [This paper is listed as her Doctoral Dissertation in the Columbia University Catalog: https://clio.columbia.edu/catalog/1503296]
Schmitt, G. K. 2018. *Ruth Sager*. Complete Dictionary of Scientific Biography. [Updated May 21, 2018]. https://www.encyclopedia.com/people/science-and-technology/genetics-and-genetic-engineering-biographies/ruth-sager
Shapiro, J. A. 1995. The discovery and significance of mobile genetic elements, *In*, D. J. Sherratt (Ed.), *Mobile Genetic Elements* [series, *Frontiers in Molecular Biology 8*], IRL Press (Oxford University Press), Oxford and New York.
Singleton, J. R. 1941. *A study of colchicine-induced tetraploidy and other colchicine effects in Xanthium chinense Miller*. Master's Thesis, University of Missouri.
Singleton, J. R. 1948. *Cytogenetic studies of Neurospora crassa*. Ph.D. Thesis, California Institute of Technology, Pasadena, CA.
Singleton, J. R. 1953. Chromosome morphology and the chromosome cycle in the ascus of *Neurospora crassa*. *American Journal of Botany* Vol. 40 (No. 3, March), pp. 124–144.
Spectre, M. 1993. *Barbara McClintock Papers. Guide to the Collection*. American Philosophical Society Library, Philadelphia, PA. (Online: Barbara McClintock Papers Mss.Ms.Coll.79, https://search.amphilsoc.org/collections/view?docId=ead/Mss.Ms.Coll.79-ead.xml) [Also see selected papers and summary of APS Library papers of McClintock available for the National Library of Medicine, Profiles in Science: https://profiles.nlm.nih.gov/spotlight/ll; The collection also reprints papers/letters from other institution's archives]
Stadler, L. J. 1944. 4. Comparison of X-ray and Ultra-violet Mutation. *Maize Genetics Cooperation News Letter 18* (January 31, 1944), pp. 21–24.
Stadler, L. J. and F. M. Uber. 1941. Genetic effects of ultra-violet radiation in maize. IV. Comparison of monochromatic radiations. *Genetics* Vol. 27 (No. 1, January 2, 1942), pp. 84–118. https://doi.org/10.1093/genetics/27.1.84
Stanford University. 1939. A Résumé of the Stanford Symposium on the Cell and Protoplasm. Stanford University June 30–July 5, 1939, 15 pp. [Published for the information of present and former students of biology at Stanford]
Stebbins, G. L. Jr. 1950. *Variation and Evolution in Plants*, Columbia University Press, New York. [Based on lectures of 1946]
Smocovitis, V. B. and F. J. Ayala. 2004. George Ledyard Stebbins 1906–2000. A Biographical Memoir. *Biographical Memoirs*, Vol. 85, The National Academies Press, Washington, DC.
Swanson, C. P. 1957. *Cytology and Cytogenetics,* Prentice-Hall, Inc. (© April 24, 1957).
University Women Honor New Yorker, Dr. McClintock, Receiving Prize at Dallas, Pleads for U.S. Aid to Science. 1947. April 17, *New York Times*, Late City Edition, p. 17.
Wallace, B. and T. L. Kass. 1974. On the structure of gene control regions. *Genetics* Vol. 77 (No. 3, July 1, 1974), pp. 541–558. https://doi.org/10.1093/genetics/77.3.541
Weil, C. 2013. Transposable controlling elements step out onto the broader scientific stage, pp. 3.11–3.21, *In*, L. B. Kass (Ed.). 2013ff. *Perspectives on Nobel Laureate Barbara McClintock's publications (1926–1984): A Companion Volume*, Vol. II, The Internet-First University Press. https://hdl.handle.net/1813/34897 [Perspective on McClintock 1950 *PNAS*]
Zinder, N. D. 1992. Forty years ago: The discovery of bacterial transduction. *Genetics* Vol. 132 (No. 2, October 1, 1992), pp. 291–294. https://doi.org/10.1093/genetics/132.2.291

8 Resignation, Renewal, and Reorganization

Belief without evidence [is] what is told by one who speaks without knowledge, of things without parallel

– Ambrose Gwinnett Bierce

RESIGNATION

McClintock's dissatisfaction with the administration at the Department of Genetics, Cold Spring Harbor, among other circumstances, led to her resignation as Staff Investigator by March 1953. In late February 1953, she wrote to the Director, Milislav Demerec, that she was considering leaving. It was her responsibility for the past 12 years to grow and maintain her plants at Cold Spring Harbor because labor was either unavailable or unreliable. The "unceasing activity" along with "constant alertness" had caused "accumulated fatigue." Such activities demanded "a decision regarding future research activities." She considered two alternatives: to change the research material to one that she could handle at Cold Spring Harbor or to resign from her position in the Institution. She wanted a few months before making a decision.[1]

Two days later, McClintock wrote to George Beadle, Chair of the Division of Biology at Caltech, who had been asking her to return to California to continue working on her *Neurospora* project. She told him of the difficulties of growing her corn at CSH. She had to do her own tractor work last summer (1952). Labor was unavailable or incompetent. She had written to Demerec to say she had two alternatives: to change the research material or to resign from her position at the institution. She told Beadle that she preferred a research associate position rather than a "regular job." She said that she had almost no one to discuss her work with after [Stanley] Stevens departed CSH, with the notable exception of Evelyn Witkin who was away for much of the past two years.[2] McClintock felt that the current staff was not capable of understanding her work and believed researchers elsewhere would fulfill her needs for "opportunity for interchange of ideas." Demerec had argued that the current CSH researchers provided a competent group of geneticists.[3] He may have been considering those who were not only staff members but also Department visitors and research associates who were quite notable but not plant geneticists.

During this time, McClintock also seemed concerned with family problems and with revising her Carnegie research reports. Her mother had been ill, which required her time and attention. Additionally, the secretary who edited the Carnegie reports had died of a heart attack on March 1, 1953. Furthermore, in responding to queries about her research, she explained to Edgar Altenburg that she had eliminated much relevant data that she wished to include in her 1952 Carnegie report.

[1] Letter B. McClintock to M. Demerec, February 20, 1953, Friday, McClintock File, Carnegie Institution of Washington, Department of Genetics, Cold Spring Harbor Laboratory Archives (henceforth McClintock DG/CSHL Archives).

[2] Stanley Stevens was in residence at CSH as a Research Associate for two years (1944–1945, 1945–1946), and Witkin was at first a Columbia University graduate student of T. Dobzhansky who encouraged her to conduct research at CSH (1944) under the guidance of Demerec and Salvador Lauria (on leave at CSH Summer 1945); after receiving her Ph.D. (1947), she was on fellowship, then appointed a Resident Investigator at CSH beginning in 1950; *see Carnegie Year Book No. 44* (1944–1945) and *Carnegie Year Book No. 49* (1949–1950). Witkin Ph.D. 1947, "Genetics of resistance to radiation in *Escherichia coli*, Reprinted from *Genetics*, Vol. 32 … [May] 1947."

[3] Letter B. McClintock to G. Beadle, February 22, 1953, Sunday, McClintock Folder No. 5.26, Beadle Collection, Caltech Archives (henceforth McClintock BC).

DOI: 10.1201/9781003332527-9

This was because she had no time to make the requested revisions due by the September 1952 deadline.[4] These additional personal challenges were not included in letters to colleagues regarding her considerations to leave Cold Spring Harbor.

McClintock's Carnegie report for 1952 confirms her omission of the details of finding another system that controlled genic expression at the *A* locus. The new mutable loci were not under the control of *Ac* (McClintock 1952, p. 214). She had observed five such mutable loci, and their description would be "postponed until further evidence is available" (McClintock 1952, p. 219 and see her Table 4). The President of the Carnegie Institution highlighted her accomplishments for the 1951–1952 research season (Report of the President 1952, p. 16):

> Barbara McClintock has reached the conclusion from her studies on maize that at least two classes of functional genetic unit are carried by the chromosomes: one of them, the gene, potentially capable of determining a particular course of cellular reactions; the other associated with the control of this potential action. Unlike the gene, which occupies a fixed position in the chromosome, the controlling units may move from one location to another. When such a unit is inserted adjacent to a known gene, it exercises its specific mode of control of the time and degree of genie action, regardless of the type of potential action of the gene itself. McClintock has been able to study the operation of the same controlling unit at a number of different gene loci, and, conversely, of different controlling units at the same gene locus.

In the March 1953 *Maize News Letter*, Peter Peterson, a student of Marcus Rhoades, had reported on a similar system for a mutable pale green locus (*pg*) and compared it with McClintock's findings (Peterson 1953a). Rhoades and Dempsey (1950) had previously reported the new mutable locus, recognized as "a mutable pale green ... the mutability ... is apparently controlled by the action of another gene which segregates independent ... is similar to the Activator type of induced mutability ... extensively studied by McClintock." Peterson named the controlling factor *En*. That year, McClintock further noted her new mutable system, which was not under *Ac* control, and initially described it as yet "unpublished" (McClintock 1953a, p. 596: Table 6, p. 595). Details followed in her Carnegie report (McClintock 1953b, p. 233). By 1954, she gave an extended description for this system that was "controlling gene expression," and "tentatively" named it *Spm*, for suppressor/mutator (McClintock 1954, p. 255). *Spm* resembled *Ac* in behavior but could be distinguished by a different phenotypic expression. At the end of 1954, McClintock submitted a summary of her "Suppressor-mutator system of control of genic action and mutation change" to the *Maize News Letter* for 1955 (McClintock 1955a, pp. 12–13). Therein, she revised her initial interpretation of the anthocyanin-producing "autonomous" mutable mutation she called $a_1^{m\text{-}1}$ and explained that *Spm,* which is an independently located factor, was responsible for the "observed behavior" [phenotypic expression] of the mutant locus.[5] The *Spm* element was a "transposable element" located elsewhere in the chromosome complement [genome] (McClintock 1955b, p. 250).[6] It could transpose in a similar manner to *Ac*. Fedoroff and colleagues chose to study the *Spm* system at the molecular level (Fedoroff 1992). Their studies were presaged by McClintock's "analysis of the 'changes in phase' of *Spm* activity and constitutes one of the earliest genetic investigations of an epigenetic regulatory system" (Fedoroff 2013a, p. 69, 2013b, p. 3.37; McClintock 1987).

---

[4] Letter B. McClintock to Edgar Altenburg, March 6, 1953; Altenburg mss. Lilly Library, Indiana University, Bloomington, Indiana (henceforth Altenburg, LL). Carnegie stenographer, Mary J. Holmes, died on March 1, 1953, personal communication, Cold Spring Harbor Archives, April 17, 2006.

[5] See McClintock (1952, p. 219: Table 4) and McClintock (1953b, p. 234), for mutation she called $a_1^{m\text{-}1}$.

[6] In this 1955 Carnegie Report, McClintock first introduced the term "transposable element" to describe a movable genetic locus. Campbell et al. (1977) accurately credited McClintock for introducing the term but mistakenly cited her 1951 Cold Spring Harbor Symposium presentation for the term, which she first introduced in 1955. See also Rieger (1991, p. 489) "transposable genetic element" and Cohen (1976, p. 734) "transposable element" defined. Neither of these authors cites McClintock (1955b, p. 250), for her first use of the term transposable element.

In early March 1953, McClintock asked Beadle for guidance on how and where to apply for funding to support her work, since she had never applied for funds except for fellowships.[7] A week later, she informed Demerec that it would be necessary for her to submit her resignation to be effective by the end of the year (1953).[8] Two days later, she submitted her letter of resignation to President V. Bush of the Carnegie Institution of Washington, effective at the end of 1953. Her reason for resigning was to associate herself again with an institution where adequate maintenance personnel was available for her maize project.[9] She immediately followed up with a handwritten personal note on the same day saying how sorry she was to be leaving the institution. She agreed with and respected Demerec's assessment that she was a "perfectionist" and "consequently untenable" and that she might be blind to her own defects. She appreciated all Bush had done to make the department a success and believed that its future was "certainly bright."[10]

Immediately upon receiving her resignation letter, Bush replied with a "sympathetic and gracious" handwritten letter that changed McClintock's mind about leaving the Carnegie Institution.[11] In response, McClintock rescinded her resignation. She wrote Bush that his expression of confidence in her had "removed all real anxiety," and "Nothing could have been more effective in restoring a badly shaken morale." She was unable to express the extent of her gratitude.[12]

She then informed Beadle about Bush's letter, which resulted in her decision to remain at Cold Spring Harbor. She told Beadle that she no longer needed to seek outside support for her work. Bush had promised that he would see that all goes well for her either at Cold Spring Harbor or elsewhere. Demerec had been called to Washington, DC, the day Bush received her letter of resignation. This had apparently settled the matter of her resigning, and she was relieved "more than I can express." She felt sorry for Demerec but believed he should have been more forthcoming with Bush about her situation.[13] In a follow-up message to Beadle, she reiterated Bush's expression of confidence in her as an investigator and an intellect, which removed some of the major causes of her difficulty. Repeated "no-confidence" votes can injure one's self-respect, she explained.[14]

McClintock kept Beadle apprised of her situation. She liked his suggestion of requesting a "roving appointment," which was exactly what she would want. It would permit her to "spend part of the time at Cal. Tech. – to grow corn, to work on *Neurospora*, to teach and not the least of all ... to be with the group, each member of which I like and admire. It would be splendid!" She would discuss the idea with President Bush of spending time at Illinois and at Caltech. In addition, she said that Ralph Singleton "came to the rescue." She will be growing her corn at Brookhaven this summer. She said it was an easy commute to the Brookhaven Laboratory.[15] Singleton had suggested that she grow her corn at Brookhaven for the summer of 1953. Her project would be a small part of his larger maize project, she told Demerec, and she wished to make plans to grow her corn at Brookhaven.[16] W. R. Singleton, formerly at the Connecticut Agricultural Research Station, had been a visiting scientist

---

[7] Letter B. McClintock to Beets [G. Beadle], March 7, 1953. Handwritten; thanks for your note [no note in file]; McClintock BC.

[8] Letter B. McClintock to M. Demerec. March 14, 1953, McClintock, DG/CSHL Archives.

[9] Letter B. McClintock to V. Bush, President Carnegie Institution of Washington, March 16, 1953 [Monday]; McClintock DG/CSHL Archives (copy of an original that was sent to Bush); original is in the Personnel file of Barbara McClintock at Carnegie Institution of Washington, DC, Archives (henceforth McClintock CIWDC Archives); note on original "received March 17, 1953."

[10] Letter handwritten B. McClintock to V. Bush, President Carnegie Institution of Washington. March 16, 1953 [Monday]; McClintock CIWDC Archives.

[11] Handwritten reply to McClintock's letter of resignation, V. Bush to B. McClintock, noted by hand on original letter of resignation "received March 17, 1953, reply mac handwritten no carbon." Bush's reply was not found among McClintock's papers, McClintock CIWDC Archives.

[12] Letter B. McClintock to V. Bush, March 19, 1953, the reply to Bush's handwritten response to McClintock's resignation letter; McClintock CIWDC Archives.

[13] Letter McClintock to G. Beadle, March 22, 1953, McClintock BC.

[14] Letter B. McClintock to Beets [G. Beadle], March 31, 1953; this is a reply to a letter from Beadle of March 25, 1953, which is not in the file; McClintock BC.

[15] Letter B. McClintock to Beets [G. Beadle], April 15, 1953; McClintock BC.

[16] Letter (handwritten) B. McClintock to M. Demerec, April 11, 1953; McClintock DG/CSHL Archives.

in Stadler's research group at the University of Missouri when McClintock was employed there in the late 1930s (see Chapter 6). Demerec promptly thanked the Assistant Director of the Brookhaven National Laboratory for making arrangements possible for McClintock to grow her corn there.[17]

In reply to an invitation from Marcus Rhoades to spend time at Illinois, McClintock summarized the obstacles she had to overcome being a woman in science. Cold Spring Harbor was the only place where she had not had to face the "anti-female bias" most of the time. Even though it existed, she had not talked about it much. It was the main reason she had not wanted to be affiliated with a regular University job. She would opt for an adjunct appointment with no tenure such as a research associate or a fellowship. She believed that her only hope was to be "foot-loose" and tied to no place permanently. She would pose this possibility to President Bush, and if he did not agree, then she would leave Carnegie. Beadle, she said, had also suggested a similar plan for spending time at Illinois, as well as Caltech. She awaited hearing from Beadle about his communications with the administration at Caltech. She concluded her letter by reporting on some interesting new mutations she was finding in her maize studies during the past year. She hoped to get together in person to discuss her findings.[18]

By mid-May, her situation at Cold Spring Harbor was still not settled in all the details, but the Washington office had assured her that plans to spend time at various institutions were acceptable to President Bush, she told Beadle. She was feeling depressed because she felt she had failed at her job at "C.S.H," meaning that she had "fouled up" the best job she could have obtained "being a woman" with no [other] job opportunities. She felt that she should have left in March, but her interest in the greenhouse crop kept her there. She inquired if President Bush had discussed the possibility with the President of Caltech of her teaching there when he visited California.[19]

Beadle contacted Bush with an offer to help regarding a place for McClintock to work away from Cold Spring Harbor on occasion, as he understood had been suggested. He made no mention that it was his original idea.[20] He mentioned that Marcus Rhoades had also invited her to spend time with him as well. He unofficially had knowledge that the Caltech administration had no reason to object to such a plan. If Bush agreed, then Beadle's Division would provide facilities for her experimental work. They would also welcome her teaching a course, if she were agreeable, and he would recommend that the Institute pay for her services. If Bush thought his suggestion had "merit," then he would be glad to discuss it with the Institute Administration and Trustees.[21] Bush was grateful for Beadle's letter, and Barbara McClintock, he said, was a "grand person." He agreed that McClintock had developed a concept of "far-reaching importance" but had not yet wrapped it up in a "conclusive and convincing fashion." Bush had told McClintock that he would do anything to help her out, and he would make it possible for her to spend time with Beadle's group. Bush also said that he had received a letter from McClintock, who said not to worry about the subject [of resigning] for it was all much clarified in her mind, and Bush believed that McClintock felt her concerns had been resolved.[22]

RENEWAL

During a visit to Cold Spring Harbor at the end of May, Bush, in consultation with Demerec, had approved her plans to spend time elsewhere than Cold Spring Harbor.[23] Demerec was grateful to Beadle for communicating with Bush, and it seemed wrong, he said, for McClintock to remain in

[17] Letter Demerec to Robert A. Patterson, Assistant Director Brookhaven National Laboratory, April 18, 1953; Letter R. Patterson to M. Demerec, April 22, 1953; McClintock DG/CSHL Archives.

[18] Letter B. McClintock to M. Rhoades, April 17, 1953; Rhoades collection, Box 12, McClintock Correspondence 1951–1989, LL.

[19] Lctter B. McClintock to Beets [G. Beadle], May 15, 1953; McClintock BC.

[20] See letter B. McClintock to G. Beadle, April 15, 1953, in replying to his letter of March 31, 1953 (the latter is not in the file).

[21] Letter G. Beadle to V. Bush, May 25, 1953, McClintock BC; original received May 25, 1953, McClintock CIWDC Archives. See also letter M. Rhoades to G. Beadle, April 15, 1953, McClintock BC.

[22] Letter V. Bush to G. Beadle, May 27, 1953; McClintock BC; carbon of this letter in McClintock CIWDC Archives.

[23] Letter B. McClintock to Beets [G. Beadle], handwritten, June 1, 1953; McClintock BC.

one place all the time. He believed the reason for her recent "low point" was complex and was due to her own makeup, the status of her research, and her physiological age.[24] It's not clear what he was implying by the latter, but it seems to be a rather chauvinistic comment.

In addition to having roving appointments at various institutions throughout the country, McClintock was encouraged to grow her corn at the Brookhaven National Laboratory, Upton, Long Island, NY. Upon receiving "P" security clearance from the "U.S. Atomic Energy Commission," she was appointed a "Guest Geneticist" in the Department of Biology effective July 3, 1953. A week later, she drove to the laboratory to begin her summer work in cooperation with W. Ralph Singleton.[25] She held Guest appointments in the Biology Department at Brookhaven from 1953 through 1972.[26]

Ralph Singleton was also collaborating with Elwood Johnson Dollinger, a student of Rhoades at Columbia University. Dollinger was conducting his research at Brookhaven Laboratory when McClintock arrived in 1953. She had proposed the experiments, and offered guidance and advice on Dollinger's doctoral research, "Studies on induced mutation in maize," and for his postdoctoral work on an "Analysis of a mutable system in maize" (Dollinger 1954, 1955).

McClintock attended the Cold Spring Harbor Symposia on "Virus," which was held during June 1953. It was organized by Max Delbruck [Caltech] in consultation with F. L. Horsfall [Rockefeller Institute] and H. M. Weaver [National Foundation Infantile Paralysis] (Demerec 1953a). James Watson, a last-minute participant in the program, presented and published more extensive details about the structure of DNA (Watson & Crick 1953c), which he and Francis Crick had announced very briefly in the April issue of *Nature* (Watson & Crick 1953a). Their symposium presentation gave additional evidence for how the DNA molecule might duplicate itself, in addition to suggesting how the nucleotide sequence can carry the "genetical information" (see also Watson & Crick 1953b). They also offered a possible molecular mechanism for "Natural Mutation," caused by a change in the sequence of nucleotide bases that might cause a tautomeric shift (a change in form) in the DNA molecule (Watson & Crick 1953c, p. 130.) Although their evidence seemed to be well received, Delbruck later questioned how DNA strands could unwind and separate without breaking the DNA double helix (Delbruck 1954). The controversy seemed to be resolved when chromosomes and subsequently DNA were later demonstrated to replicate semi-conservatively (Taylor et al. 1957; Meselson & Stahl 1958).

Only weeks before Watson and Crick's *Nature* paper appeared (April 25, 1953), McClintock had submitted (April 14, 1953) her paper on the "Induction of instability of selected loci in maize" to *Genetics*. It was published in November but did not appear until January 1954. McClintock's model was not molecular. It was a cytogenetic study, and she concluded that "mutations need not express changes in genes [and] may be the result of changes affecting the control of genic action" (McClintock 1953a, p. 594; Halpern 2016). She showed that control of gene expression (gene action) is affected by non-genic (extragenic) agents, *Ds* (Dissociation) and *Ac* (Activator), associated with a structural gene, and carried in the chromosomes. McClintock (1987, p. x) recalled receiving only three reprint requests for this paper. The molecular aspects of McClintock's extragenic agents would be isolated years later and were shown to be transposable elements (transposons) that spontaneously moved from one chromosomal locus to another (Fedoroff 1983; Fedoroff et al. 1983; see Chapters 10 and 11). McClintock's (1953a) "induction of instability" (i.e., mutation) was more than just a change in the sequence of nucleotide bases. It was an additional way of understanding how

[24] Letter M. Demerec to G. Beadle, June 2, 1953; McClintock BC.

[25] Letter R. A. Patterson, Assistant Director to B. McClintock, July 3, 1953; "New Employee Introduction" to Dr. H. J. Curtis, Director, from Employment Office, July 10, 1953, Personnel File, Barbara McClintock, Brookhaven National Laboratory, Biology Department Records, examined May 1, 2001 (henceforth McClintock BNL). Demerec (1953b, p. 209), "During the summer of 1953, McClintock grew her experimental corn at Brookhaven in cooperation with Dr. W. R. Singleton."

[26] McClintock was again appointed a research collaborator beginning November 9, 1976, for one year, and as a Guest Geneticist beginning June 1, 1979, to work in association with Dr. Benjamin Burr on experiments in Maize Genetics; McClintock BNL. See Chapter 10 for collaborations with Dr. Burr.

mutations can arise by describing how chromosome functioning can alter and regulate genotypes and phenotypes.

At the end of the summer, Demerec was a plenary speaker at the Ninth International Congress of Genetics in Bellagio, Italy. "Notable geneticists" gave plenary talks (Singleton 1954; Lerner 1953), but no women were among them. A few women presented contributed papers (Montalenti & Chiarugi 1954). McClintock did not attend this Congress, and it is not known if she had been an invited speaker. Demerec did not mention McClintock's research in his talk on the "Genetic action of mutagens" (Demerec 1954, p. 201), given in the Symposium "Genetic mechanisms and mutations." He began with the statement, "During the past ten years our group at Cold Spring Harbor had been making extensive study of both spontaneous and induced mutability and I plan to review this work here and to discuss its problems concerned with the mechanism of mutability." Demerec, however, confined his plenary talk to mutations in *E. coli* and made no reference to McClintock's contributions to the problem of induction of mutation in eukaryotic organisms.[27]

A few other plenary speakers, however, brought McClintock's work to the attention of their audience. At the same symposium, Hans Stubbe (1954, p. 226), presenting "About mutable genes in *Antirrhinum majus*" [Snapdragon], concluded that his "cytological results on *Antirrhinum* do not contradict the view developed by McClintock and [by] Goldschmidt (1951) that alleles are not qualitative changes of a gene, but can be traced back to different quantitative compositions of subunits or chromosome pieces." Stubbe and his laboratory were continuing research on Snapdragon, which had been of interest to early naturalists and geneticists because the plant had provided a means to study mutations in flower color and plant morphology. More recently, the plant has been used to study the fundamentals of plant developmental genetics (Schwarz-Sommer et al. 2003). Furthermore, research on *Antirrhinum* showed that the mutable and original "Gram locus" studied by Stubbe in *Antirrhinum* was caused by transposable elements (Golz et al. 2004).

In the 1953 symposium on "The basis of heredity," Edward Lewis (1954) mentioned McClintock's (1944) early studies of deficiencies in his presentation of "Pseudoallelism and the gene concept." Position pseudoalleles, he reiterated (Lewis 1951), "are apparently wide spread and have an important bearing on the gene concept." Pseudoallelism is a state in which two genes with similar functions are located close to one another on a chromosome and appear to be genetically linked (almost always inherited together). Such genes are said to act as a single gene. Pseudoalleles are thought to have been formed as a result of gene duplications. A duplicated gene could mutate and provide an adaptive function in a new environment. It seems odd that Lewis (1951) also had previously cited McClintock (1944) for introducing the term "pseudoalleles," and then omitted the examples reported by McClintock because her study of the pale yellow and pale green mutants in maize was only "suggestive" but did not fit his definition of "pseudoalleles." A close reading of McClintock (1944, p. 499) shows that she attributed the term "pseudo-allelic" to "Morgan, Bridges, and Schultz (1938)," and she did not originate the term. Morgan et al. (1938, p. 305) had observed that non-allelic deficiencies (mutant loci close together on adjacent loci) could explain phenotypic and chromosomal mutations in *Drosophila*. Bridges used the term to describe their observations. McClintock's results seem to indicate that she believed that her maize mutants, which were caused by homozygous deletions in the short arm of chromosome 9, did not appear to fit the concept of "pseudo-allelic" mutations (McClintock 1944, pp. 499–450).

Previously, Stephens' (1951) discussion of the function of "duplicate" linked loci had invoked McClintock's study of the pale yellow/yellow green series in corn as "representing the only case of pseudo-alleles in plants, which has been established cytologically (McClintock 1944)." However, he then concluded that "although these regions [loci] occupy adjacent positions in the ninth chromosome, ... [without chemical information] it cannot be inferred that the two regions originated as a result of duplication." Thus by definition, he rejected the idea that the duplications

[27] His Footnote: "work ... supported by ... American Cancer Society upon recommendation of the committee on growth of the National Research Council."

in maize chromosome 9 are pseudoalleles. As emphasized by Garber (1972, p. 112), even though the phenotypes McClintock observed could have been explained by invoking pseudoallelism, once McClintock (1944) had shown that the white seedling phenotype and the pale yellow seedling phenotypes were expressions of different homozygous deficiencies, "there was no need to construct speculative or unorthodox models to explain the genetic observations."

As we have previously mentioned, McClintock's (1944) study did provide a possible mechanism for inducing mutations via chromosomal deficiency and the chromosomal breakage mechanism. McClintock also recalled that it was the expansion of this experiment that she began in the summer of 1944 on the chromosome breakage mechanism and the progeny from those experiments that revealed the presence of transposable elements (McClintock 1987, pp. x–xi; see Chapter 7).

Richard Goldschmidt had opened the 1953 Congress with his Presidential address "Different Philosophies of Genetics" (Singleton 1954), in which he voiced a dynamic approach to gene action and developmental systems and their relation to regulation and integration (Goldschmidt 1954). He cited no sources and made no mention of McClintock's work. Presenting the statistical, or static, point of view vs. the physiological, or dynamic, point of view, Goldschmidt may have been attempting to downplay the criticism he had received regarding his rejection of the particulate theory of the gene (Goldschmidt 1951; Deichmann 2011).

As we have seen, in the 1940s and 1950s, McClintock and Goldschmidt held different views from the prevailing definition of a Mendelian "gene" as a single locus on a chromosome with alleles that were dominant or recessive (mutant forms). These units were long believed to be passed on from generation to generation by Mendelian inheritance. At this time, American geneticists were searching for a way to combat Anti-genetic propaganda (Selya 2012). Specifically, the Genetics Society of America was looking to fight the ideology of Russian agronomist Trofim Lysenko, who rejected Western genetics and denied the existence of genes. His influence on the decline of Russian agriculture during the Cold War has received much historical attention (DeJong-Lambert & Krementsov 2012; DeJong-Lambert 2013). Nobel Laureate (1995) Ed Lewis suggested that Lysenko's rejection of the gene may have played a part in the lack of initial acceptance of McClintock's hypothesis of gene action.[28] McClintock also believed that her system of transposing elements could direct mutations and thus change a plant's phenotype. Lysenko held a belief that he could make new plants from existing ones, but his methods and results were erroneous.

With respect to Anti-genetic propaganda and the rejection of Western genetic thought, it was at this 1953 International Congress of Genetics (ICG) that an important resolution was proposed and unanimously adopted (Singleton 1954): "The Congress asks the International Committee not to recommend that the next Congress be held in any country to which it may be expected that scientists would be refused permission to enter on grounds of race, nationality, religion, place of birth, or political association past or present."[29] At the 1952 Genetics Society of America [GSA] meeting held at Cornell, it was recommended that the next ICG be held in Canada. "Until our [U.S.] immigration laws are changed, it will be impossible to hold a really International Congress in this country," Singleton (1954) reported. He noted that of the 900 attendees at the 9th International Congress of Genetics, none were from Iron Curtain countries. The ICG resolution was "aimed directly at the United States, where the McCarran Act of 1950 allowed the State Department to deny entry visas to members of the Communist Party" (Selya 2012, p. 436).[30]

While Demerec was attending the ICG Congress, McClintock made plans to come to Caltech in the fall of 1953. She expressed how grateful she was to Beadle for all that he had done for her.

---

[28] Lee Kass interview with Ed Lewis, January 22, 1997.

[29] The resolution was announced earlier in *Nature* (News & Views 1953).

[30] "Immigration laws were a reaction to investigations by the U.S. Congress House Un-American Activities Committee of supposed Communist sympathizers during the U.S.-Russian Cold War; U.S. scientists were denied passports to travel abroad and foreign scientists were denied visas to visit the United States" (Selya 2012). See also DeJong-Lambert and Krementsov (2012) for a volume on the Lysenko Controversy and the Cold War, which includes 10 articles on the topic. See also Cassata (2017).

It had made it possible for her to "recover a healthy perspective and an enthusiastic outlook."[31] She declined a reservation, made by Beadle, for her to stay at the Athenaeum on the Caltech Campus by the first of November. She still had relatives in California who she could stay with and preferred to make living arrangements after she arrived in Pasadena. She also planned for some non-academic vacation time.[32]

At the end of October, McClintock left for California. She stopped on the way to visit Marcus and Virginia Rhoades in Illinois. R. A. Brink had asked her to visit Wisconsin for several days to discuss mutual problems and examine their material. Rhoades' and Brink's students had been working on mutable loci in maize and had found results similar to her *Ds/Ac* system in their studies. She would leave her car in Illinois and take the train to Madison, Wisconsin. Since she was driving alone this time, she might get the urge "to hurry along."[33] She informed Demerec that she also planned to visit laboratories on the west coast and would return to Cold Spring Harbor at the end of March 1954.[34] Beadle promptly notified the Caltech Biology faculty of her plans to arrive around November 15.[35]

At Illinois, McClintock conferred with Rhoades and his student Peter Peterson about his findings of a mutable pale green (*pg*) locus, as mentioned above. It caused increased mutability at the *pg* locus and was suggestive of the *Ac* effect on *Ds*-controlled loci studied by McClintock. Peterson associated his results with McClintock's concept, which considered that unstable genes represent the removal of an inhibitory locus adjacent to the dominant allele (Peterson 1953a, pp. 59–60). He hypothesized a two-element system: *I* (inhibitor of genic activity) located at the *pg* locus, and *En*, which can be located elsewhere and which causes the removal of *I* (Peterson 1960). Peterson's mutable locus was later found to be equivalent to McClintock's *Spm* transposable element system (Peterson 1977, p. 431: Table 1). At Illinois, she also had "good talks" with Salvador Luria and Sol Spiegelman, bacterial geneticists in the Department of Microbiology who had an interest in her research.[36] Previously, as a Professor at the University of Indiana, Luria had given a seminar on McClintock's work in October 1948, based on results of studies she had summarized for [Stanley] Stephens then at Texas. Luria's seminar inspired H. J. Muller to send congratulations on her "magnificent achievement in getting a solution to the ... problem of mutable genes, at least in maize." He added that it was "a relief ... for us mutation workers to know that mutable genes are in a different class from spontaneous mutations ... called 'ordinary gene mutations' ... The fact that mutations of mutable genes could apparently be directed was a source of worry to anti-Lysenkoists." Apparently, her findings that chromosomal genetic elements can spontaneously transpose resolved this issue for Muller.[37]

At the University of Wisconsin, McClintock consulted with Brink and his students as he had requested. Brink and Nilan (1952) had studied variegation at the unstable pericarp (*P*) locus, which had been earlier reported by Emerson (1914). Emerson had concluded that "The development of red in the pericarp is evidently associated with and perhaps due to a modification of some Mendelian factor for pericarp color in the somatic cells" (Emerson 1914, p. 102; Fedoroff 1998, p. 957). Emerson had suggested variegation was due to the association of some type of genetic factor that was lost and this permitted the gene to be "re-expressed" (Emerson 1914, pp. 112–113; Fedoroff 1998, p. 958). Brink and Nilan (1952, 1951) had found a genetic element that modified the expression of the variegated (*P-VV*) and red (*P-RR*) pericarp color. Their genetic breeding data in maize stocks, similar to those of Emerson's, seemed to conform to Emerson's hypothesis of a modification of some Mendelian factor for pericarp color. Their working hypothesis assumed that the unstable variegated

---

[31] Letter B. McClintock to Beets [G. Beadle], July 9, 1953; McClintock BC.

[32] Letter B. McClintock to Beets [G. Beadle] handwritten, July 30 [1953]; McClintock BC.

[33] Letter B. McClintock to Beets [G. Beadle] handwritten, October 22, 1953; McClintock BC.

[34] Letter B. McClintock to M. Demerec, October 25, 1953; McClintock DG/CSHL Archives.

[35] Memorandum from GWB [G. Beadle], October 30, 1953, to members of the Division of Biology: E. G. Anderson, A. Longley, A. Sturtevant, [Sterling] Emerson, [Max] Delbruck, [Ed] Lewis, [Norman] Horowitz, [Mary and Herschel] Mitchells; McClintock BC.

[36] Letter B. McClintock to M. Demerec, November 23, 1953, McClintock DG/CSHL Archives.

[37] Letter H. J. Muller to B. McClintock, October 27, 1948, Muller mss. LL.

allele, *VV*, was a modified form of the stable *RR* gene locus ["*VV* may be equated to *RR Mp*"]. They proposed that the difference between these loci was that *VV* is associated with a discrete element that they termed Modulator, *Mp*. The element Modulator inhibits red pigment formation. "Mutation of *VV* to *RR* consists in the loss of *Mp* from the *P* locus," via transposition (Brink & Nilan 1952, p. 539). They also hypothesized that a second modulator element caused the light variegated variation that they had observed: "Light variegated=*RR Mp* + *Mp*" pericarp color expression (Brink & Nilan 1952, p. 539). Their observations of variegated pericarp also seemed to "conform in certain respects ... to the genetic behavior of the mutable loci in the short arm of chromosome 9 in maize ingeniously analyzed by McClintock (1950) and interpreted by her in terms of chromatin material accessory to the gene" (Brink & Nilan 1952, p. 358). McClintock's explanation of mutable loci, they said, might prove applicable to their studies, but their data were "too limited to show how far the parallel extends" (Brink & Nilan 1952, p. 358). Brink and his students seem to have benefited from further discussions with McClintock during her visit to their laboratory. She told Demerec that her visit with Brink was very successful particularly with regard to *Ac* control of the *P* locus instability in maize.[38] *Mp* would soon prove to be equivalent to the activator (*Ac*) element first described by McClintock, because *Ds* responded to *Mp* in the same way as *Ac* (Barclay & Brink 1954; Peterson 1977: Table 1; Fedoroff 1998, p. 960, 1989).

At Caltech, McClintock was appointed "Visiting Professor of Biology for the period November 15, 1953, through March 31, 1954 without stipend."[39] As "Visiting Professor of Cytology" in the Division of Biology, she offered "Special Topics in Genetics," Bi 225, a required course for students majoring in Genetics, commencing on January 7, 1954, through March 8.[40] She lectured on her pioneering work on transposition and controlling units (later termed controlling elements, McClintock 1955/©1956a; Rieger et al. 1991, p. 108).[41] Caltech graduate students who enrolled in her course held her in awe and respect, and faculty regularly audited her class, which was held in room 234 of the William G. Kerckhoff Laboratories.[42] Ed Lewis' graduate student Burke H. Judd took extensive notes on this class.[43] He recalled that she "presented a lot of data from crosses designed to map *Ac* and *Ds* and to demonstrate the chromosome breakage and mobility of those elements." He noted a discussion of cell regulation: "Nuclear mechanisms for controlled segregations -- could there then be such for control at genic level-." He recalled McClintock's discussions with students of movable elements and the "roles they might play in development of the plant." Professor E. G. Anderson gave the final lecture of the term, and Judd's class notes show that Anderson reviewed the work on the "pericarp mutations" that R. A. Emerson, his Major Professor at Cornell, had studied much earlier. Anderson correlated the report of modulator [*Mp*] by Brink with the spontaneous mutations studied by Emerson. Modulator he noted was very similar to *Ac* when tested with *Ac* tester stocks. This, of course, was the discussion that McClintock had with Brink and his students at Wisconsin just a few months earlier.

Judd was completing his thesis on a direct proof of position effect at the white locus (Judd 1954) and recalled some discussion with McClintock of his work relative to chromosome breakage and rearrangement, but there is no citation to McClintock's papers in his doctoral thesis. Faculty and graduate students had multiple discussions about movable elements, and the very unusual element mobility and activity that she was studying, he said. The concept of the gene as a "unit of function (one gene-one enzyme)" was beginning to broaden probably due to work McClintock had presented earlier, he recalled. Furthermore, the Cold Spring Harbor Symposium in 1951 "shook up the

[38] Letter B. McClintock to M. Demerec, November 23, 1953; McClintock DG/CSHL Archives.

[39] Letter of appointment from President L. A. DuBridge to Dr. B. McClintock, December 8, 1953; McClintock BC.

[40] *Bulletin of the California Institute of Technology*, Catalogue 1954–1955, Vol. 63, No. 3. pp. 21, 51, 227; Recommendation for a New Faculty Appointment, December 3, 1953; McClintock BC.

[41] McClintock collection 79 Series III, Folder: California Institute of Technology, 1954 Lectures; see also Folder: unidentified Lectures 1953–1954, n.d.; McClintock Papers, American Philosophical Society Library (henceforth McClintock APS).

[42] L. Kass interview with Prof. E. B. Lewis, September 22, 1997; Former Graduate Student, R. Metzenberg, interview by LBK, February 3, 2007.

[43] Burke Judd (letter and emails, March 13, 19, and 20, 1997), Bi 225 class notes 1954, photocopies to L. Kass, March 1997.

concept quite a bit," and Goldschmidt's view of the chromosome as the unit of inheritance was not highly regarded at Caltech (in 1951). However, McClintock's work broadened the concept of genetic units toward Goldschmidt's view, and her "data were extensive and the analysis unquestionably accurate." As Professor of Genetics at the University of Texas, Austin, Judd continued to refer to his class notes for teaching, but he generally asked his students to read McClintock's original publications. In his teaching laboratories and for special projects, he used ears of corn received from McClintock, which she had tagged with "cross genotypes so that the segregation and activity of the controlling elements could be followed." His interest in studying chromosome organization "clearly gained a lot from meeting Barbara" [McClintock], he said.

Robert [Bob] Metzenberg, a graduate student of Herschel Mitchell, also enrolled in McClintock's 1954 Special Topics in Genetics class, on the subject "Aspects of Maize Genetics."[44] He was studying the biosynthesis of aromatic compounds in *Neurospora crassa* under the mentorship of Mitchell (Metzenberg 1956). Outside of class, he had many discussions with McClintock about *Neurospora*, which she was continuing to study with Mary B. Mitchell (McClintock 1954, pp. 257–260). He remembered that she spent much time preparing her class lectures, and while thinking about what to say, she walked very quickly around and around the central bench in the laboratory he was sharing with her in the Kerckhoff Laboratories. He told me he did not believe that McClintock was considered "an outsider." Caltech faculty and students were aware that she was one of the top cytologists in the country, was a member of the National Academy, the American Philosophical Society, was former President of the Genetics Society of America, was honored with the AAUW Achievement Award, and was the recipient of two honorary degrees.[45] Metzenberg recalled that McClintock suggested that he "work on something not fashionable." As a Professor of Physical Chemistry at the University of Wisconsin, Madison, he studied the genetic control of metabolism in *Neurospora*, in eukaryotic organisms, and held an interest in genome structure (Selker 2008).

Ed Lewis also attended McClintock's Special Topics class along with other faculty members in the Division of Biology.[46] He recalled that her research was "Different, unusual and strange; there was no model for it, and for a long time it didn't make sense until the molecular transposition work was done." With respect to the 1951 Cold Spring Harbor Symposium presentation, he said that she "doesn't present things in a way that was easy to understand." I asked Lewis if he thought McClintock stopped publishing because of the lack of reprint requests she received for her 1950 and 1953 publications. He felt that her excuse was a "rationalization," and was due to the difficulties she had with her writing and because she was under "no pressure to publish." He recalled that she had great difficulty writing papers on her research and that a person at Carnegie helped revise her papers and *Year Book* reports. This recollection seems to be supported by the letter she wrote to Altenburg, mentioned above, and from comments she made to Rhoades, described in earlier chapters.

While McClintock was preparing her Caltech lectures, her pioneering work was prominently featured at the December 1953 AAAS Christmas meetings, held in Boston, Massachusetts. L. J. Stadler lauded her groundbreaking research in his Presidential Address, on "The Gene," to the American Society of Naturalists.[47] Stadler questioned assumptions defining a gene: are mutations "changes in the gene itself rather than some alteration affecting its expression?" He stressed that the assumption of an intragenic alteration was contradicted by the evidence of position effects. The latter showed that

> a mutation did not necessarily represent a transformation or loss of the gene concerned; instead, it could be the result of a translocation affecting the expression of the unchanged gene. The remarkable studies of McClintock [1950a, 1951b] ... show the far-reaching importance of this limitation [assumption

[44] L Kass interview with Bob Metzenberg, February 3, 1997.
[45] Recommendation for a New Faculty Appointment for Dr. Barbara McClintock, December 3, 1953; McClintock BC; Kass and Chomet (2009: Table 2, pp. 32–33).
[46] L. Kass interview with Prof. E. B. Lewis, January 22, 1997.
[47] *American Society of Naturalists*, Presidential Address, December 30, 1953, p. 674, *Science* 118 (3075, December 4, 1953), pp. 661–682. See also published paper (Stadler 1954).

of intragenic change] in the experimental study of gene mutation. … In some cases the mutations [McClintock observed] are accompanied by detectable chromosomal aberrations at or near the locus showing instability, but in other cases no cytologically detectable chromosomal alternation is associated with the occurrence of the mutation. … [McClintock's] brilliant investigation clearly shows that expression effects may be the actual cause of apparent gene mutations, even when the mutation observed shows no indication of a change of position or of any associate chromosomal alternation.

(Stadler 1954, p. 818)

Stadler's address was presented at the society dinner. It was not a symposium presentation, and therefore, it may not have been widely heard by the geneticists, unless the latter were also members of the Naturalists' Society. His presentation on "The Gene," however, was posthumously published in the November 1954 issue of *Science* and later reprinted in Peters' (1959) anthology of *Classic Papers in Genetics*, along with McClintock's (1950a) paper on the origin of mutable loci in maize.

At this 1953 AAAS meeting, however, in the General Symposium titled "Species Which Feed Mankind," Marcus Rhoades summarized recent research by the maize geneticists, and emphasized McClintock's pioneering research. The symposium was sponsored by the American Society of Naturalists and the Genetics Society of America and had been arranged by Paul C. Mangelsdorf, the Director of the Botanical Museum, Harvard University, and M. R. Irwin, a Professor of Genetics, University of Wisconsin. In the session on "America's Principal Food Species: *Zea mays*," Rhoades presentation "Chromosomes, Mutations, and Cytoplasm in Maize" highlighted the contributions of the early maize researchers and summarized McClintock's work on *Ds/Ac* mutations. He compared McClintock's findings with those recently reported by Brink and Nilan and by his former student Peter Peterson, although no citations were listed.

Recently, Brink and Nilan, in a reinvestigation of variegated pericarp, reported data which suggest that mutation at the *P* locus involves a mechanism similar to that of *Ac-Ds*, and Peterson from his analysis of a mutable pale green locus concluded that an *Ac*-like factor was operating.

(Rhoades 1954, p. 120)

Rhoades explained that McClintock's results were "not only of the greatest importance to the problem of mutation, but they also have wide implications regarding the nature of gene action and the genetic control of differentiation" (Rhoades 1954, pp. 119–120). His presentation was given on December 27, 1953, and published in the July 1954 issue of *Science.* At this AAAS meeting, Peter Peterson also gave a contributed paper to the Genetics Society of America on his findings of transpositions at the mutable pale green locus in maize (Peterson 1953b), but he made no connection with the work of McClintock, as he had done previously (Peterson 1953a). His Abstract appeared in the same issue of *Genetics* along with McClintock's (1953a) paper on "Induction of instability of selected loci in maize."

McClintock's stay at Caltech during the winter of 1953–1954 had two apparent outcomes. In addition to teaching the class previously mentioned, McClintock also conducted research on the cytogenetics of *Neurospora* in collaboration with Mary B. Mitchell [née, Houlihan], which she described in detail in her report to the Carnegie *Year Book* upon returning to Cold Spring Harbor (McClintock 1954, pp. 257–260). She concluded that "knowledge of chromosome organization and behavior in *Neurospora* … will serve as a guide in future studies of this type" (McClintock 1954, p. 260). Indeed, such studies were continued, and in March 1961, the first "*Neurospora* Information Conference" was held at the University of California San Diego, in La Jolla, California (de Serres 1962a,b; Perkins & Davis 2002). The conference offered an opportunity for younger scientists to learn and obtain advice from some of the world's outstanding senior scientists.[48] In a session on Nuclear Cytology and Cytogenetics, chaired by McClintock's former student Patricia St. Lawrence,

[48] *Annual Report of the National Science Foundation 1961*, p. 63.

Barbara McClintock and Jessie Singleton presented "elegant cytological studies" on the different stages of meiosis. McClintock then reviewed the morphology of each of the seven chromosomes during postmeiotic mitosis (de Serres 1962a,b, p. 3). The meeting presentations were transcribed, but only brief summaries were published by the Chairs in the Conference *Proceedings* (de Serres 1962b, p. xiii). David Perkins attended this conference, and he later quoted McClintock's observations on *Neurospora* chromosomes from his copy of her unpublished transcript (Perkins 1992, pp. 33, 35).[49] Summaries of the history and progress of *Neurospora* genetics depict McClintock's legacy with respect to the future studies she had predicted in 1954 (Perkins & Davis 2002; Davis & Perkins 2002; *Fungal Genetics Reports* 1962–).

In her 1954 Carnegie report, McClintock made no mention of also teaching a class while in California. However, she did specify that other researchers were investigating "gene control systems in maize" and some of the "gene-controlling units had appeared in their materials." Without providing citations or naming their gene control unit, she reported that one of the units being investigated by Dr. Brink and his students at the University of Wisconsin behaves "much like *Ac*." As mentioned above, McClintock had discussed this work with the Wisconsin group during her travels to California.

REORGANIZATION

On her return to Cold Spring Harbor, McClintock focused her attention on "the system controlling gene expression which differs from *Ds-Ac*" (McClintock 1954, p. 255). As she had mentioned to Altenburg, this system had been observed much earlier (McClintock 1952, p. 214), but the details would be postponed "until further evidence is available" (McClintock 1952, p. 219). McClintock (1954, p. 299) tentatively named her new system that controlled genic expression *Suppressor-Mutator, Spm*. McClintock's new and unique mutable *Spm* system controlled the *A1* locus. Similar to *Ac/Ds*, the *A1* somatic instability was controlled by a two-element system that could transpose, but the mutations were not controlled by *Ac* or *Dt* loci. Also, chromosome breaks were not detected in this system. There was, however, an added layer of gene regulation that had not been observed with *Ac/Ds*. The newly controlled allele of the *A1* gene (designated *a1-m1*) was lightly colored in the kernel or plant. Only upon introduction of a fully functional *Spm* element elsewhere in the genome in conjunction with the *a1-m1* allele, did the pale pigmentation turn colorless. Additionally, deeply colored sectors appeared much like the reversion mutation that she had recognized with other transposable elements. McClintock called the suppressor that acted at a distance *Sp*. The unit responsible for the somatic reversion, she designated the Mutator (*m*). As mentioned previously, Peterson had found a similar system in his maize studies of mutable loci (Peterson 1953a,b). Peterson's *Enhancer-Inhibitor, En-I*, was later found to be equivalent to McClintock's *Spm* (Peterson 1960, 1977, 1993; Fedoroff 1992, pp. 401, 1983, 1989; Kass & Chomet 2009, p. 29).

At the June 1955 *Brookhaven Symposia in Biology*, on *Mutation*, McClintock (1956a) described elements in the maize chromosomes that control gene action; a phenomenon she described and first designated as "controlling elements." These elements "modify and control the action of the genes themselves." She presented a working hypothesis:

> there are at least two different types of basic genetic elements carried in the chromosomes: those that give rise to primary products of genes and those that control the mode of operation of these primary gene elements, the controlling elements. These latter control rates of action of the primary gene elements, their times of action in development of the organism or in the life of the cell, and also possibly, the association of a particular gene element with another such element or elements to form a unit of function.
>
> *(McClintock 1956a, p. 59)*

[49] David Perkins lent me the transcript of McClintock's talk to photocopy. McClintock's presentation is not among her papers at APS; I don't know if Perkins's papers were archived.

McClintock described various types of controlling elements and their mode of operation. The controlling elements were confined to the nucleus, which was reflected in the title of her presentation, "Intranuclear Systems Controlling Mutation." Her first examples were mutations found in the *Ds-Ac* two-element system.[50] A similar system had been observed by Nuffer (1955) and was found to be equivalent to *Ac*. She also mentioned the two-element system described by Brink and his students, which caused mutations at the *P* locus in maize and was also found to be equivalent to the *Ac* controlling element (Brink & Nilan 1952; Brink 1954; Barclay & Brink 1954). She briefly mentioned the classic example of the *Dt* (Dotted) – a_1 system originally reported by Rhoades. She went on to describe another controlling system that was not related to *Dt*-a_1 or to *Ds-Ac* and that she was currently investigating. It was the Suppressor/mutator (*Spm*) system, which she cited as "unpublished" (McClintock 1956a, p. 65). It is unclear why she did not cite her Carnegie reports (McClintock 1953b, 1954) in which she had previously identified and summarized the *Spm* controlling element system, and had already tentatively named it (McClintock 1954, p. 255), as I mentioned above.

Her talk focused on the behavior of chromosome elements that are directly concerned with the mutation process because it was the topic of the symposium. She concluded that such elements control the action of genes [phenotypic expression] and induce mutation-type responses. She considered the elements as accessory to the gene, and as such, they were called controlling elements. They can change their location in the genome by a process called transposition. Insertion of a controlling element near a gene alters its action [expression], thus producing mutations. She also suggested that controlling elements are normal components of the chromosome complement [the genome] and are responsible for controlling the time and type of activity of individual genes.

McClintock received many questions and comments on her presentation to which she patiently and carefully responded (McClintock 1956a, pp. 71–74). Most of the questions were from researchers working with organisms other than maize, but who apparently had no difficulty understanding her concepts. Such queries regarded the nature of the unstable locus, the relation of the controlling elements to heterochromatin, types of other plants that exhibit similar phenomena, and so forth. These questions reflected the topics that McClintock considered of importance but were not discussed because they were not directly concerned with the mutation process, such as mode of transposition, changes in state without change of location, inheritance patterns, interrelationships between elements of a given system, among others. Those topics were addressed by McClintock the following year at the 1956 *Cold Spring Harbor Symposia on Quantitative Biology.*

At the 1956 Cold Spring Harbor Symposia on *Genetic Mechanism: Structure and Function,* McClintock (1956b/©1957) expanded her research presentation on maize genetic elements and their controlling effect on the gene. She titled her presentation "Controlling Elements and the Gene" and considered some aspects of controlling elements that could not be discussed in her "brief review" given at the Brookhaven Symposium the previous year (McClintock 1956b, p. 197; McClintock 1956a, Figure 8.1). She repeated her suggestion that "gene elements and controlling elements represent two different classes of primary components of the chromosome and that a close relationship exists between them" (McClintock 1956b, p. 198). After reviewing the literature on maize transposable elements, she included a section on the origin and mode of operation of the *Spm*-$a_1^{m\text{-}1}$ system, which had "only been outlined in previous reports" (McClintock 1956b, p. 202). The published paper included only one comment following her talk. Once more, she believed she had failed to convince her audience of the significance of these movable elements as a general phenomenon. Franz Schrader's diary entry for that day seems to provide some confirmation of her concerns[51]:

> The lecture hall (new and with perfect acoustics) was almost full and only about 8 seats in the last row were vacant. McClintock has a mass of new stuff that seems important in various directions and is really impressive … [her] penetrating, comprehending, quick eyes are just as keen as they always have been. The discussion was largely of a 'heavens, do you really mean it,' sort.

---

[50] McClintock (1951b, 1946–1954), *Carnegie Year Books Nos. 45–53*; McClintock (1950a, 1953a).

[51] Franz & Sally Shrader's Diary entry Vol. 37, Thurs. Fri. Sat., June 6, 7, 8, 1956, Columbia University Archives.

FIGURE 8.1 Barbara McClintock and Harriett Creighton at the 1956 Cold Spring Harbor Symposia, *Genetic Mechanisms: Structure and Function*, where McClintock gave additional evidence and defined the term "controlling element" in her presentation, "Controlling Elements and the Gene." (Courtesy of Cold Spring Harbor Laboratory Archives.)

In contrast, earlier that year Demerec had highlighted McClintock's research in the *AIBS Bulletin*. The research program of the Department of Genetics, he wrote, "focuses on problems of the structure and function of genetic mechanisms. … Barbara McClintock's work with corn has identified a number of elements, or systems of interacting elements, each exercising a specific mode of control over gene action and changes in gene action" (French et al. 1956). At the end of 1956, the President of the Carnegie Institution highlighted McClintock's research:

> Several years ago McClintock demonstrated the existence of certain heritable units in the chromosome complement of maize that served to control and modify the effects of other genes, without themselves directly affecting the phenotype. Knowledge of such elements has been considerably expanded during the past year. A number of these units, now called "controlling elements," have been identified and their distinctive modes of operation examined. It is now evident that they may act not only as mutators, but also as inhibitors, suppressors, enhancers, or other kinds of modifiers of gene action. They may shift from one location to another within the chromosome complement without losing their identity in the process. A further effect, involving a structurally modified chromosome in the presence of which marked changes in chromosome organization and also in gene action take place, is also under investigation.
>
> *(Report of the President 1956, p. 16)*

Neither Rhoades nor his students attended the *Cold Spring Harbor Symposia* in June 1956 or the *Brookhaven Symposium* of 1955, yet he was well aware of her most recent discoveries. McClintock had spent time at the Universities of Illinois and Wisconsin in the spring of 1956, where she discussed with Rhoades and with Brink their common research results.[52] In late June 1956, Rhoades was invited to lecture at Cornell.[53] His lecture, "Genetic Control of Mutability," designated two kinds of mutations, intragenic and extragenic. Intragenic change has been hypothesized as possibly molecular. Extragenic mutations have been caused by four phenomena: (1) minute deficiencies,

[52] Letter B. McClintock to E. Caspari, April 5, 1956, Caspari Papers Ms. Coll. #1, File McClintock, Barbara, APS.

[53] Murphy and Kass (2011, p. 86); L. F. Randolph, File, "Rhoades Lecture Notes 1956," in "Cytogenetics Lecture Notes L.F.R."; files at Cornell to be deposited in the Cornell Archives.

which might not be visible cytologically; (2) chromosome rearrangements that result in position effects; (3) position effects involving pseudoalleles; and (4) a compound locus (e.g., *R* gene in maize) in which there is separation of component parts. He then added a fifth extragenic cause of mutation found in maize, "other elements" that affect gene action [phenotypic expression]. He mentioned three examples of the "other elements" causing extragenic mutations: (1) the *Dt* (dotted locus), and two recent studies of "unstable gene systems"; (2) *Ds-Ac*, which had recently been reported by Barbara McClintock; and (3) P^{v}*-Mp* involving Emerson's pericarp locus, reported by Brink and his students. McClintock's study showed two kinds of genes in the nucleus: "Primary genes and controlling elements [*Ds-Ac*]." The latter can transpose and determine the time and rate of action [expression] of the primary genes.

Rhoades, of course, had been one of the co-founders of maize cytogenetics (Chapter 4), and had been teaching the subject since 1941, first at Columbia University then at the University of Illinois. As we have seen, he was very familiar with McClintock's research, and as soon as her first journal article on this subject was published in 1950, he taught it to his Cytogenetics course at Illinois (Punnett 1950–1951: March 1951; Campbell 1992, p 109). Rhoades hypothesized that the controlling element *Ds* observed by McClintock at the *C* locus (*CDs*) in maize, which suppresses (controls) the expression of color, might also explain how *Dt* effects the expression of anthocyanin at the *A* locus. Later studies, however, showed that *Dt* was part of a different unstable gene system. Rhoades also described Brink's studies of Modulator (*Mp*) and explained that it had been shown to be equivalent to *Ac* because they had the same effects on causing *Ds* to transpose, but they did not affect *Dt* in a similar manner.

At Cornell, L. F. Randolph lectured on McClintock's pioneering publications in his Cytogenetics course.[54] Furthermore, students of Professor Harold H. Smith were investigating the phenomenon of broken chromosomes and mutation caused by chemical mutagens. In 1953, Jean Dolloff began studies of mutagenic agents causing mutations in maize (Kreizinger 1955, 1958, 1960). Smith encouraged her to meet with McClintock at Brookhaven National Laboratory, where he was conducting research. Smith later became a full-time member of the Brookhaven Biology Department (Murphy & Kass 2011). There, he continued to encourage students to meet with McClintock about their research problems. Seward S. Sand had also been investigating the cause of somatic instability in tobacco, with Smith as his Ph.D. advisor (Sand 1955). His continued studies of this tobacco system allowed him to correlate a mutant with variegated expression caused by a controlling element similar to that of the maize system reported by McClintock (Sand 1976).

In the summer of 1957, McClintock's colleagues in the Botanical Society of America recognized her outstanding research contributions with their Merit Award – the highest honor they can bestow on a colleague.[55] At the annual meeting banquet of the Botanical Society held on August 28 in Stanford, California, she received her certificate of merit as an "early student of radiation – induced chromosomal aberrations, pioneer in the use of such aberrations for purposes of genome analysis, important contributor to the theory of gene structure, [and] world leader in the broad field of cytogenetics."[56]

Also that year, Edgar Altenburg (1957) added a new section on "Mutable Genes" to the second edition of his popular *Genetics* text (see Chapter 7). Over half of his chapter is devoted to McClintock's work on mutable genes in maize, summarized from extensive letters he had received from McClintock and from her *CSH Symposia* publications.[57]

Yet by early 1957, McClintock again seemed to have felt a need to get away from Cold Spring Harbor (Keller 1993, p. 181). She recalled that in early spring, during a visit from Paul Mangelsdorf of Harvard University, he asked if she would train students in cytology of maize at the University in Peru.[58] She immediately accepted his invitation, and in June received an official invitation from

[54] L. F. Randolph "Cytogenetics Lecture Notes L.F.R."
[55] Since 2015 renamed "The Distinguished Fellow of the Botanical Society of America Award."
[56] *Plant Science Bulletin*, 1958, Vol. 4, No. 1, Spring, p. 6.
[57] Altenburg, LL.
[58] E. F. Keller interview with B. McClintock, January 13, 1979, APS; Provine and Sisco (1980).

Ralph Cleland, the Chairman of the National Academy of Sciences, National Research Council, Committee on Preservation of Indigenous Strains of Maize, "to help train someone who can carry on maize cytological studies for [Alexander] Grobman."[59] Mangelsdorf was a member of Cleland's Committee and for years had been conducting studies on the origin and races of maize (Mangelsdorf & Reeves 1939; Wellhausen et al. 1951, 1952; Roberts et al. 1957; Wellhausen et al. 1957; Stakman et al. 1967: Chapter 15). Grobman was working at the Escuela Nacional de Agricultura [National School of Agriculture], La Molina, Lima, Peru, on leave from his graduate school studies at Ohio State University with G. H. Stringfield (also a member of Cleland's Committee).

Mangelsdorf's collaborative efforts on the *Races of Maize in Mexico* (Wellhausen et al. 1951, 1952) inspired studies of maize races in other countries of Central and South America (Stakman et al. 1967, p. 262). As a consequence, Cleland's Committee was established in 1951 with the purpose of saving the maize diversity that had been uncovered in Mexico and that could provide opportunities to improve maize but also "to recognize, to describe, and to preserve for future use, the varieties and races which [plant breeder's] own improved productions tend to replace in some cases to extinguish" (Wellhausen et al. 1952, pp. 5–6). With a grant from President Truman's Point IV Program, the Committee established three centers with cold-storage facilities in Mexico, Colombia, and Brazil and a fourth in the United States to store standby maize samples from the others. The collections represented all the germplasm available to corn breeders (Stakman et al. 1967, p. 263; Clark 1954, 1956). In 1956, a direct grant was given to the National School of Agriculture, for the purpose of studying and classifying the races of maize in Peru (Grobman et al. 1961, p. 2).

Mangelsdorf also understood the importance of genetic and cytological studies, especially heterochromatic chromosome knobs and B-chromosomes in classifying and describing varieties of maize (Grobman et al. 1961, pp. 132, 133). Before getting McClintock involved, however, Mangelsdorf had recommended in August 1956 that the Committee propose that his Postdoctoral Associate Dr. Yu-Chen Ting make the trip to Peru to train local workers in making cytological studies of Peruvian races of Maize. Ting had been studying knobs on maize and teosinte chromosomes with Mangelsdorf at Harvard since 1954 (*Boston College Chronicle* 2005; Ting 1956, pp. 32, 35–36). Ting was unable to obtain security clearance as reported to the Committee on April 22, 1957. They then considered a new proposal to invite McClintock and called her on the phone at Cold Spring Harbor. She agreed to make the trip to Peru, and a sum of $3,000 was budgeted for her travel and subsistence expenses.[60]

McClintock left for Peru on December 5, 1957 (Cullen 2003, p. 93), spending about six weeks at the Escuela National de Agricultura, guiding Grobman, who was in charge of the Cooperative Program for Maize Research, and his assistant, Ulises Moreno, to determine the relative frequency of presence of knobs in each chromosome of maize and in each arm of the respective chromosome in three maize races of the Peruvian coast: the lowland races Perla (tropical flint), Alazan (floury, red pericarp, of high altitude origin), and Arizona (a Tuxpeno derivative introduced a good number of years ago) (Moreno & Grobman 1958b, pp. 25–26).

During this visit, McClintock conducted a preliminary examination of at least three maize collections from the Peruvian highlands and had found a maximum number of three B chromosomes per cell. This was the first report of B chromosomes from the Andean area of adaptation, which the Peruvian researchers reported in the March 1958 *Maize News Letter* (Moreno & Grobman 1958a, p. 25).

McClintock's seminar on transposable elements, Activator/Dissociation in maize, given to the Peruvian researchers and their colleagues, alerted them to the possibility of controlling elements in their maize collections.[61] In a cross of Andean sweet corn, Grobman had observed in F1 offspring

---

[59] Letter R. Cleland to B. McClintock, June 27, 1957, Folder, National Research Council, 1957–1958, McClintock APS.

[60] Division of Biology and Agriculture Records Group DNRC: Biology & Agriculture, Committee on Preservation of Indigenous Strains of Maize: Executive Committee, 1953–1960. NAS-NRC Archives, LBK email communication with NAS Archivist D. Barbiero, 11/16/01 (henceforth NAS-NRC Archives).

[61] Email A. Grobman to L. Kass, October 14, 2014.

a variegated pericarp pattern, which had been entirely absent in either parent plant. To him, it indicated the presence of a "controlling element, such as McClintock's Activator," may have been introduced in the cross from one of the parents (Grobman 1958, p. 25). They later continued to observe such mutations in Peruvian maize races (Grobman et al. 1961, pp. 54–56).

On returning to the United States in January 1958, McClintock gave a seminar at Harvard on "Chromosomal elements controlling gene action."[62] She summarized studies of the *Dt*, *Ds-Ac*, and *Spm* system of controlling elements. She mentioned that Nuffer [U of Missouri] had recently found *Dt* [a transposable element] in a stock from Peru and another from Brazil. Also that "recent experiments in Peru" on sweet corn had found mutations of *su* to *Su* in crosses of previously stable recessive *su* parents. Similar studies were obtained with genes controlling pericarp color.

The Maize Preservation Committee hoped that McClintock would continue her studies of South American maize chromosomes, thus providing an additional understanding toward the origin of indigenous maize. In July 1958, the National Academy of Sciences, Maize Preservation Executive Committee, issued a progress report from the Latin American Seed Centers during 1957–1958 (Maize Committee 1958). Grobman reported that he hoped to complete his work for the publication of *Races of Maize in Peru*, but he felt more fieldwork was needed, as he had mentioned to McClintock when she visited. He also planned to discuss this further with Mangelsdorf during his proposed forthcoming visit. McClintock had assisted in organizing the cytological studies and she planned a second trip to complete those studies (Maize Committee 1958, p. 6).

Dr. Lewis M. Roberts reported that the *Races of Maize in Colombia* had been published (Roberts et al. 1957). David Timothy reported further studies, on behalf of Director Roberts, for the Colombian-Andean Area. Collections for race studies of Bolivia, Chile, Ecuador, and Venezuela had been selected and planted at Medellín and Bogota to collect tassels for cytological studies. All tassels from lowland races had been collected, and collections for highland races were almost completed. The collections were being stored in deep freeze units with "the hope that Barbara McClintock will be able to come to Colombia to do cytological studies with these Andean collections" (Maize Committee 1958, p. 4).

Although the Peruvian researchers had reported on McClintock's visit and findings of maize chromosome studies during the winter of 1957–1958, she made no mention of that visit in her Carnegie report (McClintock 1958). Instead, she reported on continued investigations of the *Spm* system of control of gene action. She found that "The alternation of active and inactive phases during development may result in many diverse patterns of anthocyanin pigmentation, both in different parts of a plant and in different individuals of its progeny" (Demerec 1958, p. 375). She concluded that what appeared originally to be a highly complex mechanism controlling gene action had turned out to be a reflection of alternating cycles of action by the controlling element *Spm*.

In July 1958, Cleland's Committee planned for the forthcoming maize work at the Central and South American field stations (Maize Committee 1958, p. 2). At their Executive Committee meeting of July 16, 1958, "it was decided to send Dr. Barbara McClintock on a second trip to Peru for continued aid of the cytological studies of Peruvian races of maize and also to Colombia for a study of the races of Venezuela, Ecuador and Bolivia, being grown and studied there, and if possible, Chilean races being studied in Mexico."[63] Cleland then sent a formal invitation to McClintock to return to Peru and to assist the Colombian researchers with their chromosome knob studies.[64] After resolving some apparent confusion about whether McClintock would be welcome at the Colombian maize program, Director Roberts prepared for her forthcoming visit by arranging a place for her to work where good microscopes would be available.[65] She worked at the Universidad Nacional,

[62] Lecture, Series III, McClintock APS.

[63] NAS-NRC Archives.

[64] Letter R. Cleland to B. McClintock, July 27, 1958, File National Research Council, Ser. I, McClintock APS.

[65] P. C. Mangelsdorf to L. M. Roberts, October 13, 1958; Roberts to McClintock, October 15, 1958; Roberts to Mangelsdorf, October 17, 1958, all in Folder Columbian Agriculture 1958, RG 1.2, Series 311 [Colombia], Box 3, Folder 15, Rockefeller Foundation Archives.

Facultad de Agronomia e Instituto Forestal, located in Medellín, Colombia. The facilities were provided by the Dean of the School of Agronomy, Dr. Carols Garcés, and by Dr. Sánchez, in whose Plant Pathology laboratory she worked with the assistance of Señorita Rocío Díez P. [Palacio], who also contributed to the final report (McClintock 1959, p. 454). As Timothy had reported earlier (Maize Committee 1958, p. 4), he and his colleagues at Colombia collected and stored in the deep freeze maize sporocytes from Ecuador, Bolivia, Chile and Venezuela, hoping that McClintock would return to study them. As she had been doing since 1930, she applied the microscope squash technique to examine maize chromosome morphology at the pachytene stage and eventually placed knobs on the ten maize chromosomes, as she had done in her ideogram for maize chromosomes in 1933 (McClintock 1933, p. 192).[66] During her study, she only knew the country and elevation where the collections had been made. At the end of her investigations, she sent her data to Bogota, where Timothy and Hathaway plotted the data to the exact geographical locations. Upon receiving their results, it became apparent that there were "significant relations between knob constitution and geographical location, and that it would be possible to utilize knowledge of knob constitutions in attempting to trace the origins and migrations of maize races" (McClintock 1959, p. 456). McClintock's (1959) conclusions that "present day maize may have derived from several different centers" were based on the collective chromosome knob data that she and Ms. Diez-P. had observed and had been plotted by Timothy and Hathaway.[67]

The following year, additional Bolivian chromosome knob morphology and "B type" chromosome results that McClintock recorded during these studies were published in *Races of Maize in Bolivia* (Ramírez et al. 1960, pp. 25–33). Chromosome knob numbers and positions are useful characters in determining relationships in maize. When information on knob position, size, and shape is determined for a particular race or plant of maize, researchers can draw cytological relationships, even when the total number of knobs may not differ. The researchers concluded that knowledge of indigenous races "is an essential prerequisite to maize improvement," not only in local centers of maize variability like Bolivia, but also for studies of cross-breeding to improved varieties in U.S. maize.

Correlating chromosome knobs with localities was not a new idea, as McClintock (1959, p. 455) had mentioned without citing references. Those familiar with the maize cytogenetics literature would be aware that Longley (1938, 1941), as well as Wellhausen et al. (1952) and Roberts et al. (1957), had reported the importance of chromosome knobs toward an interpretation of the origin of maize. Longley (1938) had previously concluded that knobs were important to understanding the origin and distribution of maize as cited by Wellhausen et al. (1952, p. 35). Years earlier, Longley (1938, p. 195) had found that, "The number of knobs on the chromosomes of the maize of the Indians of the United States increases as the tribes approach Mexico, where the much-knobbed chromosomes of maize and maize relatives occur." His pioneering publications between 1937 and 1941 on chromosome morphology emphasized the importance of chromosome lengths, centromere positions, and heterochromatic knobs, along with numbered ideograms of the chromosomes arranged from longest to shortest to show relationships among them (Coe & Kass 2019; Dawe 2022). Over the next few years, both McClintock and Longley would continue to train and collaborate with Central and South American students on chromosome morphology for races of maize in Latin America (Figure 8.2; see Chapter 9).

During her second visit to Peru (1958–1959), McClintock continued cooperative studies with Grobman and Moreno, which were reported to the maize genetics community via the *Maize News Letter* (Moreno et al. 1959, pp. 27–28). They studied the number, position, size, and shape of chromosome knobs of the races of maize in Peru, trying to determine patterns for differentiating races.

[66] See McClintock (1933), reprinted in Kass (2013ff., pp. 1.203–1.258).

[67] See Kass 2013ff., Part IV Origin and Diversity of Maize in the Americas, for reprints and perspectives on McClintock's later maize chromosome studies; the geographic origin and place of maize domestication are controversial even today (Grobman et al. 2012; Grobman 2013; Kato 2013; Stitzer & Ross-Ibarra 2018; Katz 2018; Hirst 2020).

FIGURE 8.2 Dr. Albert E. Longley drawing chromosomes with a camera lucida, in his Bureau of Plant Industry, U.S. Department of Agriculture laboratory, Washington, DC (ca. 1930s). (A. E. Longley photos, courtesy of E. H. Coe Jr.)

They recorded frequency of knobs in chromosome arms, the shape and size of such knobs, and the frequency of the presence of the abnormal-10 chromosome [Ab10], and found that the highest number of B-chromosomes per plant was 4. Plants with B-chromosomes were associated with highland races having a low number of knobs.

Grobman soon continued his studies at Harvard, at the invitation and under the direction of Mangelsdorf (Mangelsdorf 1974). The investigations Grobman conducted and published on the *Races of Maize in Perú. Their Origins, Evolution and Classification* (Grobman et al. 1961) became the basis of his Ph.D. thesis (Grobman 1962).[68] The cytological studies of chromosome knobs and B-chromosomes reported in their monograph were conducted exclusively by Ing. Ulises Moreno (Grobman et al. 1961, pp. vii, 132–138). They acknowledged Dr. Barbara McClintock for organizing the cytological survey of Peruvian maize races and for providing technical guidance to the

[68] Email A. Grobman to L. Kass, October 14, 2014.

cytologists engaged in the work (Grobman et al. 1961, p. vii.). Maize race determinations appeared to be regulated by frequency of distribution of chromosome knobs, as observed by McClintock, as well as by geographical location (Grobman et al. 1961, p. 70). Based on tutorials from McClintock, Moreno (1960) completed a thesis on B-chromosomes and chromosome knobs, at the Agrarian University, La Molina, Lima, that led to his title of Agricultural Engineer.[69] He received more training at North Carolina State College, sponsored by the Rockefeller Foundation in 1962.[70] Eight years later, he obtained his Ph.D. (Moreno 1970) at Cornell University, funded again by a Rockefeller Foundation grant.[71]

By early 1959, McClintock was invited to initiate a program in Mexico similar to that conducted in Colombia, and to help train students at the Inter-American Maize Improvement Program Center in Mexico, under the direction of Edward Wellhausen.[72] She spent the summer of 1959 in Mexico City collecting maize sporocytes for studies of chromosome morphology and returned in the winter of 1959–1960 to complete those studies (McClintock 1960; Demerec 1960, p. 417, 420). During this time, McClintock was also concerned with the reorganization of the Cold Spring Harbor Department of Genetics and a search for a new director upon the mandatory retirement of Milislav Demerec in 1960. At the end of 1959, the Mexican Maize Improvement Program was expanded within the Inter-American Food Crop Improvement Program, which soon fostered the broader International Center for Corn and Wheat Improvement, in October 1963 (The Rockefeller Foundation 1963–1964, p. 14). The latter is known by its Spanish acronym CIMMYT for Centro Internacional de Mejoramiento de Maíz y Trigo. McClintock played a cooperative role in these programs (Stakman et al. 1967: Chapter 15, pp. 64–65). On April 12, 1966, CIMMYT was expanded as an international organization, under a new charter (Breth 1986).

Over the next few years, McClintock's continued work with the Latin American maize programs, proceeded with her investigations on maize controlling elements, and was a consultant to the reorganization of the Cold Spring Harbor Laboratory, upon the closing of the Cold Spring Harbor Department of Genetics.

LITERATURE CITED

Altenburg, E. 1957. *Genetics, Revised Edition*, Henry Holt and Company, New York.

Barclay, P. C and R. A. Brink. 1954. The relation between modulator and activator in maize. *Proceedings of the National Academy of Sciences* Vol. 40 (No. 12, December 15, 1954), pp. 1118–1126.

Boston College Chronicle. 2005. Biologist Yu-Chen Ting dies; Specialized in plant genetics. *The Boston College Chronicle* Vol. 13 (No. 15, April 14, 2005), p. 4.

Breth, S. A. 1986. *Mainstreams of CIMMYT Research: A Retrospective*, CIMMYT, Mexico, D.F. https://pdf.usaid.gov/pdf_docs/pnaay483.pdf

Brink, R. A. 1954. Very light variegated pericarp in maize. *Genetics* Vol. 39 (No. 5. September 1, 1954), pp. 724–740.

Brink, R. A. and R. A. Nilan. 1951. Studies on variegated pericarp. *Maize Genetics Cooperation News Letter* Vol. 25 (March 17, 1951), pp. 42–47.

Brink, R. A. and R. A. Nilan. 1952. The relation between light variegated and medium variegated pericarp in maize. *Genetics* Vol. 37 (No. 5, September 1, 1952), pp. 519–544.

Campbell, A. 1992. Insertion by phages and transposons, pp. 109–114, *In*, N. Fedoroff and D. Botstein (Eds.), *The Dynamic Genome: Barbara McClintock's Ideas in the Century of Genetics*, Cold Spring Harbor Laboratory Press, Plainview, NY.

[69] Email U. Moreno to L. Kass, November 18, 2016. I thank Dr. Maria Scurra, Cornell Ph.D. 1972, for help contacting Dr. Moreno.

[70] *Annual Report of the Rockefeller Foundation* 1962, p. 204.

[71] *The Rockefeller Foundation President's Five Year Review and Annual Report 1968*, p. 155.

[72] Rockefeller Foundation Archives: March 24, 1959–September 13, 1962. RG 6.13 Series 1.1, Box 32, Folder 353; McClintock, Barbara, March 195 9–September 1962. Letters to and from Wellhausen and McClintock, arrangements for doing knob studies in Mexico for Inter-American Maize Improvement Program and training students at North Carolina State College.

Campbell, A., D. Berg, D. Botstein, E. Lederberg, R. Novick, P. Starlinger and W. Szybalski. 1977. Nomenclature of transposable elements in prokaryotes, pp. 15–22, *In*, A. J. Bukhari, J. A. Shapiro, and S. L. Adhya (Eds.), *DNA Insertion Elements, Plasmids, and Episomes*, Cold Spring Harbor Laboratory, Cold Spring Harbor, NY. [See their 1979 update published in both *Gene* Vol. 5 (No. 3, March 1979), pp. 197–206) and *Plasmid* Vol. 2 (No. 3, July 1979), pp. 466–473]

Cassata, F. 2017. Lysenko in Bellagio: The Lysenko controversy and the struggle for authority over Italian genetics (1948–1956), pp. 37–72, *In*, W. deJong-Lambert and N. Krementsov (Eds.), *The Lysenko Controversy as a Global Phenomenon, Volume 2*, Palgrave Studies in the History of Science and Technology, Palgrave Macmillan, Cham. https://doi.org/10.1007/978-3-319-39179-3_2

Clark, J. A. 1954. Preventing extinction of original strains of corn. *National Academy of Science, National Research Council News Report* Vol. 4 (No. 5, September–October 1954), pp.78–81.

Clark, J. A. 1956. Collection, preservation and utilization of indigenous strains of maize. *Economic Botany* Vol. 10 (April), pp. 194–200. https://doi.org/10.1007/BF02985328

Coe, E. H. and L. B. Kass. 2019. Albert Edward Longley, 1893–1984 – Botanist, cytologist, geneticist. *Maize Genetics Cooperation Newsletter* Vol. 93, pp. 1–9. https://mnl.maizegdb.org/93/pdf/03Coe.pdf (Online October 23, 2019).

Cohen, S. N. 1976. Transposable genetic elements and plasmid evolution; Review Article. *Nature* Vol. 263 (October 28, 1976), pp. 731–738. [p. 734, "Transposable elements" defined.]

Cullen, J. H. 2003. *Barbara McClintock, Geneticist* (Women in Science), Chelsea House Publishers, New York, NY, 122 pp.

Davis, R. H. and D. D. Perkins. 2002. *Neurospora*: a model of model microbes. *Nature Reviews Genetics* Vol. 3, pp. 397–403. https://www.nature.com/articles/nrg797

Dawe, R. K. 2022. The maize abnormal chromosome 10 meiotic drive haplotype: a review. *Chromosome Research* Vol. 30 (Issue 2–3, 2 June 2022, issued September 2022), pp. 205–216. https://doi.org/10.1007/s10577-022-09693-6 [Special Issue on Non-Mendelian Inheritance and Meiotic Drive]

Deichmann, U. 2011. Early 20th-century research at the interfaces of genetics, development, and evolution: Reflections on progress and dead ends. *Developmental Biology* Vol. 357 (No. 1, 1 September 2011), pp. 3–12.

DeJong-Lambert, W. 2013. Hermann J. Muller, Theodosius Dobzhansky, Leslie Clarence Dunn, and the reaction to Lysenkoism in the United States. *Journal of Cold War Studies* Vol. 15 (No. 1, January 1, 2013, Online April 29, 2013), pp. 78–118.

DeJong-Lambert, W. and N. Krementsov (Eds.). 2012. Special issue on The Lysenko controversy and the cold war. *Journal of the History of Biology* Vol. 45 (No. 3, August 2012), includes 10 articles on the topic. https://link.springer.com/journal/10739/volumes-and-issues/45-3

Delbruck, M. 1954. On the replication of desoxyribonucleic acid (DNA). *Proceedings of the National Academy of Sciences* Vol. 40 (No. 9, September 1, 1954), pp. 783–788. https://doi.org/10.1073/pnas.40.9.783

Demerec, M. 1953a. Foreword, p. v. *Cold Spring Harbor Symposia on Quantitative Biology*, "Viruses," June 5–11 1953, Volume XVIII.

Demerec, M. 1953b. Department of Genetics, Cold Spring Harbor, Long Island, New York; M. Demerec, Director. *Carnegie Institution of Washington Year Book No. 52*, 1952–1953, pp. 205–248. [Issued December 11, 1953]

Demerec, M. 1954. Genetic action of mutagens, p. 201+, *In*, G. Montalenti and A. Chiarugi (Eds.). *Caryologia*, Vol. 6, 1953/54. Suppl.

Demerec, M. 1958. Department of Genetics, Director's Report. *Carnegie Institution of Washington Year Book 57*, 1957–1958.

Demerec, M. 1960. Department of Genetics, Director's Report. *Carnegie Institution of Washington Year Book 59*, 1959–1960.

de Serres, F. J. 1962a. Meetings, *Neurospora. Science* Vol. 135 (No. 3500, January 26, 1962), pp. 322 + 324–325.

de Serres, F. J. (Ed.). 1962b. *Neurospora information conference: a summary of the proceedings of an information conference*, [University of California, San Diego] La Jolla, California, March 2–4, 1961, National Academy of Sciences, National Research Council Publication 950, Washington, DC.

Dollinger, E. J. 1954. Studies on induced mutation in maize. *Genetics* Vol. 39 (No. 5, 1 September 1954), pp. 750–766. https://academic.oup.com/genetics/article/39/5/750/6033498 [Thesis Ph.D. Columbia University, reprinted from *Genetics* Vol. 39 No. 5, September 1954. https://clio.columbia.edu/catalog/10143975]

Dollinger, E. J. 1955. Analysis of a mutable system in maize. Abstract presented at Genetics Society of America Meeting September 1955. *Records of the Genetics Society of America* No. 24, 1955, p. 570.

Emerson, R. A. 1914. Inheritance of a recurring somatic variation in variegated ears of maize. *American Naturalist* Vol. 48, pp. 87–115.

Fedoroff, N. V. 1983. Controlling elements in maize, Chap. 1, pp. 1–63, *In*, J. A. Shapiro (Ed.), *Mobile Genetic Elements*, Academic Press, New York.

Fedoroff, N. V. 1989. About maize transposable elements and development. *Cell* Vol. 56 (No. 2, January 27, 1989), pp. 181–191. [Review]

Fedoroff, N. V. 1992. Maize transposable elements: A story in four parts, pp. 389–415, In, N. Fedoroff and D. Botstein (Eds.), *The Dynamic Genome: Barbara McClintock's Ideas in the Century of Genetics*. Cold Spring Harbor Laboratory Press, Plainview, NY. [See especially "Part IV: The *Spm* element family," pp. 402-415.]

Fedoroff, N. V. 1998. Marcus Rhoades and transposition. *Genetics* Vol. 150 (No. 3, November 1, 1998), pp. 957–961.

Fedoroff, N. V. 2013a. McClintock and epigenetics, Chapter 4, pp. 61–70, *In*, N. V. Fedoroff (Ed.), *Plant Transposons and Genome Dynamics in Evolution*, John Wiley & Sons, Inc, Oxford. First published February 4, 2013.

Fedoroff, N. V. 2013b. McClintock and epigenetics: Insights from changes in phase of transposition activity, pp. 3.34–3.39, *In*, L. B. Kass (Ed.), *Perspectives on Nobel Laureate Barbara McClintock's Publications (1926–1984): A Companion Volume*. https://hdl.handle.net/1813/34897

Fedoroff, N., S. Wessler, and M. Shure. 1983. Isolation of the transposable maize controlling elements *Ac* and *Ds*. *Cell* Vol. 35 (No. 1, November 1983), pp. 235–242.

French, C. S., M. Demerec, and G. W. Corner. 1956. Biological research in the Carnegie Institution of Washington. *AIBS Bulletin* Vol. 6 (No. 2, April 1956), pp. 17–19.

Fungal Genetics Reports (FGR). 1962–present. An online resource published by the Fungal Genetics Stock Center. https://newprairiepress.org/fgr/ [Volumes 1–32 (1962–1985) were published as *Neurospora Newsletter*. From 1986 to 2007, FGR was published as *Fungal Genetics Newsletter* (ISSN 0895-1942; eISSN: 1556-1275)]

Garber, E. D. 1972. *Cytogenetics: An Introduction*, McGraw-Hill Book Company, New York.

Goldschmidt, R. B. 1951. Chromosomes and genes, pp. 1–11, *In*, K. B. Warren (Ed.), *Genes and Mutations, Cold Spring Harbor Symposia on Quantitative Biology*, Volume XVI (June 7–15, 1951). The Biological Laboratory, Cold Spring Harbor, Long Island, NY. [(©1952]

Goldschmidt, R. B. 1954. Different philosophies of genetics. *Science* Vol. 119 (No. 3099, May 21, 1954), pp. 703–710. [Reprinted from 9th International Congress of Genetics, 1953, *Proceedings* 1954; Invited Paper, Presidential Address, Richard B. Goldschmidt, pp. 83–99]

Golz, H. F., M. Roccaro, R. Kuzoff and A. Hudson. 2004. GRAMINIFOLIA promotes growth and polarity of *Antirrhinum* leaves. *Development* Vol. 131 (No. 15, August 2004), pp. 3661–3670. https://doi.org/10.1242/dev.01221

Grobman, A. 1962. *Races of Maize in Perú. Their origins, evolution and classification*. Thesis (Ph.D.) 2 vols., Department of Biology, Harvard University, 1962 HU 90.8162

Grobman, A. 2013. Domestication and evolution of maize: new perspectives from cytogenetic, genetic and biomolecular research complementing archaeological findings, pp. 329–488, Appendix, November 2012, *In*, D. Bonavia (Ed.), *Maize, Origin, Domestication, and Its Role in the Development of Culture*, Cambridge University Press, Cambridge.

Grobman, A. and U. Moreno. 1958. 3. Tests for sugary endosperm gene in Chullpi. Andean sweet corn. *Maize Genetics Cooperation News Letter 32* (March 15, 1958), p. 25.

Grobman, A., D. Bonavia, T. D. Dillehay et al. 2012. Preceramic maize from Paredones and Huaca Prieta, Peru. *Proceedings of the National Academy of Sciences* Vol. 109 (No 5, January 31, 2012), pp. 1755–1759.

Grobman, A., W. Salhuana, R. Sevilla, with P. C. Mangelsdorf. 1961. Races of Maize in Perú. Their Origins, Evolution and Classification, Publication 915, National Academy of Sciences, National Research Council, Washington, DC, 374 pp.

Halpern, M. E. 2016. Barbara McClintock on defining the unstable genome. *Genetics* Vol. 204 (No. 1, September 1, 2016), pp. 3–4. https://doi.org/10.1534/genetics.116.194092

Hirst, K. K. 2020. The domestication of maize in America (November 20, 2020). ThoughtCo, February 16, 2021. https://www.thoughtco.com/maize-domestication-history-of-american-corn-171832

Judd, B. H. 1954. Studies of rearrangements involving heterochromatin in *Drosophila melanogaster*. I. A proof of variegated-type position effect at the white locus. II. A study of the heterochromatin of chromosome IV. Dissertation (Ph.D.), California Institute of Technology. https://doi.org/10.7907/4F9F-MV78

Kass, L. B. (Ed.). 2013ff. *Perspectives on Nobel Laureate Barbara McClintock's Publications (1926–1984): A Companion Volume*. Volumes I & II The Internet-First University Press. https://hdl.handle.net/1813/34897 [Online December 30, 2013, updates March 2014; Vol. III, Increments No. 1–3, May 2016; Vol. IV, Increment No. 4, April 2020]

Kass, L. B. and P. Chomet. 2009. Barbara McClintock, pp. 17–52, *In*, J. Bennetzen and S. Hake (Eds.), *Handbook of Maize: Genetics and Genomics*, Springer Science+Business Media LLC, New York.

Kato Y., T. A. [T. Angel Kato Y]. 2013. Significance of McClintock's pattern concept in the analysis of chromosome knob distributions in the races of maize, pp. 4.39–4.49, *In*, L. B. Kass (Ed.), 2013ff. *Perspectives on Nobel Laureate Barbara McClintock's Publications (1926–1984): A Companion Volume*. https://hdl.handle.net/1813/34897

Katz, B. 2018. Rethinking the corny history of maize, A new genetic study traces the movement of one of the world's most vital crops from Mexico to South America, *Smithsonian Magazine* [Online December 14, 2018]. https://www.smithsonianmag.com/smart-news/rethinking-corny-history-maize-180971038/

Keller, E. F. 1993. *A Feeling for the Organism. The Life and Work of Barbara McClintock*, 10th Anniversary edition, W. H. Freeman, New York.

Kreizinger, J. D. 1955. *Effects of treatments with diepoxybutane, beta-propiolactone, and ketene dimer on the endosperm in maize*. M. S. Cornell University, September 1955, Master's Thesis deposited in Library, February 1956, Thesis SB124 1956 K92.

Kreizinger, J. D. 1958. *Diepoxybutane as a chemical mutagen in maize*. Ph.D. Cornell University, June 1958, Thesis SB124 1958 K92.

Kreizinger, J. D. 1960. Diepoxybutane as a chemical mutagen in *Zea mays*. *Genetics* Vol. 45 (No. 2, February 1, 1960), pp. 143–154.

Lewis, E. B. 1951. Pseudoallelism and gene evolution, pp. 159–174, *In*, K. B. Warren (Ed.), *Genes and Mutations, Cold Spring Harbor Symposia on Quantitative Biology*, Vol. XVI (June 7–15, 1951), The Biological Laboratory, Cold Spring Harbor, NY.

Lewis, E. B. 1954. Pseudoallelism and the gene concept, pp. 100 +, *In*, G. Montalenti and A. Chiarugi (Eds.), *Caryologia*, Vol. 6, 1953/54. Suppl.

Lerner, M. I. 1953. The ninth international congress of genetics 1953. *Science* Vol. 118 (No 3076, December 11, 1953), pp. 708–709. https://doi.org/10.1126/science.118.3076.708

Longley, A. E. 1941. Chromosome morphology in maize and its relatives. *Botanical Review*. Vol. 7 (No. 5, May), pp. 263–289.

Maize Committee. 1958. Third Report - Preservation of Indigenous Strains of Maize, July 1958. Folder: National Research Council. Committee on Preservation of Indigenous Strains of Maize. McClintock Collection, Series II, APS Library.

Mangelsdorf, P. C. 1974. *Corn: Its Origin, Evolution, and Improvement*, The Belnap Press of Harvard University Press, Cambridge, MA.

Mangelsdorf, P. C. and R. G. Reeves. 1939. The origin of Indian corn and its relatives. *Texas Agricultural Experiment Station Bulletin* 574. 315pp.

McClintock, B. 1933. The association of non-homologous parts of chromosomes in the mid-prophase of meiosis in Zea mays, with 51 figures in the text and plates VII-XII. *Zeitschrift fur Zellforschung und mikroskopische Anatomie* Vol. 19 (No. 2, September 22, 1933), pp. 191–237.

McClintock, B. 1944. The relation of homozygous deficiencies to mutations and allelic series in maize. *Genetics* Vol. 29 (No. 5, September), pp. 478–502.

McClintock, B. 1946. Maize genetics: Continuation of the study of the induction of new mutants in chromosome 9. Modification of mutant expression following chromosomal translocation. The unexpected appearance of a number of unstable mutants. *Carnegie Institution of Washington Year Book No. 45*, 1945–1946, pp. 176–186. [Issued December 13, 1946]

McClintock, B. 1947. Cytogenetic studies of maize and Neurospora: The mutable *Ds* locus in maize. Continuation of studies of the chromosomes of *Neurospora crassa*. Carnegie *Institution of Washington Year Book No. 46*, 1946–1947, pp. 146–152. [Issued December 12, 1947]

McClintock, B. 1948. Mutable loci in maize: Nature of the *Ac* action. The mutable *c* loci. The mutable *wx* loci. Conclusions. *Carnegie Institution of Washington Year Book No. 47*, 1947–1948, pp. 155–169. [Issued December 10, 1948]

McClintock, B. 1949. Mutable loci in maize: The mechanism of transposition of the *Ds* locus. The origin of *Ac*-controlled mutable loci. Transposition of the *Ac* locus. The action of *Ac* on the mutable loci it controls. Mutable loci *c m-2* and *wx m-1*. Conclusions. *Carnegie Institution of Washington Year Book No. 48*, 1948–1949, pp. 142–154. [Issued December 9, 1949]

McClintock, B. 1950a. The origin and behavior of mutable loci in maize. *Proceedings of the National Academy of Sciences* Vol. 36 (No 6, June 15, 1950), pp. 344–355.
McClintock, B. 1950b. Mutable loci in maize: Mode of detection of transpositions of *Ds*. Events occurring at the *Ds* locus. The mechanism of transposition of *Ds*. Transposition and change in action of *Ac*. Consideration of the chromosome materials responsible for the origin and behavior of mutable loci. *Carnegie Institution of Washington Year Book No. 49*, 1949–1950, pp. 157–167. [Issued December 15, 1950]
McClintock, B. 1951a. Mutable loci in maize. *Carnegie Institution of Washington Year Book No. 50*, 1950–1951, pp. 174–181. [Issued December 14, 1951]
McClintock, B. 1951b [©1952, Symposium held June 1951]. Chromosome organization and genic expression, pp. 13–47, *In*, K. B. Warren (Ed.), *Genes and Mutations, Cold Spring Harbor Symposia on Quantitative Biology*, Volume XVI [June 7–15, 1951], The Biological Laboratory, Cold Spring Harbor, NY.
McClintock, B. 1952. Mutable loci in maize: Origins of instability at the *A1* and *A2* loci. Instability of *Sh1* action induced by *Ds*. Summary. *Carnegie Institution of Washington Year Book No. 51*, 1951–1952, pp. 212–219 [Issued December 12, 1952]
McClintock, B. 1953a. Induction of instability at selected loci in maize. *Genetics* Vol. 38 (No. 6, November), pp. 579–599. [Issued January 1954]
McClintock, B. 1953b. Mutations in maize: Origin of the mutants. Change in action of genes located to the right of *Ds*. Comparison between *Sh1* mutants. Change in action of genes located to the left of *Ds*. Meiotic segregation and mutation. *Carnegie Institution of Washington Year Book No. 52*, 1952–1953, pp. 227–237 [Issued December 11, 1953] [Listed in McClintock (1987, p. xiv) as a 1954 publication]
McClintock, B. 1954. Mutations in maize and chromosomal aberrations in *Neurospora:* Mutations in maize. Chromosome aberrations in *Neurospora. Carnegie Institution of Washington Year Book No. 53*, 1953–1954 pp. 254–260. [Issued December 10, 1954]
McClintock, B. 1955a. 1. Spread of mutational change along the chromosome. 2. A case of *Ac*-induced instability at the bronze locus in chromosome 9. 3. Transposition sequences of *Ac*. 4. A suppressor-mutator system of control of gene action and mutational change. 5. System responsible for mutations at *a1m-2*. *Maize Genetics Cooperation News Letter* 29 (March 17, 1955), pp. 9–13.
McClintock, B. 1955b. Controlled mutation in maize: The *a1m-1-Spm* system of control of gene action and mutation. Continued studies of the mode of operation of the controlling elements *Ds* and *Ac*. *Carnegie Institution of Washington Year Book No. 54*, 1954–1955, pp. 245–255. [Issued December 9, 1955]
McClintock, B. 1956a [Symposium held June 15–17, 1955]. Intranuclear systems controlling gene action and mutation, pp. 58–74. *In*, *Mutation, Brookhaven Symposia in Biology, No. 8.* Biology Department, Brookhaven National Laboratory, Upton, NY. [R. C. King Symposium Chairman; Issued February 1956.]
McClintock, B. 1956b [(c)1957, Symposium held June 4–12, 1956]. Controlling elements and the gene, pp. 197–216, *In*, K. B. Warren (Ed.), *Genetic Mechanisms: Structure and Function, Cold Spring Harbor Symposia on Quantitative Biology*, Volume XXI, The Biological Laboratory, Cold Spring Harbor, NY.
McClintock, B. 1958. The suppressor mutator system of control of gene action in maize: The mode of operation of the *Spm* element. A modifier element in the *a1m-1-Spm* system. Continued investigation of transposition of *Spm. Carnegie Institution of Washington Year Book 57*, 1957–1958, pp. 415–429. [Issued December 19, 1958]
McClintock, B. 1959. Genetic and cytological studies of maize: Further studies of the *Spm* system. Chromosome constitutions of some South American races of maize. *Carnegie Institution of Washington Year Book 58*, 1958–1959, pp. 452–456. [Issued December 14, 1959]
McClintock, B. 1960. Chromosome constitutions of Mexican and Guatemalan races of maize: General Conclusions. *Carnegie Institution of Washington Year Book 59*, 1959–1960, pp. 461–472. [Issued December 12, 1960]
McClintock, B. 1987. *The Discovery and Characterization of Transposable Elements.* The Collected Papers of Barbara McClintock. Genes Cells and Organisms; Great Books in Experimental Biology, J. A. Moore, series editor, Garland Publishing Co. New York. [Published December 1, 1987; Currently in the series Genes Cells and Organisms, Routledge, Taylor and Frances Group.]
Meselson, M. and F. W. Stahl. 1958. The replication of DNA in *Escherichia coli. Proceedings of the National Academy of Sciences* Vol. 44 (No. 7, July 15, 1958), pp. 671–682. https://www.pnas.org/content/44/7/671.full
Metzenberg, R. L. 1956. *Studies on the biosynthesis of aromatic compounds in Neurospora crassa.* Dissertation (Ph.D.), California Institute of Technology. https://doi.org/10.7907/WPDV-8S11
Montalenti, G. and A. Chiarugi (Eds.). 1954. *Atti del IX Congresso Internazionale di Genetica, Bellagio (Como) 24-31 agosto 1953=Proceedings of the 9th International Congress of Genetics*, Bellagio, Italy, August 24–31, 1953. *Carylogia*, Vol. 6, 1953/54. Suppl.

Moreno, U. 1960. A study of the number of B-chromosomes and the frequency and position of chromosome knobs in the races of maize in Peru. *Agronomia* (Lima), Vol. 27, pp. 39–51. [Foreign title: *Estudio del número de cromosomas tipo "B" y de la frecuencia y posición de nudos cromosómicos en razas de maíz del Perú*, Author: Moreno Moscoso, U.]

Moreno, U. 1970. *Physiological investigations on the potato plant, with special reference to the effects of different environments*. Ph.D. Cornell University, June 1970. Thesis QK82 1970 M843.

Moreno, U. and A. Grobman, A. 1958a. 1. Presence of B chromosomes in Andean maize. *Maize Genetics Cooperation News Letter* Vol. 32 (March 15, 1958), p. 25.

Moreno, U . and A. Grobman, A. 1958b. 2. Frequency of knobs in chromosomes of three races of the Peruvian coast. *Maize Genetics Cooperation News Letter* Vol. 32 (March 15, 1958), pp. 25–26.

Moreno, U., A. Grobman, and B. McClintock. 1959. 5. Study of chromosome morphology of races of maize in Peru. *Maize Genetics Cooperation News Letter* Vol. 33 (April 1, 1959), pp. 27–28.

Morgan, T. H., C. B. Bridges, and Jack Schultz. 1938. California Institute of Technology, Pasadena, California. Constitution of the germinal material in relation to heredity. Biology. *Carnegie Institution of Washington Year Book No. 37*, 1937–1938, pp. 304–309. [See p. 304 for the first use of the term pseudo-allelism]

Murphy, R. P. and L. B. Kass. 2011 (June 27). *Evolution of Plant Breeding at Cornell University: A Centennial History, 1907-2006*. Rev. edition i–x, 178 pp., Appendices, Photo Section. e-book, The Internet-First University Press, Ithaca, NY. https://hdl.handle.net/1813/23087

News and Views. 1953. International congress of genetics. *Nature* Vol. 172 (No. 4384, November 7, 1953), p. 840. https://doi.org/10.1038/172840d0

Nuffer, M. G. 1955. Dosage effect of multiple *Dt* loci on mutation of *a* in the maize endosperm. *Science* Vol. 121 (No. 3142, March 18, 1955), pp. 399–400. [Name changed to Neuffer as of 1963]

Perkins, D. D. 1992. Neurospora chromosomes, pp. 33–44, *In*, N. Fedoroff and D. Botstein (Eds.), *The Dynamic Genome: Barbara McClintock's Ideas in the Century of Genetics*, Cold Spring Harbor Laboratory Press, Plainview, NY.

Perkins, D. D. and R. H. Davis. 2002. Neurospora chronology 1843–2002. *Fungal Genetics Reports* Vol. 49, Article 2. https://doi.org/10.4148/1941-4765.1184

Peters, J. A. (Ed.). 1959. *Classical Papers in Genetics*, Prentice-Hall Inc., Englewood Cliffs, NJ.

Peterson, P. A. 1953a. University of Illinois. 6. The mutable pale green locus. *Maize Genetics Cooperation News Letter* Vol. 27 (March 17, 1953), pp. 58–60.

Peterson, P. A. 1953b. A mutable pale green locus in maize. *Genetics* Vol. 38 (No. 6, November 1, 1953), pp. 682–683.

Peterson, P. A. 1960. The pale green mutable system in maize. *Genetics* Vol. 45 (No 1, January 1, 1960), pp. 115–133. https://doi.org/10.1093/genetics/45.1.115

Peterson, P. A. 1977. The position hypothesis for controlling elements in maize, pp. 429–435. *In*, A. I. Bukhari, J. A. Shapiro and S. L. Adhya (Eds.), *DNA Insertion Elements, Plasmids, and Episomes*, Cold Spring Harbor Laboratory, Cold Spring Harbor, NY.

Peterson, P. A. 1993. Transposable elements in maize: Their role in creating plant genetic variability. *Advances in Agronomy* Vol. 51, pp. 79–124.

Provine, W. B. and P. Sisco. 1980. Interview with Barbara McClintock at Cold Spring Harbor Laboratory, Long Island, New York, August 28, 1980. Division of Rare and Manuscript Collections, Cornell University Library, Ithaca, NY. Transcripts and audiotapes.

Punnett, H. 1950–1951. *Cytogenetics Class Notebook.* [October 1950–May 1951, Prof. Marcus Rhoades Instructor]

Ramírez, E. R., D. H. Timothy, E. Diaz B., U. J. Grant, in collaboration with G. E. N. Calle, E. Anderson, and W. L. Brown. 1960. *Races of Maize in Bolivia*, Pub: 747, NAS-NRC, Washington, DC.

Report of the President. 1952. For the year ending June 30th, 1952, pp. 3–20. *Carnegie Institution of Washington Year Book No. 51*, 1951–1952. [Issued December 12, 1952]

Report of the President. 1956. *Carnegie Institution of Washington Year Book No. 55*, 1955–1956, pp. 1–28. [Issued December 14, 1956].

Rhoades, M. M. 1954. Chromosomes, mutations and cytoplasm in maize. *Science* Vol. 120 (No. 3108, July 23, 1954) pp. 115–120. [Presented in Symposium "Species that feed mankind," AAAS meeting Boston, MA, December 27, 1953]

Rhoades, M. M. and E. Dempsey. 1950. New mutable loci. *Maize Genetics Cooperation News Letter* Vol. 24 (March 17, 1950), p. 50.

Rieger, R., A. Michaelis, and M. M. Green (Eds.). 1991. *Glossary of Genetics, Classical and Molecular*, 5th edition, Springer-Verlag, Berlin.

Roberts, L. M., U. J. Grant, R. Ramírez, W. H. Hatheway, D. L. Smith, in collaboration with P. C. Mangelsdorf. 1957. *Races of Maize in Colombia*, Pub. 510. NAS-NRC, Washington, DC.

Sand, S. A. 1955. *Heritable somatic instability in Nicotiana cultures derived from a hybrid of N. Langsdorffii by N. Sanderae*. Ph.D. Cornell University February 1955. Thesis SB124 1955 S213.

Sand, S. A. 1976. Genetic control of gene expression: Independent location of FZt(3) and its interactions with the mutable *v* locus in *Nicotiana*. *Genetics* Vol. 83 (No. 4, August 1, 1976), pp. 719–736. https://doi.org/10.1093/genetics/83.4.719

Schwarz-Sommer, Z., B. Davies and A. Hudson. 2003. An everlasting pioneer: The story of *Antirrhinum* research. *Nature Reviews Genetics* Vol. 4 (August 1, 2003), pp. 655–664. https://doi.org/10.1038/nrg1127

Selker, E. U. 2008. Robert L. Metzenberg, June 11, 1930–July 15, 2007: Geneticist extraordinaire and "model human". *Genetics* Vol. 178, (No. 2, February 1, 2008), pp. 611–619. https://doi.org/10.1093/genetics/178.2.611

Selya, R. 2012. Defending scientific freedom and democracy: The Genetics Society of America's response to Lysenko. *Journal of the History of Biology* Vol. 45 (No 3, August 2012), pp. 415–442. https://doi.org/10.1007/s10739-011-9288-2

Singleton, W. R. 1954. IX International Genetics Congress. *AIBS Bulletin* Vol. 4 (No. 1, January 1, 1954), p. 26.

Stadler, L. J. 1954. The gene. *Science* Vol. 120 (No. 3125, November 19, 1954), pp. 811–819. [Presidential Address to *American Society of Naturalists*, December 30, 1953; Published November 1954]

Stakman, E. C., R. Bradfield, and P. C. Mangelsdorf. 1967. *Campaigns against Hunger*, The Belknap Press of Harvard University Press, Cambridge, MA.

Stephens, S. G. 1951. Possible significance of duplications in evolution. *Advances in Genetics* Vol. IV, pp. 247–265. [Review]

Stitzer, M. C. and J. Ross-Ibarra. 2018. Maize domestication and gene interaction. *New Phytologist* Vol. 220 (No. 2, October 2018), pp. 395–408. [First published: July 23, 2018, https://doi.org/10.1111/nph.15350

Stubbe H. 1954. Über mutable gene bi *Antirrhinum majus*, pp. 218+, *In*, G. Montalenti and A. Chiarugi (Eds.), *Caryologia* Vol. 6, 1953/54. Suppl.

Taylor, J. H., P. S. Woods, and W. L. Hughes. 1957. The organization and duplication of chromosomes as revealed by autoradiographic studies using tritium-labeled thymidine. *Proceedings of the National Academy of Sciences* Vol. 43 (No. 1, January 15, 1957), pp. 122–128. https://www.pnas.org/doi/10.1073/pnas.43.1.122

The Rockefeller Foundation 1963-1964. 1964. *The Rockefeller Foundation Program in the Agricultural Sciences, Annual Report 1963–1964* [July 1964]. Office of Publications, The Rockefeller Foundation, New York.

Ting, Y.-C. 1956. 3. Cytological studies of the *c* blotching system. 6. Cytological studies of maize-teosinte derivatives. *Maize Genetics Cooperation News Letter* Vol. 30 (March 15, 1956), pp. 32, 35–36.

Watson, J. D. and F. H. C. Crick. 1953a. A structure for desoxyribose nucleic acids. *Nature* Vol. 171 (No. 4356, April 25, 1953), pp. 737–738.

Watson, J. D. and F. H. C. Crick. 1953b. Genetical implications of the structure of deoxyribonucleic acid. *Nature* Vol. 171 (No. 4361, May 30, 1953), pp. 964–967.

Watson, J. D. and F. H. C. Crick. 1953c. The structure of DNA, pp. 123–131. *Cold Spring Harbor Symposia on Quantitative Biology, "Viruses," June 5–11, 1953, Volume XVIII* [Vol. 18], Cold Spring Harbor, NY.

Wellhausen, E. J., A. Fuentes, A. Hernández, in collaboration with P. C. Mangelsdorf. 1957. Races of Maize in Central America Pub. 511 NAS-NRC, Washington, DC.

Wellhausen, E. J., L. M. Roberts, A. Hernández, en colaboración con P. C. Mangelsdorf. 1951. *Razas de Maíz en México, su origen, características y distribución*. Programa de Agricultura Cooperativo de la Secretaria de Agriculiura y Ganaderla de México, D. F. y la Fundacion Rockefeller.

Wellhausen, E. J., L. M. Roberts, A. Hernández, in collaboration with P. C. Mangelsdorf. 1952. *Races of Maize in México: Their Origin, Characteristics and Distribution*, The Bussey Institution of Harvard University, Cambridge, MA.

9 Coming Home Again
Andrew Dickson White, Professor-at-Large, Cornell University

There is only one thing in life worse than being talked about and that is not being talked about

Oscar Wilde

ANDREW DICKSON WHITE, PROFESSOR-AT-LARGE

McClintock held the title of Andrew Dickson White Professor-at-Large, at Cornell University, during the late 1960s through the mid-1970s. She chose to remain affiliated with the Carnegie Institution of Washington (CIW/DC) Genetics Research Unit at Cold Spring Harbor (CSH), attracted by an anticipated appointment as Distinguished Service Member upon mandatory retirement at age 65.[1] Her continued affiliation with CSH did not preclude accepting the nomination by Professors Adrian Srb and Bruce Wallace to be one of Cornell's first Andrew Dickson White Professors-at-Large. McClintock's six-year appointment (1965–1971), while generally not renewable, was continued for an additional three years (1971–1974).[2] Proposed by President Perkins in 1964 and established by the Board of Trustees in 1965 to enlist the services of eminent scholars and scientists, the professorship is Cornell's highest honor, as no honorary degrees are given by the institution.[3] Published accounts during this time corroborate McClintock's recollections that a number of scientists questioned and even doubted her conclusions regarding controlling elements and gene action in the developmental process, as I will detail below.

Yet, there were researchers who had recognized and seriously considered McClintock's maize "controlling elements" toward an understanding of prokaryotic systems, before and during this time.[4] Undeterred, McClintock accepted invitations to present her research at seminars, conferences, symposia, and workshops throughout the U.S. and abroad (Spectre 1993: Series III). She shared her maize stocks with researchers and students and wrote detailed descriptions to accompany the distributed material (Spectre 1993: Series I).

Upon learning of the possibility of a forthcoming offer from CIW/DC to remain at Cold Spring Harbor upon retirement in 1967, McClintock declined a permanent faculty appointment in Cornell's new Division of Biological Sciences, with research space in both the field and laboratory, and graduate students for stimulation.[5] She preferred instead Cornell's newly established distinguished A. D. White

[1] A. Ackerman to B. McClintock, September 12, 1964, Personnel File, Barbara McClintock, Carnegie Institution of Washington, DC Archives (henceforth McClintock CIW/DC Archives).

[2] Barbara McClintock A. D. White Professor-at-Large file, A. D. White Professor-at-Large Program, Cornell University (henceforth A. D. White, McClintock, CU).

[3] "Univ. Names 7 At-Large Professors," *The Cornell Daily Sun*, April 30, 1965, p. 1.

[4] Wollman et al. (1956), Jacob and Wollman (1958a), Jacob (1958, 1960), Jacob and Monod (1961b), Monod and Jacob (1961), and Campbell (1962), etc.

[5] A. Srb, A. D. White File, courtesy of J. Srb (henceforth Srb A. D. White). McClintock CIW/DC Archives.

DOI: 10.1201/9781003332527-10

Professorship, accepting the appointment in May 1965.[6] During the next nine years, she would repeatedly contemplate leaving CSH and consider offers to permanently reside at Cornell or elsewhere.[7]

Why did McClintock accept Cornell's extended Professorship when she had declined long-term affiliations with other universities in the past? Faculty in Cornell's Department of Botany and Department of Plant Breeding embraced and respected her work, and had encouraged her to consider returning to Cornell at various times during the 1950s and 1960s. Her experience with the abrupt closing of the Carnegie Department of Genetics in June 1962 convinced her that she should look elsewhere for a place to conduct her research.[8] McClintock continued to be employed by CIW/DC when the Department of Genetics became the Genetics Research Unit on July 1, 1962.[9] However, mandatory retirement at age 65 was approaching within two years. Fortunately, by September 1964 a forthcoming offer to be a Carnegie "Distinguished Service Member" (with salary for three years) upon retirement would continue to allow her freedom to work wherever she wished.[10] Only two other members of the Carnegie Institution of Washington held such honored appointments. McClintock knew that Cornell would find a place for her as they had promised to do for Demerec upon his retirement from Cold Spring Harbor.[11] Spending time at Cornell as a Professor-at-Large gave her the resources to explore the option of remaining there should the need arise. Both these opportunities permitted continued freedom to work at her own pace. She felt wanted and needed by Cornell's faculty, staff, and students hosted by a new Section of Genetics, Development and Physiology (GDP; Kass & Cobb 2007) in the newly established Division of Biological Sciences, and by the Department of Plant Breeding. As a former Cornell student, she had received training in genetics with Plant Breeding Faculty. Upon returning from Germany in April 1934, she had completed her Guggenheim Fellowship in Cornell's Department of Botany. She then became a research assistant to R. A. Emerson, Head of the Plant Breeding Department, through 1936 (see Chapter 5). As an alumna of both departments, she was comfortable. She knew the area on and off campus. She had friends and colleagues in Srb and Wallace, former geneticists in the Department of Plant Breeding, and she had support and encouragement from many others. She knew Srb from his days as a graduate student of her good friend George Beadle at Stanford University. Wallace had been T. Dobzhansky's student at Columbia University and was a former colleague at the Long Island Biological Laboratory, until he was hired by Cornell's Plant Breeding Department as Associate Professor in 1958. Harry Stinson, the new GDP section chair and former member of Cornell's Department of Plant Breeding (hired as Professor in 1962), had previously worked with her longtime associate/colleague D. F. Jones, at the Connecticut Agricultural Research Station (Murphy & Kass 2011). Furthermore, former members (students and faculty) of Cornell's Plant Breeding Department held appointments at Brookhaven National Laboratory, where she was a guest investigator from 1953 to 1972.[12]

McClintock took her Cornell visiting professor appointment very seriously. Even though she had many other commitments, as discussed later in this chapter, she spent a minimum of one to two weeks each academic term at Cornell, presenting formal and informal lectures or seminars on her latest findings on maize controlling elements and on the origin of the corn plant (Spectre 1993: Series III).[13] She held formal and informal meetings with faculty and students in Cornell's Department of Plant Breeding, and in the newly established Section of Genetics, Development and Physiology (GDP), in the Division of Biological Sciences (Kass & Cobb 2007).

[6] A. D White, McClintock, CU.

[7] McClintock CIW/DC Archives. A. Srb files, courtesy of J. Srb.

[8] Letter B. McClintock to B. Wallace, December 1, 1963, Bruce Wallace Papers, Ms. Coll. 86, Series I, McClintock, APS Library. Report of the President 1962, p. 71.

[9] *Carnegie Institution of Washington Year Book 61*, 1961–1962, p. 438.

[10] E. Ackerman to B. McClintock, September 12, 1964; E. Ackerman to A. Hershey, January 30, 1967, McClintock CIW/DC Archives.

[11] Plant Breeding files, Cornell University.

[12] Personnel File B. McClintock, Brookhaven National Laboratory.

[13] A. D White, McClintock, CU.

On her first visit to campus as A. D. White Professor-at-Large in November 1965, she was assigned Bruce Wallace's office (he was on sabbatical leave) on the first floor of the Plant Science Building. This area had been the home to Cornell's Plant Breeding Department since 1931, when the building was first opened during the Great Depression. Upon returning to campus the following year, Wallace assigned her to Professor L. F. Randolph's room on the second floor of the Plant Science Building [Plant Science 214].[14] His room was of course a familiar place since it was close to the cytology laboratory [Plant Science 219] where she had taught with Professor Sharp in the second term of 1930–1931 and was across from the room she had used as an office and later shared with Harriet Creighton [Plant Science 220] when she returned to Cornell to complete her Guggenheim Fellowship in 1934. At Cornell, stories abound regarding McClintock climbing up the side of the Plant Science Building to get into her office when she had forgotten her key. Harriet Creighton said emphatically that it was she, herself, who had climbed to the second floor of the building, entered through the window, and unlocked the door. She added: "That building was just meant to be climbed" (Kass 2005). During the Botany Department's Centennial Celebration in 2013, I learned that the area located in the second floor wing of the Plant Science Building had been remodeled as a Plant Pathology research laboratory, thus subsuming McClintock's former office and the Cytology laboratory into that floor space (Kass 2013). The new renovations prevented public access to these rooms and precluded the building's eligibility for the National Registry of Historic Places. The building is currently being renovated again, and the area continues to be concealed from view (Figures 9.1 and 9.2).

Upon the opening of Bradfield/Emerson Hall in 1968, many faculty members were moved from the Plant Science Building to a larger space on the second and third floor of this new building. During McClintock's visits to Cornell, Srb gave her his front office in 219 Bradfield Hall as a staging area to meet with students and colleagues. It was in this office that I first met with McClintock in 1972 to discuss my thesis project, as I mentioned in the introduction to this book. Barbara McClintock was recognized by her scientific peers in Plant Biology and Genetics years before she had won the 1983 Nobel Prize for her discovery of "mobile genetic elements" and became a public figure (Fedoroff & Botstein 1992; Kass & Bonneuil 2004; Kass & Chomet 2009). At Cornell, graduate students along with time-honored Professors, recently hired Assistant Professors, and visiting faculty were greatly influenced by their discussions with McClintock during her visits in these years (Wallace 1963, 2001; Fink 1992).[15]

Professor Gerry Fink summarized McClintock's influence on his lab's research investigations of transposable elements in yeast in the chapter he contributed for the *Dynamic Genome* (Fink 1992), a *Festschrift* that Federoff and Botstein (1992) published as a tribute to McClintock's 90th birthday celebration at Cold Spring Harbor in 1992. Professors Ross MacIntyre and Peter Bruns recalled that Fink was probably one of the few faculty members in the Division that had a thorough understanding of McClintock's work on transposable genetic elements.[16] Fink and McClintock, MacIntyre recalled, spent many hours at Cornell discussing their work. Fink had a long association with Cornell. He had joined the faculty as an assistant professor of genetics in 1967, and served as a professor of genetics from 1976 to 1979 and as a professor of biochemistry from 1979 to 1982. He recalled that shortly before making his first appointment to speak with McClintock in 1970, he attended her seminar on "Maize Controlling Elements" (Fink 1992, p. 282).[17] His interest at that time was in trying to determine whether gene regulation as reported by Jacob and Monod [1960s] for *E. coli* applied to yeast. A lecture by McClintock on gene action and controlling elements certainly would be of interest. He recalled that he had difficulty understanding the critical experiments that led to the conclusions she had presented. He believed this was because of his ignorance of maize

[14] Randolph had retired in 1962 and was away from his office most of the year, Cobb 2013.

[15] A. Srb to M. Black, March 26, 1966, A. D White, McClintock, CU. Kass interviews with Cornell faculty, visiting faculty, and graduate students.

[16] Interviews L. Kass with R. MacIntyre, March 11, 1997, and P. Bruns, March 5, 1997.

[17] The lecture was titled "Programming of gene action by controlling elements" presented in March 1970, Spectre 1993: Series III.

FIGURE 9.1 Plant Science 220: Renovated former office of Barbara McClintock and Harriet Creighton on the second floor of the Plant Science Building, Cornell University, Ithaca, New York. In this office, in April 1931, Creighton and McClintock met with Thomas H. Morgan, who encouraged them to publish their cytological proof for crossing over (Chapter 5; Kass 2003). It was in this office that Crighton and McClintock drafted their manuscripts for the cytological proof for crossing over (McClintock 1931a; Creighton & McClintock 1931) and where McClintock drafted the manuscript locating genes on chromosomes (McClintock 1931b). (See Figure 9.2 for a close up of plaque on the right of the door [The area has no public access], image courtesy of Wojtek Pawlowski.)

biology. Therefore, before his first meeting with McClintock, he began reading her primary papers from the beginning since the brief summary presented in the "best genetics text of that time" was not sufficient for an understanding of her work. Other researchers in yeast genetics also felt the necessity to start at the beginning, reaching as far back as her 1932 publications, for an understanding of McClintock's studies of transposable elements and controlling elements (Herskowitz, Ira 1992, p. 294; Strathern 1992, p. 302).[18] At their first meeting together, McClintock acknowledged that Fink had acquired an understanding of her work and they discussed "the intricacies concerning chromosome movement." While discussing his work on histidine regulation in yeast, McClintock asked a crucial question: Did he think there are transposable elements in yeast? He did not because he thought that transposable elements or "McClintock elements" were something special to corn. Her question, however, led to a search for and the identification of the Ty_1 transposable elements in yeast, discovered ten years later by postdoctoral researchers working in Fink's lab at Cornell (Fink 1992, p. 285).

On one occasion during a visit to Cornell, McClintock demonstrated great sensitivity to a visiting lecturer, who she believed might be intimidated by her presence. The graduate students were

[18] See reprints of McClintock's early papers in Kass 2013ff.

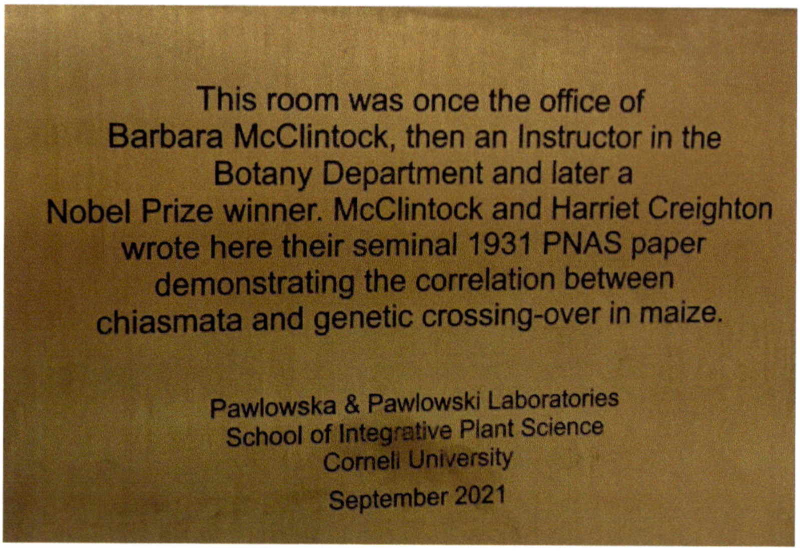

FIGURE 9.2 Wall plaque on the right of the entrance to Plant Science 220, the former office of Barbara McClintock and Harriet Creighton on the second floor of the Plant Science Building, Cornell University, Ithaca, New York. (See Figure 9.1, plaque design and image courtesy of Wojtek Pawlowski.)

planning to attend the lectures of Sydney Brenner, visiting from the Medical Research Council, Cambridge, England. Brenner had been invited by Cornell's Department of Chemistry to deliver a series of lectures on the "Problems of Gene Regulation in Higher Organisms," in the spring of 1973.[19] Years later, Brenner would share a Nobel Prize in Physiology or Medicine (2002) for his work on genetics and development in the nematode worm. Tom Kass, then a graduate student of Bruce Wallace, invited McClintock to walk with us from Bradfield Hall to the lecture hall, where Brenner was to speak.[20] "Oh no," she insisted, she would come later, enter unobtrusively and remain in the back of the room so as not to distract/disturb Sydney. The graduate students arrived early and sat in the center front rows. Cigarette in hand, Brenner delivered his lecture while walking back and forth in front and in back of the center bench, which seemed to obstruct his access to the blackboard that he used on occasion. The students hung onto his every word. Suddenly, while looking out at the audience, he stopped for a moment, seemed distracted, regained his composure, and then continued his thoughts. I realized he must have noticed McClintock's presence in the back of the lecture hall. At the end of his talk, members of the audience flocked to the front of the room to ask questions, but Brenner ignored everyone and walked directly to McClintock. "What are you doing here," he asked her. "I came to hear you speak," she responded. He took her hands in his and said, "Thank you." As a graduate student, I learned from McClintock the lesson of generosity and respect for friends and colleagues.

In 1965, her first series of lectures as Cornell Visiting Professor, and Distinguished Visitor of the New York State Science & Technology Foundation,[21] were presentations on *Aspects of gene control in Higher Organisms*, *The Controlling Elements in Maize*, and *The Origin and Migration of Races of Maize in the Americas* (Spectre 1993: Series III). Later, presentations at Cornell included a seminar on *The Stability of the Genome* (March 1967), discussions on the *Races of Maize* (January/February 1969), and lectures on *Origins of Maize*, *Origins of Gene Action* (1970), and a forward-looking seminar on ideas of the future of genetic research, *Expanding Concepts in Genetics* (1971). Stinson

[19] Cornell University, *Department of Chemistry Newsletter*, March 1973.
[20] T. Kass joined Brenner's MRC Cambridge research group as a postdoc in 1975, L. Kass diary 1975–1977.
[21] Srb A. D. White. A. D White, McClintock, CU.

would cite these as examples of why McClintock's six-year Professorship should be renewed for another four years.

I recall an informal seminar held after hours in Professor Ross MacIntyre's laboratory on the third floor of Bradfield Hall (ca. 1973). The graduate students requested that McClintock tell us about her work on the origin of maize. She sat on the lab bench, with a can of beer in one hand and a pointer in the other while tracing, on the map hanging on the wall over her head, the movement of corn as revealed by variation in chromosome knobs carried by indigenous people along the trade routes in Central and South America. At that time, we also learned that following her initial studies on the subject (McClintock 1959, 1960), much of her data on maize chromosome knobs and B-type chromosomes had been included in publications by colleagues working in Central and Latin America.

As mentioned in the previous chapter, Grobman et al. (1961) had been trained by McClintock for studies of the races of maize in Peru beginning in 1957. Subsequently, the maize chromosome knob and B-chromosome data that she had collected during her 1959–1960 visit to Colombia were published by Ramirez et al. (1960, pp. 28–29) for *Races of Maize in Bolivia*, and Timothy et al. (1961, pp. 25, 83, 1963, pp. 20–23) for *Races of Maize in Chile* and *in Ecuador*. Furthermore, in 1961, McClintock had been invited to train Latin American students in Mexico and in the U.S. to learn maize cytogenetic techniques toward practical corn improvement in maize races indigenous to their respective countries. The Rockefeller Foundation provided funding to establish a Graduate School in Chapingo, Mexico, and awarded graduate student fellowships to students from Brazil, Mexico, Chili, and Argentina to study with McClintock at North Carolina State College in Raleigh (Rockefeller Foundation Annual Reports 1961, 1962).[22]

McClintock requested that William (Bill) L. Brown of the Pioneer Hi-Bred Corn Company, then assistant director of research, join her in the project to train students in North Carolina.[23] Brown had an excellent background in taxonomy, cytology, and maize evolution, having completed his Ph.D. (1941) in cytogenetics and taxonomy with Edgar Anderson at the Washington University of St. Louis, on the recommendation of Paul Mangelsdorf. Harriet Creighton recalled teaching Anderson the corn sporocyte smear technique during the summer of 1941 at Cold Spring Harbor while working alongside McClintock and Marcus Rhoades (see Chapter 6). I found no record that Brown accompanied Anderson on that visit to Cold Spring Harbor, as reported by Duvick (1996), so it's not clear when Brown first met McClintock. Edgar Anderson was one of the first researchers to publish on the races of maize (Anderson & Cutler 1942). He had also developed the concept of introgressive hybridization (Anderson 1949). "Introgression" is the gradual infiltration of germplasm of one species into another through repeated backcrossing. Anderson was also known for his popular booklet "Corn Before Columbus" published in 1947 by the Pioneer Hi-Bred Corn Company. Brown had accepted a position as geneticist in the Corn Breeding Department at Pioneer in 1945. Four years later, he published on the numbers and distribution of chromosome knobs in maize in the U.S. (Brown 1949). Brown (1960) later published preliminary studies on chromosome knobs in *Races of Maize in the West Indies*.

McClintock and Brown mentored Latin American Scholars at North Carolina State College beginning in 1962. Local faculty members served as Chair of the student's graduate committee and were responsible for the administrative details, whereas McClintock and Brown proposed their research topics. McClintock made extended trips to Raleigh during this time to meet with and direct her students' research. Three students completed degrees within two years, Almiro Blumenschein (1964) and Warton Monteiro (1964) from Brazil, and Takeo Angel Kato [Yamakake]

[22] Rockefeller Foundation Archives: March 24, 1959–September 13, 1962. RG 6.13 Series 1.1, Box 32, Folder 353; McClintock, Barbara, March 1959–September 1962; Letters to and from Wellhausen and McClintock, arrangements for doing knob studies in Mexico for Inter-American Maize Improvement Program and training students at North Carolina State College. Archives North Carolina State University: Department of Genetics papers. Spectre 1993: Series I, North Carolina State College & The Rockefeller Foundation. Missouri Botanical Garden Archives: William Brown Papers.

[23] W. L. Brown Papers, Missouri Botanical Garden Archives.

(1964) from Mexico. Julio Safont from Argentina also studied with McClintock.[24] Ulises Moreno who McClintock trained earlier in Peru came as a Rockefeller-funded visiting scholar during this time, and Yasuo Ohata from Japan was a postdoctoral researcher who learned cytogenetics of maize from McClintock, returned to Japan in August 1965.

Blumenschein had previously collaborated with Friedrich G. Brieger et al. (1958) on *Races of Maize in Brazil and other Eastern South American Countries*. Their report, however, did not include any cytological studies. His doctoral dissertation of studies of heterochromatic chromosome knobs in maize would soon rectify that omission. Although Blumenschein already had a "Doutor em Agronomia" [1957] from the University in Sao Paulo, Brazil, concerning chromosome numbers in orchids (Blumenschein 1960) and was a member of the faculty at his home institution, he was encouraged to complete a Ph.D. at NC State College with Barbara McClintock and Bill Brown before returning to Brazil.

Prior to his studies at North Carolina State College, Kato had received his education in Mexico and learned cytology from Professor Czeslawa Prywer and from Dr. Albert Longley. Prywer was a botanist, cytologist, and teacher in the Post-Graduate School, National School of Agriculture, Chapingo, State of Mexico. Longley had recently retired from the U.S.D.A. and went to Mexico to work with the Inter-American Maize Program (later CIMMYT) of the Rockefeller Foundation and the school at Chapingo (Coe & Kass 2019). Kato was encouraged to apply to the program at North Carolina to work under the direction of McClintock on cytological studies of Mexican races of Maize (Kato 1964). Upon returning to Mexico, he and Longley pursued further studies of chromosome morphology of races of maize in Latin America, which they began during 1960–1962 (Longley & Kato 1965). Kato went on to complete a Ph.D. (1975) at the University of Massachusetts, Amherst, with maize geneticist Walton Galinat (Kato 1975, 1976).

Barbara McClintock continued to collaborate with Blumenschein and Kato on chromosome studies of Latin American maize races, which Blumenschein (1973) briefly summarized during a 1972 meeting in Cali, Colombia, organized by Adrian Srb (see below). He mentioned that their study would be published shortly; however, their major co-authored volume on *The Chromosome Constitution of Races of Maize* was published in English many years later (McClintock et al. 1981).[25]

THE LECTURE CIRCUIT AND OTHER OBLIGATIONS

Only two years from mandatory retirement in 1965, McClintock began her Cornell Visiting Professorship, which continued through 1974. During this time, she also lectured throughout the country, received invitations to speak at international symposia, worked collaboratively with former students, and contemplated her future upon mandatory retirement from CIW/DC. Srb was "delighted" that she was "willing to consider seriously" the possibility of basing herself at Cornell.[26]

In June 1965, she again presented her work at the *Brookhaven Symposia in Biology* (McClintock 1965). Ten years earlier, at a *Brookhaven symposium*, she had first introduced and defined the term "controlling element" for her audience (McClintock 1955/1956). She had reported progress on *Spm*, which had characteristics both similar to and distinct from the *Ds-Ac* system she had first described in the 1940s (Fedoroff 2013a,b). As briefly summarized by Kass and Chomet (2009), a second allele of *A1* under the control of the element she called *Spm* demonstrated a distinct suppressor regulation of gene function (see Chapter 8; see also Fedoroff 1992, pp. 402–415).

At the 1965 Brookhaven Symposia, McClintock reviewed recent work that she and her colleagues had contributed to an understanding of spontaneous mutations in maize. She further presented evidence to consider a two-element system model of regulation of gene action during development. She compared the reproduction of the maize elements with the coordinated reproduction of an incorporated

---

[24] The university has found no record of a thesis for Safont.

[25] See reprint of Timothy's (1981) review and Kato's (2013) perspective in Kass (2013ff.).

[26] Letter A. Srb to B. McClintock, January 6, 1966, Srb file, permission of J. Srb.

phage particle in the genome of bacterium, and compared her system using the terms operator and regulator, which were initially used for bacterial gene control systems (McClintock 1965).

Bacterial geneticist Francois Jacob (1958, 1960, p. 25)[27] had first proposed a hypothesis for a system of regulation acting as inducers or repressors to switch on or off the activity of genetic units. During the time Jacob's September 1958 presentation was in press, he and his colleagues had provided evidence for the operon hypothesis of gene regulation (Jacob 1960, p. 39). Via a "note added in proof," Jacob (1960, p. 39) provided a brief summary of the "experimental support" for what came to be known as the operon theory of bacterial gene regulation. They assumed that a "group of linked genes and the operator, which control their expression, constitute an integrated unit of activity or *operon*, which would in a way represent an intermediary genetic element between the gene and the chromosome. By the intermediacy of the operator, the activity of the operon ... would be submitted to the action of a specific repressor synthesized under the control of a regulatory gene not necessarily linked to the operon." Citing a paper published in French in October 1959 and another "in press" that appeared in February 1960 (Jacob & Monod 1959; Jacob et al. 1960), Jacob's "note" thus presented the first summary in English of the bacterial system of gene regulation. They did not, however, cite McClintock in these initial preliminary publications that had provided "experimental support" for their hypothesis of bacterial gene regulation (Jacob & Monod 1959; Jacob et al. 1960).

Yet between 1956 and 1958, Jacob and his colleagues had correlated McClintock's studies in maize with their findings of mutations in prokaryotic bacterial cells (Wollman et al. 1956; Jacob & Wollman 1958a). As early as 1956, at the *Cold Spring Harbor Symposia on Quantitative Biology*, Wollman et al. (1956, p. 160) suggested that mutations in donor type F+ or Hfr strains of *E. coli* bacteria were "reminiscent of the phenomena found in maize"; i.e., "the donor character could be conceived as being a 'controlling element' whose location commands ... the region of chromosome breakage ... as well as the pattern of chromosome rearrangements."[28]

With respect to their hypothetical model of gene regulation, Jacob's Harvey lecture (Jacob 1960, p. 33) did, however, make the connection between bacterial episomes and McClintock's "controlling elements" as he and Wollman had done when they originally defined the term (Jacob & Wollman 1958b; Jacob, Schaeffer & Wollman 1960):

> Episomic elements are probably not restricted to bacteria. ... In maize, the factors described as "Controlling elements" (McClintock, 1951 [*sic*, 1955 Brookhaven Symposium]) also appear to behave as episomes, since these agents are not always present they are added to certain chromosomal sites and can move from one site to another, and even from one chromosome to another.
>
> *(Jacob 1960, p. 32)*

In December 1960, Jacob and Monod submitted a "review article" of their new discovery to the *Journal of Molecular Biology* [*JMB*] written in English, but they made no comparison with McClintock's controlling elements. It does, however, mention earlier papers by Jacob and his colleagues that indeed did cite McClintock's work. Their review appeared in June 1961 (Jacob & Monod 1961a, *JMB*), concomitantly with their presentation at the Cold Spring Harbor Symposium (June 4–12, 1961) and the publication of another popular article by Jacob and Wollman (1961) published in *Scientific American*. The latter also related McClintock's controlling elements to episomes.

In the same month that their 1961 "Review Article" was published in *JMB*, Jacob and Monod presented a paper at the 1961 *Cold Spring Harbor Symposia on Quantitative Biology*, June 4–12, 1961, "On the Regulation of Gene Activity," which was published in 1962 (Jacob & Monod 1961b [©1962]). By then, McClintock had read their early reports published in 1959 and 1960 in *Comptes rendus*, and made the comparison between gene regulation in bacteria with the controlling elements she had observed in maize (Keller 1993, p. 177). She recalled sending a draft

[27] Jacob (1958), Harvey Lecture, delivered September 18, 1958, published 1960.

[28] McClintock had introduced the term "controlling element" at her 1955 presentation to the *Brookhaven Symposia* (McClintock 1956, Brookhaven).

manuscript to Jacob and Monod in the spring of 1961 (Judson 1979, p. 461). During this time, their comprehensive *JMB* Review was in press. Soon after reading their Review, McClintock (1961) submitted her manuscript to *American Naturalist* on "Some parallels between gene control systems in maize and in bacteria," which Jacob & Monod (1961b, p. 209) cited as "in press." McClintock's paper was published in the September/October 1961 issue of *American Naturalist*, prior to the publication of Jacob and Monod's 1961 Cold Spring Harbor Symposium presentation (Jacob & Monod 1961b). First acknowledging papers that had connected "controlling elements" with episomes: "recent evidence ... lends support to this interpretation" ... "that controlling elements in maize may be comparable to episomes in bacteria," McClintock then additionally compared the regulator gene-operator system in bacteria with the dual system of control in maize: "control systems in maize also resemble the operator-regulator systems of control of gene action in bacteria" (McClintock 1961, p. 266).[29]

In the published account of their June 1961 Cold Spring Harbor symposium presentation, Jacob & Monod (1961b) acknowledged that their concept of gene regulation in bacteria might be applicable to McClintock's concept of gene regulation in maize:

> Evidence for the existence of genetic systems of control has already been reported in maize (see McClintock 1956; Brink 1960). These systems appear to be composed, basically, of two elements. One is closely associated with the structural and controls its activity. The other element may be located near the first, or it may be independently located in another part of the chromosomal set. It determines the conditions to which the gene-associated element specifically responds, and therefore the change in the activity of the gene. These systems are quite specific, each gene-associated element responding only to a particular second element. As pointed out by McClintock (1961 [in press]), such dual systems of control in maize may, in some respect, be compared with the regulator gene-operator system of bacteria: it is conceivable that the maize system also operates through the intermediacy of specific, cytoplasmic molecules acting as signals like bacterial repressors.
>
> *(Jacob & Monod 1961b, p. 207)*[30]

The important difference between the maize controlling elements and bacterial regulators and operators was that the former seemed to be dispensable and able to move from one location to another. In this respect, a controlling element behaves more like a bacterial episome (Jacob & Monod 1961b, p. 208). As mentioned above, Jacob's comparison of maize controlling elements with bacterial episomes was introduced in the first definition of the episome published in French (Jacob & Wollman 1958b).

Monod and Jacob (1961) were given the opportunity to offer "General Conclusion" on the topics of the June 1961 Cold Spring Harbor Symposia. In the context of their section on "The Control of Gene Expression," they recognized McClintock's studies on gene regulation presented years earlier at a Cold Spring Harbor Symposia:

> Long before regulator genes and operator were recognized in bacteria, the extensive and penetrating work of McClintock (1956 [1956d *CSHSQB*]) had revealed the existence, in maize, of two classes of genetic "controlling elements" whose specific mutual relationships are closely comparable with those of the regulator and operator: the "Activator" of McClintock appears to work as a transmitter of signals, presumably cytoplasmic since they act both in *cis* and *trans*. By contrast, the specific *receiver* of these signals only acts in *cis* upon genes directly linked to it. Although, because of the absence of enzymological data in the maize systems, the comparison cannot be brought down to the biochemical level, the parallel is so striking that it may justify the conclusion that the rate of structural gene expression is controlled, in higher organisms as well as in bacteria and bacterial viruses, by closely similar mechanisms, involving regulator genes, suppressors, operators and operons.
>
> *(Monod & Jacob 1961, pp. 394–395)*

[29] See also W. Hayes (1968), Campbell (1962, 1969, 2007, 2013).
[30] Also quoted in Judson (1979, p. 461).

As we will see below, employing the terms for gene regulation in bacteria to the eukaryotic maize system was not appreciated and was even criticized by some non-maize geneticists. In December 1965, Jacob, A. M. Lwoff, and J. L. Monod would be awarded the Nobel Prize in Physiology or Medicine for their discoveries concerning the "genetic regulation of enzyme and virus synthesis" [in bacteria]. No comparison to the eukaryotic systems was made in their Nobel presentations, although Jacob and Monod had cited McClintock in the publication of their summer 1961 Cold Spring Harbor Symposia presentation, as mentioned above.

Years later, McClintock (1987b, p. xii) understood that the type of gene regulation in maize that she had described in her 1961 publication turned out to be a different system of control from the one she had compared with the bacterial system reported by Jacob and Monod. She explained that it was not until fifteen years after her studies seeking the answer to the regulatory aspects of gene instability that the regulation of gene action began to gain credibility. This was due to

> ... the elegant experiments of Jacob and Monod that were carried out in bacteria. ... The integrated systems of transposable elements in maize, which I called 'controlling elements' because of their distinctive modes of regulating the expression of genes, turned out to represent a quite different mode of gene regulation from that described by Jacob and Monod. Only now, more than forty years after the discovery of transposable elements, are we beginning to understand enough about the ways that they can affect genes to decipher some intriguing new aspects of gene control from their study.
>
> *(McClintock 1987b, p. xii)*[31]

At the Thomas Hunt Morgan Centennial Symposium (September 17, 1966) held in Lexington Kentucky, McClintock gave an invited talk, "The Expression of the Gene," to an audience of 150 attendees. She declined the invitation to submit a manuscript for the volume published three years later (Teas 1969), claiming that she was too busy to convert the talk (which was given from her slides) into a manuscript.[32] Many speakers who had presented talks apparently also declined or were not available to contribute a paper since the volume on *Genetics and Developmental Biology* only included 10 papers.

McClintock participated in the 26th Symposium of the Society for Developmental Biology: Control mechanisms in Developmental Processes, held at the University of San Diego, at La Jolla, California, from June 19 to 21, 1967. Her talk on "Genetic Systems Regulating Gene Expression during Development" was published in their proceedings (McClintock 1968). McClintock recalled receiving over a thousand reprint requests for this paper (Provine & Sisco 1980). The Biology Department at UCSD had been founded by David Bonner in 1960 (Chrispeels 2006). William DeWitt, a postdoc at UCSD during that time, would later return to his alma mater, Williams College, and nominate McClintock for an honorary degree, which she gladly accepted in 1972.[33]

In early 1968, the University of Texas, at Austin, held their E. B. Wilson Lecture Series (Cell Research Institute 1968). McClintock was among the eleven distinguished lecturers scheduled for their spring lecture series, named in honor of Edmund Beecher Wilson (1856–1939), one of America's foremost cytologists, and author of three editions of *The Cell in Development and Inheritance* (1898, 1900, 1925). Among the other prominent speakers was James D. Ebert, the Director of the Carnegie Institution of Washington's Department of Embryology (Baltimore, MD). Ten years later, Ebert would become President of CIW/DC. Also as speakers were Efriam Racker, who had recently been hired as a prominent biochemist in the newly established Division of Biological Sciences at Cornell, and Alfred Mirsky, recently retired from the Rockefeller University,

---

[31] McClintock (1987b), Introduction, is reprinted in Kass (2013ff.).

[32] Spectre 1993: Series III, "The expression of the Gene."

[33] Spectre 1993: Series II, file Williams College. Letter F. Foote to B. McClintock, April 27, 1972 (pdf), in Williams College Marshal Records (#AC 22), Williams College Archives & Special Collections.

but who had continued to maintain his laboratory and his research in cell biology. McClintock gave three lectures on genetic organization and control at this lecture series (Cell Research Institute 1968; Spectre 1993: Series III).[34]

CORNELL A. D. WHITE PROFESSOR REAPPOINTMENT: COMING HOME AGAIN

Encouraged by his colleagues, Harry Stinson, GDP Section Chair, nominated McClintock for reappointment in 1971.[35] A botanist/geneticist trained at Indiana University with Ralph Cleland, Stinson clearly understood McClintock's worth:

> There is no doubt in the minds of Dr. McClintock's colleagues at Cornell that she is very much an 'imaginative and daring theoretician' whose thoughts are clearly directed towards the future of genetics and biology. Nowhere is her posture better illustrated than in the seminar she gave last week as part of her current visit. This seminar, Expanding Concepts in Genetics [18 March 1971], dealt with what she terms the revolution that must take place in the minds – and experiments – of biologists who direct their attention to problems of genetics, development and evolution. As always, she performs a real catalytic function in forcing us to examine the conventional wisdom and established dogma in new lights.[36]

Stinson was convinced that McClintock genuinely enjoyed her visits to Cornell and she once described them to him as "like coming home again and seeing all my good friends."[37]

Professor Harry Stinson held the view that McClintock's decision not to reside permanently at Cornell was due to the campus unrest that plagued Universities across the nation at the time. In 1967, multiple student protests were held in opposition to the Vietnam War and anti-ROTC demonstrations were held in Barton Hall on the Cornell Campus. In 1969, Black students at Cornell also demanded equality in education, and requested a Black Studies Program at Cornell. To precipitate their request, which had been ignored, on April 19, 1969, the "Willard Straight Hall Takeover" that began the day before on the Cornell Campus was national news.[38] Cornell student Matthew Kessler published his eyewitness account of faculty and student support of this event in his hometown newspaper on April 30, 1969 (Kessler 1969). He received thank you letters for his summary that did not sensationalize the events as given in many newspapers and TV reports.[39] Professor Fink (1992, p. 281) recalled that McClintock "spent hours on the steps of Willard Straight Hall (the student union) ... debating with the various political factions that had set up booths to politicize both students and faculty. ... She often expressed surprise about how little the various ... groups had read about the subjects they were promoting." That summer, Max Black, Chairman of the Professor-at-Large Program, sent a personal note to McClintock and others in the program asking them to discount "sensational accounts" presented in the news. He attached his own eyewitness reflections of "What really happened."[40] He hoped that she would visit soon and often, and indeed, she did.

RECOGNITION BY GENETICIST COLLEAGUES

During the time McClintock was Professor-at-Large at Cornell (1965–1974), she received national recognition for her contributions to the field of genetics. The National Academy of Sciences lauded her with their highest recognition, the Kimber Genetics Award, in 1967, for achievements as an "extraordinarily

[34] Spectre 1993: Series III, February 1968, lectures; titles of these lectures differ from those in the Pamphlet distributed for the Wilson Lecture Series (Cell Research Institute 1968): Barbara McClintock – Genetic Mechanisms Influencing Gene Expression (February 12); Operation of a Control System (February 13); Versatility of Control Systems (February 15).

[35] H. Stinson to M. Black March 24, 1971, A. D White, McClintock, CU.

[36] H. Stinson to M. Black, March 24, 1971, A. D White, McClintock, CU.

[37] H. Stinson to M. Black, March 24, 1971, A. D White, McClintock, CU.

[38] *Cornell Chronicle* April 16, 2009, "A campus takeover that symbolized an era of change." *Cornell Daily Sun* April 16, 2019, "Cornell To Honor 50th Anniversary of Willard Straight Hall Takeover With Keynote Event and Commemorative Walk."

[39] All Documents in the files of Matt Kessler shared with LBK August 18, 2020, and forwarded to Cornell University Archives on same date with his permission.

[40] M. Black to B. McClintock, July 15, 1969, A. D White, McClintock, CU.

skillful and resourceful experimenter, imaginative and daring theoretician, who has produced in the past forty years a series of brilliant and significant studies on chromosome structure and function unmatched in the history of cytogenetics.'"[41] Only months later in June 1967, she reached mandatory retirement age (65), but did not suffer the same fate as Demerec. As previously arranged, she was made a Distinguished Service Member of the Carnegie Institution of Washington, with a reduced salary and retirement benefits for three years.[42] She was no longer required to make annual reports for the *Carnegie Year Books*, but "aperiodic" reports were accepted. The year she became the first female recipient of the National Medal of Science (1970; awarded May 21, 1971), the Carnegie Board of Trustees renewed her Distinguished Service Member appointment, but without compensation. McClintock again contemplated leaving CSH by June 1972 or earlier if an opportunity should arise.[43]

McClintock participated in the 1972 International Symposium on Fundamental Approaches to Plant and Animal Improvement in Cali, Colombia, South America, which Srb had organized with Alexander Hollaender. One aim of their symposium was to "foster interactions among representatives of basic and applied science ..." (Srb 1973, p. vii), and afford an "excellent opportunity to combine the basic aspects of biochemistry and genetics with the application to the improvement of plants and animals" (Hollaender, in Srb 1973, p. ix). McClintock chaired the session on Heterosis, Hybridization, and Selection.[44] Although the published Symposium volume does not reflect her participation (Srb 1973), McClintock is listed as a session chair on their funding proposal to the NSF [National Science Foundation] and AEC [Atomic Energy Commission].[45] In his 1972 presentation on mutation induction in flowering plants, R. A. Nilan (1973) mentioned the specificity of McClintock's controlling elements to inducing mutations and how they can be altered by mutagens for the purpose of stabilizing a structural gene. Adrian Srb recalled how McClintock taught him how to evade becoming ill in South America by avoiding drinks containing ice and pealing fruits and vegetables.[46]

McClintock's influence on the goal of the Cali symposium organizers is further reflected, in the contribution by Almiro Blumenschein of the Instituto de Genetica, Universidade de Sao Paulo, Piracicaba, Brazil. As mentioned above, Blumenschein was one of the students who McClintock had mentored at NC State College, along with Kato in the early 1960s. They continued their collaborations, often meeting in Mexico to review progress for the monograph they were preparing on the morphology of maize chromosomes. His presentation on "Chromosome Knob Patterns in Latin American Maize" provides a glimpse of the understanding of the origin and evolution of maize at that time. McClintock's (1959, 1960) methodology for maize chromosome knob studies provided an appreciation of geographical distributions and the derivation of modern maize from several different centers. Blumenschein reported that modern races of maize seem to be derived directly from knob complex centers, which are concentrated around southern Mexico, thus pointing to this region as the main center of the origin for modern maize varieties (Blumenschein 1973, p. 226). As mentioned above, McClintock, Kato, and Blumenschein (1981) later published a comprehensive volume of the chromosome constitutions of the races of maze in which the distribution and frequency of knobs or chromosome components are interpreted in terms of where maize developed initially and when, where, and how it was introduced into other parts of the Americas (Kato 2013).

McClintock's last Carnegie research report appeared in the *Annual Report of the Director* [Hershey's last] *of the CIW/Genetics Research Unit*, 1971, the year the Genetics Research Unit is reported to have

[41] H. Stinson to M. Black, March 24, 1971, A. D White, McClintock, CU. For details on Kimber award nomination see Comfort (2001).

[42] E. Ackerman to B. McClintock September 12, 1964; E. Ackerman to A. Hershey, January 30, 1967; Carnegie Institution to Barbara, May 19, 1967, Trustees voted appointment of Distinguished Service Member beginning July 1, 1967, for three years, with a stipend of $15,000, all letters in McClintock CIW/DC Archives.

[43] E. Ackerman to A. Hershey, May 11, 1970, "McC contemplates retiring from CIW and leaving CSH on June 30, 1972 at the latest, or earlier if an opportunity arises," McClintock CIW/DC Archives. President Richard Nixon's Daily Diary, May 21, 1971, National Medal of Science.

[44] Srb files, courtesy of J. Srb. Srb 1973.

[45] Srb files Cali Symposium, courtesy of J. Srb.

[46] L. B. Kass interview with A. Srb, October 13, 1994.

FIGURE 9.3 Barbara McClintock received the Honorary Doctor of Science degree for "distinguished service and leadership in research and education" from the University of Missouri, Columbia, Missouri, awarded June 4, 1968. (*Columbia Missourian*, June 7, 1968; image courtesy of E. H. Coe.)

closed (Report of the President 1971, p. 22), and the year McClintock was reappointed as a Cornell visiting professor. Finding other outlets for reporting her research results, she published chapters in Symposia volumes (Kass & Chomet 2009: Appendix I; McClintock 1987a),[47] and gave many lectures and seminars – outlines of which she saved in her files (Spectre 1993: Series III). Honors and awards continued to be announced in the President's reports of the *CIW Year Books*.[48] In 1968, she received her fourth honorary degree, awarded by the University of Missouri (Sc.D. 1968; Figures 9.3 and 9.4). At the White House in 1971, President Nixon recognized her award of the 1970 National Medal of Science.

[47] See Kass (2013ff.: Table I) for annotated list of McClintock's publications.

[48] For example, see *Carnegie Year Books* 71, 72; see Kass and Chomet (2009: Table 2, Awards and Recognition).

FIGURE 9.4 Barbara McClintock and colleagues at a "tea" held on June 5, 1968, given in her honor by the University of Missouri Genetics Department in celebration of her Honorary Doctor of Science degree, awarded the previous evening. From left to right: Gerry Neuffer, Chair, Department of Genetics; Lotti and Ernie Sears, faculty members; Barbara McClintock, honoree; and Dr. Albert E. Longley, Visiting Professor. (*Columbia Missourian* June 7, 1968; image and identifications courtesy of E. H. Coe.)

Her fifth honorary degree (Sc.D.) was awarded by Williams College in 1972. The following year, the research laboratory building (Animal House) where she had first worked at Cold Spring Harbor was dedicated in her name during the June 1973 CSH Symposia.[49]

ACCEPTANCE AND REJECTION OF McCLINTOCK'S IDEAS

The impact of McClintock's contributions can be judged by how and when they are accepted in the wider scientific community. This may be traced by how often her work is cited in scientific journals. The *Science Citation Index (SCI) Ten Year Cumulations* for the years 1945–1954 (1988) and 1955–1964 (1984) are useful for this purpose. The *SCI* was first issued in 1961. It lists published papers by author, and the journal where the work had been cited. The *SCI*, however, omits much work found in reference books, textbooks, and the popular science literature. Furthermore, the editor of *SCI* cautioned users of the many errors, overlooked citations, and the limitation that author's works are listed only if more than three citations had been located for an author's publications. In addition, only names of the first author of a citation were given; readers were referred to a *Source Index* to find further information for the citation author. By examining volumes listed or not in the *SCI*, and double checking cited publication dates, I found both positive and negative receptions to McClintock's ideas on the significance of the controlling elements she had found in maize during the 1940s (Fedoroff 1998), and which she first reported in the scientific journal *Proceedings of the National Academy of Sciences* in 1950 (See Weil 2013). The often repeated stories of the unenthusiastic reactions to this short initial account, and to McClintock's more detailed reports (McClintock 1951 *CSH Symposia*; 1953 *Genetics*), seem to have had a strong impact on McClintock's perception of how others understood and received her work (Keller 1983; Provine & Sisco 1980). She told Provine and Sisco (1980) that she stopped publishing accounts of the transposable elements story

[49] *CSHL Annual Report* (1973), pp. 9–10.

because few researchers requested reprints of her publications. Indeed, only three citations are listed in the *SCI Ten Year Cumulation 1945–1954* (1988) for her 1953 paper published in *Genetics*. There may have been a more compelling story behind her resolution to cease publishing around that time, however (see Chapter 8).

Before the advent of the internet, science educators were well aware of the lag time (usually 10 years) for new concepts to appear in college textbooks. For this reason, teachers supplemented texts with review articles, reference books, or original research papers; therein, students could read the primary papers, where the work first appeared. Undergraduate students often complain about the burden of such additional reading assignments, as I learned only too well. For this reason, a text that offers an accurate summary of a topic and a list of supplemental readings can provide the most current sources in a field, if only the students would read them. With this in mind, a review of the literature that includes summaries and/or references to McClintock's ideas reflects the contemporaneous views of her work and its influence at the time. In contrast, an omission to her work may reflect a rejection or a lack of comprehension of its relevance or importance to the subject. Summarizing McClintock's notable achievements in his recommendation for the 1970 National Medal of Science, Rhoades wrote that "an indication of the significance of her contributions to the fabric of genetic theory may be had from the number of times her discoveries are specifically mentioned in several recent general texts. As a rule, in such books only the pioneering, the truly significant and illuminating discoveries - the milestones - are cited." He continues by documenting the number of times McClintock's work is discussed in genetics/cytogenetics textbooks by Strickberger [1968], Swanson [1957], Herskowitz [1965], Srb et al. [1965], and Lewis and John [1964].[50]

In the 1950s, two popular textbooks (Altenburg 1957; Swanson 1957) recognized McClintock's new findings on maize controlling elements, as I mentioned in previous chapters. One of these is extremely accepting of her ideas on genic control of development (Altenburg 1957), and the other provides a caveat that her system is "of course abnormal" and the "*Ac-Ds* system is a mechanism of differentiation which … has gotten out of normal control" (Swanson 1957); later, co-authored editions would revise this idea (Swanson et al. 1967, 1981; see below). Although early editions of the widely used *Principles of Genetics* text by Sinnott and Dunn (1932, 1939) and Sinnott et al. (1950) were quick to incorporate McClintock's early contributions to cytogenetics, McClintock's research on controlling elements in maize is absent from the main text of their fifth edition (Sinnott et al. 1958). It is only in the bibliography that we find citations to the 1951 and 1956 *Cold Spring Harbor Symposia on Quantitative Biology* for "reviews of the modern state of the field of mutation studies." In these volumes, McClintock had published extensive reviews of her ideas on chromosome organization and gene expression and with controlling elements and the gene – publications she believed were not well received (Keller 1983, 1993; McClintock 1987b).

Another early textbook provided a brief summary of the contribution of Rhoades' and McClintock's research on the genetic control of mutation (Wagner & Mitchel 1955, p. 48):

> Perhaps the most interesting examples of genetic control of mutation rate have come from the investigations of Rhoades and McClintock [1951, 1953] on maize. These studies not only have confirmed the earlier finding from *Drosophila* that the rate of mutation of some genes is under the control of other genes but have also provided further data supplemented by cytological observations which may eventually lead to an understanding of spontaneous mutation.

As mentioned above, by the late 1950s and early 1960s McClintock's controlling elements were recognized by molecular biologists who compared them to episomes, "added bodies" that are either "integrated or nonintegrated" in the chromosome, which they had found and named in bacteria (Jacob &

[50] Barbara McClintock Statement of Achievements, 1969, Spectre 1993: Series II, Folder: National Medal of Science 1969, 1971. Rhoades did not include the publication dates for the textbooks that he cited in his statement recommending this award. The dates are therefore added in brackets and the volumes listed in the Literature Cited to assist readers interested in locating the primary sources.

Wollman 1958b; Jacob 1960, 1958). At a symposium held in London, in April 1960, Jacob, Schaeffer & Wollman (1960, p. 84) cited McClintock's studies in maize in relation to genetic regulation:

> In maize, the chromosomal location of the so-called controlling elements is firmly established (McClintock, 1956), but there is no experimental evidence in favor of an autonomous state although this might be postulated on the basis of the frequent movement from one chromosomal location to another.

Fedoroff (1992) noted that McClintock's pioneering research was prominent in the maize literature. Indeed, the first edition of *Corn and Corn Improvement* (Sprague 1955) includes extensive and in-depth descriptions of the work that led to McClintock's early discovery of transposable elements and mentions her then recent research on controlling elements in corn (Rhoades 1955). It also places in perspective the work of other researchers (Brink, Peterson) investigating similar phenomena in maize. The significance of McClintock's hypothesis for the genetic control of development was not part of the scope of Rhoades' chapter, but would be included in later editions of this important contribution to the maize literature (Coe & Neuffer 1977, pp. 127, 140, 144, 173–180, 195; Coe et al. 1988, pp. 139, 157, 199–203, 222–223; Carlson 1977, pp. 275–276, 1988, pp. 270–271). Detailed discussions toward an understanding of maize development was also presented by McClintock (1978) and Coe and Neuffer (1978) at a symposium on *The Clonal Basis of Development* held in Raleigh, North Carolina, in 1977.

Other volumes were positive and also critical (Bonner 1956; Sager & Ryan 1961/1962; Waddington 1962). In a symposium presentation on Genetic Unity, dedicated to the memory of Professor L. J. Stadler, D. Bonner (1956, p. 157) accepted the idea that controlling elements may play a role in growth and development, while Waddington, in his 1961 Jesup Lecture, described the "capricious action of the maize controlling elements" and thought that it was "difficult to see" how they would be adequate to control the process of development (Waddington 1962, p. 22). Later, in his book on *Heredity*, Professor D. Bonner (1961, p. 98) praised the magnificent work of Barbara McClintock with corn, which "revealed genetic elements that can induce gene mutation and direct the expression of specific genetic material." Yet he cautioned that …"the precision required for differentiation is not present." He added in this chapter on Genes and Development, "The regulator genes and 'l'operon' are also of interest to the problems of development … but they, too, lack the precision required by differentiation and development." In the second printing of their textbook on *Cell Heredity* (issued May 1962), Sager and Ryan (1961/1962: Chapter 11, p. 344) reported that the transposition phenomenon in maize shares a resemblance to episomes. They further compared the bacterial and maize systems in their summary of gene action (Sager & Ryan 1961/1962: Chapter 13, pp. 391–392). The Postscript extended the discussion of Chapters 11 and 13 by including recent work on the mechanism of gene action. Although therein they do not specifically cite the work of McClintock, the Postscript's bibliography includes the 1961 *Cold Spring Harbor Symposia*, which, as mentioned above, gave recognition to McClintock's controlling elements in genetic regulation.

W. Ralph Singleton briefly described McClintock's *Ac-Ds* system in a discussion on the nature of the gene in the first edition of his 1962 *Elementary Genetics* text (Singleton 1962, p. 409). Providing a reference to her presentation at the 1955 *Brookhaven Symposia in Biology*, Singleton (1962, p. 417) informs the reader that following her published paper is "an interesting discussion by other eminent geneticists." Singleton, a notable maize geneticist, had years before arranged for McClintock to grow her corn at Brookhaven National Laboratory (see Chapter 8), and he had an appreciation of her work on maize controlling elements. In the second edition of his textbook, Singleton (1967) cites McClintock's June 1965 Brookhaven Symposia presentation, where she summarized the work of maize researchers on control of the various systems found in maize. Here again, he informs the reader that following her presentation a lively discussion of her results ensued.

Adrian Srb, who had nominated McClintock for Cornell's A. D. White Professorship, included a description of controlling elements in maize in the second edition of his co-authored *General Genetics* text (Srb et al. 1965), which was published the year McClintock returned to Cornell. I have vivid memories of being required to understand the operon theory of Jacob & Monod, as presented in this text, while enrolled in genetics during my senior year at The City College of New York in 1968. Yet, I have a vague recollection of reading the chapter on "Extrachromosomal and Epigenetic Systems," which Srb et al. (1965) considered "a catchall for a variety of phenomena presently not referable in the ordinary way to genes in chromosomes. ... the pattern of transmission for a phenotype based on any of the various potential extrachromosomal systems cannot be predicted with certainty." Their chapter included a section on "Controlling Elements" (pp. 339–342). Conceived in terms of a regulator (*Ac*) and an operator (*Ds*), they conclude their section on "The Nature of Controlling elements" with caution: "... controlling elements are of potentially great importance in the regulation of developmental processes of organisms in which they occur. The fact that some of the manifestations of the controlling element are erratic does not preclude their participation in regular, normal processes of development" (Srb et al. 1965, p. 341). Years later, Srb lamented his limited approach to representing McClintock's ideas in his text.[51]

About a decade later, extensive descriptions of McClintock's pioneering work on maize and the two-element controlling region (*Ac-Ds*) again appeared in *Principles of Genetics* by Irwin H. Herskowitz (1973: Chapters 28 and 29). Herskowitz, a Professor at Hunter College of the City University of New York, had graduate degrees in genetics from University of Missouri (MS Zoology 1942) and had taken McClintock's Cytogenetics course at Missouri in 1941 (Herskowitz, Ira 1992, p. 293). He completed his doctorate in Genetics at Columbia University in 1948, then conducted research at Indiana University with Nobel Laureate H. J. Muller. His 1973 textbook section on gene regulation in eukaryotic systems is presented for maize, mammals, and *Drosophila*. The supplemental readings introduce the work of Brink and his students on the controlling regions in maize. Their studies also showed an *Ac* locus (initially reported as *Mp*) regulating gene expression at the pericarp locus in maize. By 1973, Herskowitz had expanded upon and updated his discussion of McClintock's two-element controlling system and the "genetic control of mutability" [gene action], which he earlier summarized in the first and second editions of his textbooks *Genetics* (Herskowitz 1962, 1965).[52]

Reviewing the literature in growth and development during this time (1965–1974), we find other examples of positive signs of the reception of McClintock's ideas on controlling elements and development: (1) reports that investigators have found in corn "controlling elements" that switch genes on and off and "can move from chromosomal site to chromosomal site, not unlike the bacterial fertility factor" (Engel 1967, p. 110); (2) "That genes exist which can control the activities of other genes as in the operator-structural gene relationship [proposed by Jacob and Monod] has been confirmed in higher organisms as well as in bacteria." In maize, the color pattern of the corn kernels is due to controlling genes that appear to affect the structural genes that are responsible for production of the characteristic pigments (Barth 1964, p. 86); (3) "elements having control of function at particular loci have become clarified by McClintock" (Williams 1964, pp. 152–157); and (4) "genic control of mutability appeared later in the proof by McClintock of genic activators and controller systems within the cell" (Dunn 1965, p. 170). By 1967, Swanson no longer referred to McClintock's controlling system as "abnormal" and paid serious consideration to condensed parts of chromosomes as a mechanism of differential expression of gene action (Swanson et al. 1967, p. 165).[53] Years later, their second edition (Swanson et al. 1981, pp. 487–494) would highlight the work on "controlling elements" to genic regulation and control, and point out their similarity to "transposable nature of episomes and insertion elements of *E. coli*."

In my first year in graduate school at Cornell (1971), we used the cytology textbook by Cohn (1969) who summarized Jacob and Monod's (1961a) theory of repression; offering an explanation of regulation of protein synthesis in the cell by regulator genes capable of switching off those genes

[51] L. B. Kass interview with A. Srb, October 13, 1994.

[52] Herskowitz (1962) includes the first publication of the 1929 photo of Emerson and his associates (p. 126), and also a discussion of McClintock's *Ac-Ds* control of mutability (pp. 213–216).

[53] Citing 1965 *Brookhaven Symposia*, especially papers by McClintock and others.

responsible for the production of specific enzymes and the possibility of a promoter gene lying adjacent to the operator genes. The control system is compared with higher organisms – "the earliest reported genetic situation compatible with the operon theory was by Barbara McClintock in corn." Within the year, I would be fortunate to meet Dr. McClintock when she visited campus on her renewed A. D. White Professorship.

When Marcus Rhoades taught cytogenetics at Indiana University during 1968, he required his students to read McClintock's (1951, 1965) symposium papers and summarized these in his lectures.[54] Rhoades also compared the *Ac-Ds* system to a regulator (*Ac*)-operator (*Ds*) and described it as a "2-element system of gene control."[55] And as early as October 1950, Charles R. Burnham, McClintock's longtime friend and colleague, introduced her pioneering work to his classes at the University of Minnesota, by reading directly from her letters, "while standing at the chalkboard."[56] As mentioned in the previous chapter, at the University of Illinois, Rhoades had required his Cytogenetics classes to read McClintock's pioneering papers beginning with Creighton and McClintock (1931). In a graduate student's cytogenetics notebook from 1950 to 1951 are five pages of notes summarizing McClintock's recent publication on the dissociation and activator system (McClintock 1950 *PNAS*).[57] He noted that McClintock's system is also associated with the *Dt* mutable expression that he had found earlier (Rhoades 1938, 1941). In his previous lecture on *Dt*, he concluded that "mutability is gene controlled." Rhoades taught his last Cytogenetics class in the spring of 1974, just before retiring from full-time teaching. By that time, he no longer included discussions of McClintock's studies of transposable elements or controlling elements. Those subjects were covered in the course on "Eukaryotic Genetics," offered by his former student Drew Schwartz.[58] Schwartz would go on to study and cooperatively identify maize controlling element mutations at the molecular level (see Chapter 11).

Yet, skepticism of McClintock's ideas of genetic control of development is apparent in both the popular and research literature on organismal growth and development. In addition to Waddington's clearly critical evaluation, two additional examples may suffice to demonstrate the veiled rejection of McClintock's hypothesis. Her 1961 publication in *American Naturalist* apparently offended some readers and one especially who had been in attendance at the 1951, 1953, and 1956 Cold Spring Harbor Symposia. In a humorous effort to depict the discord between microbiologists [molecular biologist] and embryologists [developmental biologists] of the day, Haig P. Papazian (1967) provides a popular picture of genetics in his award-winning book, *Modern Genetics*. A fictional embryologist laments the straining of a hypothesis to make it fit data that do not naturally fall into place. One of the odd phenomena that need explanation is position effect – "why do genes act differently according to their position on the chromosome, especially when next to heterochromatin?" Using an example in corn, the fictional embryologist continues, "A great deal of work has been done on certain elusive factors which seem to float about from one position to another on various chromosomes. These factors affect the functioning of the particular gene they happen to be on or near. The factors are themselves influenced by other genes, and ingenious, as well as laborious, experiments are necessary to study them. One of the discoverers of these peculiar phenomena has enthusiastically likened them to the regulator system in bacteria, but I would suggest that this is a good illustration of the hypothesis being strained. It may turn out to be fruitful, but it may lead us up the garden path and blind us to the real explanation." The microbiologist responds by complementing

---

[54] Greg Anderson, Cytogenetics notebook 1968, pp. 33–38 (McClintock 1951, 1965), courtesy of Prof. G. J. Anderson, University of Connecticut, Storrs, CT.

[55] G. Anderson Cytogenetics notebook 1968, p. 36.

[56] Ed Coe to L. B. Kass, email 27 October 2009; Letter McClintock to Burnham, January 20, 1950, Burnham collection, UMN, Box 3, file McClintock.

[57] Hope Punnett, Cytogenetics notebook, October 1950–May 1951, Rhoades lecture, "McClintock – unpublished – *PNAS* June 1950," notebook pages are not numbered, courtesy of Prof. Hope Punnett.

[58] Jim Birchler Cytogenetics notebook Spring 1974, courtesy of J. A. Birchler. Email J. A. Birchler to L. B. Kass, May 13, 2023. Carlson and Birchler (1997).

the embryologist as "progressive" and "unlike the ultraconservative kind so commonly found in the botany and zoology departments of our colleges"; and commends the embryologist for not trying to "jump on our bandwagon by borrowing from our popular streamlined vocabulary and thereby debase these new words." For those who are familiar with McClintock's (1961) publication, which compared her *Suppressor-mutator* (*Spm*) gene control system in maize with the operator-regulator systems of control of gene action in bacteria [the operon] as proposed by Jacob and Monod (see above), it would seem that she is the one under attack.

A second example of masked criticism of McClintock's work is shown in a perspective on Growth and Development found in *Gene Action*, in the Foundations of Modern Genetics series. Gene action, the authors (Hartman & Suskind 1969) state upfront, is "equated with protein formation and function." Their chapter explores some of the important problems in "biology of the future," but their critical view of hypotheses without sound experimental evidence is strong:

> Intuition tells us that some similarities will be found between regulatory systems in microorganisms and those in higher organisms and that such systems will be found to be important in guiding differentiation, but science relies upon experimental demonstration, not on intuition. It is logical to assume that controlling mechanisms peculiar to higher organisms will be discovered, we do not yet have a detailed picture of the structure of chromosomes of higher organisms; and without this, our understanding of how genic expression can be influenced and regulated is limited. Observations indicate lack of activity of the genes is due to their position next to heterochromatin and not to structural defects within the genes themselves. ... Many speculations have been advanced to explain these observations: for the most part, they are inadequately supported by experimental evidence.
>
> *(Hartman & Suskind 1969, p. 231)*

McClintock ideas are not mentioned in Sussman's (1964) *Growth and Development*, in the same Foundations series, first published in early 1960 and revised in 1964. Sussman's comment on how cells become specialized – cellular differentiation – reflects the state of this field:

> These processes are still largely unknown and present a major challenge to developmental biologists. Another mysterious aspect of cellular differentiation is the way the numbers of each cell type are carefully regulated. ... The mechanisms that control the proportions of differentiated cells are still to be discovered.
>
> *(Sussman 1964, p. 4)*

Development involves change, Sussman continues, it bears "a direct relation to genetics...we want to know if the genetic constitutions of these cells [cell types] had to change in order to permit them to look and act differently and if so, what the nature of the change was." He refers geneticist to a discussion of these matters in Bonner's *Heredity*, which as I noted above point out that controlling elements may play a role in development, but lack precision required for differentiation.

Nina Fedoroff, who later (1978–1979) began an investigation on the nature of McClintock's controlling elements[59] and co-authored a *Festschrift* for McClintock's 90th birthday (Fedoroff & Botstein 1992), tells us that the "maize genetics literature bears witness to McClintock's immediate impact." Looking for this literature, in addition to the texts mentioned above, and the many citations referred to in journal articles (*Science Citation Index*), I came across James L. Brewbaker's (1964) *Agricultural Genetics* text, which describes McClintock's studies on mutable genes. As a graduate student in the early 1950s Brewbaker, a corn geneticist who received his Ph.D. at Cornell visited McClintock at her Cold Spring Harbor Laboratory.[60] He later spent five years (1956–1961) at Brookhaven National Laboratory where he enjoyed conversations with McClintock and had attended her June 1955 symposium presentation (McClintock 1956). His text, also published in

[59] *Carnegie Institution of Washington Year Book 78*, 1978–1979.
[60] Email J. Brewbaker to R. P. Murphy & L. Kass, April 16, 2008.

the Foundation series, describes McClintock's work on transposition and its regulation in "gene action." His annotated references include R. C. King's (1962) *Genetics* text "for a most readable account of mutation." King includes an account of McClintock's study of the *Ac/Dc* controlling system (pp. 186–188, citing her 1951 Cold Spring Harbor Symposium presentation). And in King's chapter (16) on Developmental Genetics, he hypothesized (p. 270) that the intergenic regulation of genes that function to regulate the expression of neighboring genes in prokaryotic bacteria may be working in multicellular organisms, such as McClintock's *Ac-Ds* regulatory system (described p. 186). R. C. King, then a professor at Northwestern University, had been on the staff at Brookhaven and Chair of the 1955 Brookhaven Symposia in Biology titled *Mutation*, where McClintock first defined and introduced the term "controlling elements" to her audience.[61]

The examples show that McClintock's work on transposition and controlling elements in maize was often enthusiastically recognized and accepted. Nevertheless, some investigators questioned the idea that they were comparable to regulators and operators as described in bacteria. And others seemed to cast doubts due to the lack of published data. At Cornell, however, there was no lack of interest or endorsement for her ideas and perhaps for these reasons she returned year after year.

McClintock's Cornell Professorship concluded in 1974, but she was urged "to continue to come for many more years" and would be welcome to return, "... on the same terms, whenever she can."[62] Stinson later regretted that he did not pursue a way for McClintock to return on a regular basis.[63] The Genetics Research Unit at Cold Spring Harbor "closed officially" in 1974, when the Director, Al Hershey, gave up his office in June. McClintock continued working as Distinguished Service Member (without salary) of the Carnegie Institution of Washington, who paid rent for her laboratory space, administered by the Cold Spring Harbor Laboratory of Quantitative Biology (under the direction of Nobel Laureate J. D. Watson, since 1968).[64] McClintock returned to Cornell in 1982, as an honored guest for the 75th Synapsis Club reunion and the Department of Plant Breeding's Anniversary Celebration organized by William B. Provine (see Chapter 10; Murphy & Kass 2007, 2011).

LITERATURE CITED

Altenburg, E. 1957. *Genetics, Revised Edition*, Henry Holt and Company, New York.

Anderson, E. 1947. *Corn before Columbus*, Pioneer Hi-Bred Corn Co., Des Moines, IA.

Anderson, E. 1949. *Introgressive Hybridization*, John Wiley & Sons, New York, 109 pp.

Anderson, E. and H. C. Cutler. 1942. Races of *Zea mays*: I. Their recognition and classification. *Annals of the Missouri Botanical Garden* Vol. 29, pp. 69–88.

Barth, L. J. 1964. *Development, Selected Topics*, Addison-Wesley Publishing Co., Inc., Boston, MA.

Blumenschein, A. 1960. *Número de cromossomas de algumas espécies de orquídeas*. Publ Cient Univ São Paulo. Vol. 1, pp. 45–50. (Abstract in English) [Publication of his 1957 Ph.D. Thesis]

Blumenschein, A. 1964. *Identification of chromosome segments that contribute to the specificity of racial characteristics in maize* (Under the direction of Barbara McClintock and William Lacy Brown. Benjamin Warfield Smith, Major Advisor.) North Carolina State College, Raleigh, NC. (Ph.D. Thesis, Department of Genetics), pp. 1–75.

Blumenschein, A. 1973. Chromosome Knob patterns in Latin American Maize, Chapter 18, pp. 271–276, *In*, A. M. Srb, (Ed.), *Genes, Enzymes, and Populations*, Plenum Press, New York.

Bonner, D. M. 1956 [© 1957]. The genetic unit, dedicated to the memory Professor L. J. Stadler, pp. 163–170. *Cold Spring Harbor Symposia on Quantitative Biology Vol. XXI [21], Genetic Mechanisms: Structure and Function*, The Biological Laboratory, Cold Spring Harbor, NY. [K. B. Warren, Executive Ed.]

Bonner, D. M. 1961. Genes and development, Chapter 9, *In*, *Heredity*, Prentice Hall Inc., Englewood Cliffs, NJ.

[61] McClintock 1955 (1956a), Symposia on *Mutation*, published 1956, Brookhaven National Laboratory.

[62] M. Black to H. Stinson, May 30, 1974, A. D White, McClintock, CU.

[63] L. B. Kass Interview with H. Stinson, January 14, 1994.

[64] Letter M. H. Walburn to M. E. Woodring, October 16, 1974; A. Hershey to J. W. Boise, January 24, 1974, McClintock CIW/DC Archives.

Brieger, F. G., J. T. A. Gurgel, E. Paterniani, A. Blumenschein, and M. M. R. Alleoni. 1958. *Races of Maize in Brazil and Other Eastern South American Countries* Pub. 593 NAS-NRC, Washington, DC.

Brewbaker, J. L. 1964. *Agricultural Genetics*, Prentice-Hall Inc., Englewood Cliffs, NJ, 156 pp.

Brown, W. L. 1941. The cytogenetics of *Poa pratensis*. *Annals of the Missouri Botanical Garden* Vol. 28 (No. 4, November 1941), pp. 493–522. [Publication of Ph.D. Thesis]

Brown, W. L. 1949. Numbers and distribution of chromosome knobs in United States maize. *Genetics* Vol. 34 (No. 5, September 1, 1949), pp. 524–536. https://doi.org/10.1093/genetics/34.5.524

Brown, W. L. 1960. *Races of maize in the West Indies*. National Academy of Sciences Publ. No. 792, pp. 1–60.

Campbell, A. 1962. Episomes. *Advances in Genetics* Vol. 11, pp. 101–145.

Campbell, A. 1969. *Episomes* [Modern Perspectives in Biology], Harper and Row, Publishers, New York.

Campbell, A. 2007. Phage integration and chromosome structure. A personal history. *Annual Review of Genetics* Vol. 41 (December 2007), pp. 1–11. https://doi.org/10.1146/annurev.genet.41.110306.130240 [Online April 27, 2007]

Campbell, A. 2013. Comparative studies relevant to transposon function in plant development, pp. 3.87–3.90, *In*, L. B. Kass (Ed.), 2013ff. *Perspectives on Nobel Laureate Barbara McClintock's Publications (1926–1984): A Companion Volume*, Vol. II, The Internet-First University Press. https://hdl.handle.net/1813/34897

Carlson, W. R. 1977. The cytogenetics of corn, pp. 225–304, *In*, G. F. Sprague (Ed.), *Corn and Corn Improvement*, 2nd edition, American Society of Agronomy, Madison, WI.

Carlson, W. R. 1988. The cytogenetics of corn, pp. 259–343, *In*, G. F. Sprague and J. W. Dudley (Eds.), *Corn and Corn Improvement*, American Society of Agronomy, Inc. Crop Science Society of America, Inc. Soil Science Society of America, Inc., Madison, WI. ("Occasional segments of the original chapter on cytogenetics by M. M. Rhoades [1955] have been retained intact")

Carlson, W. R. and J. A. Birchler. 1997. Marcus Morton Rhoades 1903–1991. *A Biographical Memoir*, National Academies Press, Washington, DC.

Cell Research Institute. 1968. *The Cell Research Institute of the University of Texas at Austin … The E. B. Wilson Lectures … The Spring of 1968.* [E. B. Wilson Lectures, pamphlet from the Cell Research Institute vertical file, Dolph Briscoe Center for American History, The University of Texas at Austin].

Chrispeels, M. J. 2006. David Mahlon Bonner, 1916–1964. *A Biographical Memoir*. Vol. 88. National Academy of Sciences, Washington, DC, 23 pp.

Cobb, E. D. 2013. *150 Years of Botany at Cornell; A History of Botany and Plant Biology*, Department of Plant Biology, Cornell University, Ithaca, NY.

Coe, E. H. and L. B. Kass. 2019. Albert Edward Longley, 1893–1984 – Botanist, cytologist, geneticist. *Maize Genetics Cooperation Newsletter* Vol. 93, pp. 1–9. https://mnl.maizegdb.org/93/pdf/03Coe.pdf (Online October 23, 2019)

Coe, E. H. Jr. and M. G. Neuffer 1977. The genetics of corn, pp. 111–223, *In*, G. F. Sprague (Ed.), *Corn and Corn Improvement*, 2nd edition, American Society of Agronomy, Madison, WI.

Coe, E. H. Jr. and M. G. Neuffer. 1978. Embryo cells and their destinies in the corn plant, pp. 113–129, *In*, S. Subtelny and I. M. Sussex (Eds.), *The Clonal Basis of Development, The 36th Symposium of the Society for Developmental Biology* (June 1977) [Raleigh, North Carolina, U.S.A.]. Academic Press, Inc., New York.

Coe. E. H. Jr., M. G. Neuffer, and D. A. Hoisington. 1988. The genetics of corn, pp. 81–258, *In*, G. F. Sprague and J. W. Dudley (Eds.), *Corn and Corn Improvement*, American Society of Agronomy, Inc. Crop Science Society of America, Inc. Soil Science Society of America, Inc., Madison, WI.

Cohn, N. S. 1969. *Elements of Cytology*, 2nd edition, Harcourt, Brace & World, Inc., New York.

Comfort, N. 2001. *The Tangled Field*, Harvard University Press, Cambridge, MA.

Creighton, H. B. and B. McClintock. 1931. A correlation of cytological and genetical crossing-over in *Zea mays*. *Proceedings of the National Academy of Sciences* Vol. 17 (No. 8, August 15, 1931) [Communicated July 7, 1931], pp. 492–497.

Dunn, L. C. 1965. *A Short History of Genetics, the Development of Some of the Main Lines of Thought, 1864–1939*, McGraw-Hill Book Co., New York.

Duvick, D. N. 1996. William L. Brown July 16, 1913–March 8, 1991. *Biographical Memoirs*, Vol. 70, The National Academies Press, Washington, DC. https://doi.org/10.17226/5406

Engel, L. 1967. *The New Genetics*, Doubleday & Co. Inc. Discus Books paperback ed. 1968. [Manuscript dated September 17, 1962, edited by 1958 Nobel Laureate Edward L. Tatum, following death of author in 1964.]

Fedoroff, N. 1992. Maize transposable elements: A story in four parts, pp. 389–415, *In*, N. Fedoroff and D. Botstein (Eds.), *The Dynamic Genome: Barbara McClintock's Ideas in the Century of Genetics*, Cold Spring Harbor Laboratory Press, Plainview, NY. [See especially "Part IV: The *Spm* element family," pp. 402–415]

Fedoroff, N. V. 1998. The discovery of transposable elements, pp. 89–104, *In*, S.-D. Kung and S.-F. Yang (Eds.), *Discoveries in Plant Biology*, Vol. 1, World Scientific Publishing Co. Pte. Ltd., Singapore. [February 18, 1998]

Fedoroff, N. V. 2013a. McClintock and epigenetics, Chapter 4, pp. 61–70, *In*, N. V. Fedoroff (Ed.), *Plant Transposons and Genome Dynamics in Evolution*, John Wiley & Sons, Inc, Oxford. [First published February 4, 2013]

Fedoroff, N. V. 2013b. McClintock and epigenetics: Insights from changes in phase of transposition activity, pp. 3.34–3.39, *In*, L. B. Kass (Ed.), *Perspectives on Nobel Laureate Barbara McClintock's Publications (1926–1984): A Companion Volume*. https://hdl.handle.net/1813/34897

Fedoroff, N. V. and D. Botstein (Eds.). 1992. *The Dynamic Genome: Barbara McClintock's Ideas in the Century of Genetics*, Cold Spring Harbor Laboratory Press, Plainview, NY.

Fink, G. 1992. Transposable elements (Ty) in yeast, pp. 281–287, *In*, N. Fedoroff and D. Botstein (Eds.), *The Dynamic Genome: Barbara McClintock's Ideas in the Century of Genetics*, Cold Spring Harbor Laboratory Press, Plainview, NY.

Grobman, A., W. Salhuana, R. Sevilla, with P. C. Mangelsdorf. 1961. *Races of Maize in Perú. Their Origins, Evolution and Classification,* Publication 915, National Academy of Sciences, National Research Council, Washington, DC, 374 pp.

Hartman, P. E. and S. R. Suskind. 1969. *Gene Action*, 2nd edition, Prentice Hall, Englewood Cliffs, NJ. [Jointly published in Foundations of Modern Genetics and Foundations of Developmental Biology Series]

Hayes, W. 1968. *The Genetics of Bacteria and Their Viruses*, 2nd edition, John Wiley & Sons, New York.

Herskowitz, I. 1992. Controlling elements, mutable alleles, and mating-type interconversion, pp. 289–297, *In*, N. Fedoroff and D. Botstein (Eds.), *The Dynamic Genome: Barbara McClintock's Ideas in the Century of Genetics*, Cold Spring Harbor Laboratory Press. Plainview, NY.

Herskowitz, I. H. 1962. *Genetics*, Little, Brown and Company, Inc., Boston, MA.

Herskowitz, I. H. 1965. *Genetics*, 2nd edition, Little, Brown and Company, Inc., Boston, MA.

Herskowitz, I. H. 1973. *Principles of Genetics*, Macmillan Publishing Company, New York.

Jacob, F. 1958. Genetic Control of Viral Functions [Lecture delivered September 18, 1958, at the Harvey Society, New York]. Part of: Sydney Brenner Collection, Reference: SB/2/3/91. https://wellcomecollection.org/works/e895knvt

Jacob, F. 1960 [1958]. Genetic Control of Viral Functions [Lecture delivered September 18, 1958], pp. 1–39, In, *The Harvey Lectures 1958–1959*, Series LIV [54]. Academic Press, The Harvey Society, New York [January 1, 1960].

Jacob, F., D. Perrin, C. Sanchez, and J. Monod. 1960. L'opéron: groupe de gènes à expression coordonnée par un opérateur, *In, Comptes rendus de l'Académie des Sciences*, Paris, Vol. 250 (No. 9, February 29, 1960), pp. 1727–1729. [English translation in E. A. Adelberg 1960, *Papers on Bacterial Genetics*, Little Brown & Co., pp. 395–397.]

Jacob, F. and E. L. Wollman. 1958a. Genetic and physical determinations of chromosomal segments in *Escherichia coli. Symposia of the Society for Experimental Biology* No. 12, pp. 75–92. [December 1957, Published 1958]; [Reprinted in E. A. Adelberg (Ed.) 1960, *Papers on Bacterial Genetics*, Little Brown & Co, pp. 335–352]

Jacob, F. and E. L. Wollman. 1958b. Les épisomes, elements génétiques ajoutés. *Comptes rendus Hebdomadaires des Seances de L'academie des Sciences* Vol. 247 (No. 1, July 7, 1958), pp. 154–156. [English translation in E. A. Adelberg (Ed.) 1960, *Papers on Bacterial Genetics*, Little Brown & Co, pp. 398–400]

Jacob, F. and J. Monod. 1959. Gènes de structure et gènes de régulation dans la biosynthèse des protéines. *Comptes rendus de l'Académie des Sciences de Paris* Vol. 249 (No. 3, October 5, 1959), pp. 1282–1284. [Genes of structure and genes of regulation in the biosynthesis of proteins.]

Jacob, F. and J. Monod. 1961a. Genetic regulatory mechanisms in the synthesis of proteins. Review article. *Journal of Molecular Biology* Vol. 3 (No. 3, June 1, 1961), pp. 318–356.

Jacob, F. and J. Monod. 1961b. On the regulation of gene activity. *Cellular Regulatory Mechanisms, Spring Harbor Symposia on Quantitative Biology* Vol. 26 (June 4–12, 1961), pp. 193–211. https://doi.org/10.1101/SQB.1961.026.01.024 [©1962]

Jacob, F., P. Schaeffer, and E. L. Wollman. 1960. Episomic elements in bacteria, pp. 67–91, *In*, W. Hayes and R. C. Clowes (Eds.), *Microbial Genetics, Tenth Symposium of the Society for General Microbiology*, April 1960. Cambridge University Press, Cambridge.

Jacob, F. and E. L. Wollman. 1961. Viruses and genes. *Scientific American*. Vol. 204 (No. 6, June 1, 1961), pp. 93–107.

Judson, H. F. 1979. *Eighth Day of Creation*. Touchstone Book, Simon and Schuster, New York. [Based on interviews and memories, "used the voices of the scientists."]

Kass, L. B. 2005. Harriet B. Creighton: Proud botanist. *Plant Science Bulletin* Vol. 51 (No. 4, Winter), pp. 118–125. Available online, December 2005. https://www.botany.org/psbarchive/issue/2005-v51-no-4.html

Kass, L. B. 2013. Cornell University: the McClintock Years, 1919–1939 and beyond, Department of Plant Biology Centennial Celebration, June 29, 2013, Invited Symposium Presentation; Plant Science Building, College of Agriculture and Life Science, Cornell University; video recording. https://www.ecommons.cornell.edu/handle/1813/33663

Kass, L. B. (Ed.). 2013ff. *Perspectives on Nobel Laureate Barbara McClintock's Publications (1926–1984): A Companion Volume*. Volumes I & II The Internet-First University Press. URI: https://hdl.handle.net/1813/34897 [Online December 30, 2013, updates March 2014; Vol. III, Increments No. 1–3, May 2016; Vol. IV, Increment No. 4, April 2020]

Kass, L. B. and C. Bonneuil. 2004. Mapping and seeing: Barbara McClintock and the linking of genetics and cytology in maize genetics, 1928–1935, pp. 91–118, *In*, J.-P. Gaudilliere and H.-J. Rheinberger (Eds.), *Classical Genetic Research and Its Legacy: The Mapping Cultures of 20th Century Genetics*, Routledge, London. [Routledge Studies in the History of Science, Technology and Medicine]

Kass, L. B. and E. Cobb. 2007. Landmarks and milestones in American plant biology, the Cornell connection. *Plant Science Bulletin* Vol. 53 (No. 3, fall), pp. 90–101). Available online, September 2007. https://botany.org/psbarchive/issue/2007-v53-no-3.html

Kass, L. B. and P. Chomet. 2009. Barbara McClintock, pp. 17–52, *In*, J. Bennetzen and S. Hake (Eds.), *Handbook of Maize: Genetics and Genomics*, Springer, New York.

Kato Y., T. A. 1964. *Some Cytological Studies on several Mexican Races of Maize*. (Under the direction of Barbara McClintock, Lyle Llewellyn Phillips, major advisor). North Carolina State College, Raleigh, NC. (Master's Thesis, Department of Genetics), pp. 1–66.

Kato Y., T. A. 1975. Cytological studies of maize (*Zea mays* L.) and teosinte (*Zea mexicana* (Schraeder) Kuntze) in relation to their origin and evolution, University of Massachusetts, Amherst [May 1975, Advisor Walton C. Galinat], 253 pp. Doctoral Dissertation. https://scholarworks.umass.edu/dissertations_1/5931

Kato Y., T. A. 1976. Cytological studies of maize (*Zea mays* L.) and teosinte (*Zea mexicana* (Schrader) Kuntz) in relation to their origin and evolution. *Massachusetts Agricultural. Experiment Station. Bulletin* Vol. 635, 185 pp.

Kato Y., T. A. 2013. Significance of McClintock's pattern concept in the analysis of chromosome knob distributions in the races of maize, pp. 4.39-4.49, *In*, L. B. Kass (Ed.) 2013ff. *Perspectives on Nobel Laureate Barbara McClintock's publications (1926-1984): A Companion Volume*. https://hdl.handle.net/1813/34897

Keller, E. F. 1983. *A Feeling for the Organism. The Life and Work of Barbara McClintock*, 1st ed., W. H. Freeman, San Francisco, CA.

Keller, E. F. 1993. *A Feeling for the Organism. The Life and Work of Barbara McClintock*, 10th Anniversary edition, W. H. Freeman, New York.

Kessler, M. 1969. White Plains student at Cornell Tells Story of Events, *The Reporter Dispatch*, White Plains, NY, Wednesday, April 30, 1969, p. 41.

King, R. C. 1962. *Genetics*, Oxford University Press, Inc., New York.

Lewis, K. R and B. John. 1964. *The Matter of Mendelian Heredity*, J. & A. Churchill, London.

Longley, A. E. and T. A. Kato Y. 1965. *Chromosome Morphology of Certain Races of Maize in Latin America*. International Center for the Improvement of Maize and Wheat [CIMMYT]. Chapingo, State of Mexico, Mexico, Research Bulletin No. 1.

Mangelsdorf, P. C. and R. G. Reeves. 1939. The origin of Indian corn and its relatives. *Texas Agricultural Experiment Station Bulletin* 574. 315 pp.

McClintock, B. 1931a. The order of the genes *C*, *Sh*, and *Wx* in *Zea mays* with reference to a cytologically known point in the chromosome. *Proceedings of the National Academy of Sciences* Vol. 17 (No. 8, August 15, 1931) [Communicated July 7, 1931], pp. 485–491.

McClintock, B. 1931b. Cytological observations of deficiencies involving known genes, translocations and an inversion in *Zea mays*. *University of Missouri, College of Agriculture, Agricultural Experiment Station, Research Bulletin* 163 [December, authorized 23 December 1931], pp. 1–30.

McClintock, B. 1950. The origin and behavior of mutable loci in maize. *Proceedings of the National Academy of Sciences* Vol. 36 (No. 6, June 15, 1950, Communicated April 8, 1950), pp. 344–355.

McClintock, B. 1951 [©1952, Symposium held June 1951]. Chromosome organization and genic expression, pp. 13–47, *In*, K. B. Warren (Ed.), *Genes and Mutations*, *Cold Spring Harbor Symposia on Quantitative Biology*, Volume XVI [June 7–15, 1951], The Biological Laboratory, Cold Spring Harbor, NY.
McClintock, B. 1954. Mutations in maize and chromosomal aberrations in *Neurospora*: Mutations in maize. Chromosome aberrations in *Neurospora. Carnegie Institution of Washington Year Book No. 53*, 1953–1954, pp. 254–260. [Issued December 10, 1954]
McClintock, B. 1956 [Symposium held June 15–17, 1955, Issued February 1956]. Intranuclear systems controlling gene action and mutation, pp. 58–74, *In*, *Mutation, Brookhaven Symposia in Biology*, No. 8. Biology Department, Brookhaven National Laboratory, Upton, NY. [R. C. King, Symposium Chairman]
McClintock, B. 1959. Genetic and cytological studies of maize: Further studies of the *Spm* system. Chromosome constitutions of some South American races of maize. *Carnegie Institution of Washington Year Book 58*, 1958–1959, pp. 452–456. [Issued December 14, 1959]
McClintock, B. 1960. Chromosome constitutions of Mexican and Guatemalan races of maize: General Conclusions. *Carnegie Institution of Washington Year Book 59*, 1959–1960, pp. 461–472. [Issued December 12, 1960]
McClintock, B. 1961. Some parallels between gene control systems in maize and in bacteria. *American Naturalist* Vol. XCV [95] (No. 884, September–October 1961), pp. 265–277.
McClintock, B. 1965 [Symposium held June 7–9, 1965]. The control of gene action in maize, pp. 162–184, *In*, *Genetic Control of Differentiation, Brookhaven Symposia in Biology*: No. 18. Biology Department, Brookhaven National Laboratory, Upton, N.Y. [H. H. Smith Chairman of the Symposium Committee; Issued December 1965]
McClintock, B. 1968 [Symposium held June 1967]. Genetic systems regulating gene expression during development, pp. 84–112, *In*, M. Locke (Ed.), *Control Mechanisms in Developmental Processes, II. The Role of the Nucleus*, The 26th Symposium of the Society for Developmental Biology (June 1967) [La Jolla, CA, U.S.A.]. *Developmental Biology*, Supplement 1, Academic Press, New York.
McClintock, B. 1978 [Symposium held June 13–15, 1977]. Development of the maize endosperm as revealed by clones, pp. 217–237, *In*, S. Subtelny and I. M. Sussex (Eds.), *The Clonal Basis of Development*, The 36th Symposium of the Society for Developmental Biology (June 1977) [Raleigh, North Carolina, U.S.A.]. Academic Press, Inc., New York.
McClintock, B. 1987a. *The Discovery and Characterization of Transposable Elements*. The Collected Papers of Barbara McClintock. Genes Cells and Organisms; Great Books in Experimental Biology, J. A. Moore, series editor, Garland Publishing Co., New York. [Published December 1, 1987; Currently in the series Genes Cells and Organisms, Routledge, Taylor and Frances Group]
McClintock, B. 1987b. Introduction, *In*, J. A. Moore (Ed.), *The Discovery and Characterization of Transposable Elements. The Collected Papers of Barbara McClintock*, Garland Publishing Co., New York. [Reprinted in Kass 2013ff., https://hdl.handle.net/1813/34897]
McClintock, B., T. A. Kato Y., and A. Blumenschein. 1981. *Chromosome Constitution of Races of Maize. Its Significance in the Interpretation of Relationships Between Races and Varieties in the Americas*. Colegio de Postgraduados, Escuela Nacional de Agricultura, Chapingo, Edo. Mexico, Mexico, Vol. xxxi, 517 pp.
Monod, J. and J. Jacob. 1961. Teleonomic mechanism, growth and differentiation [General Conclusions], pp. 389–401. *Regulatory Mechanism, Cold Spring Harbor Symposia on Quantitative Biology*, June 4–12, 1961, Vol. 26, pp. 389–401. [©1962]
Monteiro, W. 1964. *Knob Structures in Maize*. (Under the direction of Barbara McClintock, Frank Lloyd Haynes, Jr. Major Advisor). Master's Thesis, Department of Genetics, North Carolina State College, Raleigh, NC.
Murphy, R. P. and L. B. Kass. 2007. *Evolution of Plant Breeding at Cornell University: A Centennial History, 1907–2006*, pp. 1–98, Department of Plant Breeding & Genetics, Cornell University, Ithaca, NY, Appendices A1–A98, Photo Section P1–P38. (July 2007).
Murphy, R .P. and L. B. Kass. June 27, 2011. *Evolution of Plant Breeding at Cornell University: A Centennial History, 1907-2006*. Rev. edition i–x, 178 pp., Appendices, Photo Section. e-book, The Internet-First University Press, Ithaca, NY. https://hdl.handle.net/1813/23087
Nilan, R. A. 1973. Increasing the effectiveness, efficiency, and specificity of mutation induction in flowering plants, Chapter 14, pp. 205–222, *In*, A. M. Srb (Ed.), *Genes, Enzymes, and Populations*, Plenum Press, New York.
Papazian, H. P. 1967. *Modern Genetics*, W. W. Norton & Co. Inc., New York.

President Richard Nixon's Daily Diary. Friday, May 21, 1971. The President ... went to the East Room to present the National Medal of Science Awards. For a list of recipients ... see APPENDIX "B"; WHCF: SMOF: Office of Presidential Papers and Archives RC-7. https://www.nixonlibrary.gov/sites/default/files/virtuallibrary/documents/PDD/1971/052%20May%2016-31%201971.pdf

Provine, W. B. and P. Sisco. 1980. Interview with Barbara McClintock at Cold Spring Harbor Laboratory, Long Island, New York, August 28, 1980. Division of Rare and Manuscript Collections, Cornell University Library, Ithaca, NY.

Ramírez, E. R., D. H. Timothy, E. Diaz B., U. J. Grant, in collaboration with G. E. N. Calle, E. Anderson, and W. L. Brown. 1960. *Races of Maize in Bolivia*, Pub: 747, NAS-NRC, Washington, DC.

Report of the President. 1962. *Carnegie Institution of Washington Year Book 61*, July 1, 1961–June 30 1962, Sixtieth Anniversary, pp. 1–74 [Issued December 10, 1962).

Report of the President. 1971. *Carnegie Institution of Washington Year Book 70*, 1970–1971, pp. 1–48.

Rhoades, M. M. 1938. Effect of the *Dt* gene on the mutability of the *a1* allele in maize. *Genetics* Vol. 23 (No. 4, July 1, 1938), pp. 377–397. https://doi.org/10.1093/genetics/23.4.377

Rhoades, M. M. 1941. The genetic control of mutability in maize, pp. 138–144, *In*, K. S. Brehme (Ed.), *Genes and Chromosomes - Structure and Organization, Cold Spring Harbor Symposia on Quantitative Biology*, Vol. IX [9], The Biological Laboratory, Cold Spring Harbor, NY.

Rhoades, M. M. 1955. The cytogenetics of maize, pp. 123–219, *In*, G. F. Sprague (Ed.), *Corn & Corn Improvement*, Academic Press Inc., New York.

The Rockefeller Foundation Annual Report 1961. 1961. The Rockefeller Foundation, New York.

The Rockefeller Foundation Annual Report for 1962. 1962. The Rockefeller Foundation, New York.

Sager, R and F. J. Ryan. 1961/1962. *Cell Heredity,* John Wiley & Sons, Inc. New York [1961, reprinted May 1962 with Postscript]

Science Citation Index. 1984. *SCI, Science citation index,* 1955–1964, ten year cumulation: an international interdisciplinary index to the literature of science, medicine, agriculture, technology, and the behavioral sciences. 12 Volumes. ISI, Institute for Scientific Information, Philadelphia, PA.

Science Citation Index. 1988. *SCI, Science citation index*, 1945–1954, ten year cumulation: an international interdisciplinary index to the literature of science, medicine, agriculture, technology, and the behavioral sciences. Ten Volumes. ISI, Institute for Scientific Information, Philadelphia, PA.

Singleton, W. R. 1962. *Elementary Genetics*, D. Van Nostrand Co. Inc., Princeton, NJ.

Singleton, W. R. 1967. *Elementary Genetics*, Revised 2nd edition, Van Nostrand-Reinhold Publishers, Princeton, NJ. [Issued January 1, 1967].

Sinnott, E. W. and L. C. Dunn. 1932. *Principles of Genetics*, 2nd edition, McGraw-Hill Book Co., New York, xvi+441 pp.

Sinnott, E. W. and L. C. Dunn. 1939. *Principles of Genetics*, 3rd edition, McGraw-Hill Book Co., New York, xiv+408 pp.

Sinnott, E. W., L. C. Dunn and Th. Dobzhansky. 1950. *Principles of Genetics*, 4th edition, McGraw-Hill Book Co., New York, 505 pp.

Sinnott, E. W., L. C. Dunn and Th. Dobzhansky. 1958. *Principles of Genetics*, 5th edition, McGraw-Hill Book Co., New York, 459 pp.

Spectre, M. 1993. *Barbara McClintock Papers. Guide to the Collection*. American Philosophical Society Library, Philadelphia, PA. (Online: Barbara McClintock Papers Mss.Ms.Coll.79, https://search.amphilsoc.org/collections/view?docId=ead/Mss.Ms.Coll.79-ead.xml) [Also see selected papers and summary of APS Library papers of McClintock available for the National Library of Medicine, Profiles in Science: https://profiles.nlm.nih.gov/spotlight/ll; The collection also reprints papers/letters from other institution's archives]

Sprague, G. F. (Ed.). 1955. *Corn & Corn Improvement*, Academic Press Inc., New York, 699 pp.

Srb, A. M. (Ed.). 1973. *Genes, Enzymes and Populations*. Proceedings of the Twelfth International Latin American Symposium held in Cali, Colombia, November 27–December 1, 1972. Plenum Press, New York.

Srb, A., R. D. Owen, and R. S. Edgar, 1965. *General Genetics*, 2nd edition, W. H. Freeman & Co., San Francisco, CA.

Strathern, J. N. 1992. Thinking about programmed genome rearrangements in a genomic static state of mind, pp. 299–314, *In*, N. Fedoroff and D. Botstein (Eds.), *The Dynamic Genome: Barbara McClintock's Ideas in the Century of Genetics*, Cold Spring Harbor Laboratory Press, Plainview, NY.

Strickberger, M. W. 1968. *Genetics,* The Macmillan Co., New York.

Sussman, M. 1964. *Growth and Development*, 2nd edition, Prentice-Hall, Inc., Englewood Cliffs, NJ. [Foundations of Modern Biology Series]

Swanson, C. P. 1957. *Cytology and Cytogenetics*, Prentis Hall, Inc., Englewood Cliffs, NJ. [Sixth printing 1964]

Swanson, C. P., T. Mertz, and W. J. Young. 1967. *Cytogenetics*, Prentice Hall, Inc., Englewood Cliffs, NJ. [Foundations of Modern Genetics Series]

Swanson, C. P., T. Mertz, and W. J. Young. 1981. *Cytogenetics: The Chromosome in Division, Inheritance, and Evolution*, 2nd edition, Prentice Hall, Inc. Englewood Cliffs, NJ.

Teas, H. (Ed.). 1969. *Genetics and Developmental Biology: Thomas Hunt Morgan Centennial Symposium* (1966). University of Kentucky Press, Lexington.

Timothy, D. H., B. Peña, R. Ramirez, in collaboration with W. L. Brown, and E. Anderson. 1961. *Races of Maize in Chile*. Pub 847, NAS-NRC, Washington, DC.

Timothy, D. H., W. H. Hatheway, U. J. Grant, C. Manuel Torregroza, V. Daniel Sarria, and A. Daniel Varela. 1963. *Races of Maize in Ecuador*. Pub 975, NAS-NRC, Washington, DC.

Waddington, C. H. 1962. *New Patterns in Genetics and Development*, Columbia University Press, New York and London. [Based on six Jesup Lectures given in May/June 1961 at Columbia University]

Wagner, R. P. and H. K. Mitchell. 1955. *Genetics and Metabolism*, John Wiley and Sons, Inc., New York.

Wallace, B. 1963. The annual invitation lecture. Genetic diversity, genetic uniformity, and heterosis. *Canadian Journal of Genetics and Cytology* Vol. 5 (No. 3, September 1963), pp. 239–253.

Wallace, B. 2001. (in manuscript). *On the Fringe of Glory: The Cold Spring Harbor Laboratories of the 1950s*. [Unpublished manuscript sent to W. B. Provine in May 2001 for review. Provine was asked to pass the document on to L. B. Kass and R. P. Murphy. The manuscript is now in the possession of Kass. A similar document is in the Wallace Papers at APS Library, Preface dated, July 2001.]

Williams, W. 1964. *Genetical Principles and Plant Breeding*. [Botanical Monographs Series] Blackwell Scientific Publications, Oxford.

Wollman, E. L., F. Jacob, and W. Hayes. 1956. Conjugation and genetic recombination in *Escherichia coli* K-12. *Cold Spring Harbor Symposia on Quantitative Biology, 1956, Genetic Mechanisms: Structure and Function* Vol. 12, pp. 141–162. [©1957]

10 Golden Age of Corn Genetics
Reprise

If one cannot be famous, fame might come from association with the famous

– Abe Eisenstark

THE NAME GAME: RECONCILING PROKARYOTIC AND EUKARYOTIC GENETIC SYSTEMS

In the years following her mandatory retirement at age 65 (1967) and the official closing of the Carnegie Institution's Genetics Research Unit (1974), McClintock worked at Cold Spring Harbor Laboratory.[1] She continued to be a resource person for the maize genetics community (Kass et al. 2019; Spectre 1993: Series I, Correspondence). In the mid-1970s, geneticists and molecular biologists searched for the underlying mechanism of gene action in eukaryotic organisms, including the maize plant. Concurrently, genetic elements resembling maize controlling elements were identified in prokaryotic organisms.[2]

As mentioned in the previous chapter, molecular biologists were mindful of McClintock's findings of controlling elements in maize, although they did not know the underlying molecular mechanism. As early as 1963, A. L. Taylor had reported a bacteriophage, *Mu1*, that infected *E. coli* and had its function disrupted by an insertion. He likened the ability of the phage to occupy many chromosomal sites and to suppress gene expression with the "controlling elements" of maize. His system more closely resembled maize elements than it did bacterial episomes. The bacterial phage element was similar to maize elements in that (1) both move from one position to another in the chromosome, and (2) both elements are located close to a phenotypically recognizable gene and modified its expression, sometimes completely suppressing its normal function (Taylor 1963).

Hayes (1968, pp. 221–222) pointed out that Dawson and Smith-Keary (1963) had observed instability in the *Salmonella typhimurium* bacterial system, which they believed was similar to the "controlling elements" that McClintock had observed in maize. They observed that the factor responsible for instability was shifting from one locus to another, which they called a "controlling episome," to avoid a commitment to identify their elements with those observed in maize. Hayes explained that McClintock's elements "jump" from one chromosomal locus to another. Thus, they modify the stability and expression of genes to which they are temporarily attached. He, however, warned his readers not to confuse the term controlling elements with the word "episome." The latter, he wrote, specifically defined a group of genetic elements whose "principle feature is the ability to adopt alternative cytoplasmic or chromosomal states in the cell" [Jacob & Wollman 1958]. There was no evidence in the *Salmonella* system, Hayes warned, of a physical constitution or size for the "controlling elements," nor was there evidence for an independent cytoplasmic existence [as defined by the term episome]. Further, he noted that the controlling elements described in the *Salmonella* system may turn out to be of a different nature than that of the episome as originally defined (Hayes 1968, p. 222; see Chapter 9).

---

[1] *Annual Reports 1974ff., Cold Spring Harbor Laboratory.*

[2] Starlinger and Saedler (1972, 1976), Starlinger et al. (1974), Fincham and Sastry (1974, pp. 42–43), and Bukhari (1976).

DOI: 10.1201/9781003332527-11

FIGURE 10.1 David B. Walden and Richard I. Greyson, University of Western Ontario, at the 1975 International Symposium on Maize Breeding and Genetics held at the University of Illinois, Champaign, Urbana. Walden organized and chaired the Symposium. (Courtesy of E. H. Coe & D. B. Walden.)

Relying heavily on McClintock's Carnegie Institution Reports, which she had sent to John R. S. Fincham,[3] he and G. R. K. Sastry summarized McClintock's "Controlling Elements in Maize" for the 1974 *Annual Review of Genetics*. They pointed out analogies found in prokaryotic systems such as provirus, plasmids, or transposable insertion sequence of DNA, having "a property of a controlling element." The analogies concern bacterial prophages, and the most interesting of which was *Mu-1*. Insertion sequences not associated with virus production provided an even closer analogy. The controlling elements of maize and prokaryotic systems shared possible mechanisms: 1) integration and inhibition of gene action, 2) excision of the blocking element and return to partial or full function, 3) regulation of gene activity, and 4) changes in state of the receptor of controlling elements and of bacterial phage. Molecular DNA sequencing in eukaryotes would be advantageous in demonstrating the commonality with prokaryotic systems. By 1978, such studies would begin in maize applying molecular techniques (Fedoroff & Mauvais 1979; Starlinger 1980; Burr & Burr 1980, 1981).

In 1975, at the International Symposium on Maize Breeding and Genetics, McClintock (1978a) presented an update of recent work on races of maize in Latin America that she and others were continuing to pursue (Goodman 1978a,b). McClintock was a member of the Advisory Committee for the Symposium held on September 8–12, 1975, at the University of Illinois, Champaign, Urbana. Professor David B. Waldon of the University of Western Ontario organized the meeting in honor of the "75th anniversary of the rediscovery of Mendelism" (Walden 1978, p. vii; Figure 10.1). Additionally, George Beadle (1978) and Walton Galinat (1978) presented evidence for the origin of maize (Figures 10.2 and 10.3). Concomitantly, at this 1975 Symposium, researchers communicated current work on induction of maize mutability (Peterson 1978; Neuffer 1978), gene action (Kermicle 1978; Schwartz 1978), and the usefulness of maize endosperm tissue for chemical analysis and the correlation of mutation with phenotypic analysis of development (Coe 1978, Figure 10.4; Robertson et al. 1978). Relevant to his studies on maize regulation, Peterson (1978) reminded his audience that

[3] John Fincham to B. McClintock, August 7, 1973, Spectre 1993: Series I, File, Fincham, John, 1973.

FIGURE 10.2 George W. Beadle, Denton E. Alexander, Merle T. Jenkins, and Barbara McClintock, attending the 1975 International Symposium on Maize Breeding and Genetics, organized by David B. Walden and held at the University of Illinois, Champaign, Urbana. Alexander was Symposium Executive Secretary. (Courtesy of E. H. Coe & D. B. Walden.)

FIGURE 10.3 Field trip with attendees and speakers at the International Symposium on Maize Breeding and Genetics, 1975, organized by David B. Walden and held at the University of Illinois, Champaign, Urbana. Walton C. Galinat with hat and camera bag is seated in the front row. (Courtesy of E. H. Coe & D. B. Walden.)

FIGURE 10.4 Edward H. Coe and Ron L. Phillips examining maize plants during the International Symposium on Maize Breeding and Genetics, 1975, organized by David B. Walden and held at the University of Illinois, Champaign, Urbana. (Courtesy of E. H. Coe & D. B. Walden.)

maize controlling elements were previously compared to "similar phenomena in prokaryotes ..., specifically IS [Insertion] elements and bacteriophage."

Indeed, in 1972 Starlinger and Saedler wrote, "Nothing is known at present [February 1972] at the molecular level about the occurrence of insertion mutation in higher organisms. It should be recalled, however, that McClintock established in *Zea mays* the presence of elements, which can be associated with and removed from genes, which thereby loose [*sic*] or change their function (McClintock 1956, 1961). McClintock discussed the possibility that episomes might be responsible for these observations. It would be very interesting to see whether these hypothetical episomes bear some resemblance to the insertion mutation described in microorganisms." He suggests that viruses growing in cells of higher organisms might be helpful for looking into the question. Although McClintock (1961) did recognize that controlling elements might be comparable to episomes in bacteria, she additionally compared the regulator-gene-operator system in bacteria with the dual system of control in maize (controlling elements; see Chapter 9).

Even though the maize and bacterial elements were not shown to be identical, Peterson suggested that they might provide a possible molecular model for the maize controlling element system. With current advances in plant DNA technology, he predicted that it should not be long before molecular studies would provide an "explanation of controlling and regulatory elements" in maize (Peterson 1978, p. 628). Similar to the isolation and identification of bacterial mobile genetic elements (Starlinger & Saedler 1976), maize controlling elements would eventually be cloned and identified, but reservations of their role in development remained doubtful (Starlinger 1984). In the conclusion of their 1976 review of IS-elements in microorganisms, Starlinger and Saedler wrote, "It will be interesting to see whether the newly developed techniques for the biochemical study of eukaryote chromosomes will be sufficient to characterize controlling elements in higher organisms biochemically and elucidate their role in the life cycles of these organisms."

In 1976, a summer meeting on "DNA Insertions" held at Cold Spring Harbor brought together the researchers in this field (Bukhari et al. 1976, p. 40; DNA Insertions 1976). The Director (Watson 1976, p. 12) reported that the meeting "organized by Amad Bukhari was on DNA insertion sequences, jumping genetic elements, the first of which was found by Barbara McClintock here some twenty years ago but whose universal significance has only come to be appreciated recently." Bukhari et al. (1976), while actively studying the bacterial phage Mu, had found that Mu deviates from the classic Campbell model for [lambda phage] integration. They proposed that the form of DNA active in integrative recombination is generated by the replication of Mu DNA. They thought this type of mechanism may be shared by transposable elements in general. Mu might represent a class of elements, different from most temperate phages, in which replication of the inserted molecule is a necessary step in the process of integration. They proposed that Mu is a representative of transposable elements. Their findings were consistent with the suggestion by Taylor (1963) mentioned above. McClintock prepared a lecture for this meeting on "DNA Insertions" (McClintock 1976, Abstract)[4], and the current topic was reviewed in *Science* (Kolata 1976).

It seems that the term "jumping gene" to describe transposable elements was first published in the summer of 1976 by Gina Kolata in a *Science* magazine Research News feature (Kolata 1976). Kolata, at the time a journalist at *Science* magazine and currently a *New York Times* health and science writer, did not recall why she used that term or where she might have heard it.[5] However, her news summary, titled "Jumping genes: a common occurrence in cells," refers the reader to a forthcoming book by Bukhari, Shapiro, and Adhya, who had organized the meeting on DNA Insertions at Cold Spring Harbor in May 1976. Published the following year, their book (Bukhari et al. 1977) includes contributions from the meeting's participants and an introductory chapter on terminology accurately attributes the discovery of eukaryotic transposable elements to McClintock. The term "jumping gene" is not used in their volume (and Kolata did not recall attending the meeting), but it became a popular way of describing transposable elements, originally called controlling elements by McClintock (1956a). James Shapiro, Professor at the University of Chicago, did not know when the term was first used.[6] To his knowledge, McClintock did not use it and he recalled that she also did not like using the word "gene." He reminded me that it "wasn't genes that were jumping." Sankar Adhya, Senior Investigator at the National Cancer Institute in Bethesda, MD, who also co-arranged the 1976 DNA Insertions meeting recalled that Kolata did not attend their event, but she had interviewed him by telephone following the Cold Spring Harbor meeting. His accurate recollection is that Kolata used the term "Jumping Gene" in a piece in *Science* sometime in the 70s.[7] Sydney Brenner, from the University of Cambridge, England, was invited to provide the summary at the meeting (Bukhari et al. 1976, p. ix, 1977, p. xiii), and in his detailed notes for each speaker, there is no mention of the term "Jumping Gene" (Brenner 1976), but he did note that "genes can hop." His notes also confirm that the published abstract book does not reflect all of the speakers who ultimately attended and presented talks at the meeting. For example, an abstract for the presentation by *Drosophila* geneticist Mel Green, who McClintock specifically asked (demanded) they invite to the meeting after learning that he was not on the program,[8] is absent from the Abstract book. Confirmation that Green indeed gave a presentation is found in Brenner's notes and in the chapter that he contributed to the published volume (Brenner 1976; Green 1977).

McClintock was an advisor for the DNA Insertions meeting, and she gave a talk titled "Transposable chromosomal elements in maize" in the session on "The Eukaryotic Systems"

---

4 Spectre 1993: Series III, Lecture "Transposable Chromosomal Elements in Maize," May 20, 1976.

5 Emails between L. B. Kass & Gina Kolata, *New York Times* senior writer who covers science and medicine, March 18–20, 2008.

6 Email James Shapiro to L. B. Kass, September 7, 2017.

7 Email Sankar Adhya to L. B. Kass, July 30, 2018.

8 L. B. Kass phone interview with S. Adhya, July 27, 2018.

(Bukhari et al. 1976, pp. vii; McClintock 1976; Brenner 1976).[9] Peterson spoke about "The position hypothesis for controlling elements in maize" and contributed a chapter to the published volume (Peterson 1976, 1977; Brenner 1976), and other researchers spoke about apparent controlling elements and insertions in yeast, *Drosophila*, and viruses in humans.

Brenner's (1976, pp. 1–2) unpublished meeting summary cautioned about hypothesizing functions or equivalence of eukaryotic insertion elements to those in prokaryotic systems:

> One must be careful if one wants to use the transposition elements for regulatory mechanisms. ... Concerning eukaryotic systems it can be pointed out in general, though, that the phenomena in maize or *Drosophila* cannot be a priori assumed to confirm the presence of insertion elements in these systems. The transposition elements in these systems have to be proven in their own right.

By the fall of 1976, Professor Stanley N. Cohen (1976) wrote a review article about prokaryotic transposable elements published in the journal *Nature*. Cohen had attended and was an advisor to the Cold Spring Harbor DNA Insertions meeting. He and colleagues had reported on "Promotion of insertions and deletions by translocating segments of DNA carrying antibiotic resistance genes" (Brevet et al. 1976, p. 22).[10] His review in *Nature* summarized the status of "Transposable genetic elements and plasmid evolution," and defined the term "Transposable Element" as established by a committee formed at the DNA Insertions meeting: "Transposable elements are DNA segments which can insert into several sites in a genome" (Cohen 1976, p. 374). The full report was "in press" in the forthcoming publication *DNA Insertion Elements, Plasmids, and Episomes*. It was subsequently published by the committee (Campbell et al. 1977, p. 16), who clarified that "the term 'transposable element' is derived from the work of McClintock (1952), on mobile genetic elements in maize," thus referring to her 1951 Cold Spring Harbor Symposium publication (McClintock 1951). As mentioned in previous chapters, McClintock first used the term "transposable element" in a Carnegie Institution report describing the behavior of the *Spm* element of maize (McClintock 1955, p. 250).

Cohen was a bit more optimistic than was Brenner in his conclusions about correlating eukaryotic systems with those of prokaryotic organisms. He drew a parallel between the phenomena observed by the movement of transposable "controlling" elements that regulate the phenotypic expression in maize with the similar effects of transposable genetic elements in plasmids. Movable DNA segments may have a role in the organization and control of expression in both plasmids and eukaryotic organisms, he conjectured:

> Analogies between phenomena observed in eukaryotic systems and the transposable genetic elements ... identified on plasmids are presently only speculative: nevertheless, bacterial plasmids may prove to be a highly useful model system for elucidating mechanisms involving the organization and control of genetic information on the complex chromosomes of higher organisms.
>
> *(Cohen 1976, p. 737)*

Controlling elements would soon be identified at the molecular level. Within two years, the nomenclature committee widely published an expanded definition for transposable elements in prokaryotes (Campbell et al. 1979a,b): "Transposable elements are specific DNA segments that can repeatedly insert into a few or many sites in a genome. They are classified as simple IS elements, more complex Tn transposons, and self-replicating episomes." Mobile genetic elements such as IS elements (insertion elements) were termed transposable elements or transposons and were likened to McClintock's controlling elements in contemporary textbooks of the mid-1970s (e.g., Herskowitz 1977, p. 537) and early 1980s [e.g., Avers 1980, pp. 362–363; Swanson et al. 1981, p. 490] and in popular science magazines of the day (Cohen & Shapiro 1980). Furthermore, The Rockefeller University and

[9] Bukhari et al. (1977), advisor, p. xiii. Spectre 1993: Series III, Lecture, May 20, 1976, as requested by Bukhari. Brenner (1976) notes on the speakers.

[10] See also Kretschmer and Cohen (1976, p. 26).

Harvard University recognized McClintock with honorary degrees at their June 6 and 7, 1979, commencement ceremonies, respectively (Report of the President 1979, p. 43).

In early 1980, Cohen and Shapiro summarized the significance of "Transposable Genetic Elements" toward an understanding of biological variation (Cohen & Shapiro 1980). These recently identified elements were "structurally and genetically discrete segments of DNA that have the ability to move around among the chromosomes and extra chromosomal DNA molecules of bacteria and higher organisms." Currently developed molecular techniques had revealed how DNA insertions within bacterial chromosomes and their associated plasmids (previously called episomes) could move in the bacterial genome. Such studies were important for public health, especially since antibiotic resistance can be transmitted from one bacterial cell to another by plasmids. The bacterial plasmids carried DNA sequences that included a gene for drug resistance and could move spontaneously. Such prokaryotic bacterial DNA was termed a "transposon" by R. W. Hedges and A. E. Jacob (1974), because they postulated that plasmids carrying extra DNA could be "transposed," meaning the element carrying antibiotic resistance could move from one DNA molecule to another. Transposable genetic elements have subsequently been called transposons.

Cohen and Shapiro (1980, pp. 41–42, 48–49) also reminded readers that transposable elements were originally discovered in plants. It was Barbara McClintock, they state, who first observed movable elements in the late 1940s, while conducting studies on inheritance of corn plants. The variegation that McClintock observed was the result of genetic units that moved from place to place on maize chromosomes, and could serve as biological switches to turn off or turn on the expression of genes. She therefore called these genetic elements "controlling elements."[11] Almost 20 years after McClintock first found movable elements in maize, Shapiro and others studying mutations in the bacteria *E. coli* observed phenotypic expressions that were first hypothesized to be attributable to extra DNA segments that were interrupting the continuity of the gene (Shapiro 1967, Ph.D. Thesis, pp. 94–95; Sherratt 1995; Higgins 1998) and subsequently, using molecular techniques, were demonstrated to be so (Shapiro 1968 unpublished;[12] Jordan et al. 1968; Shapiro 1969, 1995, 2009; Shapiro & Adhya 1969). These were named Insertion Sequences (IS).

Genetic studies in eukaryotic organisms seemed to provide evidence that the prokaryotic movable elements were similar to those in higher organisms. The pioneering work of McClintock on the movement of the controlling elements seemed to be analogous to transposons of prokaryotic organisms. Similar to prokaryotic systems, McClintock had shown that some controlling elements regulate the expression of distant genes carrying insertions of other controlling elements. By 1980, in addition to maize, mutations in *Drosophila* and yeast were also recognized as due to "controlling elements" (McClintock 1980; Scott et al. 1980).

In 1978, Nina Fedoroff, working at the Carnegie Institution of Washington (CIW), Department of Embryology, as a newly hired full-time staff member, began studies of McClintock's transposable elements at the molecular level. Her Department Director, Don Brown, reported that Fedoroff's "long-term goal is to describe in molecular terms the transposable elements known from Barbara McClintock's work to effect gene expression in maize" (Report of the Director 1978). Around the same time, Evelyn Fox Keller began interviewing McClintock about her life and work (Keller 1981, 1983; Wynants 2005; Keller 2014).[13] By 1983, Keller published the first popular biography of McClintock's life, considerably relying on McClintock's memories (Keller 1983, 2014; Kass 2003; see Chapter 11).

James D. Ebert, President of the CIW, reported that Nina Fedoroff planned to study Barbara McClintock's "Controlling Elements" at the DNA level (Report of the President 1979, pp. 36–37). In

[11] As noted above and in previous chapters, McClintock introduced the term at the 1955 Brookhaven Symposium and further clarified the phenomena at the 1956 Cold Spring Harbor Symposia (McClintock 1956a,b). See also Fedoroff (1983, p. 2, 1989b, p. 377).

[12] See Shapiro (1992, p. 214), "Early in 1968, positive results showed the idea was correct." Shapiro's results were submitted for publication in August 1968, but appeared in print in February 1969.

[13] Wynants (2005) and Keller (2014) report Keller's first interview with McClintock as 1977.

her first report as a staff member, she outlined her plans for studying controlling elements in maize (Fedoroff & Mauvais 1979). Ebert projected the difficulty:

> Her challenge will be to translate deductions made previously by cytogeneticists into molecular terms. The genetic systems used by Barbara McClintock to discover and describe controlling elements resulted in visible changes in the plant. The biochemical functions of these genes were not well understood. With the help of McClintock, Fedoroff hopes to produce strains of maize in which controlling elements regulate genes that are amenable to study by biochemical methods.

Fedoroff was collaborating with McClintock and with maize geneticists at the University of Missouri. She had spent time at Missouri learning how to grow and harvest maize for her forthcoming molecular studies of maize controlling elements. Professor Ed Coe recalled that she had visited Missouri in 1977 at the end of pollinating season, and in the summer of 1978 at McClintock's suggestion, she went to Missouri to learn field techniques.[14] They both recalled that in 1978 during shoot bagging time, a severe easterly storm/tornado blew plants down and Fedoroff had to rescue important plants – often as broken tassels (Fedoroff 1992, p. 401). During the summer of 1979, she worked with McClintock and Drs. Ben and Frances Burr at Brookhaven National Laboratory, where McClintock had again been appointed a guest geneticist.[15] Between 1980 and 1982 with the collaboration of technicians, postdoctoral researchers, and Brookhaven colleagues, she and her associates at the Carnegie Department of Embryology reported on the molecular techniques they had devised for studies of maize genes associated with controlling element mutations. The details are provided in the *Carnegie Institution of Washington Year Books 78* to *81* (Fedoroff & Mauvais 1979; Fedoroff et al. 1980; Chaleff et al. 1981; Fedoroff et al. 1982) and were summarized elsewhere (Fedoroff 1982).[16] The director of their department provided two reasons why movable genetic elements intrigue a developmental biologist: "First, they cause mutations and thus provide a rich source of alleles with altered gene function. Second, their special ability to move from one genetic location to another invites a detailed explanation in molecular terms, and a solution might itself suggest a powerful tool for directing genetic change" (Report of the Director 1981, p. 111). By 1983, maize controlling elements would be isolated and characterized (see Chapter 11).

In January 1980, McClintock gave an invited symposium presentation, "Modified Genic Expression by Transposable Elements" at the Miami Winter Symposium on *Mobilization and Reassembly of Genetic Information* (McClintock 1980). The purpose of her discussion was to stress the potential of controlling elements for redirecting gene activities, which she had recently reported in her invited Stadler Genetics Symposium (1978b), which emphasized mechanisms for rapidly reorganizing the genome. However, she also reminded her audience that "molecular organization of transposable CEs in maize is unknown" (McClintock 1980, p. 16). Citing many examples, she noted that the mechanism that activates silent controlling elements in maize was also suspected in other higher organisms, as others at the symposium would describe. Stabilization of the genome after reorganization by controlling elements "could give rise to new species or even new genera," she concluded.

Unaware that others had earlier proposed Barbara McClintock for the 1976 Nobel Prize in Physiology or Medicine,[17] Professor Adrian Srb of Cornell University and Dr. Robert Rabson of the Division of Biological Energy, Washington, DC, nominated Barbara McClintock for the 1980 Nobel Prize. Srb summarized McClintock's scientific creativity, and with his nomination letter, Rabson included a copy of the paper by Cohen and Shapiro (1980) that had just appeared in *Scientific*

---

[14] Email E. Coe to L. B. Kass, March 10, 2023.

[15] McClintock Personnel File, Brookhaven National Laboratory.

[16] See also Sheridan (1982), Section VI. Nucleic Acids, Cloning and Controlling Elements, for the status of national and international research laboratories working toward the molecular identification of maize genes and controlling elements as of June 1982.

[17] Letter for a joint 1976 Nobel nomination from Ken Paigen, Roswell Park Memorial Institute, Buffalo, NY, to Jim Watson, April 21, 1976, Watson Collection, CSHL Archives Repository, Reference JDW/2/2/1156/1. Also see Comfort (2001, p. 242) for the nomination by Prof. Judson John van Wyk of North Carolina.

American (see above).[18] Robert Rabson received his doctorate in 1956, at Cornell University in Plant Physiology with F. C. Steward (Cobb 2013), who was a good friend of McClintock's and always spoke highly of her.[19] This 1980 nomination, too, was unsuccessful, as were others over subsequent years (see below). However, McClintock received extensive recognition during the next few years, and despite a few additional unsuccessful Nobel nominations (see below), she was ultimately awarded the 1983 Nobel Prize in Physiology or Medicine (see Chapter 11).

In April 1980, McClintock lectured at Rutgers University on another of her favorite topics, studies of maize races. The previous year, Rutgers University had established a Professorship in her name. In a presentation titled "How chromosomes have been used to trace the origin and migration of races of maize in the Americas," she provided the background for how she became involved in these studies and provided handouts of figures from her forthcoming co-authored book with Kato and Blumenschein (McClintock et al. 1981; Kato 2013). The meticulously outlined eight pages of notes plus five pages of maps for this lecture provide an excellent summary for the background of how she became involved in research on chromosome knobs and the training of students for studies of maize races.[20]

"Movable Genetic Elements" was the theme for the 1980 *Cold Spring Harbor Symposia on Quantitative Biology* held from May 28 to June 4 (*Annual Report 1980*, pp. 115–120; *Movable Genetic Elements* 1981). The organizers were Ahmad Bukhari and Jim Hicks. Barbara McClintock and others provided suggestions for participants. McClintock declined to give a formal presentation but "gave an informal summary of her latest ideas" (Watson 1980, p. xiii; *Annual Report 1980*, pp. 115–116). H. Bentley Glass of SUNY Stony Brook summarized many of McClintock's key contributions to genetics in his Introductory remarks: "Movable Genetic Elements and Barbara McClintock" (*Annual Report 1980*, pp. 115–116; Watson 1980, p. xiii). The general focus of the Symposium was on the mechanism by which DNA rearrangements occur and may affect gene expression (*Annual Report 1980*, p. 61; McKay 1980). Three of four presentations on maize involved gene regulation. Bacterial gene regulation by transposable elements was also reported for prokaryotic organisms.

Following the Cold Spring Harbor Symposium, McClintock chaired a session on "Genetic Rearrangements" and gave a presentation (June 24, 1980) at the Gordon Conference on "Biological Regulatory Mechanisms," held in Plymouth, New Hampshire, from June 23 to 27, 1980. Early in the year, she had accepted the invitation by the co-organizers Lucy Shapiro and Ira Herskowitz (Cruickshank 1980, p. 207), and prepared a talk on "Gene Regulation in maize revealed by transposable elements." Her presentation was based on four pages of typed notes and a series of 14 slides, depicting genetic expression in maize caused by transposable elements, and comparing these with examples in other organisms, both prokaryotic and eukaryotic. She prepared an abstract for the conference that attendees received.[21] McClintock would have been quite comfortable at these conferences because no information presented was to be used without specific authorization of the individual making the contribution, and no scientific publication was expected to be prepared from the conference (Cruickshank 1980, p. 207). This meant that McClintock was under no pressure to write an account for publication, and possibly more importantly, no one could misquote her. It was a totally informal event.

RECOGNITION AND AWARDS

In the fall of 1977, Barbara McClintock, Distinguished Service Member of the Carnegie Institution of Washington, at Cold Spring Harbor Laboratory, New York, became the seventh recipient of the Lewis S. Rosenstiel Award for Distinguished Work in Basic Medical Research (Rosenstiel Award

[18] Nomination for Nobel Prize, R. Rabson to Professor Ingvar Lindquist, NFR Swedish Natural Research Council, February 5, 1980, A Srb McClintock file, courtesy of Jo Srb.

[19] L. Kass phone interview with R. Rabson, November 19, 1996.

[20] Spectre 1993: Series III, Lecture, 16 April 1980; see Chapter 9.

[21] Spectre 1993: Series III, Lecture, June [24], 1980. An abstract for the conference is included in the folder for this lecture.

FIGURE 10.5 Barbara McClintock and Jim Haber following the 1977 Rosenstiel Award dinner celebration held on April 13, 1978, at Brandeis University. McClintock is carrying the award medallion. (Courtesy of Jim Haber.)

1977). The prize, given by the Rosenstiel Basic Medical Sciences Research Center, at Brandeis University, honored McClintock's "imaginative and important contributions to our understanding of chromosome structure, behavior, and function" (Report of the President 1978, p. 31). The presentation of the award medallion and $5,000 prize were made at a dinner held on April 13, 1978 (News and Notes 1978, Figure 10.5).

At the ceremony, she was recognized as "a forerunner in the field of genetics" and her "pioneering work helped pave the way for research in modern genetics"; she had found "that chromosomes acquired changing control properties and that they are regulated differently by virtue of their attachment to a new chromosome." It was also reported that much of this work was currently "being interpreted at the molecular level." On this occasion, McClintock gave a lecture similar to the one she had recently presented at the 10th Stadler Symposium, only days before, regarding mechanisms that rapidly reorganize genomes.[22] The topic of genome restructuring was at the forefront of her thoughts during this time. She had presented a similar topic on March 13 at Downstate Medical Center in Brooklyn, New York, and at a Conference on Genetic Exchange at Pennsylvania State University in July 1978.[23] Over the next few years, she continued to lecture on the topic of reprograming genomes and posed questions and offered hypothesis about possible underlying mechanisms (see Gordon Conference mentioned above). This topic would loom large in recognition of future awards.

[22] Spectre 1993: Lecture, April 13, 1978, Brandeis University, "Mechanisms that rapidly reorganize the Maize Genome."
[23] Spectre 1993: Series III, Lectures.

In December 1978, at their 161st meeting, The New York Academy of Sciences honored McClintock with the Louis and Bert Freedman Foundation Award for Research in Biochemistry. She was cited for her imaginative and original research on the complex mechanisms involved in the transfer and expression of genetic information (Report of the President 1979, p. 42).[24] Each recipient received a monetary award and "a certificate for an outstanding paper, a body of research, instrumentation development or other distinguished contribution to biochemistry."[25] In 1978, the Academy elected the groundbreaking cancer researcher Charlotte Friend to be their first female president. McClintock was made an honorary member of their society in 1985 (Braaten 2017).[26]

McClintock attended the 49th Annual meeting of the Genetics Society of America (GSA) held at the University of Colorado, Boulder, during August 18–20, 1980. At their banquet dinner, on August 18, she was recognized with a "Salute" from the Society:

> For her brilliance, originality, ingenuity and complete dedication to research
> A skilled experimentalist, a master of interpreting details,
> a brilliant theoretician, she has played an illuminating
> and pervasive role in the development of
> cytology and cytogenetics

The certificate was signed by William K. Baker, President; Burke H. Judd, Vice President; David Suzuki, Secretary; and William J. Welshons, Treasurer.[27] Professor Jeffry B. Mitton, the local GSA representative, recalled that upon receiving the certificate, "she was so flattered and humbled that it made her feel very self-conscious." He told her that the "recognition was long overdue."[28]

Encouraged by Adrian Srb, who earlier that year had assisted with the Nobel nomination for McClintock, Professor William B. Provine, historian of science, and Plant Breeding & Genetics graduate student Paul Sisco arranged to interview Barbara McClintock at her Cold Spring Harbor Laboratory (Provine & Sisco 1980; Kass 2003).[29] They spent the day of August 28, 1980, conducting an extensive interview for the purpose of finding documentation about the "Golden Age of Corn Genetics" at Cornell University during the 1920s and 1930s. McClintock was the first of the group of maize cooperators they planned to interview, and she provided them with the names of others who might be of additional help. They did not plan to rely on the stories these researchers told but instead use their recollections to find documentation for the historical perspective they had planned to publish. Continuing this project led to a program for the 75th Synapsis Club reunion that included a celebration of the "Golden Age of Corn Genetics" at Cornell a few years later (see below).

When McClintock was interviewed by Franklin H. Portugal the week prior to the Provine & Sisco interview, she too believed that his interviews were aimed at finding documents for his forthcoming publication about the history of the Carnegie Institution of Washington.[30] Beginning in 1977 through 1980, Portugal had been hired to write about what he considered highlights in the history of the Carnegie Institution of Washington.[31] McClintock was asked to review his manuscript on "The Geneticists," part of the larger project. It was apparent to her that the account lacked

---

[24] Email Douglas Braaten, Chief Scientific Officer, The New York Academy of Sciences, to L. Kass, August 24, 2023, attachment is a list of awardees 1972–1986, Braaten mentioned that awards stopped as of 1986.

[25] Email D. Braaten to L. Kass, August 28, 2023.

[26] This honor to McClintock is missing from the Table of Honors by Kass and Chomet (2009: Table 2).

[27] GSA "Salute" Certificate, 1980. Cold Spring Harbor Laboratory Archives, Barbara McClintock Collection, Series II: Biographical Materials, 1930-2005, Box 1: Folder 21 Genetics Society Award Certificate (Photocopy), 1980.

[28] Email J. B. Mitton to L. Kass, May 7, 2014.

[29] Their interview with McClintock has been miscataloged and miscited as more than one interview because *two* individuals transcribed the four tape recordings.

[30] B. McClintock to J. D. Ebert, President CIW, October 9, 1980; Spectre 1993: Series I, File Carnegie Institution of Washington 1967–1991. Portugal (1980) Interview with Barbara McClintock, August 21, 1980, APS.

[31] See *Carnegie Institution of Washington Year Books 77, 78, 79.*

the evidence for the background knowledge to evaluate his conclusions. She had suggested that he examine documents deposited in the Cold Spring Harbor Laboratory Archives to support his story. He apparently did not do so, and the manuscript was not published.

During 1981, Barbara McClintock received national and international recognition for her contributions to Genetics. At the 50th Annual meeting of the GSA, held in Raleigh, North Carolina, Marcus M. Rhoades and Barbara McClintock were awarded "The Thomas Hunt Morgan Medal." At the meeting in Boulder the previous year, where McClintock had received her "Salute," the GSA executive committee had established the "The Thomas Hunt Morgan Award" to be given to a member for a lifetime devoted to research in genetics of the highest quality. Two medals would be awarded annually (K. Goodman 2009). Rhoades was present to accept his award, but McClintock was sent her award the following year when the medals were completed.[32] She declined to receive her medal in person at the GSA meeting at Yale in 1982, believing it was "too soon to again appear before them to receive yet another honor …," which "could engender some mutual embarrassment." She was very grateful for having the opportunity to make the decision.[33]

The October 1981 issue of the popular magazine *Science '81* featured a story about McClintock by Evelyn Fox Keller (1981). Keller had interviewed McClintock between 1978 and 1979 for an article on McClintock's work and had decided to expand the story into a book-length biography. The book's publication was delayed due to various obstructions (Keller 1983, 2014; Wynants 2005). Keller told her readers that Harvard Professor Matthew Meselson had recently speculated that history will record McClintock "as a pioneer of new and very much more subtle and complex genetic theories that are as yet only dimly understood." She concluded that "He may be right." Their prediction was prescient. In October 1981, Barbara McClintock was awarded the Wolf Prize in Medicine jointly with Stanley N. Cohen. The summary reports that she "is recognized for her imaginative and important contributions to our understanding of chromosome structure, behavior and function."[34] McClintock received the Wolf Prize at the Award Ceremony in Israel's Knesset on March 21, 1982 (*Grapevine* 1982). She gave a lecture the following day at the Weizmann Institute of Science on "The diversity of gene expression initiated by Transposable Elements."[35] Her niece, Mrs. Marjorie M. Bhavnani, and family accompanied her on this trip. As guests of the Wolf Foundation, McClintock, Mrs. Bhavnani, and her husband Ashok toured Israel (*Grapevine* 1982).

McClintock was notified on November 13, 1981, that she was the first person to be appointed as a "MacArthur Prize Fellow Laureate." The prize was awarded in recognition of her accomplishments in Genetics, which demonstrated her "originality, dedication to creative pursuits, and capacity for self-direction." The Chicago Foundation expressed hopes that the award will provide her with greater freedom from financial constraints so that she may devote herself more fully to her own creative endeavors. She had no reporting obligation of any kind, and the Laureate award was for her to use for whatever purposes she chose. Although newspaper accounts reported that the award was for $60,000 a year for life, the fact is that the amount was reduced by any other income such as pensions, retirement, or Social Security benefits to which she was entitled. Any institution where she wished to work would also receive "an unrestricted grant of $15,000 per year." They requested that their notification be kept confidential until the Foundation made the award public. At the announcement of their new program on November 17, 1981, John Corbally, President of the Foundation, said that the Laureate Program "honors a very small number of exceptionally creative and accomplished individuals … expected to be near retirement age or older, but still working." It was a "guaranteed

[32] B. H. Judd to B. McClintock, July 22, 1981, Spectre 1993: Series I.
[33] B. McClintock to B. H. Judd, President GSA, August 4, 1981.
[34] Wolf Foundation, Barbara McClintock, Wolf Prize Laureate in Medicine 1981 https://wolffund.org.il/2018/12/09/barbara-mcclintock/.
[35] Spectre 1993: Series III.

level of payment for the rest of their lives." McClintock was age 79 when she received this award, and continued working through the next decade.[36]

The following day, McClintock was honored with the Albert Lasker Basic Medical Research Award for 1981. The award is given for fundamental discovery that opens up a new area of biomedical science. The program had been established in 1945 to focus interest on fundamental biological discoveries in addition to clinical advances in the improvement of human health and to draw attention to the importance of public support of science.[37] McClintock was recognized for "Mobile Genetic Elements ... For the discovery that certain genetic elements are not static, but can move from one location to another on DNA ... transposable elements serve a fundamental role in gene control and in genetic recombination based on changes in DNA structure, which allow in turn for the development of new species."[38] *The New York Times* newspaper account emphasized McClintock's unsettled feeling of receiving yet another unsought award, which she said was "too much at once" and distracted from her work. Their story focused on her preference for privacy over prizes. She looked forward to conducting more research because she believed that most people did not yet have an understanding of what she tried to say many years ago.[39] *The Carnegie Institution of Washington Year Book 80* (Report of the President 1981, p. 49) highlighted McClintock's three major prizes for their "contributions to our understanding of chromosome structure."[40] The same volume provided the progress that was being made on the molecular aspects of controlling elements in maize (Chaleff et al. 1981, pp. 158–164).

By January 1982, Nobel Laureates Francis Crick (January 4, 1982) and Joshua Lederberg (January 18, 1982) separately nominated McClintock for the 1982 Nobel Prize. Her "Statement of Achievements" and "Biography of Barbara McClintock" were written by Marcus Rhoades to accompany both their letters. They obtained a list of her publications from the President of the Carnegie Institution of Washington. Lederberg submitted these documents with his letter, and Crick requested that Herschel Roman submit similar documents for McClintock on his behalf. Crick's nomination, however, suggested a joint award with Seymour Benzer and Sidney Brenner for work "on novel genetic systems." The nomination letters were subsequently supported by Nobel laureates, George Beadle, Alfred Hershey, Salvador Luria, and François Jacob, among others.[41] This nomination for 1982 was again unsuccessful.

The 1982 *Cold Spring Harbor Symposia on Quantitative Biology* was devoted to "Structures of DNA" (1983). Although McClintock was not on the program, she attended many talks and participated in social events during this June meeting held at the Cold Spring Harbor Laboratory.[42] At the end of June, in conjunction with her honorary degree, she presented a lecture at the University of Cambridge on "Diversity of gene expression initiated by Transposable Elements."[43] A few months earlier, she had been offered an honorary membership in the Genetical Society of Great Britain, which had been founded by William Bateson in 1919.[44] In early July, she returned

[36] Letter John Corbally President MacArthur Foundation to Dr. Barbara McClintock, November 13, 1981; Spectre 1993: Series I, File John D. and Catherine T. MacArthur Foundation 1981-1989. UPI Archives, November 17, 1981, https://www.upi.com/Archives/1981/11/17/The-John-D-and-Catherine-T-MacArthur-Foundation-Tuesday/3915374821200/; "Foundation picks geneticist as its First Fellow Laureate," *New York Times*, November 18, 1981, Section B, Page 3.

[37] The Lasker Foundation, About the awards, https://laskerfoundation.org/awards/about-the-awards/

[38] Lasker Foundation, Mobile Genetic Elements, https://laskerfoundation.org/winners/mobile-genetic-elements/

[39] Kathleen Teltsch, November 18, 1981, "Award-Winning Scientist on L. I. prizes privacy" *New York Times*, National Edition, Section B, p. 3. See also Lawrence K. Altman, November 19, 1981 "Lasker Awards go to a geneticist and a lab head," *New York Times*, National Edition, Section B, p. 2.

[40] McClintock worked as Distinguished Service Member of the Carnegie Institution of Washington, Genetics Research Unit, 1967 through 1971. Carnegie then paid for her laboratory space at the Cold Spring Harbor Laboratory, Long Island, New York.

[41] All documents from Special Collections, University of Washington Libraries, Herschel L. Roman Accession 2955-2-89-25, Box No. 9. Folder No. 3. I thank Dr. Kathleen Gale for obtaining documents on my behalf, May 5, 2006.

[42] "Structures of DNA" (1983): Front Matter, p. xi, photo of McClintock with attendees.

[43] Spectre 1993: Series III, Lecture, June 23, 1982. Honorary Dr. of Science, June 22, 1983.

[44] Spectre 1993: Series I, Genetical Society 1982.

FIGURE 10.6 Joining Barbara McClintock at the 1982 Synapsis Club reunion, organized by W. B. Provine, were former students and postdoctoral associates who had collaborated with McClintock on making the cytogenetics of maize an important project at Cornell during the 1920s and 1930s. From left to right: Marcus M. Rhoades, Charles R. Burnham, Barbara McClintock, Harriet B. Creighton, George F. Sprague, and Harold S. Perry. (Courtesy of Plant Breeding & Genetics, Cornell University.)

to Cornell for the celebration of the 75th anniversary of the Department of Plant Breeding and its Synapsis Club (Kass 2003; Murphy & Kass 2007, 2011). Graduate student members invited alumni to return for a "Round Table and Discussion" organized by Professor Provine, on the "Golden Age of Corn Genetics," where the legacy of Rollins A. Emerson and his former students and colleagues was celebrated (Burnham 1982; Morris 1982, recordings; Murphy & Kass 2007, 2011). Participating were former students and postdoctoral associates who had collaborated with McClintock on making the cytogenetics of maize an important project at Cornell during the 1920s and 1930s (Figure 10.6). McClintock was the shining star of the event (Figure 10.7), which was audiotape-recorded for posterity.[45] Marcus Rhoades presented a handout on their accomplishments at the July 1st event and noted that most of the pioneering discoveries were made by McClintock (Murphy & Kass 2011, p. 23; see also Rhoades 1984). On July 2, 1982, McClintock also gave an invited keynote symposium lecture on "Potentials of the Plant Genome," which unfortunately was not published.[46]

During this year, McClintock was again recognized with two additional major awards, and received honorary degrees from Yale University during commencement in May 1982, and from the University of Cambridge the following month. In June, she was notified that she was to be honored with the Louisa Gross Horwitz Prize of Columbia University. She and Susumu Tonegawa were both recognized for their research on movable gene segments. They were "cited for their contributions to the understanding of the evolution of genetic information and the control of its

[45] Recordings in Cornell University Archives. Spectre 1993: Series III, IV. Files of Professor Pardee: Synapsis Reunion, Plant Breeding & Genetics, Cornell University.

[46] Spectre 1993: Series III, Lecture, July 1982.

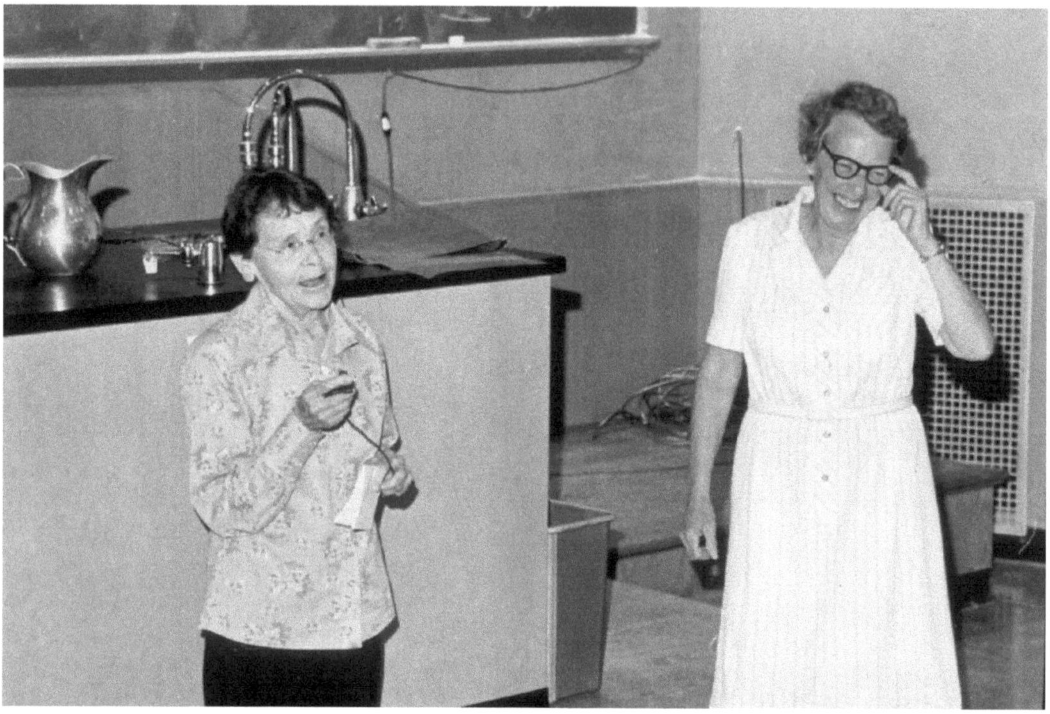

FIGURE 10.7 Barbara McClintock and Harriet Creighton at the 1982 Synapsis Club reunion Round Table and Discussion on the "Golden Age of Corn Genetics," at Cornell University. (Courtesy of Plant Breeding & Genetics, Cornell University.)

expression" (People & Places 1983, p. 228; Horwitz Prize 1982). The award was given at a formal dinner at the University. On Wednesday evening, October 6, she was honored and gave a lecture on the "Diversity of gene expression initiated by transposable elements." Therein, she also spoke about her experiences at Cornell that influenced her current research.[47] In November, McClintock was notified that she and Armin Braun, The Rockefeller University plant pathologist, were to share the Charles Leopold Mayer Prize from the French *Academie des Sciences*, for outstanding work in the biological sciences, specifically in the area of "cell biology" (the study of the structure, function, and behavior of cells). She accepted the invitation to receive the prize at the session of the *Academie* on Monday, December 6, 1982. The next day, she presented a lecture at the Pasteur Institute titled "Some unexpected results of a simple experiment." She explained that the start of the experiment went back to the summer of 1931, when she had been working at the University of Missouri.[48] As I have noted in earlier chapters, the experiences that McClintock gained over her long career led her to notice exceptional genetic characters (factors, phenotypic expressions) in her studies of maize. For this reason, the "*Companion Volume*" (Kass 2013ff.) that includes reprints of these early papers and provides relevant perspectives should be of value in comprehending how McClintock arrived at her later understanding of maize movable elements.

As this chapter has shown, well past retirement age, McClintock continued to give many lectures and symposia presentations.[49] She published her last Carnegie research report in 1971, and thereafter primarily published in Symposium volumes (Kass 2013ff.: Table I; Kass & Chomet 2009:

[47] Spectre 1993: Series I, Columbia University; Series III, Lectures.

[48] McClintock began her National Research Council Fellowship at Missouri in 1931, see Chapter 5, McClintock (1931), Kass and Birchler (2014/2016). Spectre 1993: Series I, Charles Leopold Mayer Prize 1982. Spectre 1993: Series III, Lecture at Pasteur Institute. See Braun's National Academy of Sciences Memoir by Chrispeels and Meins (2014).

[49] Spectre 1993: Series III, Lectures.

Appendix I). She received many honors and awards for her fundamental pioneering research (Kass & Chomet 2009: Table 2). Colleagues from Cornell and elsewhere unsuccessfully nominated her for the Nobel Prize (1980–1982). McClintock's greatest recognition was yet to come.

LITERATURE CITED

Annual Report 1980. *Cold Spring Harbor Laboratory*. Cold Spring Harbor, New York.

Avers, C. 1980. *Genetics*, D. Van Nostrand & Co., New York.

Beadle, G. W. 1978. Teosinte and the origin of maize, pp. 113–128, *In*, D. B. Walden (Ed.), *Maize Breeding and Genetics*, John Wiley & Sons, New York. [1975 International Symposium]

Braaten, D. 2017. New York Academy of Sciences – A concise history. *Newsletter*. https://www.nyas.org/magazines/celebrating-200-years/new-york-academy-of-sciences-a-concise-history/

Brenner, S. 1976. Concluding Remarks – CSHL DNA Insertions Meeting, May 1976 [includes notes on meeting] Sydney Brenner Collection (1927–2010), SB/3/9, CSHL Archives Repository. https://libgallery.cshl.edu/items/show/74161

Brevet, J., P. Nisen, D. J. Kopecko, and S. N. Cohen. 1976. Promotion of insertions and deletions by translocating segments of DNA carrying antibiotic resistance genes, p. 22, *In*, A. I. Bukhari, J. A. Shapiro, and S. L. Adhya (Eds.), *DNA Insertions*, Abstracts book. Cold Spring Harbor, New York.

Bukhari, A. I. 1976. Bacteriophage *Mu* as a transposition element. *Annual Review of Genetics* Vol. 10 (December 1976), pp. 389–412. https://doi.org/10.1146/annurev.ge.10.120176.002133

Bukhari, A. I., L. Ambrosio, F. deBruijn, S. Froshauer, H. Khatoon, and E. Ljungquist. 1976. Insertion elements and plasmids, pp. 37–40, *Annual Report 1976*, Cold Spring Harbor Laboratory, Cold Spring Harbor, Long Island, NY.

Bukhari, A. L., J. A. Shapiro and S. L. Adhya (Arrangers). 1976. *DNA Insertions; Abstracts of Papers Presented at the Meeting on DNA Insertions, May 18–21, 1976*, Cold Spring Harbor Laboratory. Cold Spring Harbor, Long Island, NY.

Bukhari, A. I., J. A. Shapiro, and S. L. Adhya (Eds.). 1977. *DNA Insertion Elements, Plasmids, and Episomes*, xiii, 782 p. ill., Cold Spring Harbor Laboratory, Cold Spring Harbor, NY.

Burnham, C. 1982. The golden age of corn genetics. Panel Discussion, July 1, 1982. Recording, two sound cassettes. Cornell University Library. https://catalog.library.cornell.edu/catalog/3601898

Burr, F. A. and B. Burr. 1980. The cloning of Zein sequences and an approach to Zein genetics, pp. 227–231, *In*, C. J. Leaver (Ed.), *Genome Organization and Expression in Plants*, NATO Advanced Study Institutes Series, Vol. 29. Plenum Press, New York. [Meeting of July 11–21, 1979, Edinburgh, Scotland]

Burr, B. and F. Burr. 1981. [1980 Symposium] Detection of changes in maize DNA at the shrunken locus due to the intervention of *Ds* elements. *Cold Spring Harbor Symposia on Quantitative Biology* Vol. 45, *Movable Genetic Elements*, 1980. Pt. 2, pp. 463–465 [Published February 1981]. https://doi.org/10.1101/SQB.1981.045.01.061

Campbell, A., D. Berg, D. Botstein, E. Lederberg, R. Novick, P. Starlinger, and W. Szybalski. 1977. Nomenclature of transposable elements in prokaryotes, pp. 15–22, In, A. J. Bukhari, J. A. Shapiro, and S. L. Adhya (Eds.), *DNA Insertion Elements, Plasmids, and Episomes*, Cold Spring Harbor Laboratory, Cold Spring Harbor, NY. [See their 1979 update published in both *Gene* Vol. 5, March 1979, and *Plasmid* Vol. 2, July 1979]

Campbell, A., D. Berg, D. Botstein, E. M. Lederberg, R. P. Novick, P. Starlinger, and W. Szybalski. 1979a Nomenclature of transposable elements in prokaryotes. *Gene* Vol. 5 (No. 3, March 1979), pp. 197–206.

Campbell, A, P. Starlinger, D. E. Berg, D. Botstein, E. M. Lederberg, R. P. Novick, and W. Szybalski. 1979b. Nomenclature of transposable elements in prokaryotes. *Plasmid* Vol. 2, (No. 3, July), pp. 466–473.

Chaleff, D., J. Mauvais, S. McCormick, M. Shure, S. Wessler, and N. Fedoroff. 1981. Controlling elements in maize. *Carnegie Institution of Washington Year Book 80* (1980–1981), pp. 158–174. [Issued December 1981]

Chrispeels, M. and F. Meins Jr. 2014. Armin Charles Braun 1911–1986. A Biographical Memoir of the National Academy of Sciences, January 2014. National Academy of Sciences Press. https://www.nasonline.org/member-directory/deceased-members/57430.html

Cobb, E. D. 2013. *150 Years of Botany at Cornell; A History of Botany and Plant Biology*, Department of Plant Biology, Cornell University, Ithaca, NY.

Coe, E. J. Jr. 1978. The aleurone tissue of maize as a genetic tool, pp. 447–459, *In*, D. B. Walden (Ed.), *Maize Breeding and Genetics*, John Wiley & Sons, New York. [1975 International Symposium]

Cohen, S. N. 1976. Transposable genetic elements and plasmid evolution; review article. *Nature* Vol. 263 (October 28, 1976), pp. 731–738.
Cohen, S.N. and J. A. Shapiro. 1980. Transposable genetic elements. *Scientific American* Vol. 242 (No. 2, February), pp. 40–49. [Bibliography p. 172. Author biographies p. 12.]
Cruickshank, A. M. 1980. Gordon research conferences. *Science* Vol. 207 (No. 4436, March 14, 1980), pp. 1229–1263.
Dawson, G. W. P. and P. F. Smith-Keary. 1963. Episomic control of mutation in *Salmonella typhimurium. Heredity* Vol. 18 (February 1, 1963), pp. 1–20. https://doi.org/10.1038/hdy.1963.1
DNA Insertions. 1976. Program. Arranged by Amad I Bukhari, Cold Spring Harbor Laboratory; James Shapiro, University of Chicago; Sankar Adhya, National Institutes of Health. 126 Participants, pp. 83–86, Summer Meetings, *Annual Report 1976*, Cold Spring Harbor Laboratory.
Fedoroff, N. 1982. Introduction to transposable controlling elements in maize, pp. 203–211, *In*, W. F. Sheridan (Ed.), *Maize for Biological Research*, Plant Molecular Biology Association, Charlottesville, VA.
Fedoroff, N. V. 1983. Controlling elements in maize, pp. 1–63. In, J. Shapiro (Ed.) *Mobile Genetic Elements*. Academic Press, Inc., New York.
Fedoroff, N. V. 1989b. Maize transposable elements, pp. 375–411, *In, Mobile DNA*, D. E. Berg and M. M. Howe (Eds.). American Society for Microbiology, Washington, D. C.
Fedoroff, N. V. 1992. Maize transposable elements: A story in four parts, pp. 389-415, *In*, N. Fedoroff and D. Botstein (Eds.), *The Dynamic Genome: Barbara McClintock's Ideas in the Century of Genetics*. Cold Spring Harbor Laboratory Press, Plainview, NY.
Fedoroff, N. V., D. Chaleff, J. Mauvais, M. Shure, and S. Wessler. 1982. Molecular studies on maize genes with controlling element mutations. *Carnegie Institution of Washington Year Book 81* (1981–1982), pp. 161–171. [Issued December 1982]
Fedoroff, N. and J. Mauvais. 1979. Controlling elements in maize. *Carnegie Institution of Washington Year Book 78* (1978–1979), pp. 61–65.
Fedoroff, N., S. McCormick, and J. Mauvais. 1980. Molecular studies on the controlling elements of maize. *Carnegie Institution of Washington Year Book 79* (1979–1980), pp. 51–62.
Fincham, J. R. S. and G. R. K. Sastry. 1974. Controlling elements in maize. *Annual Review of Genetics* Vol. 8, pp. 15–50.
Galinat, W. C. 1978. The inheritance of some traits essential to maize and teosinte, pp. 93–111, *In*, D. B. Walden (Ed.), *Maize Breeding and Genetics*, John Wiley & Sons, New York. [1975 International Symposium]
Goodman, K. 2009. *An Abridged History of the Genetics Society of America*. Compiled from the Records of the Genetics Society of America and the *GENETICS* Journal, 1931–2008 by staff member Karen Goodman, Spring 2008. https://www2.genetics-gsa.org/pdf/GSAhistoryscrapbook.pdf
Goodman, M. M. 1978. A brief survey of the races of maize and current attempts to infer racial relationships, pp. 143–158, *In*, D. B. Walden (Ed.), *Maize Breeding and Genetics*, John Wiley & Sons, New York. [1975 International Symposium]
Grapevine. 1982. Tour coordinator goes to Israel for Award Ceremony. *Grapevine, American Museum of Natural History* Vol. 39 (No. 2, March/April 1982), p. 2. https://archive.org/details/grapevine3639amer/page/n97/mode/2up
Green, M. M. 1977. The case for DNA insertion mutations in *Drosophila*, pp. 437–445, *In*, A. I. Bukhari, J. A. Shapiro, and S. L. Adhya (Eds.), *DNA Insertion Elements, Plasmids, and Episomes*, Cold Spring Harbor Laboratory, Cold Spring Harbor, NY.
Hayes, W. 1968. *The Genetics of Bacteria and Their Viruses*, 2nd edition, John Wiley & Sons, New York.
Hedges, R. W. and A. E. Jacob. 1974. Transposition of ampicillin resistance from RP4 to other replicons. *Molecular and General Genetics* Vol. 132 (No. 1, August 30, 1974), pp. 31–40.
Herskowitz, I. H. 1977. *Principles of Genetics* (2nd edition). Macmillan Publishing Company, New York.
Horwitz Prize. 1982. 1982 Louisa Gross Horwitz Prize. Barbara McClintock and Susumu Tonegawa, for research in the evolution of genetic information and the control of its expression. https://awardsandwinners.com/category/louisa-gross-horwitz-prize/1982/
Higgins, N. P. 1998. Book Review: *Mobile Genetic Elements*/Frontiers in Molecular Biology [Series]. *The Quarterly Review of Biology* Vol. 73 (No. 1, March 1998), pp. 68–69.
Jacob F. and E. L. Wollman. 1958. Les épisomes, elements génétiques ajoutés; *Comptes Rendus Hebdomadaires des Seances de L'academie des Sciences*, Paris Vol. 247 (No. 1, July 7, 1958), pp. 154–156. [French translated to English in E. A. Adelberg 1960, *Papers on Bacterial Genetics,* pp. 398–400]
Jordan, E., H. Saedler and P. Starlinger. 1968. Oo and strong-polar mutations in the *gal* operon are insertions. *Molecular and General Genetics* Vol. 102 (No. 4, December 1968), pp. 353–363. [Received September 1968]

Kass, L. B. 2003. Records and recollections: A new look at Barbara McClintock, Nobel Prize-Winning geneticist. *Genetics* Vol. 164 (No. 4, August 1, 2003), pp. 1251–1260. https://www.genetics.org/content/164/4/1251

Kass, L. B. (Ed.). 2013ff. *Perspectives on Nobel Laureate Barbara McClintock's Publications (1926–1984): A Companion Volume.* Volumes I & II The Internet-First University Press. https://hdl.handle.net/1813/34897 [Online December 30, 2013, updates March 2014; Vol. III, Increments No. 1–3, May 2016; Vol. IV, Increment No. 4, April 2020]

Kass, L. B. and J. A. Birchler. 2014/2016. Setting up a prepared mind: Summary and analysis of "Cytological observations of deficiencies involving known genes, translocations and an inversion in *Zea mays*," Vol. III, Increment #1, pp. 1.47.1–1.47.15, *In*, L. B. Kass (Ed.), *Perspectives on Nobel Laureate Barbara McClintock's Publications (1926–1984): A Companion Volume*, The Internet-First University Press. https://ecommons.cornell.edu/bitstreams/9711d284-f6c1-4f0a-88a0-68eab1c8d7c8/download (15.27 MB) [Online June 7, 2016]

Kass, L. B., E. H. Coe, M. N. Cook, M. E. Smith, and J. L. Singer (Eds.). 2019. *Founding of the Maize Genetics Cooperation News Letter at Cornell University: A 90th Anniversary Tribute.* (Foreword by E. S. Buckler) [e-book] Internet-First University Press, Ithaca, NY. https://ecommons.cornell.edu/handle/1813/66550. [Online October 10, 2019]

Kass, L. B. and P. Chomet. 2009. Barbara McClintock. pp. 17–52, *In*, J. L. Bennetzen, and S. C. Hake (Eds.), *Handbook of Maize: Genetics and Genomics*, Springer Science, New York

Kato Y., T. A. [T. Angel Kato Y]. 2013. Significance of McClintock's pattern concept in the analysis of chromosome knob distributions in the races of maize, pp. 4.39–4.49, *In*, L. B. Kass (Ed.), 2013ff. *Perspectives on Nobel Laureate Barbara McClintock's Publications (1926–1984): A Companion Volume.* https://hdl.handle.net/1813/34897

Keller, E. F. 1981. McClintock's maize, *Science* Vol. 2 (No. 6, October), pp. 54–58; p. 102.

Keller, E. F. 1983. *A Feeling for the Organism. The Life and Work of Barbara McClintock*, 1st edition, W. H. Freeman, San Francisco, CA.

Keller, E. F. 2014. Pot-holes everywhere: How (not) to read my biography of Barbara McClintock, pp. 33–42, *In*, P. Govoni and Z. A. Franceschi (Eds.), *Writing About Lives in Science: (Auto) Biography, Gender, and Genre*, V&R Unipress, Gottingen.

Kermicle, J. L. 1978. Imprinting of gene action in maize endosperm, pp. 357–371, *In*, D. B. Walden (Ed.), *Maize Breeding and Genetics*, John Wiley & Sons, New York. [1975 International Symposium]

Kolata, G. B. 1976. Jumping genes: A common occurrence in cells. *Science* Vol. 193 (No. 4251, July 30), pp. 392–394.

Kretschmer, P. J. and S. N. Cohen. 1976. Selection for replicons containing the TnA translocation segment-general method for selected translocation, p. 26, *In*, A. I. Bukhari, J. A. Shapiro & S. L. Adhya, 1976. *DNA Insertions*, Abstracts book. Cold Spring Harbor, New York.

McClintock, B. 1931. Cytological observations of deficiencies involving known genes, translocations and an inversion in *Zea mays. University of Missouri, College of Agriculture, Agricultural Experiment Station, Research Bulletin 163* [Authorized 23 December 1931], pp. 1–30.

McClintock, B. 1951 [©1952, Symposium held June 1951]. Chromosome organization and genic expression, pp. 13–47, *In*, K. B. Warren (Ed.), *Genes and Mutations, Cold Spring Harbor Symposia on Quantitative Biology*, Vol. XVI [Vol. 41, June 7–15, 1951], The Biological Laboratory, Cold Spring Harbor, Long Island, NY.

McClintock, B. 1955. Controlled mutation in maize: The *a1m-1-Spm* system of control of gene action and mutation. Continued studies of the mode of operation of the controlling elements *Ds* and *Ac*. *Carnegie Institution of Washington Year Book No. 54*, 1954–1955, pp. 245–255. [Issued December 9, 1955]

McClintock, B. 1956a [Issued February 1956, Symposium held June 15–17, 1955]. Intranuclear systems controlling gene action and mutation, pp. 58–74. *In, Mutation, Brookhaven Symposia in Biology, No. 8.* Biology Department, Brookhaven National Laboratory, Upton, NY. [R. C. King, Symposium Chairman.]

McClintock, B. 1956b [©1957, Symposium held June 1956]. Controlling elements and the gene, pp. 197–216, *In*, K. B. Warren (Ed.), *Genetic Mechanisms: Structure and Function, Cold Spring Harbor Symposia on Quantitative Biology*, Volume XXI [June 4–12, 1956], The Biological Laboratory, Cold Spring Harbor, Long Island, NY.

McClintock, B. 1961. Some parallels between gene control systems in maize and in bacteria. *American Naturalist* Vol. XCV [95] (No. 884, September–October 1961), pp. 265–277.

McClintock, B. 1976. Transposable chromosomal elements in maize, p. 41, *In*, A. I. Bukhari, J. A. Shapiro and S. L. Adhya (Eds.), *DNA Insertions*, Abstracts book, Cold Spring Harbor, New York.

McClintock, B. 1978a [Symposium held September 1975]. Significance of chromosome constitutions in tracing the origin and migration of races of maize in the Americas. Chapter 11, pp. 159–184, *In*, D. B. Walden (Ed.), *Maize Breeding and Genetics*, John Wiley & Sons, Inc., New York. [1975 International Symposium]

McClintock, B. 1978b. [Symposium held April 7–8, 1978] Mechanisms that rapidly reorganize the genome. *Stadler Genetics Symposia*, Vol. 10, pp. 25–48. University of Missouri, Columbia, MO. [Pg. 48 is a plate of photographs of Symposium participants. Symposium held at Columbia, Missouri, U.S.A.. Proceedings published by University of Missouri, Agricultural Experiment Station.]

McClintock, B. 1980. Modified gene expressions induced by transposable elements, pp. 11–19, *In*, W. A. Scott, R. Werner, D. R. Joseph, and Julius Schultz (Eds.), *Mobilization and Reassembly of Genetic Information. Proceedings of the Miami Winter Symposium*, January 1980. Miami Winter Symposium Vol. 17. Academic Press, New York. [Lecture given on 7 January 1980].

McClintock, B., T. A. Kato Y., and A. Blumenschein. 1981. *Chromosome Constitution of Races of Maize. Its Significance in the Interpretation of Relationships Between Races and Varieties in the Americas.* Colegio de Postgraduados, Escuela Nacional de Agricultura, Chapingo, Edo. Mexico, Mexico, xxxi, 517 pp.

McKay, R. 1980. Movable genes. *Nature* Vol. 287 (No. 5779, September 18, 1980), pp. 188–189. https://doi.org/10.1038/287188a0

Morris, R. 1982. Golden age of corn genetics. Panel Discussion. July 1, 1982. Recording, one cassette. Files of R. Morris, duplicate sent to L. B. Kass.

Movable Genetic Elements. 1981. [1980 Symposium] *Cold Spring Harbor Symposia on Quantitative Biology, Vol. 45, Movable Genetic Elements, 1980.* Cold Spring Harbor Laboratory, Cold Spring Harbor, NY. [Issued February 1981]

Murphy, R. P. and L. B. Kass. 2007. *Evolution of Plant Breeding at Cornell University: A Centennial History, 1907–2006*, pp. 1–98, Department of Plant Breeding & Genetics, Cornell University, Ithaca, NY, Appendices A1–A98, Photo Section P1–P38. (July 2007).

Murphy, R. P. and L. B. Kass. 2011. *Evolution of Plant Breeding at Cornell University: A Centennial History, 1907–2006.* Revised edition. https://hdl.handle.net/1813/23087 (Online June 27, 2011).

Neuffer, G. 1978. Induction of genetic variability, pp. 579–600, *In*, D. B. Walden (Ed.), *Maize Breeding and Genetics*, John Wiley & Sons, New York. [1975 International Symposium]

News and Notes. 1978. Barbara McClintock receives Brandeis Science Award. *Journal of Heredity* Vol. 69 (No. 2, March–April 1978), p. 106.

People and Places. 1983. Columbia University's 1982 Louisa Gross Horwitz Prize has been awarded to two scientists for their research on movable gene segments. *BioScience* Vol. 33 (No. 3, March 1983), pp. 228–231.

Peterson, P. A. 1976. The position hypothesis for controlling elements in maize, p. 43, *In*, A. I. Bukhari, J. A. Shapiro, & S. L. Adhya, *DNA Insertions*, Abstracts book. Cold Spring Harbor, New York.

Peterson, P. A. 1977. The position hypothesis for controlling elements in maize, pp. 429–435. *In*, A. I. Bukhari, J. A. Shapiro and S. L. Adhya (Eds.), *DNA Insertion Elements, Plasmids, and Episomes*, Cold Spring Harbor Laboratory, Cold Spring Harbor, NY.

Peterson, P. A. 1978. Controlling elements: The induction of mutability at the *A2* and *C* loci in maize, pp. 601–631, *In*, D. B. Walden (Ed.), *Maize Breeding and Genetics*, John Wiley & Sons, New York. [1975 International Symposium]

Provine, W. B. and P. Sisco. 1980. Interview with Barbara McClintock at Cold Spring Harbor Laboratory, August 28, 1980. Rare and Manuscript Collections, Cornell University Library. Transcripts and audiotapes. [Miscatalogued as two interviews]. https://catalog.library.cornell.edu/catalog/2093697

Report of the Director. 1978. Department of Embryology, Introduction, Donald D. Brown, pp. 9–11. *Carnegie Institution of Washington Year Book 77* (1977–1978).

Report of the Director. 1981. Department of Embryology, Introduction, Donald D. Brown, pp. 109–112. *Carnegie Institution of Washington Year Book 80* (1980–1981).

Report of the President 1977–1978. 1978. *Carnegie Institution of Washington Year Book 77*, 1977–1978, pp. 1–43.

Report of the President 1978–1979. 1979. *Carnegie Institution of Washington Year Book 78, 1978–1979*, pp. 1–43.

Report of the President. 1981. *Carnegie Institution of Washington Year Book 80, 1980–1981*, pp. 1–50.

Rhoades, M. M. 1984. The early years of maize genetics. *Annual Review of Genetics* Vol. 18 (December 1984), pp. 1–29.

Robertson, D. S., I. C. Anderson, and M. D. Bachmann. 1978. Pigment -deficient mutants: Genetic, biochemical, and developmental studies, pp. 461–494, *In*, D. B. Walden (Ed.), *Maize Breeding and Genetics*, John Wiley & Sons, New York. [1975 International Symposium]

Rosenstiel Award. 1977. Lewis S. Rosenstiel Award for Distinguished Work in Basic Medical Research, Brandeis University, Division of Science, Rosenstiel Basic Medical Sciences Research Center, Waltham, MA, Past Winners. https://www.brandeis.edu/rosenstiel/rosenstielaward/past.html

Schwartz, D. 1978. A model for control of the level of gene expression in maize, pp. 551–557, *In*, D. B. Walden (Ed.), *Maize Breeding and Genetics*, John Wiley & Sons, New York. [1975 International Symposium]

Scott, W. A., R. Werner, D. R. Joseph, and J. Schultz (Eds.). 1980. *Mobilization and Reassembly of Genetic Information*. Proceedings of the Miami Winter Symposium, January 1980. Miami Winter Symposium Vol. 17. Academic Press, New York.

Shapiro, J. A. 1967. *The structure of the galactose operon in Escherichia coli K-12*. Ph.D. Thesis, University of Cambridge, Cambridge, UK. [July 1967]

Shapiro, J. A. 1969. Mutations caused by the insertion of genetic material into the galactose operon of *Escherichia coli*. *Journal of Molecular Biology* Vol. 40 (No. 1, February 28, 1969), pp. 93–105. [Received August 1, 1968]

Shapiro, J. A. 1992. Kernels and colonies: The challenge of pattern, pp. 213–221, *In*, N. Fedoroff and D. Botstein (Eds.), *The Dynamic Genome: Barbara McClintock's Ideas in the Century of Genetics*. Cold Spring Harbor Laboratory Press, Plainview, NY.

Shapiro, J. A. 1995. The discovery and significance of mobile genetic elements, pp. 1–17, *In*, D. J. Sherratt (Ed.), *Mobile Genetic Elements* [series *Frontiers in Molecular Biology 8*], IRL Press (Oxford University Press), Oxford and New York.

Shapiro, J. A. 2009. Letting *E. coli* teach me about genome engineering. *Genetics* Vol. 183, pp. 1205–1214.

Shapiro, J. A. and S. L. Adhya. 1969. The galactose operon of *E. coli* K-12. II. A deletion analysis of operon structure and polarity. *Genetics* Vol. 62 (No. 2, June), pp. 249–64. https://doi.org/10.1093/genetics/62.2.249. [Received October 18, 1968, Published June 1969]

Sheridan, W. F. (Ed.) 1982. *Maize for Biological Research*, Plant Molecular Biology Association, Charlottesville, VA. [June 1982]

Sherratt, D. J. (Ed.). 1995. *Mobile Genetic Elements* [series *Frontiers in Molecular Biology* 8], IRL Press (Oxford University Press), Oxford and New York.

Spectre, M. 1993. *Barbara McClintock Papers. Guide to the Collection*, American Philosophical Society Library, Philadelphia, PA. (Online: Barbara McClintock Papers Mss.Ms.Coll.79, https://search.amphilsoc.org/collections/view?docId=ead/Mss.Ms.Coll.79-ead.xml) [Also see selected papers and summary of APS Library papers of McClintock available for the National Library of Medicine, Profiles in Science: https://profiles.nlm.nih.gov/spotlight/ll; The collection also reprints papers/letters from other institution's archives]

Starlinger, P. 1980. A reexamination of McClintock's "controlling elements" in maize in view of recent advances in molecular biology, pp. 537–551, *In*, C. J. Leaver (Ed.). *Genome Organization and Expression in Plants*, NATO Advanced Study Institutes Series, Vol. 29. Plenum Press, New York. [Meeting of July 11–21, 1979, Edinburgh, Scotland]

Starlinger, P. 1984. Transposable elements. *Trends in Biochemical Sciences* Vol. 9 (No. 4, April 1984), pp. 125–127.

Starlinger, P. and H. Saedler. 1972. Insertion mutations in microorganisms. *Biochemie* Vol. 54 (No. 2, February 1972), pp. 177–185.

Starlinger, P. and H. Saedler. 1976. IS-elements in microorganisms, pp. 111–152, *In*, W. Arber et al. (Eds.), *Current Topics in Microbiology and Immunology/Ergebnisse der Microbiologie und Immunitätsforschung*, Vol. 75, Springer, Berlin, Heidelberg. https://doi.org/10.1007/978-3-642-66530-1_4

Saedler, H., H. J. Reif, S. Hu, and N. Davidson. 1974. IS2, A genetic element for turn -off and turn-on of gene activity in *E. coli*. *Molecular and General Genetics* Vol. 132 (No. 4, 16 September), pp. 265–289. https://pubmed.ncbi.nlm.nih.gov/4610339/

Structures of DNA. 1983. [1982 Symposium] *Cold Spring Harbor Symposia on Quantitative Biology* 1982, Vol. *XLVII* [Vol. 47], *Structures of DNA, 1982*: Front Matter, pp. v–xx, The Cold Spring Harbor Laboratory, Cold Spring Harbor, NY.

Swanson, C. P., T. Mertz, and W. J. Young. 1981. *Cytogenetics. The Chromosomes in Division, Inheritance and Evolution*, 2nd edition, Prentice-Hall, Inc., Englewood, Cliffs, NJ.

Taylor, A. L. 1963. Bacterio-phage induced mutations in *Escherichia coli*. *Proceedings of the National Academy of Sciences* Vol. 50 (No. 6, December 1, 1963), pp. 1043–1051. https://doi.org/10.1073/pnas.50.6.1043
Walden, D. B. (Ed.). 1978. *Maize Breeding and Genetics*, John Wiley & Sons, New York. [1975 International Symposium]
Watson, J. D. 1976. Director's report, pp. 5–15. *Annual Report 1976*, Cold Spring Harbor Laboratory, Cold Spring Harbor, NY.
Watson, J. D. 1980. Foreword, p. xiii. *Cold Spring Harbor Symposia on Quantitative Biology, Vol. 45 "Movable Genetic Elements"* [May 28th to June 4th 1980]. Cold Spring Harbor Laboratory [©1981].
Wynants, M. 2005. Interview with Dr. Evelyn Fox Keller. *BioEssays* 27 (No. 7, July), pp. 748–758. [Interview of November 29, 2004; Online 13 June 2005]. https://onlinelibrary.wiley.com/doi/10.1002/bies.20251

11 Unshared Nobel Prize and Inclusive Recognition

I took the one less traveled by, and that has made all the difference

– Robert Frost

THE NOMINATIONS FOR THE 1983 NOBEL PRIZE

As reviewed in the previous chapter, Herschel Roman and colleagues began discussions the previous summer regarding nominating McClintock for the 1982 Nobel Prize. In August 1981, Herschel Roman asked François Jacob about nominating McClintock for the Nobel Prize, and if there was any chance of her being considered by the Nobel Committee. It seemed to him that she should share in the prize if it were given for transposable elements.[1]

I mentioned in Chapter 10 that their 1982 nomination for McClintock was unsuccessful. Marcus Rhoades had summarized McClintock's work on "controlling elements" emphasizing their function, and he did not use the term "transposable elements." He wrote that her outstanding contribution is her analysis of the control of gene action in maize and the discovery of the two-unit interacting system. This concept, he said, was the precursor of the regulator-operator theory of gene regulation that won for its promulgators, Jacob and Monod, the Nobel Prize in 1965.[2] The Prize in Physiology or Medicine was awarded to F. Jacob, A. Lwoff, and J. Monod "for their discoveries concerning genetic control of enzyme and virus synthesis" in prokaryotic cells.

Jacob believed that a Nobel Prize should be given for transposable elements and that McClintock should be included.[3] He had no doubt that she had discovered such elements before anyone else. He believed the difficulty would be in communicating her ideas. He had read her Cold Spring Harbor paper[4] without understanding what she meant, and he assumed the same for Roman. Jacob recalled that it was only at the end of the 1960s, when bacterial transposons were found, that a reasonable interpretation could be made for her published paper reporting on the maize eukaryotic system.[5] His reservation was that it would be difficult to explain McClintock's work to the Swedish surgeons who vote for the prize (see below). He told Roman that, as a former Nobel Laureate, he would certainly support her nomination if the Nobel Committee asked for his opinion.

Jacob had cited McClintock's work in his Harvey Lecture of 1958 (Jacob 1958, 1960). He and Monod had credited her research and compared it to regulatory systems in bacteria, in their published papers resulting from their participation in the 1961 Cold Spring Harbor Symposium (Jacob & Monod 1961b; Monod & Jacob 1961: Chapter 9):

> Long before regulator genes and operator were recognized in bacteria, the extensive and penetrating work of McClintock (1956) had revealed the existence, in maize, of two classes of genetic "controlling

---

[1] Letter H. Roman to F. Jacob, August 27, 1981, University of Washington Libraries, Herschel L Roman Accession 2955-289-25, Box 3 (henceforth Roman Papers UW).

[2] Barbara McClintock: Statement of Achievements written by Marcus Rhoades for the 1982 Nobel Nomination in Physiology or Medicine, Roman Papers UW; see Chapter 10.

[3] François Jacob to Herschel Roman, September 11, 1981, Roman Papers, UW.

[4] Probably implying McClintock's (1951) *CSHSQB* Symposium paper.

[5] The term transposon was introduced in 1974 by R. W. Hedges and A. E. Jabob. Bacterial transposable elements were first hypothesized by Jim Shapiro (1967) and reported in the literature in 1968ff, as insertion sequences; see Chapter 10.

DOI: 10.1201/9781003332527-12

elements" whose specific mutual relationships are closely comparable with those of the regulator and operator [in bacteria].

(*Monod & Jacob 1961, pp. 394–395*)

The question posed by Roman was not about nominating McClintock for work on gene control systems, but it was for studies of transposable elements. Crick, as you will recall (Chapter 10), had nominated her in 1982 to share a prize for "novel genetic systems."

Because their 1982 Nobel nomination for McClintock was unsuccessful,[6] Roman suggested that Lederberg forward the nomination again in early 1983.[7] Roman also asked if the supporting evidence, and letters supporting the nomination needed to be brought up-to-date. When Lederberg did not respond, Roman wrote to him again in May asking if he would nominate McClintock for the 1983 Nobel Prize.[8] In June, Lederberg replied by jotting a quick note by hand to the bottom of Roman's letter; he would be glad to renominate Barbara again for the 1984 prize, which had a deadline date of February 1, [1984].

Meanwhile, another team of McClintock supporters had already nominated her for the 1983 Nobel Prize on January 26, 1983. Ira Herskowitz and Bruce Alberts, then on the faculty in the Department of Biochemistry and Biophysics of the University of California, San Francisco, nominated McClintock for her discovery of "Transposable Genetic Elements."[9] They emphasized that McClintock had discovered that pieces of genetic information are capable of movement from one chromosomal location to another and that such movement can lead to a major alteration in gene expression. They made no mention of gene regulation or control of development as previous nominees had done. The lack of a discussion of regulation or control has been offered as a rationale for why McClintock was seriously considered for the 1983 award for transposable elements and not for gene regulation (Comfort 2001a,b). However, this justification seems limited for several reasons.

Significantly, Nobel Prizes for Physiology or Medicine are only given for a Discovery (Norrby 2022a; see below). There is no doubt that McClintock was the first to discover movable genetic elements. Nonetheless, descriptions of her previous major awards (Chapter 10) illuminate the breadth of her discoveries. The 1970 National Medal of Science, awarded in 1971, recognized her achievements "For establishing the relations between inherited characters in plants and the detailed shapes of their chromosomes, and for showing that some genes are controlled by other genes within chromosomes."[10] The 1981 Horwitz Prize was for research on movable gene segments, which contributed to an understanding of the evolution of genetic information and the control of its expression [gene regulation]. The 1981 Lasker Prize was for mobile genetic elements: the discovery that certain genetic elements are not static and serve a fundamental role in gene control. The 1981 Wolf Prize was awarded for an understanding of chromosome structure, behavior, and function. These studies were made by examining maize kernels cytologically and morphologically (McClintock 1984a, p. 797). McClintock was continually observing organismal development as she examined the mitotic and meiotic cell divisions in breakage-fusion-bridge cycles. These, she believed, had activated silent elements that were potentially transposable. Transposition events can be observed phenotypically during the development of the maize kernel. McClintock showed that maize displays clonal development. This can be detected by the occurrence and timing of a transposition event when observed phenotypically as variegated sectors of a dominant or recessive character (e.g., expression of color or no color), during somatic tissue development in the maize kernel (see McClintock 1978b; Coe & Neuffer 1978).

[6] Lederberg had submitted his Nomination on January 18, 1982; see Chapter 10.

[7] H. Roman to J. Lederberg, February 23, 1983, Roman Papers, UW, Box No. 9. Folder No. 3.

[8] H. Roman to J. Lederberg, May 7, 1983. Roman Papers, UW.

[9] Email, Professor Bruce Alberts to L. Kass, September 15, 2023; Comfort 2001b.

[10] President Richard Nixon's Daily Diary, May 21, 1971, Appendix B.

Furthermore, by 1980 transposable elements were included in textbook discussions of regulation of gene expression and differentiation during development (e.g., Avers 1980, p. 336ff). The main purpose of studies seeking to understand how gene expression is regulated was to answer the questions posed regarding differentiation and development in multicellular organisms. If all cells have a complete set of genes, then phenotypic differences should arise due to differential gene action [gene expression]. The latter had been observed as genes are turned on and off in different cell types, as shown by the operon model of gene control in prokaryotic cells. At this time, knowledge of the molecular organization of chromosomes in eukaryotic organisms seemed essential for understanding the interactions that regulate gene expression. Transposable elements found in prokaryotic and eukaryotic species (yeast, *Drosophila*, and corn) appeared to contribute to an understanding of the control of gene expression in these organisms. Barbara McClintock was noted as the first person to describe genetic studies showing altered gene expression as seen in variegated phenotypic patterns in kernels on the ears of corn, and "brilliantly attributed the complex phenomena to the behavior of transposable elements" (Avers p. 363).

Also, three significant volumes were published in early 1983 that featured the importance of transposable genetic elements. In January 1983, James Shapiro's (1983) edited volume, *Mobile Genetic Elements*, included a chapter on "Controlling Elements in Maize," authored by Nina Fedoroff (1983). This first chapter summarized McClintock's genetic studies of controlling elements [transposable elements] and detailed current molecular studies identifying maize transposable elements (see below). A second important volume, "*Genes*" issued in April 1983, highlighted Shapiro's book and the recent work on mobile elements in eukaryotes. Authored by Benjamin Lewin, the founding editor of the journal *Cell* in 1974, this first volume of *Genes*, was intended to "discern general principles and describe the state of the art" in an exciting new field "that has all but overwhelmed the traditional discipline of genetics" (Lewin 1983, p. v, see below).

In addition, Bruce Alberts, who had co-nominated McClintock for the 1983 Nobel Prize, co-authored the first edition of another important textbook issued in early 1983: *Molecular Biology of the Cell*. Therein, the authors drew attention to the importance of transposable elements in evolution, plant and animal diversity, and genetic control. They referenced McClintock's (1956) Cold Spring Harbor Symposium presentation for the potential importance of maize "controlling elements" toward an increase in biological diversity (Alberts et al. 1983, pp. 473–475; McClintock 1956b). Like Lewin's *Genes*, Albert's texts have educated many students over the years and have continued to be revised and updated. The most recent edition was published in 2022 (Alberts et al. 2022).

Another volume that may have played a role in bringing McClintock's contributions to the attention of the Nobel Committee's "in-depth" reviewers (see below) was a new biography by Evelyn Fox Keller (1983) featuring McClintock's life and work. It appeared in the summer of 1983, just as the deadline for reviews was due to the Nobel Committee (see below). As I mentioned previously, Keller had interviewed McClintock and published a featured article about her that appeared in October 1981, just before the announcement of the Wolf Prize to McClintock and Stanley N. Cohen.

WHO DECIDES?

As I discussed in a perspective regarding H. J. Muller's (1946) Nobel Prize (Kass 2018), many people who should have shared the prize have not – this decision is influenced by many factors, and in some cases, this award is quite political, as was shown by István Hargittai in his 2002 book titled *The Road to Stockholm*. The decisions are dependent on the nominations (and nominators)[11] and on the members of the Swedish Academy and/or the Karolinska Institute who review them. While bringing the reader's attention to the limitations of the Nobel Award process, Hargittai (2002: Chapters 3 and 12) explains why some have received the prize and others have not. He details that

[11] See Hargittai (2002, p. 22), for the six groups who may nominate for the prize in physiology or medicine. See also E. Norrby (2010).

the review process is a considerable task that is bound to produce mistakes, since the committees are not and have not been the most informed on the weight of discoveries in subfields in which they may have no extensive experience. They very much rely on the information submitted by the nominators (see Norrby 2010 below). This was Jacob's concern when he wrote to Roman regarding the nomination they proposed to submit for McClintock.

Another view of the politics of the Nobel Prize is offered in a book by Friedman (2001). Even though it focuses on the prizes in chemistry and physics, it depicts how the members of the committees make decisions on who gets a Nobel Prize, and how many of those judgments included political, national, self-interest, and resentful, negotiated, and closed-minded agendas. It shows extensive interpersonal clashes between committee members and how downplaying of important nominees and their works was a frequent practice.

Erling Norrby (2010), a member of the Karolinska Institute's Committee for Nobel Prizes in Physiology or Medicine, has provided additional clarification of the selection procedure. Nobel Prizes for Physiology or Medicine are only given for a Discovery (Norrby 2022a). The selection process begins once all nominations have been submitted by the January 31 deadline (Hargittai 2002, pp. 22–24; Norrby 2010, pp. 17–21).[12] In February, the long list of candidates is first reviewed by a Committee (5 members and 10 adjuncts) of the Karolinska Institute. All nominations are scrutinized by the committees, but 80%–90% of them had already been encountered in work during preceding years (Norrby 2010, p. 20). The list of nominees is shortened in March. Individual members of the committee may be "in-depth" reviewers, but outside reviewers of both national and international origins are used.[13] All "in-depth" reviews are submitted to the full committee by the end of August for discussion, and are collected in a paper-bound form (a booklet). The booklets provide the background material for final discussions that take place with the 50-member Karolinska Institute Nobel Assembly in mid-September. Occasionally, there is a need for two meetings. The announcement by tradition occurs in the first complete week of October.

All documents supporting McClintock's nominations may have been brought to the attention of the full committee for consideration, as part of the "in-depth" outside reviews due by the August deadline, as noted above. The Nobel Committee would certainly have had much current information to ponder as they deliberated, over a period of eight months, on nominations for Barbara McClintock that were previously and recently submitted for the 1983 Nobel Prize in Physiology or Medicine.

The files on Nobel Prize winners are closed for 50 years (Tønnesson 1999; Hargittai 2002, p. 16; Norrby 2010, p. 98), and McClintock's documents will not be available to researchers until 2034.[14] The archives of the committee for physiology or medicine, in contrast to the committees for physics and chemistry, do not present a discussion about candidates. It only notes which candidates are worthy of a prize and reviews those that have not yet been nominated or not yet found worthy. Hence, even when the archives are opened for the 1983 prizes, procedures clarifying the committee's decision for an unshared prize in Physiology or Medicine for 1983 may not be a rich source of information for details on McClintock's Nobel Prize.[15]

THE DYNAMIC GENOME AND CELL BIOLOGY

As mentioned above, *Genes*, by Benjamin Lewin (1983), may have lent further support to the importance of transposable genetic elements and their broader implications. The text asked, "What is a gene, how is it reproduced, how is it expressed, what controls its expression?" The book is divided

---

[12] Additional details kindly provided by Professor Erling Norrby to L. B. Kass, emails, September 19 & 20, 2023. Norrby was an adjunct member of the Committee for the 1983 Nobel Prize for Physiology or Medicine; I thank Dr. Bruce Alberts for introducing me to Dr. Norrby.

[13] The latter since 1970.

[14] Email Professor E. Norrby to L. B. Kass, September 19, 2023. For more information on the Nobel Prizes, see books by Erling Norrby (2010–2022b).

[15] Email Professor E. Norrby to L. B. Kass, September 19, 2023.

into ten parts. "Part 10" is titled "The Dynamic Genome: DNA in Flux," and the introduction to Part 10 featured various quotes from Barbara McClintock's (1956) Cold Spring Harbor Symposium presentation, describing "controlling elements" in maize (McClintock 1956b [©1957]). Chapters that followed "Part 10" summarize the most recent studies of transposable elements in prokaryotic and eukaryotic organisms. One important point noted by Lewin was that "We can now recognize the 'controlling elements' [of maize] as 'transposable elements' with several functions, analogous to those described for other systems." However, the molecular biology of McClintock's "controlling elements" awaited sequence information to further understand the three families of maize controlling elements most investigated at that time: *Ds-Ac*, *Spm/En*, and *Dt* (Lewin 1983, p. 612). By late 1983, additional molecular studies would be published in *Cell* (see below) providing further understanding for the controlling element families in maize. *Genes* became one of the most important textbooks for the current understanding of these topics, and over the years, it has continued to be updated and new editions issued. The most recent is *Lewin's Genes XII* (Krebs et al. 2018).[16]

MOLECULAR STUDIES IDENTIFY TRANSPOSABLE ELEMENTS IN MAIZE

As mentioned in the previous chapter, by 1983 maize controlling elements were being investigated at the molecular level,[17] and soon would be isolated and sequenced. In a book chapter published in January 1983, Fedoroff (1983) summarized the status of maize "controlling elements." She provided the background for genetic and molecular studies of the maize system that McClintock had initiated in the 1940s. McClintock had genetically identified changes in the genome [chromosomes] that were observed in maize kernels and plants during somatic cell division. These changes were brought about by "transposable elements" [McClintock 1955, p. 250], which came to be called "controlling elements" (McClintock 1956a,b). The elements were recognized by their ability to move from one site to another [on a chromosome]. The insertion of a controlling element at a given site on a chromosome appeared to control the activity of adjacent loci (genes). The frequency and timing of events associated with the controlling elements were subject to developmental regulation, and their actions [phenotypic expressions] were observed to be tissue-specific.

Insertion of a controlling element on a chromosome [DNA molecule] could generate an unstable allele at a chromosomal locus. The instability was originally referred to as a mutable allele (*m*). Loss of the controlling element itself or loss of its ability to transpose changes an unstable mutable allele (*m*) to a stable allele. Deletions, duplications, inversions, and translocations were observed at sites where controlling elements were present. Chromosome breakage was also seen in the presence of some elements. Fedoroff also reported how the selective interaction among controlling elements allowed different families of those elements to be distinguished. There appeared to be more than one family of controlling elements. Each family had two types of elements. There were those capable of independent excision and transposition, and those that are not capable of transposition. As mentioned above, three families of maize "controlling elements" were known at that time: *Ds-Ac*, *Spm/En*, and *Dt*.

Fedoroff (1983) reported that controlling element insertion mutations are of two basic types. One type is inherently unstable (*m* alleles) and is associated with the insertion of an element capable of autonomous excision and transposition. Such controlling element insertion mutations had been termed autonomous by McClintock (1950 [*PNAS*]), and the elements causing such mutations were later called "autonomous" controlling elements. The second type of controlling element mutation is not intrinsically unstable, but it becomes unstable when an autonomous element is present elsewhere (in *trans*) in the genome (chromosomes). The elements causing such mutations do not appear to be

[16] See also Krebs et al. (2021).

[17] Fedoroff and Mauvais (1979), Fedoroff et al. (1980), Chaleff et al. (1981), Burr and Burr (1981), Fedoroff (1982), Fedoroff et al. (1982), Fedoroff et al. (1983a) (presented August/September 1982), and Fedoroff et al. (1983b) (submitted July 1982).

capable of autonomous transposition and came to be called "non-autonomous" controlling elements. McClintock did not use the term non-autonomous in her 1950 *PNAS* article with respect to the mutant locus that is not inherently unstable. Instead, McClintock described the "mutable locus" as one that "requires a separate activator for instability to be expressed," and she termed it "the activator requiring class." The terms for autonomous and non-autonomous controlling elements continue to be applied to transposable elements (transposons) governed by a two-element interacting system in maize and other organisms.

According to Fedoroff (1983), it appeared likely that non-autonomous elements (e.g., *Ds*) were derived from *trans*-acting autonomous elements (e.g., *Ac*) by loss of functions needed for transposition, or another function such as influencing gene expression (Chaleff et al. 1981; Fedoroff et al. "in press" [published November 1983b]). Forthcoming molecular studies permitted the isolation and structural analysis of maize controlling elements and provided evidence that *Ds* was derived by a deletion in the *Ac* element (Courage-Tebbe et al. 1983; Fedoroff et al. 1983d; Doring & Starlinger 1984).

The September 1983 issue of *Cell* featured a variegated maize kernel on its cover. The image depicted sectors with revertant gene expression caused by the controlling element *Ds* at the *Shrunken* (*Sh*) locus. It accompanied a report on the structure of the *Sh* locus in a maize strain containing an unstable recessive mutation caused by the transposable controlling element *Dissociation* (*Ds*) (Courage-Tebbe et al. 1983). By November, the transposable maize controlling elements *Ac* and *Ds* had been isolated and sequenced (Fedoroff et al. 1983d). Evidence was reported that the genetically defined *Ac* element corresponds to a 4.3-kb insertion, while the two *Ds* elements correspond to 4.1- and 2.0-kb insertions, all near the 3′ end of the *Waxy* (*Wx*) transcription unit. Accompanying this study was a report that the *Wx* locus had been identified and isolated by molecular analysis (Shure et al. 1983; Fedoroff et al. 1983c). It had been demonstrated that a mutation caused by the controlling element *Dissociation* (*Ds*) was attributable to an insertion of approximately 2.4 kb at the *Wx* locus. Furthermore, by October 1983 Peacock's group in Australia had submitted a report that molecular studies of *Ds* controlling element mutations at the alcohol dehydrogenase locus (*Adh1*) had been isolated and sequenced (Sutton et al. 1984).

The ubiquitous nature of transposable genetic elements and studies of their molecular structure, which began years before, found a new paradigm deserving of recognition. The discovery of movable genetic elements was original to Barbara McClintock and justified consideration by the Nobel Committee.

McCLINTOCK'S ENDEAVORS DURING 1983

During 1983, McClintock continued to engage with colleagues via correspondence, invited lectures, and was recognized as an outstanding alumna by her alma mater, Cornell University. Replying to a letter of February 18, 1983, from Elizabeth Blackburn requesting reprints or preprints of articles dealing with telomeres for a review to be published in the *Annual Review of Biochemistry*, McClintock wrote that her studies related to the behavior of newly broken ends of chromosomes had commenced in the early 1930s, and continued into the 1940s.[18] She no longer had reprints for those papers.[19] She also wrote that, "In those early days one did a thorough job before publishing. Thus my articles are long and detailed – the type of article no one wants to read in this day of avalanching mini-reports." She enclosed a list of her publications relating to fusion of broken ends of chromatids (the chromatid type of breakage-fusion-bridge cycle) and of chromosomes (the chromosome type of breakage-fusion-bridge cycle), and checked ones relevant to the topics. She added, "At least you will have received references to my studies of the behavior of broken ends of chromosomes, even though you may have no desire to examine these reports." She explained that in the embryo,

[18] Letter B. McClintock to E. Blackburn, March 11, 1983, Spectre 1993: Series I: Correspondence Blackburn, Elizabeth H. 1983.

[19] See Kass (2013ff.) for reprints of McClintock's early papers from the 1930s.

the broken end "heals." The healing process provides a new telomere. She also wrote that she had observed chromosomes with broken ends that did not heal and hypothesized that such ends lack an enzyme required for the formation of new telomeres. She also thought that telomere repair enzymes were not found in the gametophyte or endosperm in her maize stocks. It's interesting to note that McClintock did not use the term telomere in her original publications, but she applied the term in this communication with Blackburn.[20] A co-authored review article with Jack W. Szostak was published the following year (Blackburn & Szostak 1984). They had read the papers recommended by McClintock and cited her research for broken and fused chromosomes.[21] Later, Blackburn and Szostak discovered that telomeres have a particular DNA that prevents chromosomes from breaking down. As a graduate student, Carol Greider, together with Blackburn (in 1984), discovered telomerase, which helps produce protective DNA "caps" on the ends of telomeres. Blackburn, Greider, and Szostak were awarded the 2009 Nobel Prize in Physiology or Medicine "for the discovery of how chromosomes are protected by telomeres and the enzyme telomerase."[22]

In April, McClintock accepted invitations to lecture at Wayne State University in Detroit, Michigan, and to the Graduate Genetics Course at SUNY Stony Brook, Long Island, New York. In these lectures, she reviewed the history of genetics and her career within the subject. She provided background on how she first identified maize transposable elements, which she called controlling elements, and described their contribution to the action [expression] of genes.[23]

By May, she returned to thoughts of reprograming genomes in a lecture at the University of Chicago. She had emphasized the subject at her Stadler Genetics Symposium, "Mechanisms that rapidly reorganize the genome" and further presented those ideas in this presentation on "Genome shock as a means of reprogramming genomes."[24] The subject of "trauma" (genome shock) as another mechanism for initiating change in genome organization would be further explored in an invited talk to the Tissue Culture Association held in Orlando, Florida, in mid-June (McClintock 1983).[25] The challenges facing genomes would continue to be McClintock's focus as she prepared later presentations, and it became the subject of her Nobel Prize lecture (McClintock 1984a,b; see below).

Graduate degrees were showered on McClintock at the end of the 1982–1983 academic year. In mid-May, she received an honorary Doctor of Science from Bard College, Annandale-on-Hudson, NY.[26] SUNY Stony Brook presented her with what appeared to be an earned degree (not honorary) with the following decree: "The Degree of Doctor of Science, with all the honors, rights and privileges belonging thereto in testimony whereof this Diploma is granted bearing the seal of the university given at The State University Center at Stony Brook this twenty second day of May, nineteen hundred eighty three." Accompanying the award was a tribute to her accomplishments: "… you made fundamental genetic discoveries long before the larger scientific community could appreciate their true value. Now, the molecular biology community acknowledges the foresight and imagination embodied in your work. For your patience and persistent labor in a lifetime of dedication to science, The State University of New York is honored to confer upon you the degree of Doctor of Science."[27] In early June, she was awarded an honorary Doctor of Science degree from New York University and attended the pre-commencement dinner with family members.[28]

[20] The term telomere was suggested by H. J. Muller (1938) in a paper that mentioned the recently published work of McClintock.

[21] See also Blackburn (1992) for recollections of her correspondence with McClintock about this topic.

[22] See the Nobel Prize in Physiology or Medicine (2009), https://www.nobelprize.org/prizes/medicine/2009/summary/

[23] Spectre 1993: Series III, Lectures April 15, 1983, "Some unexpected results of a simple experiment"; April [28], 1983, State University of New York at Stony Brook ["History of origin of controlling elements: their contribution"].

[24] Spectre 1993: Series III, Lecture [10 May] 1983, University of Chicago.

[25] Spectre 1993: Series III, Lecture, June 16, 1983; see reprint of McClintock's published abstract in Kass (2013ff., p. 3.141).

[26] Spectre 1993: Series I, Bard College, December 7, 1982, Bard College to Barbara McClintock, Honorary Doctor of Science, May 18, 1983, https://families.bard.edu/commencement/honorary-degrees/.

[27] Spectre 1993: Series II, State University of New York 1983, May 22, 1983.

[28] Spectre 1993: Series I, New York University. Doctor of Science June 3, 1983, pre-commencement dinner, June 2, 1983, attended with her niece and niece's husband.

A few weeks later, McClintock gave a public symposium presentation on June 29, 1983, at the Genetics Society of Canada annual meeting held at the University of Ottawa.[29] The topic of the symposium was "Unraveling Heredity: Views on Research in Genetics," and her lecture was titled, "Lessons learned from a simple experiment." The aim of her topic was to review experiments conducted during her career beginning in 1931, and which suggested some aspects of gene regulation that "the future must unravel." Each experiment had been a lesson that led to another experiment. She reviewed her experimental results on ring chromosomes, healing of broken chromosomes, and her surprise at finding mobile elements, during studies of broken chromosomes. She mentioned that current molecular analysis had revealed much basic information for transposable elements "with more basic information expected in the very near future." She recalled that she had stopped growing maize for continued [genetic] analysis in 1968, because it was time for the molecular analysis and also because "another major research project, long needing attention, became her focus." The major research project was not identified but it was probably the work on the races of maize, which had been published only a few years earlier (McClintock et al. 1981).

McClintock was honored with an "Outstanding Alumni Award" during Cornell's College of Agriculture and Life Sciences "Roundup" on September 24, 1983. Although she was not available to receive her award in person, Dean David Call sent her the "Outstanding Alumni Award" plaque, which had been presented to the other four winners the previous fall.[30] A few weeks later, she received international and definitive recognition for her pioneering discovery of the dynamic genome.

THE 1983 NOBEL PRIZE IN PHYSIOLOGY OR MEDICINE

Alfred Nobel's initial "Will" (1893) provided funds for a committee originally formed from the faculty of medicine of the Karolinska Institute [Medical School] to award Nobel Prizes for new discoveries in the fields of physiology and medicine. The prizes would be given to the most worthy, be they men or women, with no consideration to their nationality. This "Will" was superseded in 1895 with one having new guidelines for "prizes to those who, during the preceding year, shall have conferred the greatest benefit to mankind." Five percent of his estate was left to prizes for "the person who shall have made the most important discovery within the domain of physiology or medicine" (Bergström 1984).[31]

The Assembly of the Karolinska Institute announced on "10 October 1983 the award of the Nobel Prize in Physiology or Medicine to Barbara McClintock of the Carnegie Institution of Washington, Cold Spring Harbor, N.Y. United States, for her discovery of mobile structures in the genome" (*Les Prix Nobel 1983*, p. 11; Figure 11.1).[32] It is usually reported "for the discovery of Mobile Genetic Elements." All winners were expected to deliver a lecture prior to receiving their prize. McClintock delivered her lecture on December 8, 1983 (McClintock 1984b).[33] His Majesty the King of Sweden presented the award at the ceremony held on December 10, 1983, in the "grande salle du Palais des Concerts" (Soderlind 1983, Figure 11.2). For each winner, a speech was first presented that

---

[29] Letter B. McClintock to D. B. Walden, December 24, 1982, Spectre 1993: Series I, Genetics Society of Canada. Spectre 1993: Series III, Lecture, June 29, 1983.

[30] *Cornell Chronicle*, September 1, 1983. Cornell *Alumni News*, October 1983. Letter Dean Call to B. McClintock, January 13, 1984, B. McClintock Alumni File, Cornell College of Agriculture & Life Sciences, Office of Alumni Affairs and Development, 270 Roberts Hall, Ithaca, NY.

[31] Opening address by the President of the Nobel Foundation, K.S.D. Bergström at the 1983 Nobel Celebration on the 150th anniversary of Alfred Nobel's death, December 10, 1983. Bergström had shared the 1982 Nobel Prize in Physiology or Medicine.

[32] The published version is "pour sa découverte des structures mobiles de la masse génétique," *Les Prix Nobel 1983*, p. 11 (1984); I thank Jim Shapiro for the best translation from the French.

[33] A text of McClintock's speech of December 8, 1983, can also be seen at this link: https://www.nobelprize.org/uploads/2018/06/mcclintock-lecture.pdf, and the recorded presentation of December 8, 1983, at this link, https://www.nobelprize.org/prizes/medicine/1983/mcclintock/lecture/

FIGURE 11.1 Karolinska Institute announced the award of the Nobel Prize in Physiology or Medicine to Barbara McClintock (center) on October 10, 1983, for the "Discovery of Transposable Genetic Elements." On her way to the press conference held at Cold Spring Harbor Laboratory, she celebrated the news with CSHL colleagues, Joseph (Joe) F. Sambrook (left), Assistant Director for Research, and staff scientist Terri Grodzicker (right), who was then investigating "the biological role of inserted gene products and their controlling elements." (Courtesy of Cold Spring Harbor Laboratory Archives.)

summarized the importance and significance of the award. Dr. Nils G. Ringertz, Professor of Cell Biology and Vice President of the Committee for the prize for physiology or medicine gave the summary of McClintock's discovery and work (*Les Prix Nobel 1983*, p. 8; Ringertz 1984, pp. 27–29).[34] The 1983 Nobel Prize in Physiology or Medicine recognized "a great discovery about the organization of genes on chromosomes and how these genes, by changing places, can alter their function." Her discovery resulted in "new knowledge of great medical importance," he added. Ringertz reported that McClintock found variegated plants that were caused by a small piece of a chromosome that had moved to a place on the chromosome close to a gene coding for pigment. The genes were believed to undergo mutations during the development of the kernel. The effect was to switch off the gene, and often, the chromosome showed a break at the site of integration. McClintock called these types of genetic material "control elements" [*sic*] since they altered function of neighboring genes. She [genetically] mapped several families of these elements between 1948 and 1951. She predicted that "mobile genetic elements" were probably present in other organisms. Her work "received little attention," he told the audience, because when it was first presented it was overshadowed by

[34] See text of Ringertz speech presenting McClintock to the King of Sweden on December 10, 1983, at this link, https://www.nobelprize.org/prizes/medicine/1983/ceremony-speech/, and recorded presentation speech of Ringertz December 10, 1983: https://ww w.nobelprize.org/prizes/medicine/1983/mcclintock/lecture/. Soderlind (1983) reported that each award was accompanied by a $190,000 cash stipend.

FIGURE 11.2 Barbara McClintock received the Nobel Prize in Physiology or Medicine for the "Discovery of Transposable Genetic Elements" at the award ceremony held on December 10, 1983, in Stockholm, Sweden. Here, she holds her Nobel diploma and gold medal, which had been awarded to her by Sweden's King Carl XVI Gustaf. (Courtesy of Cold Spring Harbor Laboratory Archives.)

the discovery that the DNA molecule stores the genetic information in its structure. Any change in one of the DNA building blocks [nucleotides] could have serious effects. Ringertz may have been thinking of the paper McClintock published in November 1953 in *Genetics* (issued 1954), eight months after Watson and Crick (1953a) published their model of DNA in the journal *Nature*, and after Watson had presented it at the June Cold Spring Harbor Symposium on Viruses (Watson & Crick 1953b). Ringertz emphasized that, at the time, few geneticists were prepared to consider that genes could move in the "irresponsible manner that McClintock proposed for controlling elements." In the mid-1960s, mobile genetic elements were found to play an important role in the spread of antibiotic resistance, which is a serious medical problem in hospitals. Additional medical significance was attributed to mobile genetic structures when they were discovered to underlie the formation of antibodies. He explained that human chromosomes carry a set of mobile building blocks for antibody genes. By combining these blocks in the cell, the body can generate "millions of genes for antibodies." He further added that in recent years, similar structures have been found to transform a normal cell into a cancer cell.

Ringertz summarized that maize mobile genetic elements have been found to have counterparts in bacteria, humans, and other animals. "Working completely on her own," he emphasized that she carried out these studies that demonstrated the "heredity information is not as stable as had previously been thought." Her discovery led to new insights into "how genes change in evolution, and how mobile genetic structures on chromosomes can change the properties of cells."

It seems clear from Ringertz's summary that the Committee was familiar not only with the nomination, as submitted by Herskowitz and Alberts, but also with the documents that had been submitted previously with nominations for her Nobel Prize. Those nominations had included extensive explanations for the significance of transposable elements, which McClintock and others had called controlling elements. Ringertz's review providing background for the discovery of transposable elements demonstrates an understanding of the significance of these elements in reorganizing the genome and its regulation during development of an organism.

McCLINTOCK'S NOBEL LECTURE

McClintock's Nobel lecture did not focus on the "discovery of mobile genetic elements" for which she had been awarded the prize in physiology or medicine (McClintock 1984a,b; Shapiro 2013). It did, however, provide a brief summary of her career and the lessons she had learned through a series of experiments she had conducted during that time. Her lecture focused on the challenge that cells face to reorganize their genomes in response to trauma or genome shock (Shapiro 2013). McClintock had recently lectured on these topics (see above), and she focused on the experiments as lessons that "revealed how a genome may react to conditions for which it is unprepared." She had learned that broken chromosomes were "responsible for the activation of potentially transposable elements that are carried in a silent state in the maize genome." The elements' mobility allowed them to associate with different gene loci and to "take over control of action [expression] of the gene" wherever it would enter. In addition to "modifying gene action," the elements could "restructure the genome." Such changes could involve "a few nucleotides, to gross modifications involving large segments of chromosomes, such as duplications, deficiencies, inversions, and other more complex organizations." Such effects of the first traumatic event alerted her to anticipate unusual responses of a genome to various shocks that it might receive. Her examples showed the importance of stress in initiating genome modifications by mobilizing cell mechanisms that could restructure genomes. She concluded, "We know about the components of genomes that could be made available for such restructuring. We know nothing, however, about how the cell senses danger and instigates responses to it that often are truly remarkable." This was a challenge for future research.

McClintock's lecture did not concentrate on any evidence for her discovery of transposable elements, which was not as exceptional as one might think (Shapiro 2013). Norrby (2022a) explained that other Nobel Laureates had focused their lectures on future work rather than the subject of their discovery. For example, the 1962 Prize in Physiology or Medicine was given to Watson, Crick, and Wilkins "for their discoveries concerning the molecular structure of nucleic acids and its significance for information transfer in living material" (Nobel Prize 1962). Crick's Nobel lecture asked certain general questions about the genetic code and how far they could be answered. For example: Do codons overlap? How is the base sequence divided into codons? How many bases are there in any one codon?[35] Watson's lecture was an historical perspective of RNA and protein synthesis, with many questions remaining to be answered about the functional significance of ribosomal RNA.[36] Later, RNA would be found to play an important role in gene regulation and in the replication of types of movable elements classified as Class I transposons (Mattick & Amaral 2023; see below).

I believe it is mistaken to perpetuate the story that McClintock's emphasis and insistence that controlling elements played a role in differentiation and development delayed the recognition of her Nobel Award (Mattick & Amaral 2023, p. 82). Nobel prizes for physiology or medicine "are only

[35] Francis Crick Nobel Lecture, December 11, 1962, On the Genetic Code, https://www.nobelprize.org/prizes/medicine/1962/crick/lecture/

[36] James Watson, Nobel Lecture December 11, 1962, The Involvement of RNA in the Synthesis of Proteins, https://www.nobelprize.org/prizes/medicine/1962/watson/lecture/

awarded for discoveries" (Norrby 2022a). McClintock won the Nobel Prize "pour sa découverte des structures mobiles de la masse génétique" (*Les Prix Nobel 1983*, p. 11): "for her discovery of mobile structures in the genome"; for the "discovery of transposable genetic elements," which indeed she was the first to identify genetically (Chapter 7).

When McClintock spoke with graduate students, including me, about her research, she often assumed we already had the background to understand the discussion. Many details would be omitted, and therefore, as David Botstein told me, "McClintock spoke in shorthand."[37] McClintock had long ago demonstrated to her satisfaction that parts of chromosomes were moving in the genome, which she had termed "transposable elements" in 1955 (Chapter 8; McClintock 1955). She called these elements "controlling elements" because when the element moved (transposed) next to a genetic locus with a known function, such as color expression in a maize kernel, the element seemed to control the action (expression) of that locus (McClintock 1956a,b).

During a discussion with McClintock about controlling elements in maize, she explained to Professor Provine and graduate student Paul Sisco, in August 1980, that the controlling element is something that can give you dramatic changes (Provine & Sisco 1980). She did not know exactly where they were in the genome because the *Ds* and *Ac* elements moved, but she believed that the change in the DNA was subtle. She said that maize researchers were getting the enzymes for some of the loci and were looking for the DNA sequences.[38] One of her major questions was "what is the nature of the control?" McClintock mentioned that some people had commented that she must be pleased "that all this transposition is going on"; meaning that transposable elements have now been found in many eukaryotic organisms in addition to prokaryotes. She cautioned that they should not take that for granted since "The real point is control." What she was emphasizing is that the "real secret of all this [mobile genetic material] is control" and she believed "it was subtle." In other words, we now needed to understand how genes controlled development. Further, she wanted to know the molecular mechanism underlying the genetic control of the evolutionary process and for understanding development. Her collaborations with the molecular geneticist working to solve these questions would provide some answers (see above).

RATIONALE FOR DELAY OF HER RECOGNITION

There are several reasons why it took so long for mainstream biologists to recognize McClintock's studies of movable genetic elements as ubiquitous. As I briefly summarized (Kass 2000), McClintock worked with maize and many believed these findings were confined to that organism. The phenomena were often reported as anomalous and had only been observed in a few other organisms. McClintock's findings of movable genetic units, published in a major scientific journal (McClintock 1950), and followed by a presentation at the Cold Spring Harbor Symposia in 1951 (McClintock 1951), were reported as "making no impact" (R. Lewin 1983). As mentioned previously, this lack of recognition had been attributed to a dearth of understanding of the maize life history and the insistence that genes on chromosomes were static. After all, even McClintock had mapped the location of genes on chromosomes under such an assumption (McClintock 1931b) and provided the cytogenetic proof for crossing over (McClintock 1931a; Creighton & McClintock 1931) that had been hypothesized and shown genetically by Thomas H. Morgan and his research colleagues. Morgan had been awarded the 1933 Nobel Prize in Medicine or Physiology "for his discoveries concerning the role played by the chromosomes in heredity." The chromosomes harbored the genes in a linear array, and such genes corresponded to phenotypic traits. Morgan also provided a hypothesis and genetic evidence for crossing over for which McClintock and her student Harriet Creighton had demonstrated cytogenetic proof (see Chapter 5; Kass 2003).

---

[37] L. B. Kass interview with David Botstein, February 4, 1997.

[38] These would soon be isolated and sequenced by researchers in the U.S., Germany, and Australia (Doring & Starlinger 1984); see above.

McClintock herself attributed "the dogma of the constancy of the genome" as a reason that her idea for a dynamic genome was questioned (R. Lewin 1983, p. 404). Certainly, those in the genetics community who had found and published on similar phenomena embraced her studies of transposable elements, yet some still questioned their functions (Chapters 9 and 10; Starlinger 1984). Others accepted with little question, taught her work to their classes, and included her findings in their textbooks (Chapter 10).

Furthermore, a system other than the classical gene might be responsible for macromutations as postulated by Richard Goldschmidt. Such sudden mutations could be at the root of evolutionary changes. McClintock seemed to be in full support of Goldschmidt's hypothesis (McClintock 1951; Goldschmidt 1951) and he of hers (Chapter 7; see also Goldschmidt 1955). Both questioned the currently accepted paradigm of evolutionary genetic thought. They emphasized reorganization of the genetic material rather than "point mutations" within a gene. For this reason, their ideas of a genetic system that did not exclusively rely on the classical stationary gene on a chromosome may not have been seriously considered.[39] New ideas in science, which contradict the prevailing model, dogma, or then current paradigm, are not easily accepted (Kuhn 1962, 1970, p. 24; Allen 1975, p. 72; R. Lewin 1983; Kass & Chomet 2009; Shapiro 2023). The discovery of anomalies was considered revolutionary and led to a paradigm shift. Such "scientific revolution," Kuhn believed, was the phase in which underlying assumptions in a subject are reexamined and a new paradigm is established (Kuhn 1962, 1970: Section XII).

Additionally, as Professor Ed Lewis suggested, geneticists may have associated the erroneous ideas of Lysenko with McClintock's belief that her system of transposing elements could direct mutations and thus change a plant's phenotype (Chapter 8).[40] Lysenko's denial of the existence of genes and belief that he could make new plants from existing ones may have played a role in some researcher's skepticism of McClintock's new model for the units of heredity.

Students of maize and snapdragon genetics had observed occurrences similar to those reported in maize by McClintock in their own field studies. These mutations were also considered anomalous and confined to those eukaryotic organisms. Additional data published in 1953 and two symposium presentations by McClintock in 1955 (McClintock 1956a) and 1956 (McClintock 1956b) brought attention to her work. The latter presentation stressed the idea of genetic regulation by extragenic elements that McClintock called "controlling elements" (McClintock 1956a,b). During this time, some researchers gave more than a nod to her studies but considered they might be restricted to maize (Chapter 9), while others would later seek a molecular explanation for such regulatory systems. McClintock later recognized that gene regulation in maize "turned out to represent a quite different mode of gene regulation" from that represented in bacteria (McClintock 1987: Chapter 9).

As described in Chapter 9, between 1959 and 1960 molecular biologists provided experimental support for the operon hypothesis of gene regulation in prokaryotic cells, published first in French. Their review article, published in English (Jacob & Monod 1961a), brought attention to this work but did not make any correlation to studies in eukaryotic cells. McClintock (1961a) recognized the prokaryotic regulatory system as similar to the controlling elements system she had been studying for many years and soon published her reasoning in *American Naturalist.* Campbell (2013) also believed that McClintock's scientific article did not have "a substantial impact on the broader molecular genetics community, except for those who had previously followed it." Even though her paper had been cited 35 times between 1962 and 1964 (according to the 1984 *Science Citation Index*), his assessment was that "few if any investigators changed the course of their work because of it." The discovery of a molecular model for DNA during the early 1950s also may have overshadowed

[39] See also Steven M. Stanley (1979) *Macroevolution*, in support of the idea of punctuated equilibria as initially proposed by N. Eldredge and S. J. Gould (1972) also may have played a part in accepting an alternative outlook on heredity and evolution. The theory states that evolution occurs primarily through short bursts of intense speciation, followed by lengthy periods of stasis or equilibrium. The model postulates discontinuous evolution; i.e., that species can change very quickly over evolutionary time rather than by continuous, gradual, and incremental changes.

[40] L. B. Kass interview with Ed Lewis, January 22, 1997.

McClintock's early investigations and focused the efforts of the molecular geneticist in a different direction (Chapter 8). Jacob and Monod's recognition of McClintock's studies on gene regulation (Jacob & Monod 1961b; Monod & Jacob 1961) seems to have had limited influence. Delayed recognition would come when mobile genetic elements were discovered in other organisms.

As early as 1969, repetitive non-coding DNA had been identified and proposed as a gene regulator for higher cells (Britten & Davidson 1969). The maize system of gene regulation was described as an example that would fit this model of gene regulation:

> Genes are known in maize which display control over producer genes and are located in the genome at sites distant from the producer genes that they control. ... Control of the expression of the producer gene is accomplished through the interaction of the distant regulatory gene with the contiguous regulatory gene. This point has been demonstrated by insertion of the contiguous regulatory genes at different sites in the genome, near known genes, which respond to the same control system governed by the distant regulatory unit.
>
> *(Britten & Davidson 1969, p. 355)*

By 1972, repetitive non-coding DNA had been regarded as "Junk DNA" and by 1980 was considered "Selfish DNA" (Orgel & Crick 1980) "... DNA sequences which appear to have little or no function, such as much of the DNA in the introns of genes and parts of the DNA sequences between genes." Decades later, a new assessment provided the view that transposable elements can now be perceived as "... deeply integrated into the physiology, development and evolution of species. Controlling and controlled, the [transposable] elements are intertwined with virtually every facet of biology" (Feschotte 2023).[41]

Once movable genetic elements were discovered in bacteria, molecular geneticists familiar with McClintock's controlling elements correlated the prokaryotic transposable elements with those found in eukaryotic systems (Fincham & Sastry 1974; Bukhari et al. 1977). The general focus of the 1980 Cold Spring Harbor Symposia, *Movable Genetic Elements*, was on the mechanism by which DNA rearrangements occur and may affect gene expression (Chapter 10).

In the late 1970s, investigators began extensive studies seeking to identify and sequence maize controlling elements (reviewed in Fedoroff 1982, 1983). Movable elements were eventually identified in many eukaryotic systems.[42]

In the early 1980s, McClintock received major awards recognizing her contribution to an understanding of chromosome structure and to her discovery that genetic elements are not static, "mobile genetic elements," and to their role in gene control (Chapter 10; Kass & Chomet 2009: Table 2). Popular accounts of the 1981 Lasker Prize and the 1982 Horwitz Prize brought attention to "mobile genetic elements," and "movable gene segments," respectively (Chapter 10).

Months after McClintock was awarded the Nobel Prize "for her discovery of mobile structures of the genome," reviews of transposable elements appeared in the scientific literature in April, June, and December 1984 (Starlinger 1984; Freeling 1984; Fedoroff 1984; Doring & Starlinger 1984). None paid tribute to McClintock's Nobel Prize, which had been announced the previous October; possibly because her Nobel lecture (McClintock 1984a) was not published until November 1984, or because their reviews were in press before the award was announced. Starlinger (1984) provided a perspective on when transposable elements were discovered at the genetic level by McClintock and summarized the current understanding at the biochemical level for using transposons "to tag genes for isolation procedures."[43] Starlinger's review of "Transposable Elements" questioned the function

[41] See also Makalowski (2003, 2007a,b), Fedoroff (2012), Mhiri and Grandbastien (2022), Marchant et al. (2022), and Mattick and Amaral (2023, p. 85). A correction to Feschotte (2023) was intended as a change from "scientist emeritus" to "scientist emerita", not to "professor emerita," email, C. Feschotte to L. Kass February 20, 2024. Note, however, that I have found no documentation that McClintock was a "professor emerita" at CSHL.

[42] Scott et al. (1980), Bukhari et al. (1977); see also reviews in Starlinger (1984) (April) and in Doring and Starlinger (1984) (December).

[43] See Hannah and Schwartz (2009) for a "personal account of how of how maize molecular genetics came into existence."

of such elements in development: "Do the mobile elements play a role in development. Here I am doubtful at present." Plants devoid of the elements by genetic criteria develop perfectly well, he said. He asked if there were "copies of the element that function in a different way which we have not yet discovered?" Starlinger believed that answers to this question would eventually be found and that transposable elements by that time would "no longer be thought of as 'selfish DNA' devoid of any function but their own propagation."

Freeling (1984) provided an extensive review of plant transposable elements and insertion sequences, including comparisons found in other eukaryotic systems. He stressed the importance of getting molecular sequence data for a complete understanding of how transposons move and for an interpretation of their function. He also criticized the idea that transposable DNA was selfish and had no function. Fedoroff's (1984) article was published in the more popular venue *Scientific American*, and a composite image of maize kernels appeared on its cover. Fedoroff reviewed the genetic evidence for McClintock's studies of transposable genetic elements in maize and introduced the studies that had recently been conducted at the molecular level in her laboratory and in laboratories in Germany (Starlinger) and Australia (Peacock) but provided only a limited bibliography.[44] Doring and Starlinger (1984) reported on recent studies being conducted in several laboratories, for plant transposable elements at the DNA level. They hoped to find a structural basis for McClintock's genetic observations of transposable elements and for the underlying basis for the "concept that the expression of genes can be regulated (controlling elements)."

"TIME FOR THE MOLECULAR ANALYSIS" OF TRANSPOSABLE ELEMENTS

Following initial identification, isolation, and sequencing of maize transposable elements in the 1980s, the question of how transposons move in the genome (transpose) and their role (function) in organismal development and in species evolution became the focus of future research (Nelson 1987, 1988; Fedoroff 1989a–c; Berg & Howe 1989; Gierl et al. 1989). McClintock's maize stocks continued to be resources for molecular analysis (Fedoroff 1992; Fedoroff & Botstein 1992). Many publications detailed their genomic nature and possible functions (e.g., Saedler & Gierl 1996; Feschotte et al. 2002). Epigenetic systems were introduced as mechanisms for genome regulation (Holliday 1990; Haig 2004, 2012). Although defined first by Waddington (1942) to describe internal and external interactions between the environment and the genes, which lead to the development of the phenotype, the meaning of the term epigenetics has changed with time (Haig 2012). The current understanding is that epigenetic changes modify the activation of certain genes, but not the genetic code sequence of DNA (Berger et al. 2009; Haig 2012). DNA methylation is one type of epigenetic modification that has been found to explain the underlying molecular mechanism responsible for controlling elements and especially the *Spm* mutations first reported by McClintock (Fedoroff 2013b,c; Weil 2013). Furthermore, there is evidence that DNA methylation is associated with paramutation, imprinting, and other transposable element inactivation (Noshay et al. 2018; Liu et al. 2022; Mhiri & Grandbastien 2022). Also, as new mobile elements were discovered, suggestions for a universal system for naming them were proposed (Roberts et al. 2008).

As mentioned above, transposons are classified as autonomous or non-autonomous. In brief, molecular analysis has revealed that transposons encode their own transposase enzyme, which enables the mobility of the elements, while non-autonomous transposons require the transposase activity of another transposable element. Autonomous DNA transposons move via a "cut and paste" mechanism (Class II transposons). DNA transposons are delineated by flanking terminal repeats that mark the location where the transposase excises the DNA. These DNA elements then reintegrate at a different location within the genome. The excision from DNA leaves marks of these

[44] A bibliography listed McClintock's (1951) Cold Spring Harbor Symposium paper, McClintock (1965), Fedoroff (1983), and Fedoroff et al. (1983d).

flanking repeats that can be used to study the rate and level of DNA transposition events within a genome. The insertion of these transposons can affect the phenotypic expression of nearby genes and may completely disrupt genes near or into which they enter.

RNA transposons are called retrotransposons because they are transcribed into an mRNA and require a reverse transcription to integrate into the genome (Class I transposons). They work via a "copy and paste" mechanism and are believed to be responsible for increasing eukaryotic genome size (genomic expansion). The most common mobile elements in the human genome are the Long Interspersed Nuclear Elements (LINEs) and the Short Interspersed Nuclear Elements (SINEs). These retrotransposons are most abundantly represented by the autonomous LINE1 (L1) and non-autonomous Alu elements, respectively. Alu elements rely on the expression of the L1 in order to be reverse-transcribed and integrated into the genome. As their classifications signify, LINEs are longer than SINEs. This is due to the presence of a second DNA sequence that encodes the transposase. An emerging body of evidence indicates that LINEs and SINEs function to regulate gene expression by affecting chromatin structure, gene transcription, pre-mRNA processing, or aspects of mRNA metabolism.[45]

Mobile genetic elements are major components of most eukaryotic genomes, yet the magnitude of their impact on genome evolution and function remains a matter of investigation and controversy (Fedoroff 2013a; Bourque et al. 2018; Wells & Feschotte 2020; Fueyo et al. 2022; Pandita & Pandita 2023).

A NOBEL LAUREATE

McClintock was active and received continued recognition for many years following the award for which she is most remembered. During 1984, she lectured at national universities, and at Cold Spring Harbor's Banbury Center. She presented her Nobel lecture as a keynote address at a Symposium on Genome Rearrangements in Colorado, and traveled to Mexico for a special meeting at CIMMYT on strengthening agricultural research in Latin America and the Caribbean. After the meeting, she gave a special lecture on "Response of the genome to shock."[46] Rutgers University awarded her a Doctor of Science for "recognition decades before others that chromosomes contain 'controlling elements' capable of undergoing transposition and of influencing the expression of genes and that genetic systems are fluid, dynamically regulated, and subject to controlled reorganization," which had profoundly "transformed our view of life. … free of the constraint of prevailing scientific dogma, and, above all, uniquely creative in the way you see and think about biology, you are the very herald [precursor, forerunner] of the new era of molecular biology."

In 1985, she lectured and met with faculty and students at the University of Minnesota, presented a Plenary lecture at the First International Congress of Plant Molecular Biology held in Savannah, Georgia, then traveled to Cali, Colombia, to lecture on "The significance of challenge in revealing the potentials of the genome."[47] That year, she was elected an honorary member of the New York Academy of Sciences (Braaten 2017).

On March 8, 1986, McClintock was inducted into the National Women's Hall of Fame in Seneca Falls, New York. She was recognized for her work in cytogenetics and honored for her achievements in science and genetics. Susan Cooper, director of Libraries and Marketing at the Cold Spring Harbor Laboratory, accepted the award on McClintock's behalf.[48] McClintock originally declined

[45] Summarized from Deragon and Zhang (2006), Mattick and Amaral (2023), and from General Biology: Transposons, https://bccbio.commons.gc.cuny.edu/origins/transposons/

[46] Spectre 1993: Series III, Lectures 1984.

[47] Spectre 1993: Series III, Lectures 1985.

[48] "Hall to honor Harriet Beecher Stowe," *The Post-Standard*, Thursday, February 27, 1986, p. B-6(A). L. B. Kass interview with Susan Cooper, March 18, 1997. See B. McClintock Alumni File, Cornell University, for many Newspaper articles reporting this event.

the invitation, but Cooper assured her that the award would not interfere with her work. Cooper had assumed responsibility, on behalf of McClintock, for declining the many invitations to lecture and to personally receive awards. Cooper volunteered to accept the award on McClintock's behalf. McClintock said, "Tell them I am honored and delighted to be in such good company." Charlotte Williams Conable, Chairwoman of the Woman's Rights National Historical Park Advisory Commission, Cornell alumna (1951), and author of *Women at Cornell: The Myth of Equal Education* (1977), was the keynote speaker at the event.[49]

McClintock was invited to give the Page Laboratory Dedication at Cold Spring Harbor in 1987. She gave her last formal lecture at Brookhaven National Laboratory in May 1988, speaking about "The extraordinary range of DNA content and organization in plants and animals."[50] At the annual Carnegie Evening of May 4, 1989, Carnegie Institution of Washington, Distinguished Service Member Barbara McClintock was honored for her extraordinary discovery of transposable genetic elements in maize (Singer 1989, pp. 3–4). The featured speaker, Nina Fedoroff of the Department of Embryology, described her studies at the level of DNA molecules of the moveable genetic elements in the chromosomes of maize, which McClintock had discovered using classical genetic and cytogenetic techniques, decades before.[51] McClintock then spoke informally, paying tribute to the Institution's dedication to scientific freedom. She further lauded earlier Carnegie administrations for supporting her during years when most biologists failed to comprehend the significance of her work (Singer 1989, p. 4).

On June 16, 1992, Barbara McClintock celebrated her 90th birthday at the Cold Spring Harbor Laboratory. Many of her close friends and colleagues were present for the event (Figure 11.3). Nina Fedoroff and David Botstein (1992) presented her with *The Dynamic Genome*, a Festschrift of contributed memories written especially for the volume (Fedoroff 1994, pp. 7–8, 1995, pp. 230–231).

FIGURE 11.3 Barbara McClintock celebrated her 90th birthday with friends, Harriet Creighton (left) and Nina Fedoroff (right), among many others, at Cold Spring Harbor Laboratory on June 16, 1992. McClintock holds her major gift, *The Dynamic Genome*, a Festschrift edited by Fedoroff and David Botstein for the special occasion. (Courtesy of Cold Spring Harbor Laboratory Archives.)

[49] "Cornellians have a part in Women's Hall event," *Cornell Chronicle*, Thurs. March 6, 1986. "3 women inducted in hall of fame," *Finger Lakes Times*, Monday, March 10, 1986, p. 9. "Women's Hall Inducts 3," *Cayuga/Seneca, The Post Standard*, Monday, March 10, p. A-7 (A). *The Cornellian 1951*, Charlotte Elizabeth William [Conable] Vol. 83, p. 358, https://catalog.library.cornell.edu/catalog/314011

[50] Spectre 1993: Series III, Lecture, May 10, 1988.

[51] See *Carnegie Year Book 87* for Fedoroff's essay summarizing some of her maize molecular studies.

Friends and colleagues described how McClintock's pioneering research motivated their own investigations. Reviews of more recent literature reveals that McClintock's legacy continues to stimulate genomic studies (Feschotte 2023) and her name has been memorialized by posthumous honors, awards, and recognition (Table 11.1).

A few months later, on September 2, 1992, McClintock died at Huntington Hospital on Long Island. Her obituary appeared on the first page of the *New York Times* two days later (Kolata 1992). The story recounted memories that McClintock worked alone, chose not to publish her revolutionary findings for years, had no telephone, and never gave lectures! The illustrations of McClintock's Nobel-winning work on movable genetic elements were labeled "Jumping Genes," a term coined by Kolata in 1976 (see Chapter 10). The term continues as the popular name for transposons. A memorial service was held at Cold Spring Harbor on November 17, 1992. McClintock's friends and colleagues spoke at this celebration of her life (Peterson 1993).[52] Nina Fedoroff (1995) wrote McClintock's obituary for the National Academy of Sciences *Memoirs*. It was preprinted with permission in the January 1994 issue of *Genetics* (Fedoroff 1994). The summary was based on McClintock's publications, her own friendship with McClintock, and memories recorded in Keller's (1983) biography of McClintock.

In 2016, I brought to the attention of the maize genetics community the inaccuracy of the report that Barbara McClintock was buried at the Huntington Rural Cemetery of Long Island, New York (Kass 2016). Cold Spring Harbor Archivist, Clare Clark, then reached out to McClintock's niece and nephew, Marjorie M. Bhavnani and Peter McKinley, for clarification on where their aunt Barbara McClintock was buried. With their permission, I provide here the summary that I previously shared

TABLE 11.1
McClintock's Legacy: Posthumous Honors, Awards, and Recognition

1993	Benjamin Franklin Medal, The Carnegie Institution of Washington, Washington, DC
2004	Barbara McClintock Professorships established at Cornell University, Ithaca, NY
2005	McClintock Graduate Student Award established at Cornell University by Robert Rabson.
	Barbara McClintock U.S. Postal Service American Scientists Commemorative Postage Stamp
2007	Barbara McClintock Society [A philanthropic society], established by the Carnegie Institution of Washington, Washington, DC.
	Department of Plant Breeding & Genetics Centennial Celebration recognized McClintock's contributions to maize genetics in their Centennial book and symposium (Murphy & Kass 2007a,b).
2008	Barbara McClintock inducted into the Connecticut Woman's Hall of Fame
2013	Department of Plant Biology Centennial recognized McClintock's contributions to the former Departments of Botany and of Plant Breeding (Kass 2013; Cobb 2013)
2014	Barbara McClintock Prize for Plant Genetics and Genome Studies, established by the Maize Genetics Cooperation
2018	Cornell University renamed their Biology Without Borders series the Barbara McClintock Life Sciences Lecture Series committed to inspire, and stimulate the Cornell community to engage in interdisciplinary research.
2021	A dormitory at Cornell University was named for Barbara McClintock
	A full-day symposium celebrating McClintock's legacy was held at the Memorial Sloan Kettering Institute in New York City in November.
2022	Barbara McClintock Women Innovator Award (one of two Women Innovator Awards launched by the Center for Technology Licensing at Cornell University) to support female faculty, staff, and students in technology innovation and commercialization in entrepreneurship.

[52] Many speaker recollections were quite accurate. However, readers should note that it was Lester W. Sharp (not R. A. Emerson) with whom McClintock did her Ph.D. at Cornell.

with the maize community of their knowledge that McClintock's ashes were deposited at Cold Spring Harbor Laboratory (Kass 2017):

> Peter McKinley recalled that a woman on the Cold Spring Harbor Laboratory administrative staff, whose name he has forgotten [probably Susan Cooper], suggested he might spread the ashes of their aunt on McClintock's "experimental lands," but Mr. McKinley chose not to. Probably within "a couple of weeks," but no more than a month after her death, Mr. McKinley deposited his aunts ashes directly off the trail that led along the water from where McClintock lived on the Cold Spring Harbor Laboratory campus toward the Sound. He recalled that he wanted to find a place that was "quiet" and "with a view of the Sound." He looked for "a quiet place with a view." He thought of how "she liked to be alone" and he remembered that "she enjoyed being on the shore." Mr. McKinley recalled that about 1/8–1/4 mile in, the path rises and looks out over the water. At that spot he walked four to five yards off the trail up into the "brush and vegetation" off the path and left the ashes there. He found the site to be "quiet, beautiful." McClintock's nephew said a prayer of his own devising drawing from his being an Episcopalian with "what I thought were appropriate words, showing respect." He remarked that "it was a solitary event."

LESSONS FROM STORIES AND LEGENDS

McClintock has been portrayed as a lone maverick. Yet, throughout her long career, McClintock was a member of the maize genetics community. In the mid-1920s, Rollins A. Emerson founded this group, which successfully precluded isolation, as can be demonstrated by reports from those who contributed unpublished research to their annual publication (Kass et al. 2005, 2019). Since its beginning, the maize genetics cooperation and the founding of the *Maize Genetics Cooperation News Letter* in 1929 were a cooperative venture (Kass et al. 2019). Students of maize genetics unselfishly shared unpublished data in their annual reports and exchanged corn stocks when requested (Emerson 1932). Though many of their journal publications or institutional reports were not co-authored, a careful reading of the acknowledgments reflects the cooperative efforts of these researchers (e.g., see Coe & Kass 2005; McClintock 1939). Investigators visited and gave lectures and seminars nationally and internationally, followed by extensive discussions of current research problems. Subsequently, long before the days of faxes or emails, long handwritten letters would continue the consultation. Often these included suggestions for new experiments that might lead to answers for hypotheses previously considered.[53] From its inception, Barbara McClintock was an active member of Emerson's cooperative enterprise, which she embraced throughout her lifetime. As I have shown in the previous chapters of this narrative, McClintock mentored students, shared results via personal correspondence, and visited friends and colleagues nationally and internationally. In addition, she shared unpublished results in seminars, lectures, and symposium presentation, as well as in contributions to the *Maize Genetics Cooperation News Letter.*[54] It is therefore difficult to reconcile the documents and correspondence in McClintock's files, as cited in previous chapters, with recollection and newspaper reports that McClintock was a lone and isolated researcher.

Yet, Rhoades had emphasized that McClintock, herself, and without assistance, conducted the experiments that led her to discover movable elements in her corn stocks.[55] Indeed, while working as a full-time Research Investigator at Cold Spring Harbor beginning in 1943, and until 1953, she took responsibility for planting, harvesting, and examining the corn plants and kernels that led to her interpretation of movable elements as the underlying mechanism for spontaneous mutations in maize (Chapters 7 and 8). It was in the early 1950s that long hours of experimental work had led her to submit her resignation and seek to work somewhere other than Cold Spring Harbor. Yet,

[53] Correspondence among investigators is found in various archival collections at Universities and other Institutions nationally and internationally. An excellent example is Nelson's (1992) memoir published in *The Dynamic Genome*. See also Spectre 1993: Series I, Correspondence, Nelson, Oliver E. Jr., Five folders of letters, 1961–1990, between McClintock & Nelson; also see correspondence, 1962–1988, between McClintock & Coe, Edward H. Jr.

[54] See Spectre 1993: Series I, II, III.

[55] Rhoades 1982: Barbara McClintock Statement of Achievements, Roman Papers UW.

we have seen that McClintock collaborated on other projects such as studies of *Neurospora* and investigations on races of maize. At Missouri, she had mentored students and colleagues of Stadler, among others (Chapter 6), and fostered colleagues both on and off campus (Crouse 1992). She mentored Rhoades' students at Columbia University and discussed her findings with them when they worked closely with her at Cold Spring Harbor. When Rhoades left Columbia for Illinois in 1948, they corresponded about her work, and she visited his laboratory and discussed her project with his students. When the latter found similar phenomena in their studies, they cited her results (and she theirs), and Rhoades presented these findings at national meetings. She visited colleagues at other institutions, gave lectures, and attended meetings where she discussed her experimental results.

Why then did Rhoades emphasize that McClintock had worked alone? As previously discussed, McClintock was distrustful of assistants and sometimes students because she believed they did not pay careful attention to their responsibilities. When re-examining work that she had trusted to a student, she found the data may have been falsified and she subsequently reinvestigated the material (McClintock & Hill 1931; Provine & Sisco 1980). In the 1920s and 1930s, she almost lost important plants due to lack of attention from those she had depended upon to oversee her experiments. While a graduate student at Cornell, she believed she deserved priority of authorship for a project for which she had done most of the work (Randolph & McClintock 1926; Kass 2003; see Chapter 3). She, therefore, distanced herself from the project and worked alone, with limited exceptions, over the rest of her career. This early experience seems to have played a part in her decision to encourage students to work independently. McClintock was resourceful. She was quick to explore and modify new experimental techniques, which those around her sometimes found uncomfortable (see Chapters 3 and 4, Beadle & McClintock 1928; Pawlowski 2013). Rhoades told Roger Lewin that "She wanted to be on top of her research. She wanted to be very close to her research material" (R. Lewin 1983, p. 403). By working without a research group, Lewin speculated, "she was reflecting her scientific approach, rather than just a solitary personality." McClintock was considered to be a very private person. She rarely discussed her personal life with anyone but her very close friends, whom she trusted. Trusted friends do not reveal information that one does not wish to share. For this reason, many stories have circulated about McClintock, with little documentation.

McClintock shunned interviews, and therefore, many journalist's accounts of her accomplishments and her life have relied on undocumented stories portrayed by colleagues and friends. Such a narrative featuring McClintock's then recently awarded Lasker prize, among others, appeared in the summer of 1982 in the Long Island, New York, *Newsday Magazine* (Kessler, 1982). Exaggerated stories and much misinformation about McClintock's family and her work experiences at Cold Spring Harbor and elsewhere were publicized therein, and continued to be repeated.

Individuals who had read and studied McClintock's papers and had heard her presentations believed she was worthy of a Nobel award. This was shown by the documented summaries included in their nominations for this highest recognition, as mentioned above. Furthermore, they had instilled this enthusiasm in graduate students who had taken their courses. One colleague recalled being laughed at during a dinner following an academic job interview in late September 1983, when he predicted Barbara McClintock might win the Nobel Prize in Physiology or Medicine that year. The reactions were something like "you wish" and "never happen." Of course, he was thrilled when shortly thereafter his prophesy was fulfilled. Remarkably, he never heard any more from any member of the department who had expressed amusement at his vision of Nobel Laureate Barbara McClintock.[56]

Biographical material about McClintock (Keller 1983) was publicized throughout the world after the announcement of her unshared 1983 Nobel Prize for Physiology or Medicine (e.g., Altman 1983). National news stories about McClintock's life were embellished and interpreted in the context of contemporaneous views about women's role in society and on academic policy (see Kass 2003;

[56] Letter U Mass, to J. Birchler, September 21, 1983, File of Prof. J. Birchler, University of Missouri; emails J. Birchler to L. Kass, October 23 & 26, 2023.

Wilford 1983). Misrepresented accounts of McClintock's life have resulted in many legends, which continued to be repeated over the years, without documentation. Keller's book, as I mentioned, chiefly relied on interviews as she "let [her interviewees] words speak for themselves." Interviews with Cornell faculty also presented misinformation about McClintock's Major Professor and when she grew corn at Cornell (Segelken 1983).[57]

In a book detailing *Postwar Cornell*, the author, Edmundson, relied on interviews for information that happened decades previously (Edmundson 2015; Acriche 2015). The interviews were "emotionally great material" but not reliable in terms of accuracy. He, like Keller (1983), "let [his interviewees] words speak for themselves," but he additionally searched through archives until he "found the paper that corresponded to the event that the person had described to [him] in the interview."

Keller (2014) wrote that published biographies of McClintock seemed to rely on stories. McClintock's story …"from the start, was shaped, and reshaped, by the memories of our informants, by the interest of the author (or authors), and the needs of the various kinds of readers." … She added, "we constantly rewrite our memories" and asked, "Can we say which of the [published] accounts are the truer one?" She answered, "Probably not." Yet, as an historian of science, V. B. Smocovitis (1999) aptly summarized with regard to "selective memories" displayed in interviews: "Common sense, detective-work or triangulation involving multiple checking of statements of fact with other sources is … a requirement" for obtaining more precise and accurate information about one's biographical subject.

I suggest that finding the documentation to support the stories told about McClintock, and not relying on memories, may result in a more accurate historical account. This book, therefore, is intended to fulfill that goal.

LITERATURE CITED

Acriche, S. 2015. New book details life at Cornell Post World War II. *The Cornell Daily Sun*, November 1, 2015. https://cornellsun.com/2015/11/01/new-book-details-life-at-c-u-post-world-war-ii/

Alberts, B., D. Bray, J. Lewis, M. Raff, K. Roberts, and J. D. Watson. 1983. *Molecular Biology of the Cell*, Garland Publishing, New York.

Alberts, B., R. Heald, A. Johnson, D. Morgan, M. Raff, K. Roberts, P. Walter; Problems by J. Wilson, T. Hunt. 2022. *Molecular Biology of the Cell*, 7th edition. W. W. Norton & Co., New York.

Allen, G. E. 1975. *Life Science in the Twentieth Century*, John Wiley & Sons, Inc., New York.

Altman, L. K. 1983. Long Island Biologist wins Nobel in medicine. *The New York Times*, October 11, 1983, Section A, p. 1, Section C, p. 6.

Beadle, G. W. and B. McClintock. 1928. A genetic disturbance of meiosis in *Zea mays*. *Science* Vol. 68 (No. 1766, November 2, 1928), p. 433. [This became George Beadle's dissertation research project]

Bergström, S. 1984. Opening address, by the President of the Nobel Foundation [K. Sune D. Bergström, December 10, 1983]. *Les Prix Nobel 1983*, pp. 18–20.

Berg, D. E. and M. M. Howe (Eds.). 1989. *Mobile DNA*, American Society for Microbiology, Washington, DC.

Berger, S. L., T. Kouzarides, R. Shiekhattar, and A. Shilatifard. 2009. An operational definition of epigenetics. *Genes & Development* Vol. 23 (No. 7, April 1, 2009), pp. 781–783. https://doi.org/10.1101/gad.1787609

Blackburn, E. H. 1992. Broken Chromosomes and telomeres, pp. 381–388, *In*, N. Fedoroff and D. Botstein (Eds.), *The Dynamic Genome: Barbara McClintock's Ideas in the Century of Genetics*, Cold Spring Harbor Laboratory Press, Plainview, NY.

Blackburn, E. H. and J. W. Szostak. 1984. The molecular structure of centromeres and telomeres. *Annual Review of Biochemistry* Vol. 53 (July), pp. 163–194.

Bourque, G., K. H. Burns, M. Gehring, V. Gorbunova, A. Seluanov, M. Hammell, M. Imbeault, Z. Izsvák, H. L Levin, T. S. Macfarlan, D. L. Mager, and C. Feschotte. 2018. Ten things you should know about transposable elements. *Genome Biology* Vol. 19 (No. 1, November 19, 2018), Article Number 199. https://doi.org/10.1186/s13059-018-1577-z

[57] As documented in earlier chapters, Lester W. Sharp was McClintock's Major Professor (not R. A. Emerson) and she last grew corn at Cornell in 1939 (not in the 1940s).

Braaten, D. 2017. New York Academy of Sciences – A concise history. *Newsletter*. https://www.nyas.org/magazines/celebrating-200-years/new-york-academy-of-sciences-a-concise-history/

Britten, R. J. and E. H. Davison. 1969. Gene regulation for higher cells: A theory: New facts regarding the organization of the genome provide clues to the nature of gene regulation. *Science* Vol. 165 (No. 3891, July 25, 1969), pp. 349–357. https://doi.org/10.1126/science.165.3891.349

Bukhari, A. I., J. A. Shapiro, and S. L. Adhya (Eds.). 1977. *DNA Insertion Elements, Plasmids, and Episomes*, xiii, 782 p. ill., Cold Spring Harbor Laboratory, Cold Spring Harbor, NY.

Burr, B. and F. Burr. 1981. [1980 Symposium] Detection of changes in maize DNA at the shrunken locus due to the intervention of *Ds* elements. *Cold Spring Harbor Symposia on Quantitative Biology* Vol. 45, *Movable Genetic Elements*, 1980. Pt. 2, pp. 463–465 [Published February 1981]. https://doi.org/10.1101/SQB.1981.045.01.061

Campbell, A. M. 2013. Comparative studies relevant to transposon function in plant development, pp. 3.87–3.90, *In*, L. B. Kass (Ed.), 2013ff. *Perspectives on Nobel Laureate Barbara McClintock's Publications (1926–1984): A Companion Volume*, Vol. II, The Internet-First University Press. https://hdl.handle.net/1813/34897

Chaleff, D., J. Mauvais, S. McCormick, M. Shure, S. Wessler, and N. Fedoroff. 1981. Controlling elements in maize. *Carnegie Institution of Washington Year Book 80* (1980–1981), pp. 158–174. [Issued December 1981]

Cobb, E. D. 2013. *150 Years of Botany at Cornell; A History of Botany and Plant Biology*, Department of Plant Biology, Cornell University, Ithaca, NY.

Coe, E. and L. B. Kass. 2005. Proof of physical exchange of genes on the chromosomes. *Proceedings of the National Academy of Sciences* Vol. 102 (No. 19, May 10), pp. 6641–6656. [Online May 2, 2005]. https://www.pnas.org/content/102/19/6641.full

Comfort, N. C. 2001a. *The Tangled Field*, Harvard University Press, Cambridge, MA.

Comfort, N. C. 2001b. From controlling elements to transposons: Barbara McClintock and the Nobel Prize [Historical Perspective]. *Trends in Biochemical Sciences* Vol. 26 (No. 2, July 1, 2001), pp. 454–457.

Conable, C. W. 1977. *Women at Cornell: The Myth of Equal Education*, Cornell University Press, Ithaca, NY.

Courage-Tebbe, U., H.-P. Döring, N. Fedoroff, and P. Starlinger. 1983. The controlling element *Ds* at the *Shrunken* locus in *Zea mays*: Structure of the unstable *sh-m5933* allele and several revertants. Vol. 34 (No. 2, September 1983), pp. 383–393. [Received April 29, 1983, Revised June 20, 1983, Published September 1983].

Creighton, H. B. and B. McClintock. 1931. A correlation of cytological and genetical crossing-over in *Zea mays*. *Proceedings of the National Academy of Sciences* Vol. 17 (No. 8, August 15, 1931) [Communicated July 7, 1931], pp. 492–497.

Crouse, H. V. 1992. Barbara McClintock: Recollections of a graduate student, pp. 25–32, *In*, N. Fedoroff and D. Botstein (Eds.), *The Dynamic Genome: Barbara McClintock's Ideas in the Century of Genetics*, Cold Spring Harbor Laboratory Press, Plainview, NY.

Deragon, J.-M. and X. Zhang. 2006. Short Interspersed Elements (SINEs) in plants: Origin, classification, and use as phylogenetic markers. *Systematic Biology*, Vol. 55 (No. 6, December 1, 2006), pp. 949–956. https://doi.org/10.1080/10635150601047843

Döring, H.-P. and P. Starlinger 1984. Barbara McClintock's controlling elements: Now at the DNA level. *Cell* Vol. 39 (No. 2, Part 1, December 1984), pp. 253–259. https://doi.org/10.1016/0092-8674(84)90002-3

Edmundson, B. 2015. *Postwar Cornell: How The Greatest Generation Transformed A University 1944–1952*, Cayuga Lake Books, Ithaca, NY. [Issued December 11, 2015]

Eldredge, N. and S. J. Gould. 1972. Punctuated equilibria: An alternative to phyletic gradualism, pp. 82–115, *In*, T. J. M. Schopf (Ed.), *Models in Paleobiology*, Freeman, Cooper and Co., San Francisco, CA.

Emerson, R. A. 1932. The present status of maize genetics. *Proceedings of the Sixth International Congress of Genetics* Vol. 1, pp. 141–52. [Paper presented in August, published in December 1932]

Fedoroff, N. 1982. Introduction to transposable controlling elements in maize, pp. 203–211, *In*, W. F. Sheridan (Ed.), *Maize for Biological Research*, Plant Molecular Biology Association, Charlottesville, VA.

Fedoroff, N. V. 1983. Controlling elements in maize, pp. 1–63, *In*, J. Shapiro (Ed.), *Mobile Genetic Elements*, Academic Press, Inc., New York. [Issued January 1983]

Fedoroff, N. V. 1989a. About maize transposable elements and development, review. *Cell* Vol. 56 (No. 2, January 27, 1989), pp. 181–191. https://doi.org/10.1016/0092-8674(89)90891-x

Fedoroff, N. V. 1989b. Maize transposable elements, pp. 375–411, *In*, D. E. Berg and M. M. Howe (Eds.), *Mobile DNA*, American Society for Microbiology, Washington, DC.

Fedoroff, N. V. 1989c. Maize transposable elements in development and evolution. *American Zoologist* Vol. 29 (No. 2, May 1989), pp. 549–555. [From the Symposium on Science as a Way of Knowing-Cell and Molecular Biology presented at the Annual Meeting of the American Society of Zoologists, December 12–16, 1988, at San Francisco, CA]

Fedoroff, N. V. 1992. Maize transposable elements: A story in four parts, pp, 389–415, *In*, N. Fedoroff and D. Botstein (Eds.), *The Dynamic Genome: Barbara McClintock's Ideas in the Century of Genetics*, Cold Spring Harbor Laboratory Press, Plainview, NY.

Fedoroff, N.V. 1994. Barbara McClintock (June 16, 1902–September 2, 1992). *Genetics* Vol. 136 (No. 1, January 1, 1994), pp. 1–10. https://doi.org/10.1093/genetics/136.1.1

Fedoroff, N. V. 1995. Barbara McClintock 1902–1992. A biographical Memoir. *Biographical Memoirs* Vol. 68. The National Academy Press, Washington, DC.

Fedoroff, N. V. 2012. Transposable Elements, Epigenetics, and Genome Evolution. *Science* Vol. 338 (No. 6108, November 9, 2012), pp. 758–767. [AAAS Presidential Address, given February 2012]

Fedoroff, N. V. (Ed.). 2013a. *Plant Transposons and Genome Dynamics in Evolution*, John Wiley & Sons, Inc, Oxford.

Fedoroff, N. V. 2013b. McClintock and epigenetics, Chapter 4, pp. 61–70, *In*, N. V. Fedoroff (Ed.), *Plant Transposons and Genome Dynamics in Evolution*, John Wiley & Sons, Inc, Oxford. [First published February 4, 2013]

Fedoroff, N. V. 2013c. McClintock and epigenetics: Insights from changes in phase of transposition activity, pp. 3.34–3.39, *In*, L. B. Kass (Ed.), *Perspectives on Nobel Laureate Barbara McClintock's Publications (1926–1984): A Companion Volume*. https://hdl.handle.net/1813/34897

Fedoroff, N. V. and D. Botstein (Eds.). 1992. *The Dynamic Genome: Barbara McClintock's Ideas in the Century of Genetics*, Cold Spring Harbor Laboratory Press, Plainview, NY.

Fedoroff, N. V., D. Chaleff, U. Courage-Tebbe, H.-P. Döring, M. Geiser, P. Starlinger, E. Tillmann, E. Weck, and W. Werr. 1983a. Mutations at the *Shrunken* locus in maize caused by the controlling element *Ds*, pp. 61–72, *In*, O. Ciferri and L. Dure (Eds.), *Structure and Function of Plant Genomes* NATO Science Series A: (NSSA Vol. 63). Plenum Press [Proceedings of the meeting of NATO Advanced Study Institute held at Porto Portese, Italy, August 23–September 2, 1982, published October 1983.]

Fedoroff, N. V., D. Chaleff, J. Mauvais, M. Shure, and S. Wessler. 1982. Molecular studies on maize genes with controlling element mutations. *Carnegie Institution of Washington Year Book 81* (1981–1982), pp. 161–171. [Issued December 1982]

Fedoroff, N. V., S. McCormick, and J. Mauvais. 1980. Molecular studies on the controlling elements of maize. *Carnegie Institution of Washington Year Book 79* (1979–1980), pp. 51–62.

Fedoroff, N. V. and J. Mauvais. 1979. Controlling elements in maize. *Carnegie Institution of Washington Year Book 78* (1978–1979), 61–65.

Fedoroff, N. V., J. Mauvais and D. Chaleff. 1983b. Molecular studies on mutations at the Shrunken locus in maize caused by the controlling element *Ds*. *Journal of Molecular & Applied Genetics* Vol. 2 (No. 1, November 1983), pp. 11–29. [Submitted July 1982, accepted November 1982] https://pubmed.ncbi.nlm.nih.gov/6221058/

Fedoroff, N. V., M. Schwartz, M. Shure and S. Wessler. 1983c. Molecular analysis of mutations at the Shrunken and Waxy loci of maize caused by the controlling elements Ac and *D;* Structure of the *sh-m5933* allele of the *Shrunken* locus (Fedoroff); The *pcWx* cDNA plasmids are homologous to the genetically identified *Waxy* locus (Shure & Wessler); Cloning of the *Waxy* locus and the *Ac & Ds* elements (Fedoroff); Direction of transcription of the *Wx* locus and approximate extent of the transcription unit (Schwartz); Preliminary Analysis of the RNAs transcribed in endosperms carrying the *Ac wx-m9*, *wx-m9* alleles (Shure). *Carnegie Institution of Washington Year Book 82* (1982–1983), pp. 171–180. [Issued December 1983]

Fedoroff, N. V., S. Wessler, and M. Shure. 1983d. Isolation of the transposable maize controlling elements *Ac* and *Ds*. *Cell* Vol. 35 (No. 1, November 1983), pp. 235–242. https://doi.org/10.1016/0092-8674(83)90226-x

Feschotte, C. 2023. Transposable elements: McClintock's legacy revisited, Perspective. *Nature Reviews Genetics* Vol. 24 (September 18, 2023), pp. 797-800. https://doi.org/10.1038/s41576-023-00652-3

Feschotte, C., N. Jiang, S. Wessler. 2002. Plant transposable elements: Where genetics meets genomics. *Nature Reviews Genetics* Vol. 3 (No. 5, May 1, 2002), pp. 329–341. https://doi.org/10.1038/nrg793

Fincham, J. R. S. and G. R. K. Sastry. 1974. Controlling elements in maize. *Annual Review of Genetics* Vol. 8 (December 1974), pp. 15–50.

Friedman, R. M. 2001. *Politics of Excellence: Behind the Nobel Prize in Science*, W. H. Freeman & Co., New York.

Fueyo, R., J. Judd, C. Feschotte, and J. Wysocka. 2022. Roles of transposable elements in the regulation of mammalian transcription. Review article. *Nature Reviews Molecular Cell Biology* Vol. 23 (February 28, 2022), pp. 481–497. https://www.nature.com/articles/s41580-022-00457-y

Gierl, A., H. Saedler, and P. Peterson. 1989. Maize transposable elements. *Annual Review of Genetics* Vol. 23 (December 1989), pp. 71–85.

Goldschmidt, R. B. 1951. Chromosomes and genes, pp. 1–11, *In*, K. B. Warren (Ed.), *Genes and Mutations, Cold Spring Harbor Symposia on Quantitative Biology*, Volume XVI (June 7–15, 1951). The Biological Laboratory, Cold Spring Harbor, Long Island, NY. [(©1952]

Goldschmidt, R. B. 1955. *Theoretical Genetics*, University of California Press, Berkeley.

Hannah, C. and D. Schwartz. 2009. The birth of maize molecular genetics, pp. 53–62, *In*, J. L. Bennetzen and S. Hake (Eds.), *Handbook of Maize: Genetics and Genomics*. Springer Science+Business Media LLC, New York.

Haig, D. 2004. The (dual) origin of epigenetics. *Epigenetics, Cold Spring Harbor Symposia on Quantitative Biology* Vol. 69, pp. 67–70. https://doi.org/10.1101/sqb.2004.69.67

Haig, D. 2012. Commentary: The epidemiology of epigenetics. *International Journal of Epidemiology* Vol. 41 (No. 1, February 2012), pp. 13–16. https://doi.org/10.1093/ije/dyr183

Hargittai, I. 2002. *The Road to Stockholm: Nobel Prizes, Science, and Scientists*, Oxford University Press, Oxford.

Hedges, R. W. and A. E. Jacob. 1974. Transposition of ampicillin resistance from RP4 to other replicons. *Molecular and General Genetics* Vol. 132 (No. 1, August 30, 1974), pp. 31–40.

Holliday R. 1990. DNA methylation and epigenetic inheritance. *Philosophical Transactions of the Royal Society of London. Series B, Biological Sciences*. Vol. 326 (No. 1235, January 30, 1990), pp. 329–338. https://doi.org/10.1098/rstb.1990.0015

Jacob, F. 1958. Genetic Control of Viral Functions [Lecture delivered September 18, 1958, at the Harvey Society, New York]. Part of: Sydney Brenner Collection, Reference: SB/2/3/91. https://wellcomecollection.org/works/e895knvt

Jacob, F. 1960 [1958]. Genetic Control of Viral Functions [Lecture delivered September 18, 1958], pp. 1–39, *In, The Harvey Lectures 1958–1959*, Series LIV [54]. Academic Press, The Harvey Society, New York [January 1, 1960].

Jacob, F. and J. Monod. 1961a. Genetic regulatory mechanisms in the synthesis of proteins. Review article. *Journal of Molecular Biology* Vol. 3 (No. 3, June 1, 1961), pp. 318–356.

Jacob, F. and J. Monod. 1961b. On the regulation of gene activity. *Cellular Regulatory Mechanisms, Cold Spring Harbor Symposia on Quantitative Biology* Vol. 26 (June 4–12, 1961), pp. 193–211. https://doi.org/10.1101/SQB.1961.026.01.024 [©1962]

Kass, L. B. 2000. Barbara McClintock, (1902–1992). American Botanical Geneticist, pp. 66–69, *In*, R. Robinson (Ed.), *Plant Sciences*, Macmillan Science Library, New York.

Kass, L. B. 2003. Records and recollections: A new look at Barbara McClintock, Nobel Prize-Winning geneticist. *Genetics* Vol. 164 (August), pp. 1251–1260. https://www.genetics.org/content/164/4/1251

Kass, L. B. 2013. Cornell University: the McClintock Years, 1919–1939 and beyond, Department of Plant Biology Centennial Celebration, June 29, 2013, Invited Symposium Presentation; Plant Science Building, College of Agriculture and Life Science, Cornell University; video recording. https://www.ecommons.cornell.edu/handle/1813/33663

Kass, L. B. (Ed.). 2013ff. *Perspectives on Nobel Laureate Barbara McClintock's Publications (1926–1984): A Companion Volume*. Volumes I & II The Internet-First University Press. https://hdl.handle.net/1813/34897 [Online December 30, 2013, updates March 2014; Vol. III, Increments No. 1–3, May 2016; Vol. IV, Increment No. 4, April 2020]

Kass, L. B. 2016. *Requiescat in pace*, Barbara McClintock. *Maize Genetics Cooperation Newsletter* 90 [Online January 2017]. https://mnl.maizegdb.org/90/pdf/03kass.pdf

Kass, L. B. 2017. Barbara McClintock rests at Cold Spring Harbor, New York, *Maize Genetics Cooperation Newsletter* 91 [Online]. https://mnl.maizegdb.org/91/pdf/03Kass.pdf

Kass, L. B. 2018. Questions and reflections on H. J. Muller's Nobel Prize winning research. *Drosophila Information Service* Vol. 101 (December 2018), pp. 93–106. https://www.ou.edu/journals/dis/DIS101/DIS101.html

Kass, L. B., C. Bonneuil, and E. Coe. 2005. Cornfests, cornfabs and cooperation: The origins and beginnings of the Maize Genetics Cooperation News Letter. *Genetics* Vol. 169 (No. 4, April), pp. 1787–1797. https://www.genetics.org/content/169/4/1787 [Online May 6, 2005]

Kass, L. B., E. H. Coe, M. N. Cook, M. E. Smith, and J. L. Singer (Eds.). 2019. *Founding of the Maize Genetics Cooperation News Letter at Cornell University: A 90th Anniversary Tribute*. (Foreword by E. S. Buckler) [e-book] Internet-First University Press, Ithaca, NY. https://ecommons.cornell.edu/handle/1813/66550. [Online October 10, 2019]

Kass, L. B. and P. Chomet. 2009. Barbara McClintock, pp. 17–52, *In*, J. Bennetzen and S. Hake (Eds.), *Handbook of Maize: Genetics and Genomics*, Springer, New York.

Keller, E. F. 1983. *A Feeling for the Organism. The Life and Work of Barbara McClintock*, 1st edition, W. H. Freeman, San Francisco, CA.

Keller, E. F. 2014. Pot-holes everywhere: How (not) to read my biography of Barbara McClintock, pp. 33–42, *In*, P. Govoni, and Z. A. Franceschi (Eds.), *Writing About Lives in Science: (Auto)Biography, Gender, and Genre*, V&R Unipress, Gottingen.

Kessler, R. E. 1982. The corn lady who stood genetics on its ear. *The Newsday Magazine*, Long Island, New York, August 22, 1982, pp. 1, 8–12, 28–29.

Kolata, G. 1992. Dr. Barbara McClintock, 90, Gene researcher pioneer, dies. *New York Times* National Editon, Friday, September 4, 1992, p. 1, C 16.

Krebs, J. E., E. S. Goldstein, and S. T. Kilpatrick. 2018. *Lewin's Genes XII*, 12th edition, Jones & Bartlett Learning, LLC, Burlington, MA.

Krebs, J. E., E. S. Goldstein, and S. T. Kilpatrick. 2021. *Lewin's Essential Genes*, 4th edition, Jones & Bartlett Learning, LLC, Burlington, MA.

Kuhn, T. S. 1962. *The Structure of Scientific Revolutions*, University of Chicago Pres, Chicago, IL.

Kuhn, T. S. 1970. *The Structure of Scientific Revolutions*. 2nd edition, Enlarged, University of Chicago Press, Chicago, IL.

Les Prix Nobel 1983, Nobel Prizes, Presentations, Biographies and Lectures. 1984. Almqvist & Wiksell International, Stockholm-Sweden. © By the Nobel Foundation.

Lewin, B. 1983. *Genes*, John Wiley & Sons, New York. [Issued April 27, 1983]

Lewin, R. 1983. A Naturalist of the Genome: The Nobel Prize for Physiology or Medicine recognizes the prescient work of a geneticist whose proposals contradicted the prevailing dogma, *Science* Vol. 222 (No. 4622, October 28, 1983), pp. 402–405. https://www.science.org/doi/10.1126/science.6312562

Liu, P., D. Cuerda-Gil, S. Shahid, R. K. Slotkin. 2022. The epigenetic control of the transposable element life cycle in plant genomes and beyond. *Annual Review of Genetics* Vol. 56 (November 2022), pp. 63–87. https://doi.org/10.1146/annurev-genet-072920-015534

Makalowski W. 2003. Genomics. Not junk after all. *Science* Vol. 300 (No. 5623, May 2, 2003), pp. 1246–1247. https://doi.org/10.1126/science.1085690

Makalowski, W. 2007a. What is junk DNA, and what is it worth? *Scientific American Newsletter*, February 12, 2007. https://www.scientificamerican.com/article/what-is-junk-dna-and-what/

Makalowski, W. 2007b. Ask the Experts: "What is "junk" DNA, and what is it worth?" *Scientific American*, Vol. 296 (No. 5, May 2007), p. 104. [A reply to A. Khalavinia, from Iran]

Marchant, D. B., G. Chen, S. Cai et al. 2022. Dynamic genome evolution in a model fern. *Nature Plants* Vol. 8 (No. 9, September) pp. 1038–1051. https://doi.org/10.1038/s41477-022-01226-7

Mattick, J. and P. Amaral. 2023. *RNA, the Epicenter of Genetic Information*, CRC Press. https://www.taylorfrancis.com/books/oa-mono/10.1201/9781003109242/rna-epicenter-genetic-information-john-mattick-paulo-amaral

McClintock, B. 1931a. The order of the genes *C*, *Sh*, and *Wx* in *Zea mays* with reference to a cytologically known point in the chromosome. *Proceedings of the National Academy of Sciences* Vol. 17 (No. 8, August 15, 1931) [Communicated July 7, 1931], pp. 485–491.

McClintock, B. 1931b. Cytological observations of deficiencies involving known genes, translocations and an inversion in *Zea mays*. *University of Missouri, College of Agriculture, Agricultural Experiment Station, Research Bulletin* 163 [December, authorized December 23, 1931], pp. 1–30.

McClintock, B, 1939. The behavior in successive nuclear divisions of a chromosome broken at meiosis. *Proceedings of the National Academy of Sciences* Vol. 25 (No. 8, August) [Communicated July 7], pp. 405–416. [Inversion 9]

McClintock, B. 1950. The origin and behavior of mutable loci in maize. *Proceedings of the National Academy of Sciences* Vol. 36 (No. 6, June 15, 1950), pp. 344–355. [Communicated April 8, 1950]

McClintock, B. 1951 [©1952, Symposium held June 1951]. Chromosome organization and genic expression, pp. 13–47, *In*, K. B. Warren (Ed.), *Genes and Mutations, Cold Spring Harbor Symposia on Quantitative Biology*, Volume XVI [June 7–15, 1951], The Biological Laboratory, Cold Spring Harbor, Long Island, NY. [©1952. [This reference has been cited as 1951 or 1952- see *Science Citation Index*; McClintock (1987) lists it as 1951]

McClintock, B. 1955 [July 1, 1954–1955]. Controlled mutation in maize: The a1m-1-Spm system of control of gene action and mutation. Continued studies of the mode of operation of the controlling elements *Ds* and *Ac*. *Carnegie Institution of Washington Year Book No. 54*, 1954-1955 [Issued December 9, 1955, submitted June 1955]: 245–255.

McClintock, B. 1956a [Issued February 1956, Symposia held June 15–17, 1955]. Intranuclear systems controlling gene action and mutation, pp. 58–74. *In, Mutation, Brookhaven Symposia in Biology, No. 8.* Biology Department, Brookhaven National Laboratory, Upton, NY. [R. C. King, Symposium Chairman.]

McClintock, B. 1956b [(c)1957, Symposium held June 1956]. Controlling elements and the gene, pp. 197–216, *In*, K. B. Warren (Ed.), *Genetic Mechanisms: Structure and Function, Cold Spring Harbor Symposia on Quantitative Biology*, Volume XXI [June 4–12, 1956], The Biological Laboratory, Cold Spring Harbor, Long Island, NY.

McClintock, B. 1965 [Symposium held June 7–9, 1965]. The control of gene action in maize, pp. 162–184, *In*, *Genetic Control of Differentiation, Brookhaven Symposia in Biology*: No. 18. Biology Department, Brookhaven National Laboratory, Upton, N.Y. [H. H. Smith Chairman of the Symposium Committee; Issued December 1965]

McClintock, B. 1983. [Abstract]. Trauma as a means of initiating change in genome organization and expression. *In Vitro* Vol. 19 (3, Part II) [March 1983], pp. 283–284. [Paper presented at 34th annual meeting of the Tissue Culture Association, June 12–16, 1983, Orlando, FL]

McClintock, B. 1984a [Nobel lecture, presented December 8, 1983]. The significance of responses of the genome to challenge. *Science* Vol. 226 (No. 4676, November 16, 1984), pp. 792–801.

McClintock, B. 1984b [Nobel lecture presented, December 8, 1983]. The significance of responses of the genome to challenge, pp. 174–193. *Les Prix Nobel 1983*. Almqvist & Wiksell International, Stockholm. © 1984. By the Nobel Foundation.

McClintock, B. 1987. Introduction, pp. vii–xi, *In*, J. A. Moore (Ed.), *The Discovery and Characterization of Transposable Elements. The Collected Papers of Barbara McClintock. Genes Cells and Organisms; Great Books in Experimental Biology*, Garland Publishing Co., New York.

McClintock, B. and H. E. Hill. 1931. The cytological identification of the chromosome associated with the R-G linkage group in *Zea mays*. *Genetics* Vol. 16 (No. 2, March 16, 1931), pp. 175–190.

McClintock, B., T. A. Kato Y., and A. Blumenschein 1981. *Chromosome Constitution of Races of Maize. Its Significance in the Interpretation of Relationships Between Races and Varieties in the Americas*. Colegio de Postgraduados, Escuela Nacional de Agricultura, Chapingo, Edo. Mexico, Mexico, xxxi, 517 pp.

Mhiri, C. F. B and M. A. Grandbastien. 2022. Specificities and dynamics of transposable elements in land plants. *Biology* (Basel) Vol. 11 (No. 4, April), Number 488. https://doi.org/10.3390/biology11040488

Monod, J. and J. Jacob. 1961. Teleonomic mechanisms, growth and differentiation [General Conclusions], pp. 389–401. *Cold Spring Harbor Symposia on Quantitative Biology,* June 4–12, 1961, Vol. 26, pp. 389–401. [©1962]

Muller, H. J. 1938. The remaking of chromosomes. *The Collecting Net* Vol. 13 (No. 8, September 3, 1938), pp. 181, 183–195, 198. [Lecture given September 1, 1938 at Woods Hole Biological Laboratory, Falmouth, Cape Cod, MA]

Murphy, R. P. and L. B. Kass. 2007a. *Evolution of Plant Breeding at Cornell University: A Centennial History, 1907–2006*, pp. 1–98, Department of Plant Breeding & Genetics, Cornell University, Ithaca, NY, Appendices A1–A98, Photo Section P1–P38. (July 2007).

Murphy, R. P. and L. B. Kass. 2007b. "The Evolution of Plant Breeding at Cornell University." Keynote address, Department of Plant Breeding and Genetics, Centennial Symposium, July 26, 2007. [Written by Kass & Murphy; Presented by Murphy, assisted by Kass]. Video no longer available, slide presentation images at this link: https://www.ctl1.com/publicaccess/calscentennial/murphy/rnh.htm; [posted February 27, 2014].

Nelson, O. E. 1987. The waxy locus in maize twenty-five years later. [Genetics Perspective] *Genetics* Vol. 116 (No. 3, July 1, 1987), pp. 339–342, https://doi.org/10.1093/genetics/116.3.339

Nelson, O. E. (Ed.). 1988. Transposable Elements (*Basic Life Sciences*, Vol. 47), Plenum Press, New York. [Proceedings of an International Symposium on Plant Transposable Elements held August 22–26, 1987, at the University of Wisconsin-Madison, Madison, Wisconsin]

Nelson, O. E. 1992. The mutable *waxy* and *bronze1* alleles of maize, pp. 123–131, *In*, N. Fedoroff and D. Botstein (Eds.), *The Dynamic Genome: Barbara McClintock's Ideas in the Century of Genetics*, Cold Spring Harbor Laboratory Press, Plainview, NY.

Nobel Prize. 1962. The Nobel Prize in Physiology or Medicine 1962. https://www.nobelprize.org/prizes/medicine/1962/summary/

Norrby, E. 2010. *Nobel Prizes and Life Sciences*, World Scientific Publishing Co. Pte. Ltd, Singapore. [September 24, 2010]

Norrby, E. 2022a. The changing concept of the gene through the lens of Nobel Prizes awarded in the past century. Office of NIH History lecture series, Monday, February 13, 2023, 11:00:00 AM. https://videocast.nih.gov/watch=49150

Norrby, E. 2022b. *Nobel Prizes: Genes, Viruses and Cellular Signaling*, World Scientific Publishing Co. Pte. Ltd, Singapore. [September 9, 2022]

Noshay, J. M., P. A. Crisp, and N. M. Springer. 2018. The maize methylome, pp. 81–96, *In*, J. Bennetzen, S. Flint-Garcia., C. Hirsch, and R. Tuberosa (Eds.), *The Maize Genome. Compendium of Plant Genomes*, Springer, Cham. https://doi.org/10.1007/978-3-319-97427-9_6

Orgel, L. E. and F. H. Crick. 1980. Selfish DNA: The ultimate parasite. Review Article. *Nature* Vol. 284 (No. 5757, April 17, 1980), pp. 604–607.

Pandita, D. and A. Pandita (Eds.). 2023. *Plant Transposable Elements: Biology and Biotechnology*, Apple Academic Press co-published with CRC Press. https://doi.org/10.1201/9781003315193

Pawlowski, W. P. 2013. Genetic analysis of meiosis using the *asynaptic* 1 mutant: A perspective on George W. Beadle and Barbara McClintock's 1928 contribution, pp. 1.9–1.13, *In*, L. B. Kass (Ed.), *Perspectives on Nobel Laureate Barbara McClintock's publications (1926–1984): A Companion Volume*, Vol. I, The Internet-First University Press. https://hdl.handle.net/1813/34897

Peterson, T. 1993. A celebration of the life of Dr. Barbara McClintock. *Probe* Vol. 3 (No. 1/2, January to June 1983), pp. 18–19. [Celebration at Cold Spring Harbor Laboratory, November 17, 1992.]

President Richard Nixon's Daily Diary. Friday, May 21, 1971. The President ... went to the East Room to present the National Medal of Science Awards. For a list of recipients ... see APPENDIX "B"; WHCF: SMOF: Office of Presidential Papers and Archives RC-7. https://www.nixonlibrary.gov/sites/default/files/virtuallibrary/documents/PDD/1971/052%20May%2016-31%201971.pdf

Provine, W. B. and P. Sisco. 1980. Interview with Barbara McClintock at Cold Spring Harbor Laboratory, Long Island, New York, August 28, 1980. Division of Rare and Manuscript Collections, Cornell University Library, Ithaca, NY.

Randolph, L. F. and B. McClintock. 1926. Polyploidy in *Zea mays* L. *American Naturalist* Vol. LX (No. 666) [January/February 1926, received, no date given], pp. 99–102.[first report of a triploid corn plant & the behavior of its chromosomes, reprinted in Kass 2013ff.]

Ringertz, N. 1984. The Nobel Prize for Physiology or Medicine [1983]. *Les Prix Nobel 1983*, pp. 27–29.

Roberts, A. P., M. Chandler, P. Courvalin, G. Guédon, P. Mullany, T. Pembroke. J. L. Rood, C. J. Smith, A. O. Summers, M. Tsuda, and D. E. Berg. 2008. Revised nomenclature for transposable genetic elements. *Plasmid* Vol. 60 (No. 3, November 2008), pp. 167–173. https://doi.org/10.1016/j.plasmid.2008.08.001

Saedler, H. and A. Gierl (Eds.). 1996. *Transposable Elements*, Current Topics in Microbiology and Immunology 204, Springer-Verlag, Berlin, Heidelberg.

Scott, W. A., R. Werner, D. R. Joseph, and J. Schultz (Eds.). 1980. Mobilization and Reassembly of Genetic Information. Proceedings of the Miami Winter Symposium, January 1980. *Miami Winter Symposium* Vol. 17. Academic Press, New York.

Segelken, R. 1983. Nobel Laureate Barbara McClintock did her early research at Cornell. *Cornell Chronicle* Vol. 15 (No. 8, October 13, 1983), pp. 1, 16.

Shapiro, J. A. 1967. *The structure of the galactose operon in Escherichia coli K-12*. Ph.D. Thesis, University of Cambridge, Cambridge, UK. [July 1967]

Shapiro, J. (Ed.). 1983. *Mobile Genetic Elements*, Academic Press, Inc., New York. [Issued January 1983]

Shapiro, J. A. 2013. The special character of Barbara McClintock's Nobel prize address, pp. 3.133–3.140, *In*, L. B. Kass (Ed.), 2013ff., *Perspectives on Nobel Laureate Barbara McClintock's Publications (1926–1984): A Companion Volume*. https://hdl.handle.net/1813/34897

Shapiro, J. A. 2023. Evolution without accidents [original essay title: "Random mutations have nothing to do with origin of species"]. *Aeon* July 6, 2023. https://aeon.co/essays/why-did-darwins-20th-century-followers-get-evolution-so-wrong

Singer, M. 1989. President's Commentary. *Carnegie Institution of Washington Year Book* 88, 1988–1989, pp. 1–14.

Shure, M., S. Wessler and N. Fedoroff. 1983. Molecular identification and isolation of the Waxy locus in maize, *Cell* 35 (No. 1, November 1983), pp. 225–233. https://pubmed.ncbi.nlm.nih.gov/6313224/

Smocovitis, V. B. 1999. Living with your biographical subject: Special problems of distance, privacy, and trust in the biography of G. Ledyard Stebbins Jr. *Journal of the History of Biology* Vol. 32 (No. 3, Winter 1999), pp. 421–438.

Soderlind, R. 1983. Sweden's King Carl XVI Gustaf presented the 1983 Nobel science and literature prizes Saturday to five American scientists and the British novelist William Golding. UPI Archives, December 10, 1983. https://www.upi.com/Archives/1983/12/10/Swedens-King-Carl-XVI-Gustaf-presented-the-1983-Nobel/2786439880400/

Spectre, M. 1993. *Barbara McClintock Papers. Guide to the Collection*, American Philosophical Society Library, Philadelphia, PA. (Online: Barbara McClintock Papers Mss.Ms.Coll.79, https://search.amphilsoc.org/collections/view?docId=ead/Mss.Ms.Coll.79-ead.xml) [Also see selected papers and summary of APS Library papers of McClintock available for the National Library of Medicine, Profiles in Science: https://profiles.nlm.nih.gov/spotlight/ll; The collection also reprints papers/letters from other institution's archives]

Stanley, S. M. 1979. *Macroevolution, Pattern and Process*, W. H. Freeman & Co., San Francisco, CA.

Starlinger, P. 1984. Transposable elements. *Trends in Biological Sciences* Vol. 9 (No. 4, April 1984), pp. 125–127. https://doi.org/10.1016/0968-0004(84)90115-4

Sutton, W. D., W. L. Gerlach, D. Schwartz and W. J. Peacock. 1984. Molecular analysis of *Ds* controlling element mutations at the *Adh1* locus of maize. *Science* Vol. 223 (No. 4642, March 25, 1984), pp. 1265–1268. [Submitted October 1983] https://pubmed.ncbi.nlm.nih.gov/17759347/

Tønnesson, Ø, 1999. Sources of the History of the Nobel Peace Prize: [The Archive of the Norwegian Nobel Committee; "After 50 years the Director of the Norwegian Nobel Institute may give access to these archival sources, primarily for the purpose of historical research."], December 1, 1999. https://www.nobelprize.org/prizes/themes/sources-of-the-history-of-the-nobel-peace-prize/

Waddington, C. H. 1942 The epigenotype. *Endeavour* Vol. 1, pp. 18–20. [Reprinted in *International Journal of Epidemiology* 2012 Vol. 41, pp. 10–13.]

Watson, J. and F. Crick. 1953a. A structure for desoxyribose nucleic acids. *Nature* Vol. 171 (No. 4356, April 25, 1953), pp. 737–738.

Watson, J. D. and F. H. C. Crick. 1953b. The structure of DNA, pp. 123–131. *Viruses, Cold Spring Harbor Symposia on Quantitative Biology*, June 5–11 1953, Volume XVIII [Vol. 18].

Weil, C. 2013. Transposable controlling elements step out onto the broader scientific stage, pp. 3.11–3.21, *In*, L. B. Kass (Ed.). 2013ff. *Perspectives on Nobel Laureate Barbara McClintock's publications (1926–1984): A Companion Volume*, Vol. II, The Internet-First University Press. https://hdl.handle.net/1813/34897

Wells, J. N. and C. Feschotte. 2020. A field guide to eukaryotic transposable elements. *Annual Review of Genetics* Vol. 54 (November 2020), pp. 539–561.

Wilford, J. N. 1983. Brillant loner who loves genetics, Barbara McClintock. *The New York Times*, October 11, 1983, Section C, p. 6.

Index

Note: **Bold** page numbers refer to tables; *italic* page numbers refer to figures and page numbers followed by "n" denote footnotes.

AAAS Symposia 23
AAAS Toronto Meeting, 1921 22–24, *24*, 53, 62, 65, 93
AAUP Statement of Principles for Academic Tenure and Freedom 120
"Above all nations is humanity," Club's motto 18
"aceto-carmine smear" technique 34, 60, 66
"aceto-orcein smear" technique 146, 148
Achievement Award of the American Association of University Women 136
Adhya, S. 212
Alberts, B. 230, 239
  Molecular Biology of the Cell 231
Alexander, D. E. *210*
Allen, C. E. 39
Allen, G. 75, 79
Allen, S. M. 59n34, *95*
Altenburg, E. 140, 156, 170
American Association of University Professors (AAUP) Achievement Award 119, 150, 165
American Journal of Botany 23, 106, 117
American Men of Science 18, 92, 116, 136, 149
American Naturalist 22, 23, 38, 139, 190, 199, 241
American Society of Naturalists 23, 150, 165, 166
Anderson, E. G. 19, *24*, 33, 37, 39, 51, *52*, 56, 57, 65, 66, 72, 73, 87, 90, 94, 164
  National Research Council Fellowships 81
Anderson, E. S. 95, 122, 122n119, 123, 127, 187
  "Corn Before Columbus" 187
Andrew Dickson White Professor-at-Large 1, 2n5, 182–182, 182n2
antibiotic resistance 214
Anti-genetic propaganda 162
anti-Jewish laws 83
Antisemitism in Academia 18, 19, 114
A1 somatic instability 167
Assembly of the Karolinska Institute 236
Atomic Energy Commission (AEC) 193
Auerbach, C. 83
"autonomous" controlling elements 233
autonomous DNA transposons 243

Babcock, E. B. 56, 80
bacterial gene control systems 189
bacterial plasmids 214
Bailey, L. H. 37
Baker, W. K. 218
Bartley, M. A. **104**
Bateson, W. 21, 22n32, 23, *24*, 220
  Mendel's Principles of Heredity 22
Bauer, H. 113
Beadle, G. W. 2, 2n3, 36, 39, 40, 40n31, 48, 51, 52, 56, 58, 62, 66, 86, 100, 101, 129, 136, 145–147, 150, 156, 158–159, 162, 163, 183, 209, *210*, 220
  doctoral research 57
  hypothesis 141
  maize investigations with E. G. Anderson 72
  NRC Fellowship 73
Belling, J. 33–35, 47, 60, 65
Benzer, S. 220
Berger, M. G. 136
Beveridge, W. 89
Bhavnani, M. M. 6, 219, 246
Bierce, A. G. 156
Biological Bulletin 22, 47
Birchler, J. A. 49n15, 79–81, 81n34, 118n99, 199n58, 222n48, 248n56
Black, B. 2n5
Blackburn, E. 234, 235
Black, M. 192
Blakeslee, A. F. 33, 37, 39, 56, 65, 108, 110, 123, 125, 126
  Annual Research Report 122
  International Botanical Congress Cytology *40*
Bliss, C. 91
Blumenschein, A. 83, 187, 188, 193
Bodger, E. M. 30
Bonner, D. 142, 191, 197
  Heredity 197, 200
Bonneuil, C. 65
Boston University School of Medicine (BUSM) 8
Botanical Gazette (Morgan) 19
Botany Department's Centennial Celebration 184
Botstein, D. 184, 240, 245, *245*
Braun, A. 222
Bregger, T. 48, 51, 53
Brenner, S. 186, 186n20, 212, 213, 213n9, 220
Brewbaker, J. L.
  Agricultural Genetics text 200
Bridges, C. B. 23, *24*, 47, 75, 94–95, 161
Brieger, F. G. 83, 91, 91n92, 91n93, 188
Brink, R. A. 58, 63, 66, 141, 163, 164, 166, 167
  studies of Modulator (*Mp*) 170
Brookhaven National Laboratory 158, 160
Brookhaven Symposia in Biology 167, 188, 201
Brookhaven Symposium 137, 168, 169, 188, 189, 214n11
Brooklyn Daily Eagle newspaper 7, *11*
Brown, D. 214
Brown v. Board of Education 115
Brown, W. L. (Bill) 187, 188
  Races of Maize in the West Indies 187
Brown, S. W. **104**
Bruns, P. 184
Bukhari, A. 212, 216
  DNA Insertion Elements, Plasmids, and Episomes 212
Bulletin of the AAUP 120
Burr, B. 160n26, 215, 233n17
Burr, F. 215, 233n17

Burnham, C. R. 49, 57–58, *60*, 62, 66, 81, 87, 101, 107, 121, 136, 199
maize investigations with Anderson 72
semisterile-2 63–64, 73
semisterile-2 maize stock 76
Synapsis Club reunion *221*
Bush, V. 108, 124, 128, 145, 150, 158, 159
BUSM *see* Boston University School of Medicine (BUSM)
Bussell, F. B. *24*

California Institute of Technology (Caltech) 19, 65, 73, 81, 82–91, 94, 95, 107, 110, 113, 123, 129, 136, 138, 147
Division of Biological Sciences, Caltech 86, 86n60
Caltech Executive Council 73
Cameron, D. R. 80
Cameron, J. W. *102*, **104**
Campbell, A. 212, 241
Carnegie Board of Trustees 193
Carnegie Institution of Washington (CIW) 22, 108, *109*, 121, 124, 126, 128–130, 145, 158, 182, 183, 191, 193, 201, 214–218, 220
Carnegie Institution of Washington Year Book (McClintock) 1, 121, 124, 137, 145, 193, 215, 220
Caspari, E. 142, 142n16
Castle, W. E.
Genetics and Eugenics 22
cell biology 192, 222, 232–233, 237
The Cell in Development and Inheritance (Wilson) 191
Child, C. M. 85
Chomet, P. 76, 188
"chromosomal type" breakage-fusion-bridge cycle 103, 105, 119, 124, 144
"chromosome-breakage mechanism" 145
The Chromosome Constitution of Races of Maize (McClintock et al.) 83, 188
chromosome knobs 64, 171–175, 187, 188, 193, 216
CIMMYT (Centro Internacional de Mejoramiento de Maiz y Trigo) 175, 244
CIW *see* Carnegie Institution of Washington (CIW)
Clark, C. 246
Clark, F. J. *102*, **104,** 116, 117
Clark, J. A. 171
Clarke, A. E. 86
Clausen, R. E. 80n27, 80n28
Cleland, R. E. 39, *40*, 86, 95, 171, 172, 192
Coe, E. H. 76, 197, *211*, 215
Cohen, S. N. 213–215, 219, 231
Cohn, N. S. 198
colchicine-induced tetraploidy 147
Cold Spring Harbor Laboratory 3, 6, 175, 200, 208, 216, 218, 220, *237*, 244, *245*, 247
Cold Spring Harbor Symposia on Genes and Mutations 136
Cold Spring Harbor Symposia on Quantitative Biology 108–110, 119, 122, 168, 169, 189, 220
Cold Spring Harbor Symposium 119, 122, 123, 139, 141, 164, 165, 189, 190, 201, 213, 216, 229, 231, 233, 238
Comfort, N. xi, 14n3, 118n99, 136n2, 139n8, 141, 193n41, 215n17
Companion Volume (Kass) xi, xiii, 49n15, 143n17, 144n18, 188n25, 191n31, 222, 234n19
Conable, C. W.
Women at Cornell: The Myth of Equal Education 245
Conference Statement on Academic Freedom and Tenure 120
Congress *Proceedings* (International Congress of Genetics) 38, 85
Connecticut Agricultural Experiment Station (CAES) 117
Cooper, S. 244
Copeland, R. S. 11, 11n23
Corbally, J. 219
Corn and Corn Improvement (Rhoades reports on work of McClintock) 197
Cornell Synapsis Club 33
Cornell University 92–95, 115
AAAS meeting in Toronto *24*
Cornell University Archives 14
Cornell University Messenger Lectureship Series 75
Cytology and Advanced Cytology courses 2n4
Department of Botany, Cornell 1n1, 2, 2n2, 25, 31, 32, 36, 38, 72, 183
Department of Plant Breeding 2, 16, 22, 23, 27, 31, 36n25, 39, 40n31, 48, 50, 60n35, 72, 110, 148, 148n35, 148n36, 183, 201, 221, **246**
Division of Biological Sciences, Cornell 3, 182, 183, 191
Doctoral Committee members 2, 20, 25
Dr. Barbara McClintock, Instructor in Botany *60*
"Golden Age of Corn Genetics" 218, *222*
Graduate School, Biological Sciences course announcements 3
"Holding Court" 1, 3
map of *55*
"McClintock Shed" *61*
McClintock's postgraduate instructor at 46–48, *60*
microscope slides 2, 2n4
New York State College of Agriculture 14–16, 25, *26*
Pachytene chromosomes in maize, first report 26, 35, 49n4, 58, 61, 62, 64
Plant Science Building *185*, *186*
researchers from 23
Section of Genetics, Development and Physiology 1, 1n1, 2, 183
women in plant breeding 30–31
women in science 16, 38, 114
cornfest (cornfab) 53
Correns, C. 83
Corson, D. R. 114, 114n72
Creighton, H. 2, 2n2, 12, 31, 49, 51, 57, 58, *59*, 72, *74*, 76, 105, 106, 184, 187, 240
The Annual Report of The Biological Laboratory 122, 122n119
Cold Spring Harbor Symposia *169*
on cytological evidence 82
direct and indirect cytological evidence 80
experimental maize cross and progeny types *77*
McClintock's 90th birthday celebration *245*
memories 75
Morgan's "References to Literature" 79
Plant Science 220 *185*, *186*
Synapsis Club reunion *221*, 222
Crepis 35, 35n23, 80, 81
Crick, F. 160, 220
crossing-over theory 75–79
cross-shaped configuration (synaptic complex) 64
Crouse, H. **104**, 107n32
Crozier, W. J. 48
c-sh-wx linkage group 73–75
Curtis, W. C. 100, 101, 103, 110–113, 116, 118, 120, 126–128

cytogenetics of maize 2, 24, 34, 36, 46, 48, 49, 54, 58, 65–66, 72, 83, 84, 92, 94, 106, 116, 117, 121, 123, 127, 142, 148, 151, 166, 170, 171, 172, 173, 187, 188, 193, 199
Cytology and Cytogenetics (Swanson) 117, 143

Dachau concentration camp 88
Darwin, C. 22n33
  evolutionary theory 14
  On the Origin of Species by Means of Natural Selection or the Preservation of Favoured Races in the Struggle for Life 22
Davenport, C. B. 33
Dawson, G. W. P.
  Salmonella typhimurium bacterial system 208
Delbruck, M. 123, 160
Demerec-Dayton, Z. (Hartman, Z) 148, 148n36, 148n37, 149
Demerec, M. 16, *17*, *24*, 33, 37, 51, 56, 65, 95, 106, 108, *109*, 122–126, 141, 145, 148, 149, 156, 158, 159, 161, 162, 164, 169, 175, 193
Demerec, M. A. Z. 18; *see also* Ziegler M. A.
Demerec, R. (Demerec Dyson-Hudson, Dyson-Hudson) 16, 16n16, *17*, 18n23, *109*
Dempsey, E. 157
DeVries, H. 21
  mutation 34
  Plant Breeding 22
DeWitt, W. 191
Diez-P., S. R. 173
Dobroscky, I. D. (I. Dobroczki, Mrs. Carleton van deWater) 18, 18n19
  International Botanical Congress Cytology *40*
Dobzhansky, T. 86, 106, 183
Dollinger, E. J. 160
Dolloff, J. 170
Doring, H. -P. 243
Dorsey, E. 32
Dorsey, M. J. *24*
dotted (*Dt*) gene 93, 94, 137, 168
double-strand crossing over theory 80
Dreiser, T. 10
Drosophila chromosomes 21–23, 34, 36, 37, 47, 49, 51–54, 57, 58, 65, 66, 75–77, 82, 84, 86–87, 89, 94, 113, 123, 138, 140, 141, 144, 146, 161, 196, 198, 212–214, 231
Ds-Ac two-element system 168, 188
Duggar, B. M. 47
Dunn, L. C. 196
  "usable textbook in cytogenetics" 121
Duvick, D. N. 187

Eames, A. J. 36, 38, 47
East, E. M. 36, 39, 48–50, 95
  Inbreeding and Outbreeding 22
Ebert, J. D. 191, 214
Edmundson, B.
  Postwar Cornell 249
Emerson College of Oratory 6, 7
  College by the Commonwealth of Massachusetts 7
Emerson, R. A. 2, 16, *17*, 19, 20, 23, 24, *24*, 25, 30, 31, 33–36, 36n24, 48, *60*, 66, 72, 93, 94, *95*, 100, 101, 110, 119, 121, 137, 163, 164, 183, 221, 247
  assigning linkage groups, to cytologically identifiable chromosomes 60–64
  career in meteorology 92
  genetics of maize 53–54
  maize and genetics 50–52
  maize cooperation 52–53
  Maize Genetics Cooperation group 65
  pericarp locus 170
  school at Cornell 49
  submitted manuscripts to *PNAS* 77, 81
Emerson, S. E. *24*, 57
endosperm 24, 54, 58, 124, 143, 144, 209, 235
epigenetic systems 243
Etheridge, W. C. 50
eukaryote chromosomes 211
"Eukaryotic Genetics" 199
eukaryotic organisms 139, 161, 165, 208, 213, 214, 231, 233, 240, 241
Eyster, L. A. *24*
Eyster, W. H. 37, 51

Fedoroff, N. V. 137–139, 146, 197, 200, 214, 215, 231, 233, 234, 243, 245, *245*
A Feeling for the Organism (Keller) 6
Ferguson, M. 57
Fincham, J. R. S. 209
Fink, G. 184
Fisher, H. E. *95*
flu pandemic of 1918 10, 11
The Flying Dutchman III (McClintock) 9, 12
Forbes Magazine 16
Francis, G. *102*
Franklin, B. 150
Fraser, A. C. 21, 25, 30, 31, 33, 35, 39, 48, 51, *95*
Fraser, C. V. (Fotheringham, A. C.) 25, 25n36
Friedman, R. M. 232
Fritch, C. von 88

Gaines, L. 115
Gajdusek, C. 18
Galinat, W. C. 188, 209, *210*
Galston, A. 122n117
Garber, E. D. 162
Garcés, C. 173
"gene control systems in maize" 167, 189, 190, 200, 230
Genes (Lewin) 232, 233
Genetical Society of Great Britain 220
genetic code sequence of DNA 243
genetic map (linkage map) 50, 65–66
Genetics 105, 118, 119, 139, 144, 149, 160, 196
Genetics and Eugenics (Castle) 22
Genetics Building (Curtis Hall) 102, 108, 116, 117n88
genetics of maize 14, 36, 53–54
Genetics Research Unit, CIW 182, 183, 193, 201, 208, 220n40
Genetics Society and Society of Naturalists Symposium 148
Genetics Society executive committee 150
Genetics Society of America (GSA) 106, *106*, 118, 151, 162, 218
Genetics Society of America Nominating Committee 150
Genetics Society on the Inter-Society Committee on Science Foundation Legislation 150
Gentleman's Agreement (Hobson) 16
Gilbert, W. M. 126

Glass, H. B. 89, 90, 90n81, *102, 106,* 216
Goebbels, J. 89
"Golden Age of Corn Genetics" 49, 218, 221, *222*
Goldring, W. 38
Goldschmidt, R. B. 83, 84, 89, 90, 139, 140, 143, 161, 162, 241
Gordon, M. 123
Green, M. M. 212
Greider, C. 235
Greyson, R. I. 209
Grobman, A. 171, 187
    Races of Maize in Perú. Their Origins, Evolution and Classification 174
Grodzicker, T. *237*
Growth and Development (Sussman) 200
GSA *see* Genetics Society of America (GSA)
Guggenheim Foundation 8, 91, 92
Guthrie, M. J. *102*, 114, 116, 120

Haber, F. 89, 89n79
Haber, J.
    Rosenstiel Award dinner celebration *217*
Haig, D.
    "The meaning of the term epigenetics has changed with time" 243
Haldane, J. B. S. 89
Handy, S. R. 7, 8
Hanson, F. B. 92–94, 102, 111, 116, 123
Hargittai, I.
    The Road to Stockholm 231
Hartmann, M. 84, 89–90
Hayes, H. K. 50, 208
Hedges, R. W. 214
Heitz, E. 91, 113, 116
Heredity (Bonner) 200
Hershey, A. 201, 220
Herskowitz, I. H. 196, 230, 239
    Principles of Genetics 198
Herskowitz, Ira 216, 230
Hesse, R. 84
Hicks, J. 216
Hill, H. E. 58, 59n34, 62, 63n38
Hindenburg, P. von 83
Hitler, A. 83, 87
Hobson, L. Z.
    Gentleman's Agreement 16
Hollaender, A. 193
homeopathic medicine 8, 11
Horowitz, N. 141, 148
Horsfall, F. L. 160
Houlahan, M. B. 146
Hunt Institute for Botanical Documentation 2n2, 2n4, 60n35
Hutchison, C. B. 1, 20–22, 24, *24*, 25, 50–51, 85
hypothetical model of gene regulation 189

IBC *see* International Botanical Congress (IBC)
ICG *see* International Congress of Genetics (ICG)
Immigration laws 162n30
Inbreeding and Outbreeding (East and Jones) 22
insertion sequences (IS) 209, 212, 214, 243
Inter-American Maize Improvement Program Center in Mexico 175, 188
International Botanical Congress (IBC) 37, *40*, 47
International Congress of Genetics (ICG) 47, 53, 65, 81, 82, 84, 85, 87, 90, 162
International Congress of Plant Sciences Sections of Genetics and Cytology 38
International Education Board (IEB) Fellowship 39, 48, 85
International Education Board of the Rockefeller Foundation 84
International Research Fellow in Genetics of the Rockefeller Foundation 86
International Symposium on Maize Breeding and Genetics 209, *210*
"Intranuclear Systems Controlling Mutation" 168
Introduction to Cytology (Sharp) 2
introgressive hybridization 187
"iron-acetocarmine" smear technique 34
Irwin, M. R. 166

Jackson, G. 38, *42*
Jacob, A. E. 137, 214
Jacob, F. 182n4, 184, 189, 189n27, 190, 191, 197, 198, 200, 220, 229, 229n1, 232, 242
Jenkins, M. T. *210*
Jesup Lecture 197
Jewish Telegraphic Agency (*JTA*) 88
John Innis Horticultural Institute 89, 91
Joint Genetics Sections of the American Society of Zoologists 149
Jollos, V. 83
Jones, D. F. 39, 50, 51, 93, 183
    Inbreeding and Outbreeding 22
Journal of the Washington Academy of Sciences 22
Judd, B. H. 164, 218
jumping genes *see* mobile genetic elements

Kaiser Wilhelm Institute for Biology in Berlin-Dahlem (KWIB/BD) 82–84, 86, 89, 90
Kaiser Wilhelm Institute for Brain Research in Berlin-Buch (KWIBR) 89, 90
Kaiser Wilhelm Society (KWG) 83, 89, 90
Kass, L. B. 2n2, *52*, 65, 76, 79, 81, *102*, 188
Kass, T. 186
Kato Y., T. A. 83, 173n67, 187, 188, 188n25, 193, 216
    Chromosome Constitution of Races of Maize 83, 188
Kaufmann, B. 86, *106*
Keim, F. D. 39
Keller, E. F. 7, 8n8, 129, 142, 214, 219, 231, 249
    A Feeling for the Organism xi, 6
Kerckhoff, W. G. 147
Kessler, M. 192
Kimball, R. F. 123
Kimber Genetics Award 192
Kimmelman, B. 51
King, R. C.
    Genetics text "for a most readable account of mutation" 201
Kittridge, M. 6, 142
knob interchange chromosome 73
Kohler, R. E. 86
Kolata, G. B. 11n23, 212, 246
Kühn, A. 90
Kvakan, P. 51
KWG *see* Kaiser Wilhelm Society (KWG)

Lasker Prize 230, 242, 248
Law for the Restoration of the Professional Civil Service 83, 88, 89

Lederberg, J. 6, 220, 230
Lewin, B. 231
Genes 232, 233
Lewin, R. 248
Lewis, E. B. 161, 162, 165, 241
Lewis, E. M. (Mrs. Seth Ames Lewis) 9n10, 9n11
Lewis, I. F. 46
Lewis S. Rosenstiel Award 126, *217*
LIBA *see* Long Island Biological Association (LIBA)
Li, H. W. 51
Lindstrom, E. W. 50
linkage groups 24, 33–35, 41, 48–53, 56–58, 60–66, 73, 76, 79, 80, 94
Livermore, J. R. *95*
Long Island Biological Association (LIBA) 108, 122
Long Island Biological Laboratory (Bio-Lab), Cold Spring Harbor 108, *109*, 122, 142, 183
ashes deposited 246, 247
Longley, A. E. 173, 188, *195*
drawing chromosomes with camera lucida *174*
Love, H. H. 36, 39, *95*
Luria, S. 123, 123n120, 163, 220
Lwoff, A. M. 191, 229
Lysenko, T. 162, 241

MacCracken, H. N. 10
MacDowell, E. C. 122, 125n131, 128
MacIntyre, R. 184, 187
MacLean, H. A. 8n10
MacLean, M. M. 8n10
maize ear carrying translocation 8-9, *78*
maize chromosome morphology 62, 173
maize eukaryotic system 229
Maize Genetics Cooperation group 65, 92
Maize Genetics Cooperation News Letter 136, 143, 157
Maize Preservation Committee 172
maize system of gene regulation 242
Mangelsdorf, P. C. 166, 170, 187
Races of Maize in Mexico 171
Mangold, O. 89
Marine Biological Laboratory (MBL) 46, 47, 106, *106*, 116, 118
Marshak, A. G. 48, 95
Massachusetts Homeopathic Medical Society 8
Mavor, J. W.
work on X-rays and mutation 34
MBL *see* Marine Biological Laboratory (MBL)
McCarran Act (1950) 162
McClintock, B. *4*, 51, *102*, 193, *194*
AAAS Toronto Meeting, 1921 22–24
Ac/Dc controlling system 201
"*Ac-Ds* position effect" in maize 143
Ac-Ds system 197
affiliation 19
Albert Lasker Basic Medical Research Award 220
Annual Review of Biochemistry 234
Annual Review of Genetics 209
antisemitism in academia 18–19, 114, 114n73
"aroused particular interest" 142
associates at Columbia University 127
birthday celebration with friends *245*
Botanical Society of America Merit Award 170
as Botany Scholar 31
Brookhaven Symposia in Biology 167
in Brooklyn, New York 9–12
from Caltech to Germany 82–84
Carnegie Institution of Washington Year Book 1, 121, 124, 137, 145, 193, 215, 220
Carnegie Institution Reports 145, 156–157, 209
Cold Spring Harbor, 1941 108–110, 121–123, *169*, 231
The College of Agriculture Announcement for 1919–20 15, *15*, 19–22
comprehensive undergraduate curriculum and encouragement 27
Comptes rendus 189
Connecticut Women's Hall of Fame 246
considerations for changing jobs 113–115
"controlling element" first defined 167
controlling elements in genetic regulation 189, 197, 233, 241, 242
Cornell Alumna and instructor, in Botany 41
Cornell Plant Breeding Synapsis Club *95*
Cornell Professorship 201
Cornell University, 1934–1936 14, 92–95
Corrigenda 79
with Creighton (HBC) *74*, 76
Curt and Evelyn Stern meet McClintock, in America 84–86
cytogenetics of maize plant 46–47
Cytology class 57
cytology meets corn genetics, at Cornell 54, 56–59
delay of her recognition 240–243
direct and indirect cytological evidence for crossing-over 79, 80
discovery of transposable elements 1
discovery of transposition 136–137, 139
Dynamic Genome 130, 232–233
elected to American Philosophical Society 150
elected to National Academy of Sciences 136, 149
"Emerson School" at Cornell 49
endeavors during 1983 234–236
in Erasmus Hall High School 9–12, *11*
experimental maize cross and progeny types *77*
extracurricular activities 18
family life of 6
final draft for pioneering paper 80
findings of controlling elements in maize 208
The Flying Dutchman III 9, 12
friends with foreign and different faith students 16
with Gemma Jackson *42*
"Gene-enzyme relationships in *Neurospora*" 142
and geneticists *106*
genetics literature 137
Genetics text 170, 197, 201
Gordon Conference on "Biological Regulatory Mechanisms" 216
graduated with Bachelor of Science in Agriculture 26, *26*
growing up in unconventional family 7–8
Guest Investigator, Department of Genetics, CIW 123–125
Guggenheim Research Proposal 86–87
hypothesis, for genetic control of development 139, 197, 199
ideas, acceptance and rejection of 195–201
IEB application 48
"Induction of instability of selected loci in maize" 160, 166
innovative thoughts on "nuclear organization and gene function" 142

International Botanical Congress, Ithaca, New York *39*, *40*
investigations 79
leave of absence for 1941–1942 110–113
life and contributions of 1–3
"MacArthur Prize Fellow Laureate" (first) 219
maize chromosome knob studies 193
Maize Genetics Cooperation News Letter 75, 143, 247
maize stocks 243
Master's Degree with Sharp 32
"McClintock effect" 144
"McClintock Shed" *61*
Mendelian "gene" 162
modified staining techniques 60
Molecular Biology of the Cell 231
Morgan's "References to Literature" 79
names of 8–9
National Research Council Fellowships 19, 38, 81, 90
National Women's Hall of Fame 244
Nature 139, 160, 213, 238
Neurospora chromosomes 66, 145–148, 158, 166, 167, 248
Nobel Committee 234
Nobel lecture 239–240
Nobel Prize in Physiology or Medicine, 1983 236–239, *237*, *238*, 248
nominations for 1983 Nobel Prize 229–231
Nucleolar Organizer in maize, first named and described 48n7, 88
originally named Elenore 8
"The origin and behavior of mutable loci in maize" 139
participation in Thursday "Evening Lecture" series 122
permanent appointment, CIW/CSH's Department of Genetics 125–126
pioneering research 166, 246
pioneering study 143
pioneering work on maize 198
Plant Science 220 184, *185*, *186*
PNAS paper, 1950 1, 139
postgraduate instructor, at Cornell University 46–48
presentations and collaborations 141, 145–149
problems with University administration 115–117
prokaryotic regulatory system 241
Races of Maize in Bolivia 173
"Recent Investigations in the Cytogenetics of Maize" 106
recognition and awards 216–223
recognition and discrimination, 1928 48–49
recognition by geneticist colleagues 192–195
records and recollections of 129–131
research and publication responsibilities 118–119
research and teaching assistant, in Botany 32–40
research on genetic control of mutation 196
Resident Investigator at the Department of Genetics 136, 142, 150, 247
resignation as Staff Investigator, in Department of Genetics 156–159
ring chromosome paper for publication 82
"ring fragments" in maize 81
Rise of National Socialism, in Germany 87–92
Rosenstiel Award dinner celebration *217*
seminar on transposable elements 171
significant contributions, 1928–1931 49
"smallest and fourth smallest" chromosomes 64
Spm transposable element system 163
with Stadler, in Genetics Research Group 100, **101**
"Staff of Instruction and Extension Work" 16
studies and teaching of maize in Perú 171
studies in maize 189
studies of genetics and first course in cytology 24–27
studies with broken chromosomes 143–145
Suppressor-mutator (*Spm*) gene control system 200
Synapsis Club Razzberry meeting 27, 27n40, *27*
Synapsis Club reunion *221*, 222
teaching responsibilities, University of Missouri, 1936–1943 103, **104**
"The Thomas Hunt Morgan Medal" 219
"transposable element" first used the term 157, 213
transposable elements in maize 233–234
transposition and controlling elements 201
triploid corn plant in Cornell stocks 60
two-element controlling system 198
undergraduate curriculum and encouragement 14
undergraduate education in Cornell University 14
using trisomic methods 63
The Witan 10
Wolf Prize to 219, 230, 231
with women from class of 1923 *41*
in Women's Cosmopolitan Club 18
(work that led to) early discovery of transposable elements 197
work on "mutable loci" 140
works at Cornell University 1
works with maize cytologists 72
Year Books of the Carnegie Institution of Washington 136
McClintock, E. Mignon 9, 11
McClintock, Malcom R. (Tom) 8, 9, 9n11
McClintock, S. H. 7, 7n5, 8, 8n10, 9n11, 12, 13
McClintock, S. Marjorie 6, 8, 8n10, 9, 9n12, 10, 10n15, 10n16, 10n17
Washington Square Players 10
McClintock, T. H. 7–9, 8n8, 8n10
McClung, C. E. 95
McKinley, P. 8n10, 246, 247
The Mechanism of Mendelian Heredity (Morgan) 22
meiosis 61, 62, 64, 93
Mendel, G. 14, 21, 83
factor 163
laws 50
Mendel's Principles of Heredity (Bateson) 22
Messenger, H. 75
Metz, C. W. 23, *24*, 33, 47
Metzenberg, R. 141, 165
Mexican Maize Improvement Program 175
microsporocyte (pollen mother cell) 64
Middlebush, F. 126
Mills, K. O. *102*
Mirsky, A. 123, 191
Missouri Agricultural Experiment Station 101
Missouri State Legislature 103
Mitchell, H. K. 146, 165
Mitchell, M. B. 165, 166
mitosis 1, 57, 61, 62, 64, 146, 167
Mitton, J. B. 218
mobile genetic elements 1, 212, 213, 237
Modern Genetics (Papazian) 199
modifiers 140, 169
Moe, H. A. 86

Monod, J. L. 184, 190, 191, 197, 198, 200, 229
Monteiro, W. 187
Moreno, U. 171, 174, 175
Morgan, T. H. 21, 23, 33, 47, 56, 65, 66, 73, 75, 85, 92, 107, 147, 161, 240
Botanical Gazette 19
The Mechanism of Mendelian Heredity 22
Nobel Prize in Physiology or Medicine 82
The Physical Basis of Heredity 22, 36
"References to Literature" 79
research group at Columbia University 84
mosaic mutants 37, 86
Muller, H. J. 19, 47, 75, 84, 89, 123, 144, 150, 163, 198, 231
cytogenetic work on *Drosophila* 49
"The gene as the basis of life" 37
with Timofeef-Ressovsky, in Guggenheim Fellowship 90
X-ray work 34
Munger, H. 2
Murphy, R. P. v, xiii, xiv, 2n3, 16n16, 30, 40n31, 49n11, 49n12, 49n15, 50n16, 50n17, 107n26, 111n55, 116n84, 169n53, 200n60

National Academy of Sciences 76, 136, 143, 149–151, 172, 192, 195, 246
National Medal of Science 3, 193, 194, 196, 230
National Registry of Historic Places 184
National Research Council (NRC)
committee 73, 149, 171
Fellowships 38, 60, 72, 73, 81, 86, 89, 90, 100
National Science Foundation Act (1950) 150
Navashin, M. 35, 35n23, 80, 80n28, 81, 118n99
Nazi Civil Service Law for non-Aryans 90
Neuffer, M. G. 111n55, *195*, 197; *see also* Nuffer, M. G.
Neurospora chromosomes 66, 145–148, 158, 166, 167, 248
Neurospora crassa 146, 148n33, 165
New York State College of Agriculture Announcement **15,** 15–16
Nilan, R. A. 163, 166, 193
19th Amendment 14
Nixon, R. M. 3, 193n43, 194, 230n10
Nobel, A. 236
"non-autonomous" controlling elements 234
normal Mendelian ratios 62
Norrby, E. 232, 239
Nuffer, M. G. 168, 172; *see also* Neuffer, M. G.
Nuremberg Laws 83, 84

Oelkers, F. 91
Olmo, H. P. 80n28, **104**
O'Mara, J. G. *102*
O'Neill, E. 10
operon theory of bacterial gene regulation 189
Orthogenesis 23

Page Laboratory Dedication at Cold Spring Harbor 245
Painter, T. S. 75, 94, 113
Panel of the Division of Biology of the Committee on Growth of the National Research Council 149
Papazian, H. P.
Modern Genetics 199
Patterson, J. T. 47
Pearl, R. 33
Perkins, D. 167, 167n49
Perry, H. S.
Synapsis Club reunion *221*
Peters, J. A.
Classic Papers in Genetics 166
Peterson, P. A. 57n29, 157, 163, 166, 167, 209, 211, 213
phenotypic genetic changes 79
Phi Kappa Phi College of Agriculture Honor Society 27
Phillips, R. L. *211*
Phipps, I. F. 57
The Physical Basis of Heredity (Morgan) 22, 36
Planck, M. 83, 89
Plant Breeding (DeVries) 22, 54
Plant Sciences Congress (Plant Congress, international Congress of Plant Sciences) 38, *40,* 47
Plessy v. Ferguson 115
PNAS (scientific journal) 1, 47, 77, 81, 94, 105, 125, 139, 233, 234
Portugal, F. H. 218
Postwar Cornell (Edmundson) 249
Principles of Genetics 196, 198
Prize, C. L. M. 222
Proceedings of the National Academy of Sciences 22, 139, 195
Program for the American Society of Naturalists 23
prokaryotic bacterial DNA 214
prokaryotic regulatory system 241
Provine, W. B. 6, 57, 113, 195, 201, 218, *221*, 240
Prywer, C. 188
"pseudoalleles" 161, 162, 170
"pseudo-allelic" mutations 161
pseudoallelism 161, 162

Rabson, R. 215–216
Racker, E. 191
Raffel, D. *106*
rainbow map 52, *52*, 53
Ramírez, E. R.
Races of Maize in Bolivia 187
Randolph, L. F. 32, 32n8, 34, 35, 35n20, 35n21, 36, 38, 47, 49, 51, 56, 60, 72, 170, 184
Carnegie Year Book 33
International Botanical Congress, Ithaca, New York *39*, *40*
recombination frequency data 50
reconciling prokaryotic and eukaryotic genetic systems 208–216
Rèdei, G. *102*
Redfield, H. 47
Reed, H. D. 20, 20n31
regulator-gene-operator system in bacteria 211
regulator-operator theory of gene regulation 229
repetitive non-coding DNA 242
retrotransposons 244
R–G linkage group 62
Rhoades, M. M. 2n4, 51, 53, 56–58, *60*, 62, 65, 66, 72, 75, 79, 90, 93, 101, 103, 107–109, 111, 113, 117–119, 121–123, 125, 127, 136, 137, 145n19, 149, 157, 159, 166, 168, 170, 187, 199, 220, 221, 229, 247, 248
"Genetic Control of Mutability" 169
research on genetic control of mutation 196
Synapsis Club reunion *221*
"The Thomas Hunt Morgan Medal" 219
using trisomic methods 63

Rhoades, V. 123, 163
Richey, F. D. 32n8, 34n17, 34n18, 34n19, 35n20, 35n21, 36n24, 38, *39*, 53, 53n21, 56n27, 102n10
Ringertz, N. G. 237, 238–239
Rise of National Socialism, in Germany 87–92
Ris, H. 148
The Road to Stockholm (Hargittai) 231
Robbins, W. M. 47, 100, 115
Roberts, L. M. 172, 173
Rockefeller Foundation Archive Center (RF) 39, 84
Roman, H. *102*, **104**, 220, 229, 230
RR gene locus 164
Ryan, F. J.
  Cell Heredity 197
Ryan, J. F.
  The Story of the 27th Division 10

Saedler, H. 211
Safont, J. 188
Sager, R. 149
  Cell Heredity 197
Salmonella typhimurium bacterial system 148, 208
Sambrook, J. F. *237*
Sand, S. S. 170
Sastry, G. R. K. 209
Sawicki, J. *4*
Sax, K. 39
Schaeffer, P. 197
Schrader, F. 127, 140, 141, 148, 168
Schrader, S. 127
Schramm, J. R. 47
Schultz, J. 19, 161
Schwartz, D. 149, 199
Science Citation Index (SCI) Ten Year Cumulations 195, 196
Sears, E. R. **101,** *102*, 116n84, 118n96, 129n151, 195
Shapiro, J. A. 212, 214, 215
  Mobile Genetic Elements 231
Sharp, L. W. 1, 33, 36, 38, 40n33, 41, 47–49, 56, 57, 110, 184
  cytology class 60n35
  Cytology texts 34
  IEB committee 49
  International Botanical Congress Cytology *40*
  Introduction to Cytology 2, 25
  lab in Agronomy Building *59*
  "Report Regarding Graduate Work" 32
  Roll Book 2n3, 16n15, 18n21, 40n33, 46n2
  satiric monograph on *Eoörnis pterovelox gobiensis* (Whiffenpoof) 25, 25n36
  The Summer School of Biology 46
Shull, G. H. 39
Shultz, J. 47
Sigma Delta Epsilon 18, 38, 114
Singer, M. 19n29, 40n31, 100n2, 245
Singh, S. 51
Singleton, J. R. **104**, 121n111, 129, 129n50, 135, 147, 147n27, 148, 154, 155, 167
Singleton, W. R. *102*, **105**, 158, 160, 160n25, 161, 162, 181, 206
  Brookhaven Symposia in Biology 197
  Elementary Genetics text 197
Sinnott, E. W. 196
Sisco, P. 6, 113, 195, 240
sister mitotic cells 103
"smallest and fourth smallest" chromosomes 64
Smith, H. H. 170
Smith-Keary, P. F.
  Salmonella typhimurium bacterial system 208
Smith, L. **101**, *102*
Smocovitis, V. B. 249
somatic reversion 167
Spm element 157, 167, 213
Sprague, G. F. 51, **101**, *102,* 116
  Synapsis Club reunion *221*
Srb, A. 182, 188, 193, 196–198, 215, 218
Stadler, L. J. 36, 36n24–26, 37, 39, 47, 47n6, 56, 58, 66, 72, 79, 81, 90, 93, 95, *102*, 110–113, 117, 119, 120, 123, 126, 128, 129, 144, 149, 165, 166, 197
  genetics research group 88, 103–108
  with McClintock, in Genetics Research Group 100, **101**
  Stadler Genetics Symposium 129, 215, 235
  X-ray-induced mutations 105
Starlinger, P. 211
  review of "Transposable Elements" 242–243
State and the Land Grant College's Experiment Stations 54
Stebbins, G. L. 146
Steinitz, L. (Sears) 146, *195*
Stephens, S. G. 161
Stern, C. 58, 75, 77, 78, 95, 106, 150
  McClintock and Curt meet, in America 84–86
  Messenger Lectures 79
  Pittsburgh meeting 93
  Rise of National Socialism, in Germany 87–92
  Rockefeller Foundation Fellowship 82
  work with McClintock 82–84
Stern, E. S. 85
Stevens, S. 156, 156n2
Steward, F. C. 216
Stinson, H. 2, 2n5, 3, 183, 192, 201
Stone, W. 75, 75n19
The Story of the Maize Plant (Weatherwax) 32
Strickberger, M. W. 196
Stubbe, H. 161
Sturtevant, A. H. 47, 86, 95, 123, 147
The Summer School of Biology (Sharp) 46
Suppressor-mutator (*Spm*) gene control system 157, 168, 200
Sussman, M.
  Growth and Development 200
Sutherland, J. P. 8
Suzuki, D. 218
Swanson, C. P. 196
  Cytology and Cytogenetics 117, 143
Szostak, J. W. 235

tabula rasa 65
Tai, S. E. *95*
Tatum, E. 141, 148
Taylor, A. L. 208, 212
Taylor, W. R. 46
Teas, H.
  Genetics and Developmental Biology 191
theory of crossing-over 75–79
Thomas Hunt Morgan Centennial Symposium 191
Timofeef-Ressovsky, N. V. 90
Timothy, D. H. 173
  Races of Maize in Chile and *in Ecuador* 187
Ting, Yu-Chen 171
T6-4 interchange chromosome 80

traditional plant geneticists 65
Trajkovich, H. A. Z. *17*, 18, 30, 51; *see also* Ziegler, H. A.
trans-acting autonomous elements 234
transposable elements 240
  Antirrhinum 161
  bacterial gene regulation 216
  in Carnegie Institution report 213
  "controlling elements" 229
  in evolution 231
  induction of 105
  "jumping gene" to 212
  lab's research investigations of 184
  in maize 233–234
  McClintock's discovery of 125, 136, 143, 145, 191, 197, 199
  McClintock's genetic observations of 243
  P (pericarp) locus 138
  reviews of 242
  Spm element 157, 163
  "time for the molecular analysis" of 243–244
triploid endosperm 54
trisomic methods 63
Truman, H. S. 150
  Point IV Program 171
Tsao, C. Y. *95*
Tucker, C. M. 115–117, 120, 127
two-element interacting system 234
two-element system model of regulation 188
Tyler, A. (A. Titlebaum) 19n29, 47, 86, 86n61

Uber, F. M. *102*, 144
Uhl, C. H. 2
University of Berlin (Kaizer Willhelm University, Friedrich Willhelm University) 49, 82, 84
University of Missouri 112
  Anderson's copy archived at *52*
  career at 129
  cytogenetics lectures in 2
  Department of Botany, Missouri 100, 103, **104**, 129, 148
  faculty appointment at 120, 127
  maize geneticists at 215
  McClintock (Assistant Professor) at 100–103
  McClintock's experiences at 130
  McClintock's request for a leave 112–113
  policy 113
  postdoc with Stadler at 49
  promotion, tenure and AAUP 119–121
  research consultant **105**
  with Stadler's genetics research group 103–108, 159
  work of 117

U.S. Agricultural Experiment Stations 22
U.S. Atomic Energy Commission 160
U.S. Congress House Un-American Activities Committee 162n30

Vodkin, M. *4*

Waddington, C. H. *106*, 243
Walden, D. B. 209, *210*, *211*
Wallace, A. R. 23
Wallace, B. 141, 142, 182, 184, 186
Warren, K. B. 8
Watson, J. D. 139, 160, 201, 215n17, 238, 239, 239n36
Weatherwax, P.
  The Story of the Maize Plant 32
Weaver, E. A. **104,** 122
Weaver, H. M. 160
Weaver, W. 92, 107n30
Weil, C. 139
Weinstein, A. 19
Weinstein, E. 16
Weismann, A. 20
Wellhausen, E. J. 173, 175, 175n72
Welshons, W. J. 218
Wessler, S. 160, 215, 220, 234, 243
Wettstein, F. von 83, 90
White, A. D. 182, 184, 192, 197, 199
Wiggins, R. G. *95*
Wilson, E. B. 191
  The Cell in Development and Inheritance 191
  E. B. Wilson Lecture Series 191–192
Witkin, E. 156
Wollman, E. L. 189, 197
Women at Cornell: The Myth of Equal Education (Conable) 245
women, in plant breeding at Cornell University 30–31
Woods Hole Marine Biological Laboratory 46, 116, 118
Wright, S. 23, *24*, 123

Year Books of the Carnegie Institution of Washington 22, 136, 145, 193, 194, 215

Zametkin, L. K. 16; *see also* Hobson
Zea chromosomes 49n14, 58, 81, 93
Zea mays 34, 38, 51, 211
Zeleny, C. 84, 85
Ziegler, H. A. 16; *see also* Trajkovich, H. A. Z.
Ziegler, M. A. 16, *17*; *see also* Demerec, M. A. Z
Zimmer, C. 84
Zinder, N. D. 141